FARM LAND EROSION
IN TEMPERATE PLAINS ENVIRONMENT AND HILLS

Cover Photo by S. Wicherek.
Effects of sudden rainfalls: sheet erosion in a field beets provokating a loss of soil fertility and water pollution. Foreground: flood on a departmental road. Soissonnais plateau (Parisian basin), May 1989.

FARM LAND EROSION

IN TEMPERATE PLAINS ENVIRONMENT AND HILLS

Proceedings of the International Symposium on Farm Land Erosion
Paris, Saint-Cloud, France, 25-29 May 1992

Edited by

STANISLAS WICHEREK

Centre de Biogéographie-Ecologie
U.R.A. 1514 C.N.R.S.
E.N.S. Fontenay/Saint-Cloud
Saint-Cloud, France

ELSEVIER

AMSTERDAM . LONDON . NEW YORK . TOKYO 1993

ELSEVIER SCIENCE PUBLISHERS B.V.
Sara Burgerhartstraat 25
P.O. Box 211, 1000 AE Amsterdam
The Netherlands

Library of Congress Cataloging-in-Publication Data

International Symposium on Farm Land Erosion (1992 : Paris and Saint-Cloud, France)
Farm land erosion : in temperate plains environment and hills : proceedings of the International Symposium on Farm Land Erosion, Paris, Saint-Cloud, France, 25-29 May 1992 / edited by Stanislas Wicherek.
p. cm.
Editorial also in French.
ISBN 0-444-81466-3 (acid-free paper)
1. Soil erosion--Congresses. 2. Soil conservation--Congresses. 3. Agriculture--Environmental aspects--Congresses. I. Wicherek, Stanislas. II. Title.
S622.2.I615 1992
631.4'5'0912--dc20 93-7318
CIP

ISBN 0-444-81466-3

This book is printed on acid-free paper

Printed in The Netherlands

EDITORIAL

Changes to agricultural systems in France (and Europe in general) over the last two decades have brought on a spectacular worsening of soil erosion and degradation; part of the agronomical asset and water quality are threatened. "How can risk be evaluated? What solutions should be adopted without radically disturbing the socio-economic orientation of major agricultural regions?" were the principal questions for debate at the International Symposium on "*Farm land erosion in temperate plain and hill environments*", organized by the Centre de Biogéographie-Ecologie, E.N.S. – C.N.R.S. – U.R.A. 1514, at the Ecole Normale Supérieure de Fontenay-Saint Cloud, Paris, which took place May 25-29, 1992, in cooperation with the "European Society for Soil Conservation (E.S.S.C.), the "International Geographical Union (I.G.U. – COMTAG) and the "Réseau Erosion" (Erosion Network of the O.R.S.T.O.M.) under the auspices of the French Minister of Research and Space and the French Minister of Agriculture and Forests.

The Scientific Organization Committee was made up of the following Professors: *P. Arnould*, E.N.S. Fontenay-Saint Cloud, France; *R.B. Bryan*, University of Toronto, Canada; *J. De Ploey**†, University of Louvain, Belgium; *W. Froehlich*, Academy of Sciences, Cracow, Poland; *E. Roose*, ORSTOM, Erosion Network, Montpellier, France; *Y. Veyret*, University of Paris VII, France; *S. Wicherek*, Chairman, C.N.R.S and E.N.S. Fontenay-Saint Cloud, France. Secrétariat de rédaction (sub-editing) was assumed by Mrs. *M.O. Boissier*, with help of the technical staff: M. *G. Chêne*, Mrs. *M. Le Berre* and Mrs. *M. Mekharchi*.

Our international symposium was attended by 160 participants from 25 different countries. Within the framework of the *année européenne* or "European Year", all member countries of the EEC were represented; we were also able to invite scientists from the Eastern Euopean countries, thus gaining access to their knowledge in this area, which was highly appreciated and enriching. The major agricultural countries (U.S.A., Canada) were also strongly represented.
As for France, every major French research organization was represented at this bilingual symposium (French/English): The *Centre National de la Recherche Scientifique* (CNRS – The National Center for Scientific Research), *Institut National de la Recherche Agronomique* (INRA – National Institute for Research in Agronomy), *Bureau de Recherches Géologiques et Minières* (BRGM – The Geological and Mining Research Office), *Centre National du Machinisme Agricole, du Génie Rural, des Eaux et des Fôrets* (CEMAGREF – National Center for Agriculture, Rural development, Water and Forests), *Organisme de Recherche Scientifique des Territoires d'Outre-Mer*

---

† Shortly before the date of the Symposium *Professor Jan De Ploey* passed away. We have lost an internationally reknown scientist, one of the pioneers who succeeded in heightening awareness of the importance of the soil asset. He was particularly appreciated for his immense scientific competence and his remarkable human qualities.

(ORSTOM – Organization for Scientific Research in the Overseas French Territories), as well as the French Universities and "Grandes Ecoles".

In this publication, you will encounter a number of papers chosen by the Scientific Organization Committee. Sixty-five of the best papers were selected for oral presentation, along with ten posters, covering the following five major themes:

1. From the parcel to the watershed. Role of station studies. Processes analysis and establishing partial balances. Section Chairman: *Professor R.B. Bryan*, University of Toronto. Rapporteur: *Professor F. Papy*, INRA, Paris-Grignon.
2. From the geosystem level to the regional level: Essay on typology, to define erosion sensitiveness and potential. Section Chairman: *Professor G. Bertrand*, University of Toulouse. Rapporteur: *Professor Y. Veyret*, University Paris VII / E.N.S.
3. Methods and tools. Evaluation of the respective contributions of field studies, remote sensing, GIS, modellization and Caesium 137. Section Chairmen: *Professor J. Boardman*, Oxford University; *Professor J. Poesen*, University of Louvain. Rapporteurs: *Professor Cl. Bernard*, Ministry of Agriculture, Québec. *Professor W. Froelich*, Academy of Sciences, Cracow.
4. Other examples of intensified erosion. Section Chairman: *Professor P. Sanlaville*, C.N.R.S., University of Lyon II. Rapporteur: *Professor R. Neboit*, University of Clermont Ferrand.
5. Recent and past changes in agricultural structures: incidence on erosion. Traditional strategies for water management. Proposals for better and more efficient soil conservation. Section Chairman: *Professor E. Roose*, ORSTOM, Montpellier. Rapporteur: *Professor P. Arnould*, E.N.S. de Fontenay-Saint Cloud.

During this symposium, considerable progress made in fundamental and applied research at the center of environmental studies was reported. The forty-seven papers selected for this publication corroborate this statement. The majority are in English but we have made a point of including a French summary and three articles written in the French language. Some presented papers have not been considered for this book by reason of their objectives, results published elsewhere, language or delay; they will be published later in the *Cahiers no. 6* of our Research center.

Both of the excursion days in the Parisian Basin to major agricultural lands (Laonnois-Soissonnais regions) and Champagne vineyard, led to direct contact with the "users" of the land, the decision-makers, the managers of the environment, which is fairly uncommon. The experience was highly appreciated by all concerned especially since such strong ties exist between researchers/scientists and those who are directly involved with managing the agricultural milieu, the users. Vast problems were discussed: the relationship between Science and experience, the needs of Society, while stressing the rapport existing between the physical, biological and human milieus, as well as those between farmers and the EEC.

Stanislas Wicherek
C.N.R.S. – E.N.S.
Paris, October 1992

EDITORIAL

Les mutations des systèmes de cultures en France et en Europe depuis une vingtaine d'années ont entraîné une spectaculaire aggravation de l'érosion des sols et une partie du capital agronomique est menacée ainsi que la qualité des eaux. La question posée "Comment évaluer les risques et quelles solutions adopter sans bouleverser radicalement les orientations socio-économiques des régions de grandes cultures ?" a été au coeur des débats du Symposium international intitulé "*Erosion des terres agricoles en milieu tempéré de plaines et de collines*", organisé par le Centre de Biogéographie-Ecologie, E.N.S.-C.N.R.S. U.R.A. 1514, à l'Ecole Normale Supérieure de Fontenay - Saint-Cloud, Paris, du 25 au 29 mai 1992, en coopération avec "European Society for Soil Conservation (E.S.S.C.), "International Geographical Union (I.G.U-COMTAG) et le Réseau Erosion (O.R.S.T.O.M.), sous le haut patronage du Ministre français de la Recherche et de l'Espace et du Ministre français de l'Agriculture et de la Forêt.

Le Comité Scientifique d'Organisation était composé par les professeurs suivants : *P. Arnould*, E.N.S. Fontenay - Saint-Cloud, France ; *R.B. Bryan*, Université de Toronto, Canada ; *J. De Ploey* (†)*, Université de Louvain, Belgique ; *W. Froehlich*, Académie des Sciences, Cracovie, Pologne *E. Roose*, Orstom, Réseau Erosion, Montpellier, France ; *Y. Veyret*, Université Paris VII, France ; *S. Wicherek* : Coordinateur et Organisateur, C.N.R.S. - E.N.S. Fontenay - Saint-Cloud, France.Le Secrétariat de rédaction a été assuré par Mme M.O. Boissier avec l'aide de l'équipe technique : M.G. Chêne, Mme M. Le Berre et Mme M. Mekharchi.

Le Symposium a remporté un vif succés, 160 personnes sélectionnées, venant de 25 pays ont participé. On peut remarquer que dans le cadre de cette année européenne tous les pays de la CEE étaient représentés et nous avons pu également inviter des scientifiques des Pays de l'Est et accèder à leurs connaissances dans ce domaine, ce qui a été très apprécié et fort enrichissant. Il ne faut pas oublier pour autant une forte représentation des grands pays agricoles (U.S.A., Canada).

En ce qui concerne la France tous les grands Organismes français de Recherche ont participé à ce symposium bilingue : Centre National de la Recherche Scientifique (C.N.R.S.) , Institut National de la Recherche Agronomique (I.N.R.A.), Bureau des Recherches Géologiques et Minières (B.R.G.M.), Centre National du Machinisme Agricole, du Génie Rural des Eaux et des Forêts (C.E.M.A.G.R.E.F.), Organisme de Recherche Scientifique des Territoires d'Outre-Mer (O.R.S.T.O.M.), Universités et Grandes Ecoles.

* Peu de temps avant le Symposium, Mr le Professeur J. De Ploey est décédé. Nous avons perdu un scientifique de renommée internationale ; un des pionniers qui a su mettre en évidence l'importance du "capital sol". Nous l'avons tout particulièrement apprécié pour ses immenses compétences scientifiques ainsi que pour ses remarquables qualités humaines.

Le Comité Scientifique d'Organisation avait sélectionné 65 communications pour une présentation orale, ainsi que 10 posters, sous les 5 thèmes suivants :
1 - De la parcelle au bassin versant. Place des études stationnelles. Analyse des processus et établissement de bilans partiels. Président de Séance : *Pr. R.B. Bryan*, Université de Toronto ; Rapporteur : *Pr. F. Papy*, INRA, Paris, Grignon.
2 - Du géosystème à la région : essai de typologie visant à définir les sensibilités et les potentialités de l'érosion. Président de Séance : *Pr. G. Bertrand*, Université de Toulouse, Rapporteur : *Pr. Y. Veyret*, Université Paris VII-E.N.S.
3 - Méthodes et outils : évaluation des apports respectifs des études de terrain, télédétection, SIG, modélisation et Césium 137. Présidents de Séance *Pr. J. Boardman*, Université d'Oxford ; *Pr. J. Poesen*, Université de Louvain ; Rapporteurs : *Pr. Cl. Bernard*, Ministère de l'Agriculture, Québec ; *Pr. W. Froehlich*, Académie des Sciences, Cracovie.
4 - Autres exemples d'érosion exacerbée. Président de Séance : *Pr. P. Sanlaville*, C.N.R.S., Université Lyon II ; Rapporteur : *Pr. R. Néboit*, Université de Clermont-Ferrand.
5 - Appréciations des modifications actuelles et passées des structures agraires : incidence sur l'érosion. Stratégies traditionnelles paysannes et gestion de l'eau. Propositions pour une meilleure gestion de la ressource sol et lutte anti-érosive. Président de Séance : Pr. *E. Roose*, Orstom, Montpellier ; Rapporteur : *Pr. P. Arnould*, E.N.S. de Fontenay - Saint-Cloud.

Pendant ce Symposium, on a pu constater dans ce domaine un progrès considérable des recherches fondamentales et appliquées qui font partie intégrantes des études environnementales. Dans cet ouvrage vous trouverez 47 communications sélectionnées qui traduisent bien ces propos. Les articles sont présentés en majorité en anglais à la demande de l'Editeur Elsevier, mais nous avons pu obtenir d'inclure un résumé en français ; ainsi que 3 contributions. Certaines communications n'ont pu être retenues pour cette publication en raison de leurs objectifs, de résultats déjà publiés, de la traduction et des délais non respectés ; elles seront publiées ultérieurement dans "Cahiers n° 6" de notre Centre de Recherche.

Les 2 journées de travail sur le terrain en Bassin parisien, sur les terres de grandes cultures (Laonnois-Soissonnais) ainsi que dans les vignobles de Champagne ont permis de rencontrer des utilisateurs, des décideurs, des gestionnaires des milieux, ce qui est peu habituel dans ce type de rencontre et cela a été apprécié. Ces liens entre les scientifiques et les utilisateurs de l'Espace sont indissociables et indispensables.
Les problèmes au sens large ont été posés : relations entre Science et ses acquis, besoins de la Société dans ce domaine.
L'étude des articulations entre les milieux physiques, biologiques et humains ainsi que les rapports entre exploitants agricoles et la CEE ont été privilégiés durant tout ce Symposium, qui s'est révélé opportun dans le contexte actuel européen et mondial.

Stanislas Wicherek
C.N.R.S.-E.N.S.
Paris Octobre 1992

Acknowledgments to organizations participating in the Symposium
Organismes remerciés pour leur participation à l'organisation du Symposium

- Ministère de la Recherche et de la Technologie.
- *French Ministry of Research and Technology.*

- Ministère de l'Agriculture et de la Forêt.
- *French Ministry of Agriculture and Forest.*

- Ministère des Affaires Etrangères.
- *French Ministry of Foreign Affairs.*

- E.N.S. (*) de Fontenay - Saint-Cloud.
- *E.N.S. (*) Fontenay - Saint-Cloud.*

- C.N.R.S. (**) : Département des Sciences de l'Homme et de la Société.
- *C.N.R.S. (**) : Department of Human and Social Sciences.*

- Université de Reims.
- *University of Reims.*

- I.N.R.A. (***), Station d'Agronomie de Châlons-sur-Marne.
- *I.N.R.A. (***) Châlons-sur-Marne Agricultural Station.*

- I.N.R.A. (***) Station Agronomique de Laon.
- *INRA (***) Laon Agricultural Station.*

- Agence de l'Eau Seine-Normandie, Service Agriculture.
- *Seine-Normandie Water Works, Agricultural Service.*

- Chambre d'Agriculture de Laon et Syndicat du Pays de Serre (Aisne).
- *Laon Chamber of Agriculture and Pays de Serre Association (Aisne) .*

- Comité Départemental du Tourisme de l'Aisne (Laon).
- *Aisne Regional Tourism Committee (Laon).*

- Conseil Général de l'Aisne.
- *Aisne General Council.*

- Conseil Régional de Picardie.
- *Picardie Regional Council.*

- Mairie de Saint-Cloud.
- *Municipality of Saint-Cloud.*

- Mairie de Coulommes La Montagne.
- *Municipality of Coulommes La Montagne.*

- Mairie d'Erlon (Laonnois).
- *Municipality of Erlon (Laonnois).*

- Mairie de Cessières (Laonnois).
- *Municipality of Cessières (Laonnois).*

- Mairie de Vierzy (Soissonnais).
- *Municipality of Vierzy (Soissonnais).*

- Centre d'Etudes Techniques Agricoles du Soissonnais (CETAS), association privée fondée par 70 agriculteurs.
- *Center for Technical Agricultural Studies of the Soissonnais region (CETAS), private association founded by 70 farmers.*

- Propriétaire récoltant de champagne à Passy sur Marne.
- *Passy sur Marne champagne property farmer.*

(*) E.N.S. : Ecole Normale Supérieure / *Upper College of Education.*
(**) C.N.R.S. : Centre National de la Recherche Scientifique / *National Center for Scientific Research.*
(***) INRA : Institut National de la Recherche Agronomique / *National Institute for Agronomical Research.*

CONTENTS / TABLE DES MATIERES

The Soil Asset : Preservation of a natural resource

S. Wicherek

Centre de Biogéographie-Ecologie, URA 1514 CNRS, Ecole Normale Supérieure de Fontenay - Saint-Cloud, Le Parc, 92211 Saint-Cloud, France.

1. CONCERNING THE TEMPERATE PLAIN ENVIRONMENT : GOING FROM THE "NORM", FROM BIOLOGICAL TO SOCIAL

The temperate plain environment has long been considered as reference environment. Even if its "normality" has been questioned by knowledge of other natural environments, the separatist approach of classical physical geography has always been predominant in fundamental research. The soaring interest in bio-geographical study has however, created a privileged domaine where we can integrate different physical parameters (soil, water, climate) and construct a code for "reading" the landscape.

Curiously the temperate plain environment is a geographical sector that has been neglected by research in general, and yet this area is the center of the contemporary issues on environment and development. Thus over the past twenty years our research center has oriented its perspective on the study of relationships between physical, biological and human environments, the problem of soil degradation for example.

The temperate plain environment, located in the middle central Europe is not simply a northern European great plain, it is also a referential environment "under influence". It appears well-known to us, because it is such a familiar environment. These plains, plateaus, hills and valleys represent a "Douce France", and extends to encompass a "Douce Europe", whose bioclimatic values have been exploited by an ancient system of values. Generations of farmers have finely adjusted their agricultural systems to the local variations of climate and soil : biological, economic and social elements have functionned on a local level over numerous centuries. Economic imperatives have been satisfied by exploiting all the nuances of the temperate plain ; a differentiated rural society with specific rules has closely managed this environment and its space.

Except for a few remarkable exceptions to the rule, relative simplification of agricultural systems and transformation of the rural society have erased from the landscape the finest elements of this agricultural mosaic, the concrete expression of environmental diversity (1).

The large fields and forests alternate over the landscape, masking the intricate expression of bioclimatic nuances. Over many decades, these nuances have been found on hillsides and valleys, on lands left vacant by a "small farming population". The planting of trees, both spontaneous and volontary, has progressively made these nuances disappear. Today the patience, the vision and the knowledge of phyto-geographers are needed to recognize, in species or vegetal groups, a living tribute to the diversity of temperate plain environments. Villages, surrounded by housing developments, have sprung up between the large forests and ever-expanding agricultural lands. A new village society is being set up on tree-covered hillsides.

Any in-depth study of the physical and biological components of the temperate plain environment must encounter past and present societal phenomena. While a researcher may begin his study by recording and observing physical and biological processes, correlating these processes implies that he eventually takes "human action" into account. *Human action* is not only a passive filter, it is also a major protagonist in environmental production.

II. PROBLEMS BECOMING MORE AND MORE ACUTE

Mutations within the European agricultural network, and in particular on the loamy lands of plains and plateaux in oceanic temperate climates, have brought on a spectacular worsening of erosion, even in situations which had previously been considered as mildly or non- erosive (2 - 3). This refers to the major agricultural regions which have undergone intense expansion over the past 25 years without forgetting changes in agricultural techniques and regrouping of lands (Photo 1).

Photo 1. Damages after erosion sheet in a field of onions. Several hectares of cultivated lands carried away. Damages on the road. (Marlois région, Parisian basin) S. Wicherek, May 1990.

These lands are, however, fragile and easily affected by surface sealing or by settling, a problem increased by the use of larger and heavier farm machinery. The economic consequences are impressive : on the interfluve level, soil loss, and as this concerns a non-renewable resource, affect seriously production costs and profitability, since natural fertilizers have to be replaced in other way ; off-farm damages are no less severe : blocking of water/rain collection networks, muddy flooding of tillages increased in turbidity and water pollution... (Photo 1).

Objectives and needs are clear How can risks be evaluated and what solutions should be chosen without radically disturbing the socio-economic orientation of these regions ?

By distancing ourselves from the work completed and the results obtained, it becomes evident that the foundation must be laid for a "human physics" of the countryside. Climatic low atmospheric traits, particular means of transfer of mineral and vegetal matters (soil erosion) the water cycle in function of vegetal cover (forests, lawns and farms) must all be placed in a harmonious relationship with the way land is used. This use of the land can not be seen as a simple "disturbance" that should be reduced in order to see better results. These "disturbances", modulated over time and within a given space, are the trademark of society's conscious or unconscious management of the physical environment.

III. SOIL DEGRADATION IN FRANCE ON A TEMPORAL SCALE

Over the past few decades, the theme of soil erosion has been one of the main concerns of some French researchers who have developed considerable research, information, techniques and results of noteworthy importance. This work, however, is characterized by its extreme diversity, and notably by 1) a variety of research sites ; 2) its multiple approaches, and ; 3) the complexity of the techniques used.

In France, critical periods for soil erosion involving different areas may be identified, with the affected sites in highly differentiated areas. Schematically speaking, these different periods and areas can be grouped under four major categories :

1. Since the 19th Century, mountain environments.
2. The Mediterranean region, in a limited climatic context.
3. Vineyards and sometimes orchards, on hillsides and steep slopes.
4. Since 1960-1970, the large agricultural plains of the Parisian Basin and also the South-West.

Since the 1980's, French researchers have gradually become aware of the fact that soil erosion could affect any regions of France (Fig. 1A, 1B). They have also evaluated the planetary dimension of these issues and have seen gradual development of research devoted to these issues in other European countries. Even though soil erosion receives less attention from the media than other environmental phenomena, like acid rain or forest fires, it poses a real threat to agriculture, as a natural risk with serious consequences (4).

On the national level, major research organizations like the CNRS, INRA , CEMAGREF, BRGM and ORSTOM, have shown a growing interest in the soil erosion problem and have implemented research programs.

Thus, carrying out fundamental research in the hope of finding concrete solutions to the needs of local societies and communities is essential.

Agricultural land erosion in France
Erosion des sols cultivés en France

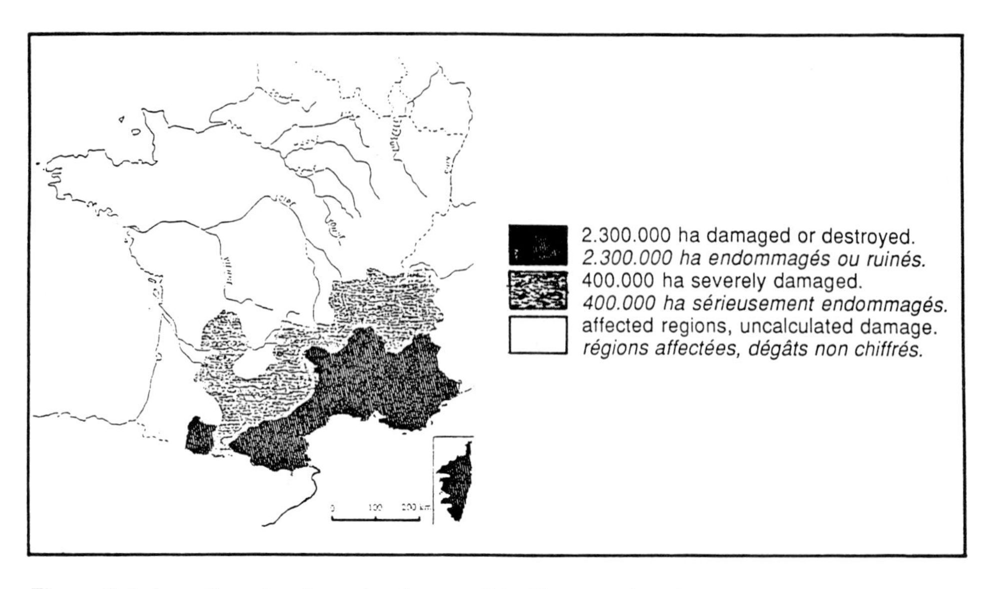

Figure 1-A. Location of soil erosion by runoff in France, situation in 1950.
Localisation de l'érosion des sols par ruissellement en France, situation en 1950.
(S.Hénin et T. Gobillot, 1950).

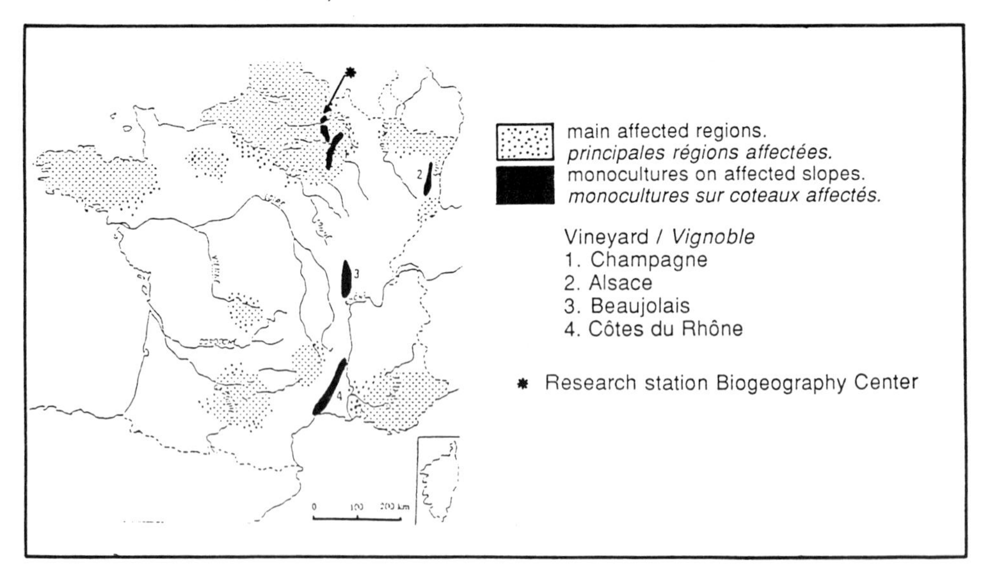

Figure A-B. Water erosion on cultivated soils in France, from existing bibliography.
Erosion hydrique des sols cultivés en France, d'après la bibliographie existante 1990.

IV. PRIORITIES OF FUNDAMENTAL RESEARCH AND WORK SCALES : SUGGESTING METHODS OF SOIL UTILIZATION IN ORDER TO LIMIT DEGRADATION

1. Fundamental research at the station level seeks a better understanding of the mechanisms and functioning of soil erosion, taking into consideration the climatic (for example, rainfall intensity) and physical parameters (for example, humidity). This leads to determining a threshold and what triggers this threshold (5). In this approach, different types of vegetal cover and their influence on erosion play a major role. From a methodological point of view, it is only at this level that precise measurements under maximum conditions of homogeneity (soil, climate...) can be made. These measurements concern climatic parameters, evolution of physical, hydric and mechanical conditions of the soil, evolution of the surface state, runoff and sediment loss.

Use of certain measurement stations chosen as "surface-tests" responds to a feasability criterion : only at this level can a continuous follow-up be envisaged (for example, surface state evolution and formation of runoff during rainstorms). This is also essential for understanding the mechanisms and simulation protocol of laboratory research.

However, the station environment can not reproduce the diversity of the slope or the watershed, nor situational variety into account. This level is therefore necessary, but still insufficient: the number of stations and their distribution should permit all functional units, whose inter-relationships are necessarily studied at the catchment area and slope levels, to be represented.

2. Farm land erosion at the level of the elementary catchment area or group of catchment areas, allows us to follow the evolution of the erosion process and the different forms it takes, to determine the physical and chemical composition of eroded substances and to subsequently propose more immediate applications for anti-erosive protection. Problems concerning water management and quality (ex. nitrates) are also examined. This spatial unit, with its autonomous hydric networks reaching hundreds of hectares, with water flowing from a crest line toward a permanent or temporary evaculation canal, is an elementary unit for the study of soil degradation, and in a broader sense, for environmental studies in general. Recording movement and amounts of liquid and solid flow is essential to understand phenomena in correlation with vegetal cover and soil tillage, and to define better management for technique large-scale agricultural land (6 - 7).

3. Catchment area typology and risk cartography. Forecasting the forms and risks of soil erosion on agricultural lands under moderate climatic and topographical conditions, as well as preventing it has been hindered by insufficient knowledge of the mechanisms involved in runoff formation and concentration mobilization and transportation of solid particles.

Actually we are unable, to objectively and properly evaluate and classify the diversity of the situations encountered, the reasons for this apparent worsening and consequences on superficial hydrology, such as : - modifications in plot structure, - evolution of practiced agricultural systems : progressive disappearance of prairies, evolution of agricultural techniques, degradation of the organic status of the soil and its properties (Fig. 2 A, 2 B).

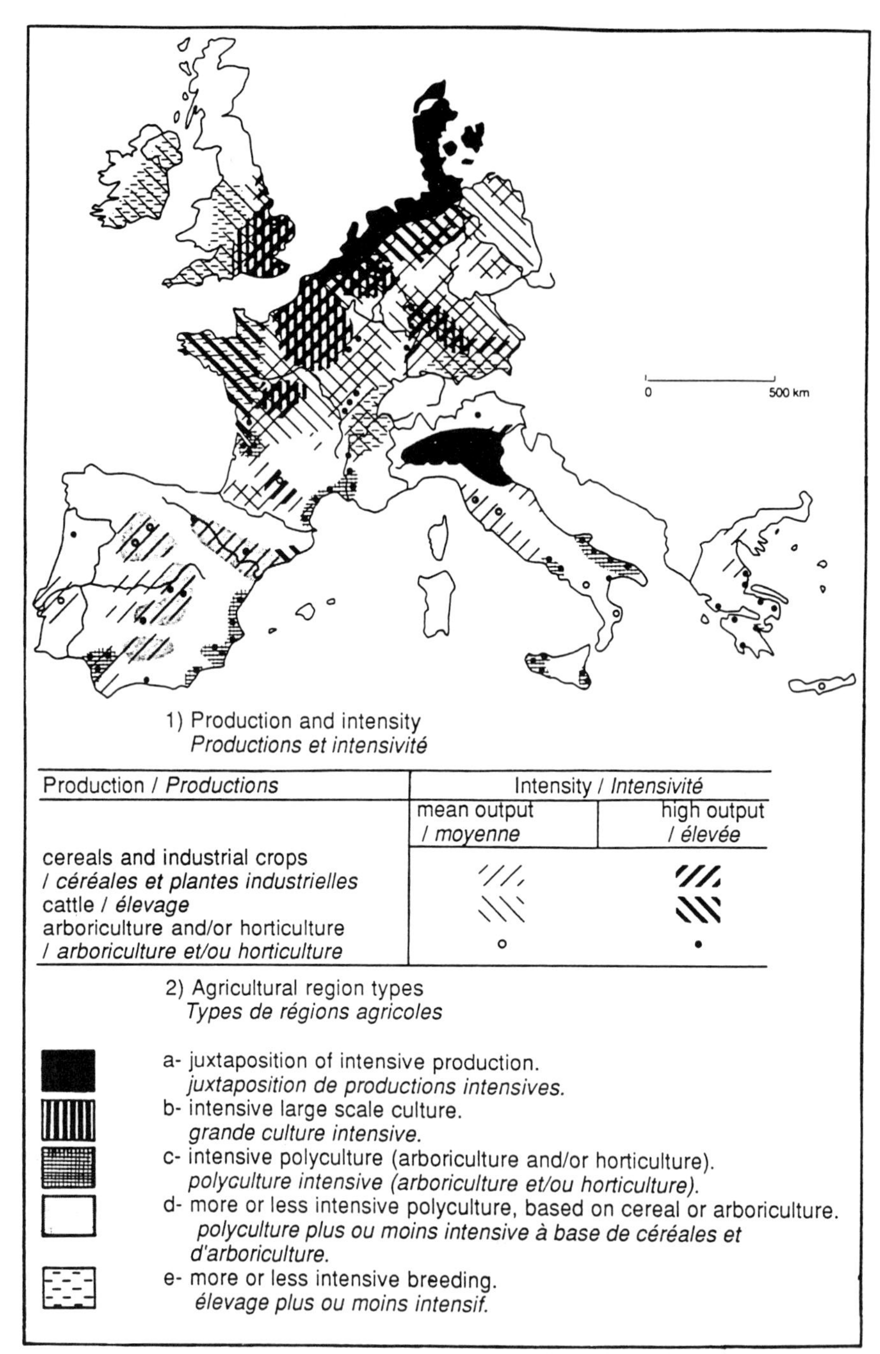

Figure 2-A. Agricultural region typology in U.E. countries
Typologie des régions agricoles de la C.E.E.

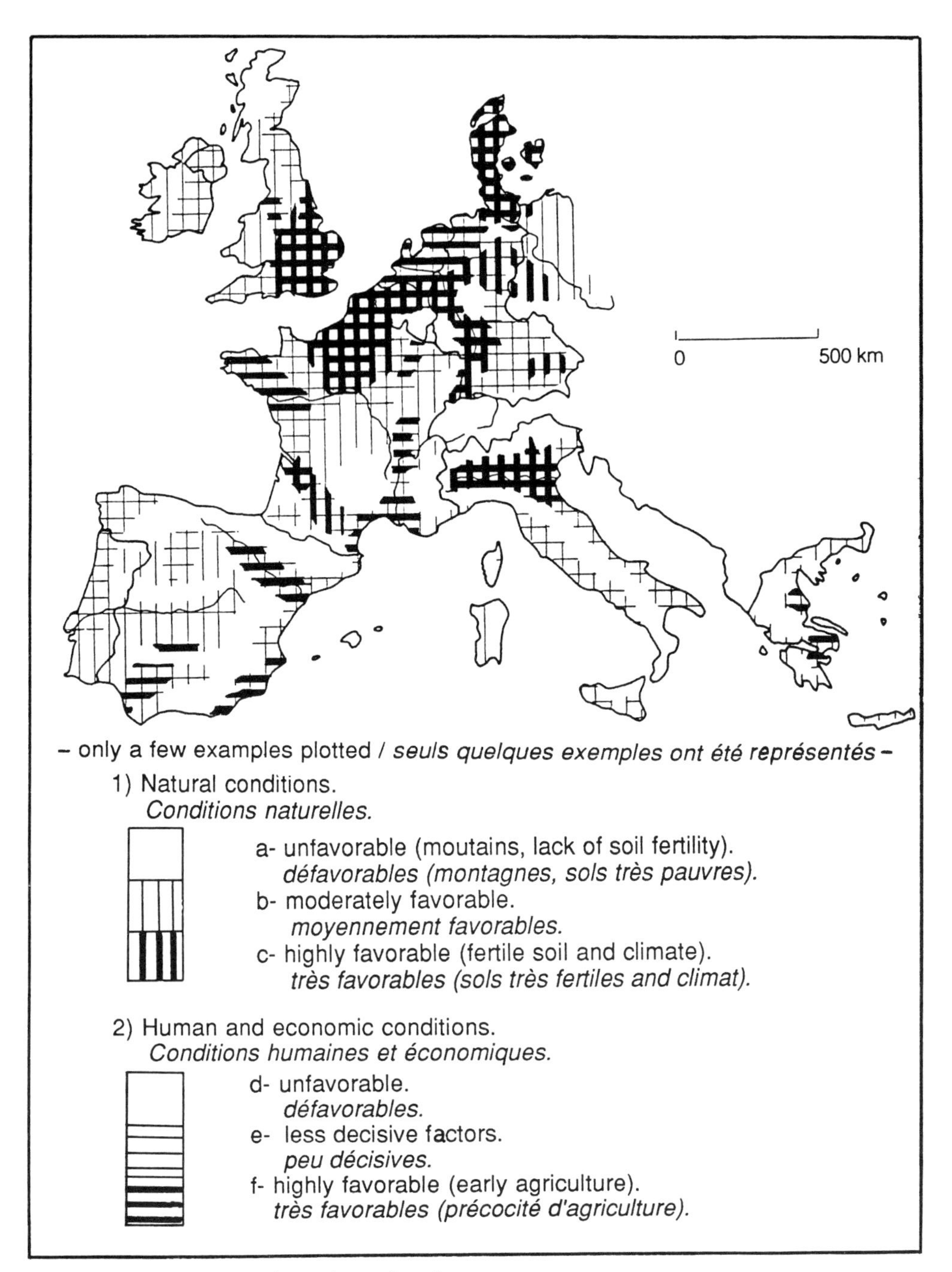

Figure 2 B. Nature's role and man's role
La part de la nature et la part de l'homme
According to Charvet J.P. and Brunet P., 1991.

Permanent factors should be considered. In order to do so, a catchment area typology should be established (8 - 9) largely based on a "morphopedological diagnosis". This should help in drawing up a cartography of erosion risk potential. Morphopedological characteristics, largely a function of morphostructural inheritance, seem to be permanent elements, and their role in triggering the erosion process depends on climatic and anthropic factors.

V. SOIL DETERMINES WATER QUALITY AND MUST BE CONSERVED

Physical, chemical and biological degradation of large-scale farm lands affects the natural fertility of the soil and also influences the water quality.
-*Physical degradation*. - This takes the form of compaction (farming practices and utilization techniques) and occurs on reserves of water which begin to move, generating hydric erosion, and in particular sheet erosion, or more rarely rill erosion, as well as hypodermic runoff, worsened by "tillage soles" formed by increasingly frequent passages of heavy farm machinery (ex. in the Parisian Basin, 18-20 annual passages). A recent discovery in France by Rhone-Poulenc of a vaccination concerning cereal crops in particular, will lead to a decrease in the number of necessary passages and therefore a decrease in compaction, which is an essential improvement as far as infiltration levels of the soil and erosion are concerned.
-*Chemical degradation* - Liquid and solid flows bring on chemical impoverishment, acidity, and salinization, which have then to be corrected chemical fertilizers, N, P, K (Nitrogen, Phosphorus, Potassium). Only two-thirds of this is absorbed by vegetation, with one-third remaining available for runoff movement, directly joining up the evacuation network or the water table (10). If one adds to this the use of pesticides, it is hardly surprising that the natural quality of the soil and water is in a state of continuous degradation. Furthermore, chemical degradation causes :
-*Biological impoverishment* of the fauna and microfauna, indispensable for proper soil balance, impoverishment of organic matter content, currently in constant decrease, and especially on loamy soils due to diffuse runoff, reaching the critical threshold of 1.8 to 2 % (while it was about 3 % twenty years ago).
If we add to this the problem of atmospheric pollution in the form of sulfur dioxide (SO_2), nitrogen oxide (NOx) or waste (for ex. heavy metals), it is no surprise that over the past few years, the natural quality of water and soil have been diminished on the large-scale farm lands of Europe and North America (Fig. 2 bis).

According to the works of pedologists (11) water and pollutant percolation rate is about 50 cm per year. Taking into account that, in our regions, water tables are located even up to several dozen meters in depth, a gap of a few years, or even a few decades, exists between the actual pollution of the water table and the effects of farming practices. Consequently, pollution should really not be more of a problem than it was, say, in the 1970's, but there is, however, a "snow-ball" effect of accumulation, as well as the time factor. The same is true for a brutal change in use of the soil : years or even decades must go by before water and soil quality can improve. On the other hand, reconstituting the natural fertility of the soil will take centuries ; this explains the dire need for good management of the soil asset.

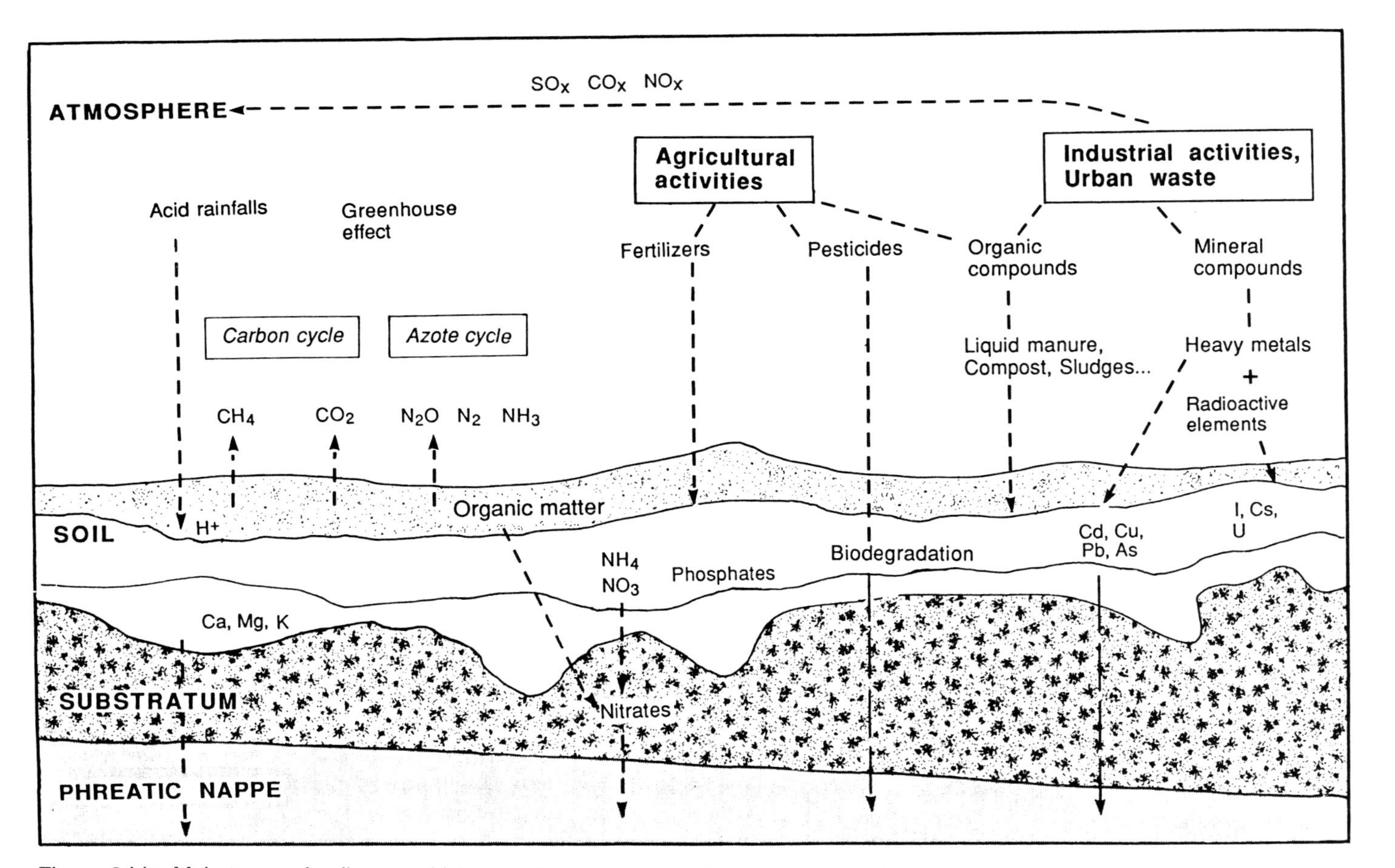

Figure 2 bis. Main types of polluants which pass through the soil, from (12) modified.
Principaux types de polluants traversant le sol, d'après (12) modifié.

VI. SOIL AND FARMING IN GREAT TEMPERATE PLAINS REGIONS OF THE NORTHERN HEMISPHERE

Available arable land (S.A.U. - utilizable agricultural surface) without major constraints and apt to receive all types of agriculture, represents 22 % of the total global area, of which only one-half is currently being farmed (Fig. 3.) (12). The majority of fertile farm land is located in the Northern Hemisphere ; these lands are also those which are the most vulnerable to erosion resulting, for the most part, from agricultural practices and land use (13 - 14). This concerns, in particular, the great plains of North America running up to the Canadian prairies, and the North-Western European plains, extending out to the Germano-Polish and Russian plains and the plains of Northern China. Consequently, most of the world agricultural potential can be found in the world's most industrialized countries : the U.S., Canada, the EEC, Eastern Europe, the USSR, Japan (15 - 16) Tab. 1.

Agricultural statistics (1990) / *Agriculture en chiffres (1990)*

Countries	Population (millions hab)	Surface (1 000 km2)	S.A.U. / Total effective area (%)	Arable lands / S.A.U. (%)	Part of the working population in agriculture (%)
EEC (12)	328	2260	58	52,5	7,4
Eastern Europe	137	1246	69	65,0	18,0
USSR	280	22275	27	62,0	15,0
USA	248	9372	46	48,0	2,9
Japan	124	372	15	51,0	7,9
Canada	26	9959	5	85,0	6,0
France	57	551	59	56,0	3,6

Table 1. Reprinted from : J. Beaujeu Garnier, Images économiques du Monde, 1991, S. Wicherek, 1992.

All these regions have an abundance of arable land, which explains the question of land freeze over the next years ; in ten years' time, this could reach 50% of all European farmland, because of surplus production (17 - 18).

For this reason, these regions (countries) are incontestably privileged, because up till now the world's food supply has depended upon them. This is especially crucial when we consider that world population will probably double over the next two decades and that the problem of food supply is linked to the North-South balance.

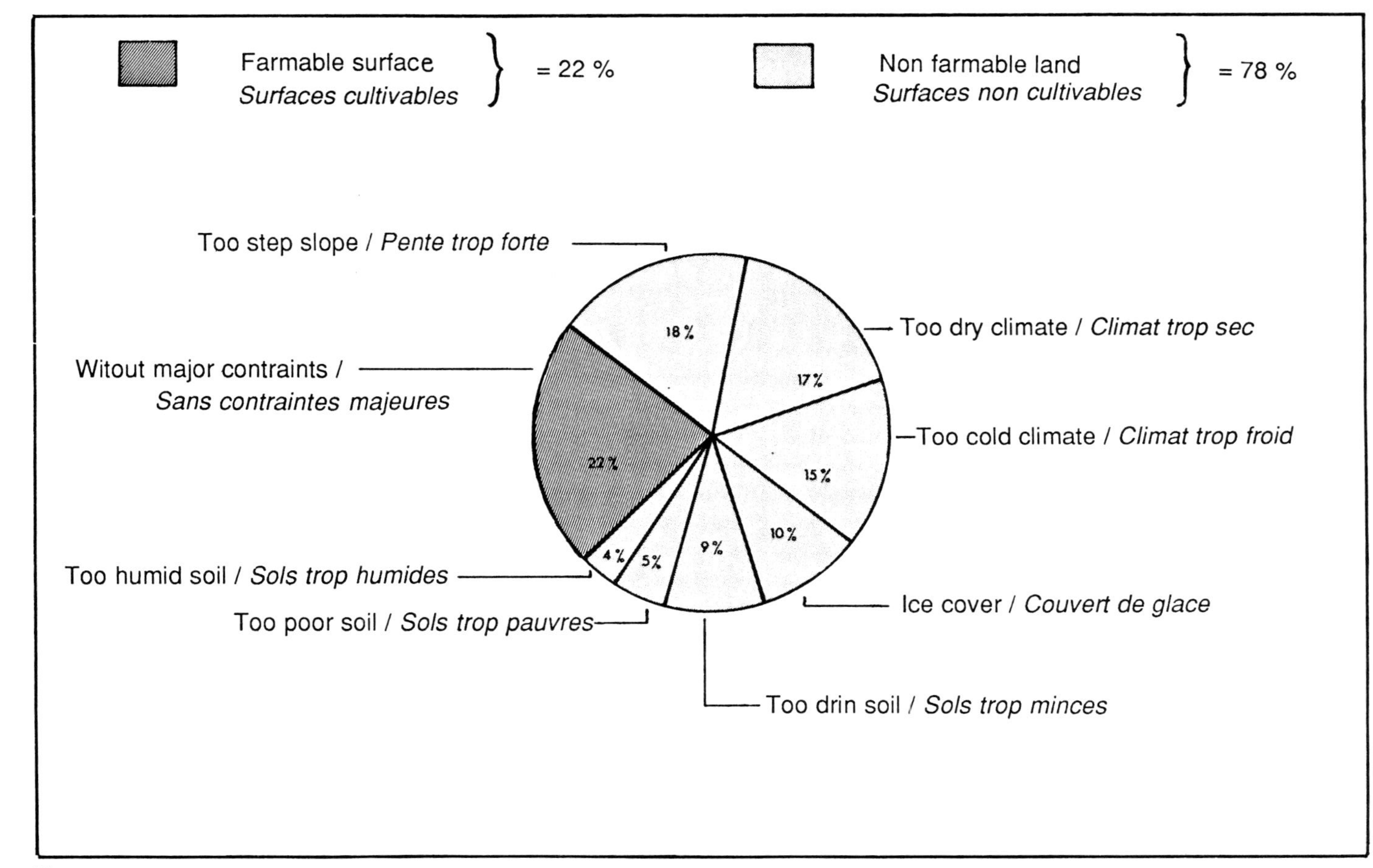

Figure 3. World soil ressources / *Ressources mondiales en sols.*
According to Robert M., 1992.

Our duty as scientists is to propose which types of soil should be frozen, under what form and in what contexts, in order to better protect them, and to keep them in reserve as one of humanity's most coveted assets for the future (Fig. 4).

For lands which remain in use, degradation of their natural qualities (fertility) must be surveyed - especially since the quality of water and food depends on the state of the soil. Substratum relationships, vegetation, the health of entire populations and the quality of the environment should be correlated and economic cost should be evaluated. An exceptional case within these regions is that of Canada whose agricultural soil reserves are practically non-existant (due to the climatic limitations) (19 - 20). Soil, degradation however, is relatively high, which explains the problem of their agricultural competitiveness world-wide (this represents 20 % of the total surplus of Canada's commercial balance). Fig. 5.

In comparison, in France, the situation is totally different, because major agricultural potential and production represent a quarter of all agricultural production for the EEC countries. It is also interesting to stress the role played by history and the behavior of farmers and political decision-makers vis-à-vis soil degradation.

The American "Dust Bowl" of the 1930's and a resulting heightened awareness of problems had little impact in other countries. Europe did not feel concerned by these types of problems, and in particular, of the temperate plain environment, until the introduction of intensive and extensive large-scale farming in the 1970's, occurring at the same time as regrouping of the land in France, (the "remembrement" policy), resulted in an increase of erosion. The scientific community and political decision-makers have since then become more aware of this dilemna.

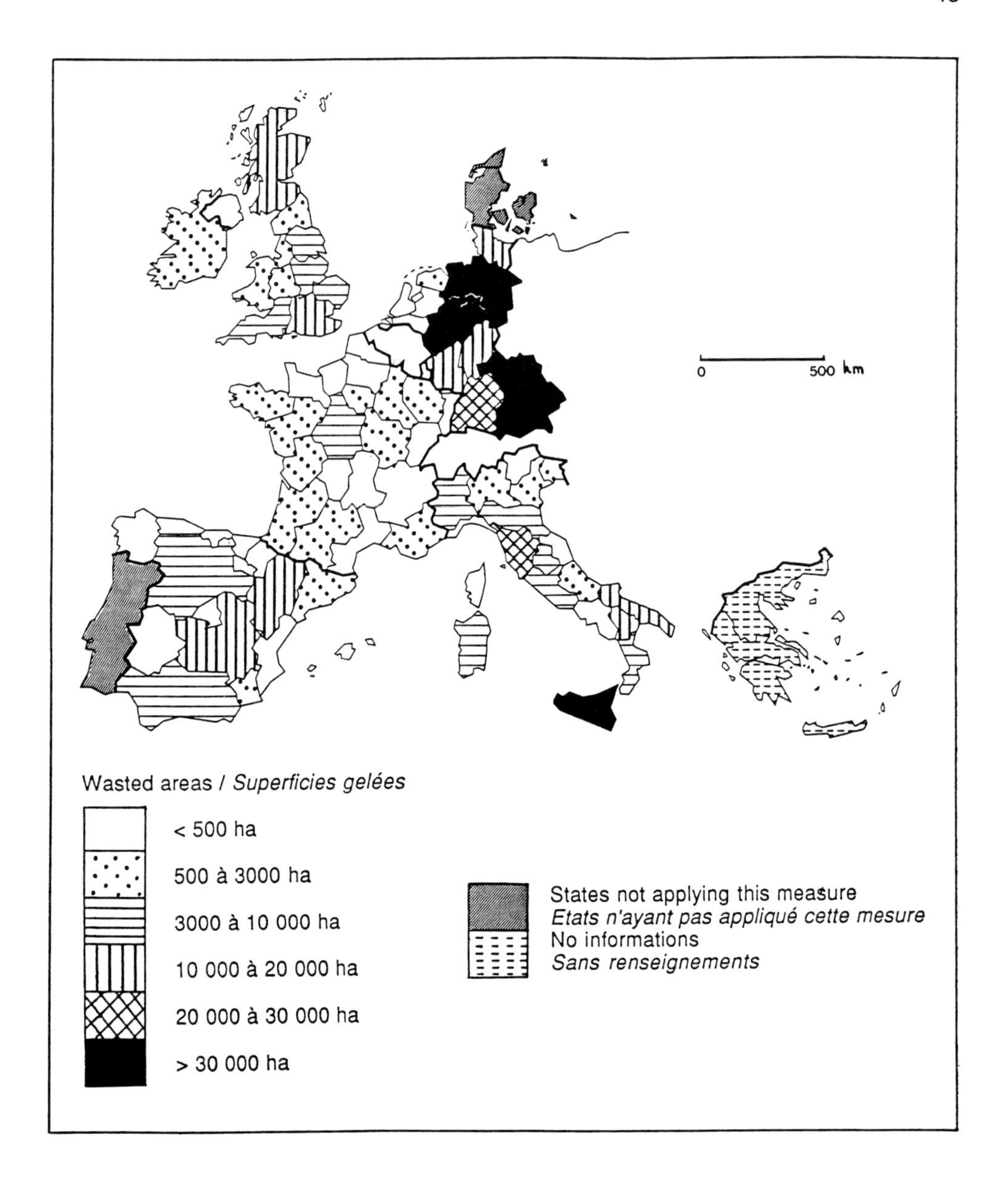

Figure 4. Fallow lands / Gel des terres. 1988-1989. According to Charvet and Brunet P., 1991.

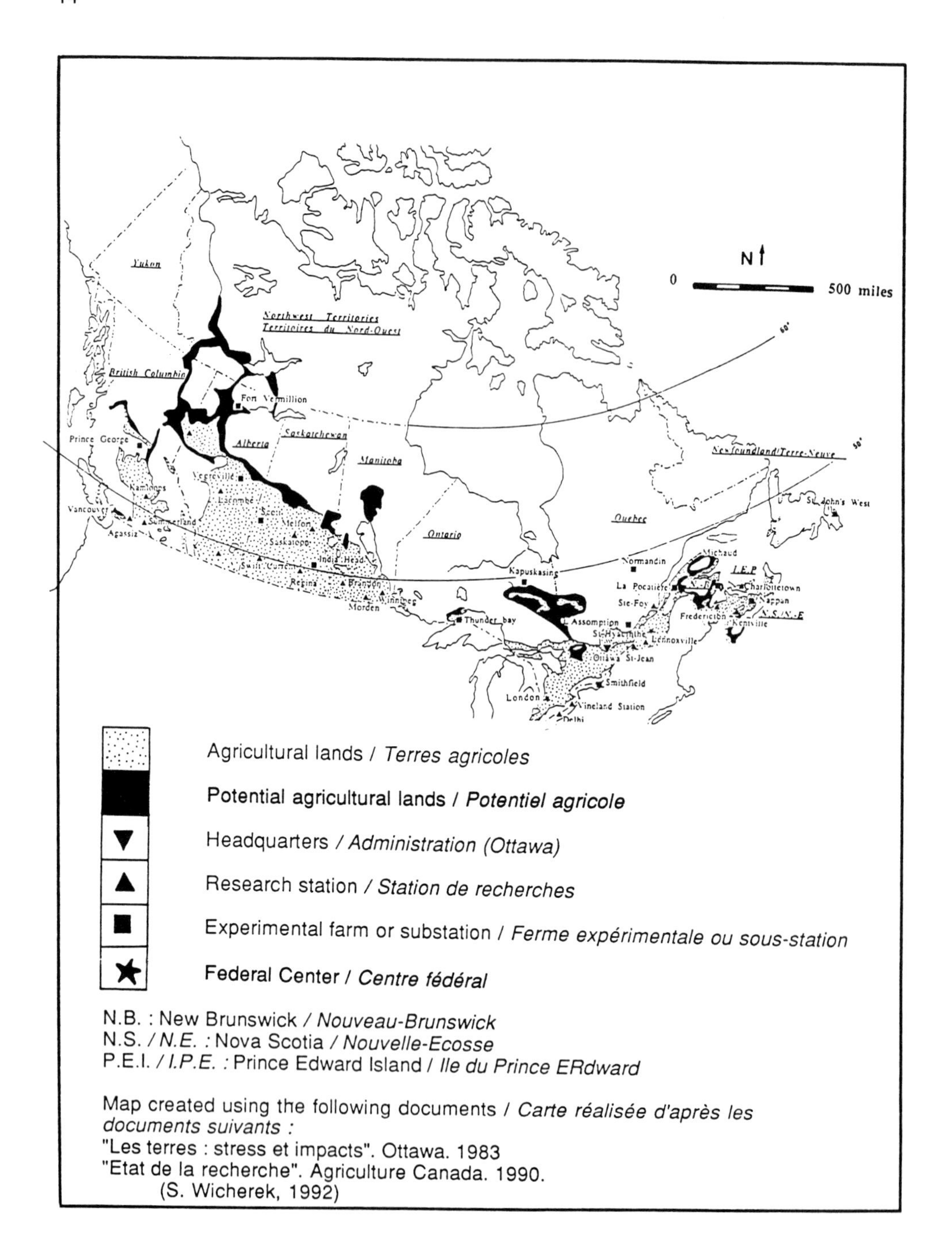

Figure 5. Canadian agricultural lands and research Establishments. Terres agricoles et Etablissements de recherche.

CONCLUSION : The Necessary International Dimension

The Northern European large-scale farming plains have long been considered to be stable environments, with no noteworthy erosion. However, over the past twenty years or so, soil degradation has been taking place. Who is responsible ? Man. It all began around 1970 with the regrouping policy. The result was that, instead of stagnating and infiltrating, water ran off and carried soil and natural fertilizers in its wake, and in particular organic matter. This explains why chemical fertilizers needed to be used, and they too were carried off by rain, polluting rivers and water tables... Furthermore, organic matter is what holds loamy fragile soil together. New techniques and farming methods have worsened the situation, and incessant passages of machinery have packed down the soil leading to formation of a "tillage sole", a layer of compacted, waterproof soil, thirty centimeters thick, which in turn causes leaching of the superficial layer of the soil. In consequence : there is a decrease in natural fertility, and productivity, there are flooding, mud slides, superficial crusting, gullies... all kinds of catastrophies fairly unheard of twenty years ago.

A better understanding of the soil erosion phenomenon, to which classical criteria for evaluation are poorly adapted, is of major scientific interest because of its scientific implications as well as those in the areas of the Environment, and Agriculture, where people in charge are taking interest in advances made in understanding erosion.

Any in-depth study of physical and biological components of the temperate plain environment will encounter past or present societal phenomena. If, at first, the researcher may limit himself to recording and observing the physical and biological processes at work, correlating these processes implies his taking "human action" into account. Human behavior is not only a passive filter, it is in itself a protagonist participating in the production of the environment.

Most of the countries with temperate plain environments must absolutely take control over their lands. Mastering the soil means two different things : on one hand, conserving the quality of the land by fighting erosion, impoverishment of organic matter or abusive chemical fertilization which influence water quality. On the other hand, a grass-roots policy allowing soil occupation to be controlled and land freezing, concerning Europe in particular, must be put into place, in the hopes of avoiding the savage urbanization of the rural environment.

REFERENCES

1. Wicherek, S. (1990) : Paysages agraires, couverts végétaux et processus d'érosion en milieu tempéré de plaine de l'Europe de l'Ouest. (Agrarian landscapes and canopies : erosion processes on western european temperate plains environment). *Soil Technology, Catena,* 3, 2, 199-208.
2. Wicherek, S. (1992) : Sols : l'érosion en temps réel. *Journal du CNRS,* 25, p. 24.
3. De Ploey, J. (1990) : La conservation des sols. *La Recherche, Suppl.* 227, 38-41.
4. Neveu, A. (1991). Agriculture. Economie de l'Agriculture française en Europe. Forces et faiblesses. *Dunod,* 192 p.
5. Wicherek, S. (1988) : Les relations entre le couvert végétal et l'érosion en climat tempéré de plaines. Exemple : Cessières (Aisne - France). (Relationship between the vegetal cover and erosion under temperate climates of plains. Example of Cessières (02 Aisne - France)). *Zeitschrift für Geomorphologie,* 32, H.3, 339-350.

6. Wicherek, S. (1991) : New approach to the study of erosion in cultivated lands. (Nouvelle approche pour l'étude de l'érosion des sols cultivés). *Soil Technology, Catena,* 4, 2, 99-100.
7. Boardman, J., Foster, I.D.L. and Dearing, J.A. (1990) : Soil erosion on agricultural land British geomorphological Research group Symposia Series, *Wiley,* Chichester, 687 p.
8. Veyret, Y., Wicherek, S. and Arnould, P. (1991) : Terres de grandes cultures : l'érosion des sols; exemples pris dans le Bassin Parisien, 27 p., 20 diapositives. Publication dans la Collection dirigée par J.C. Miskovsky, Université Paris VI, Dept. geodynamique des milieux continentaux et concours du Centre de Recherches Archéologiques du CNRS, *Géopré,* 6, 27 p., 20 diapositives.
9. Arnould, P., Veyret, Y. and Wicherek, S. (1992) : Influence des modifications des structures agraires sur l'érosion des sols (Changes in rural structures and soil erosion). *B.A.G.F.,* 69, 2, 100 p.
10. Mousset, S. (1987) : A propos de la fertilité des sols français. *Information géographique,* 51, 5, 190-195.
11. Duchaufour, Ph. (1988) : Abrégé de pédologie, *Masson, Paris,* 2e ed., 224 p.
12. Robert, M. (1992) : Le sol, ressource naturelle à préserver pour la production et l'environnement. *Cahiers, Agricultures,* 1, 1, 20-34.
13. Chisci, G. and Morgan, R.P.C. (Eds) (1986) : Soil erosion in the European Community. Impact of changing agriculture. Balkema, Rotterdam, 233 p.
14. Neboit, R. (1991) : L'homme et l'érosion. L'érosion des sols dans le monde. *Univ. Clermont-Ferrand II,* Fasc. 34, N.S., 2e Ed., 270 p.
15. Beaujeu-Garnier, J. and al. (1991) : Images économiques du Monde,*Sedes,* 236 p.
16. Wicherek, S. (1992) : Excursion guide. Internation Symposium "Farm land erosion in temperate plains environments, 25-29 May 1992, Saint-Cloud, Paris, 148 p.
17. Politique Agricole Commune (P.A.C.) (1990) : Le marché unique de 1993. *Ministère de l'Agriculture et de la Forêt,* 1, 30 p.
18. Brunet, P. and Charvet, J.P. (1991) : L'agriculture de la C.E.E. *Documentation photographique, Documentation française,* 7004, 40 p.
19. Dumanski, J., Coote, D.R. and Luciuk Gand Lok, C. (1986) : Soil conservation in Canada. *Journal of soil and water conservation,* U.S.A., 41, 4, 204-210.
20. Wicherek, S. (1993) : Les terres agricoles du Canada et leur dégradation. (Agricultural lands in Canada : their degradation).*Cahiers, Agricultures* (sous presse).
21. Wicherek, S. and Boissier, M.O. (1991) : Viticulture and soil erosion in the North of Parisian Basin. Example : the mid Aisne region, France. (Viticulture et érosion des sols dans le nord du Bassin parisien. Exemple : région de l'Aisne médiane, France). *Zeitschrift für Geomorphologie N.F. Suppl.,* 83, 115-126.
22. Monnier, G., and Boiffin, J. and Papy, F. (1986) : Réflexions sur l'érosion hydrique en conditions climatiques et topographiques modérées des des systèmes des grandes cultures d'Europe de l'Ouest. Cah. ORSTOM, Série Pédologique, XXII, 2, 123-131.
23. Vogt, H. and Vogt, T. (Eds) (1979) : Colloque sur l'érosion des sols en milieu tempéré non méditerranéen, Strasbourg-Colmar, sept. 1978, 251 p.
24. Godard, A. and Rapp, A. (1987) : Processus et mesure de l'érosion (XXVème Congrès International. Géographpie UGI, Paris, 1984. Editions du CNRS, 576 p.

Farm Land Erosion: In Temperate Plains Environment and Hills
S. Wicherek (Editor)

Simple methods of characterizing erosive rainfall with reference to the South Downs, southern England

J. Boardman [a], and D.T. Favis-Mortlock [b]

[a] School of Geography, University of Oxford, Mansfield Road, Oxford OX13TB, United Kingdom.

[b] Departement of Building, Brighton Polytechnic, Lewes Road, Brighton BN2 4AT, United Kingdom.

SUMMARY

Erosion monitoring has been carried out since 1982 on an area of about 36 km^2 of agricultural land on the South Downs, southern England. The dominant crops are winter cereals and most erosion occurs from September to January. In this area land use and management have changed little from year to year on a majority of fields. Rainfall is therefore the main control on the total amount of erosion occurring each year in the monitored area. Erosion in fields of winter cereals is initiated in the autumn by rainfall events of about 30 mm in two days. Erosive rain falling in the growing season is characterized by use of a Rainfall Index (RI). The RI for each year shows a good relationship to total and median soil loss in the monitored area. In the years 1988-91 two problems emerged in the use of the RI: (1) it is a poor predictor of low amounts of erosion, and (2) it fails to take account of late winter erosion on ploughed land. There are, therefore, problems in the application of the RI as originally proposed. Other methods of characterizing erosive rainfall are explored. These include using a RI weighted according to the timing of the rainfall event in the growing season. The highest daily or the total rainfall in the period of erosion risk may also be used. All methods have their limitations but some provide a relatively satisfactory means of predicting total soil loss, median soil loss or the number of sites of erosion, under current conditions, in the monitored area.

RESUME

Un suivi de l'érosion a été effectué depuis 1982 sur une surface de terre arable d'environ 36 km^2, dans les South Downs, au Sud de l'Angleterre. Les céréales d'hiver constituent la culture dominante. L'érosion est la plus active entre septembre et janvier. Dans cette région, l'utilisation et la gestion du sol, demeurent inchangées d'une année à l'autre. La pluie constitue le facteur principal qui commande l'importance de l'érosion survenant dans la région étudiée.

L'érosion sur les céréales d'hiver commence à l'automne par des précipitations d'environ 30 mm en 2 jours. L'ampleur des évènements pluvieux, facteurs d'érosion, durant la saison végétative est estimée à l'aide d'un Index de Pluviométrie (Rainfall Index - R.I.). De façon générale, le R.I. révèle un bon rapport à la perte de sol totale et médian dans la région de référence. Le rapport est moins bon si l'on inclut l'année 1989-1990. Ceci est dû à la chute de fortes pluies à la fin de l'hiver ce qui a pour effet d'agir sur les terres labourées plus que sur les céréales d'hiver. Cependant la perte de sol totale reste relativement faible. Des problèmes existent donc pour utiliser le R.I. ainsi conçu. D'autres méthodes sont proposées pour apprécier l'érosion hydrique. Eles incluent l'utilisation d'un R.I. établi en fonction de la période au cours de la saison végétative où survient l'épisode pluvieux.

Il est possible de prendre en compte le taux le plus élevé de pluviosité quotidienne ou le taux de pluviométrie total durant la période de risque d'érosion. Toutes les méthodes ont des limites, certains fournissent cependant de façon relativement satisfaisante, un moyen de prévoir la perte de sol totale, des sites d'érosion dans les conditions actuelles caractérisant la région de référence. Ces approches fournissent aussi la base de prévisions.

INTRODUCTION

Studies of the relationship between rainfall and erosion have emphasised the ability of raindrops to detach soil particles and cause splash erosion. In western Europe rainfall amounts and intensities are much lower than in tropical areas and many parts of the USA and maps based on R values as defined by Wischmeier and Smith (1), tend to suggest that there should be little risk of erosion (e.g. 2). However, field studies have shown that erosion occurs frequently and may be associated with relatively low rainfall amounts and intensities (3) ; topography, crop type and management are also important. Other factors include the susceptibility of soils to crusting (4) and compaction (5), linear elements in the landscape (6) and the concentration of flow in valley bottoms (7 - 8 - 9). Rainfall is recognised as one element interacting with others in the erosional process. For example, Papy and Boiffin (10) recognise the importance of cumulative rainfall in controlling crusting and use this parameter to define periods of erosion risk.

A re-examination of the character of erosive rainfall is necessary for several reasons.

The recognition of the central importance of rill and gully processes with less emphasis on detachment, splash and wash (e.g. 11 - 9), forces us to consider rainfall events that produce runoff and any form of concentrated flow. The few measurements of flow in rills suggest that even under unexceptional rainfall conditions large amounts of material are transported (12). De Ploey (6) recognises situations where there is runoff but little detachment and entrainment on slopes but gullying in the valley bottom. The relationship of rainfall to these processes has to be studied in the field rather than on experimental plots.

There is therefore a need for empirically based, easy to calculate methods for characterizing erosive rainfall. These indices should correlate with measured amounts of erosion on farmers' fields and thus be able to predict erosion under similar conditions in the future.

RAINFALL AND EROSION IN BRITAIN

Several studies in Britain have shown that erosion occurs with relatively low rainfall amounts. Evans and Nortcliff (13) in a study of erosion in north Norfolk, use 7.5 mm as a measure of daily rainfall necessary to produce erosion. This was based on Evans and Morgan (14) who described erosion caused by a thunderstorm of 7.4 mm. Boardman (15) also compared the number of days when more than 7.5 mm of rain fell during the period of erosion with a longer term average. On compacted sandy soils in the West Midlands, Reed (16) states that erosion occurs with rainfall intensities of 1 mm hr^{-1} in events totalling 10 mm. Evans (17) suggests that 'rainfalls which cause erosion range from storms of about 10 mm in the summer months to about 20 mm falling over three days on to wet soils in the winter months' (p 524)

Frost and Speirs (18) report cases of serious erosion as a result of daily rainfall amounts as low as 12.7 mm.

There are clear discrepancies between the rainfall values listed above. However, these are explained by differences in soil, topography, crops and management of the areas of observation. They are all relatively low values compared to those that appear to cause erosion in other countries (19). This is probably because the values were obtained from field investigations rather than plot studies. Erosion occurs on fields at lower rainfall threshold values than on plots because of the large size of fields and the concentration of water along linearities such as valley bottoms and vehicle wheel tracks. Most plot experiments fail to reproduce the conditions found on agricultural fields. Rainfall threshold values obtained from plots are therefore not appropriate for field situations.

In this paper we consider simple methods by which erosive rainfall may be characterized and related to amounts of erosion. These are based on observations of erosion in the field over a ten-year period. We attempt to relate rainfall to total and median amounts of erosion per year within a specified area. An approach to predicting soil loss on individual fields using an Expert System is considered elsewhere (20). We have also used north American computer-based erosion models for the same end ; these, however, require long runs of rainfall data or use a measure of rainfall energy such as the R factor from the USLE (21).

THE STUDY AREA

The South Downs in southern England rise to just over 200 m and are composed of Chalk, a soft Cretaceous limestone, with patches of sandy and clayey deposits of Tertiary and Quaternary age on some interfluves. Dry valley networks dissect the landscape giving rise to valley-side relief of up to 150 m. The principal soils are classified as Andover 1 association (22) and are typically thin, stony rendzinas containing 60-80 % silt of loessial origin. They are silty clay loams and silt loams rarely thicker than 25 cm and frequently with A horizons of only 15 cm over bedrock.

Farming on the Downs is now dominated by the growing of autumn planted wheat and barley ('winter cereals'). In many areas over 55 % of the arable area is under winter cereals and locally it is higher. Slopes of up to 25° are under winter cereals.

Most erosion occurs in winter cereal fields between September and January. It is caused by runoff from smooth, bare or relatively bare, slopes. Other factors such as the large size of fields, the presence of wheel tracks and the concentration of runoff in valley bottoms are also important (23).

Between 1982 and 1992 an area of about 36 km^2 of agricultural land in the eastern South Downs was monitored. Soil loss was estimated by measurement of the volumes of rills, gullies and areas of deposition. Data relating to soil, morphometry, crop, farm practice and rainfall were collected for each site of erosion.

Mean annual rainfall on the eastern South Downs is between 750 and 1000 mm with an autumn peak (24). There are seven rain gauges in or very close to the study area but these vary in type and length of record. Most use is made of the Southover, Lewes, autographic gauge (23, Figure 7.3). The mean annual rainfall at Southover is 759 mm (1950-1989).

RAINFALL AND EROSION ON THE SOUTH DOWNS

In the study area the amount of land under winter cereals remains roughly constant from year to year. A few fields are in rotations with oil seed rape, grass ley or spring cereals but many remain under winter cereals. Farming practices change little from year to year. Erosion occurs at predictable topographic locations mainly in winter cereal fields.

Over the ten-year period of monitoring, erosion appears to have been closely related to rainfall because other factors have changed little. Early results suggested this relationship and therefore the need to characterize erosive rainfall. Plot experiments in 1985-86, and observations on adjacent fields, showed that rilling began at a rainfall threshold of about 30 mm in two days. In the critical period before vegetation cover inhibits runoff this rainfall amount is frequently exceeded : between 1 September and 30 November in the years 1982-91 at the Southover gauge there were 22 such events.

Rainfall intensity appears to be less important than quantity (cf. 13). In the autumn and winter of 1987 the maximum short-period intensity recorded at Southover was 40 mm $hr^{-1}$for 15 minutes (12, Table 1). However, erosion is frequently the result of low intensity rainfall.

As a result of the recognition of a threshold for the initiation of rilling, a Rainfall Index (RI) was proposed which would characterize each winter cereal growing season in terms of its erosive rainfall. This procedure is applied only to rainfall which occurs during an 'erosion season' defined for the South Downs as between 1 September and 1 March. Each rainfall event is allotted a value (Table 1) and these are then summed to give an annual total (Table 2). Annual RI values in Table 2 differ slightly from those published in Boardman (23) because they are based on the rainfall record from one rain gauge, that at Southover, rather than incomplete records from several.

A log-linear relationship was observed between the RI and soil loss and total soil loss for each of the seven erosion seasons between 1982 and 1989 (Figure 1), and in terms of median soil loss for the same period in Boardman et al. (24). The regressed line for 1982-89 in Figure 1 fits the measured data reasonably well, with an R^2 value of 0.747 ; similarly the fit for median soil loss values in Boardman et al. (24) is 0.797. However when data for subsequent years is included in the analysis (Figure 1), a log-linear fit to the data is less good with the R^2 value falling to 0.555 (Table 3). Examination of this later data, suggested that the RI is a less reliable indicator of very low amounts of erosion and that the timing of the rainfall events is important.

In 1989-90 a different pattern of rainfall with a late winter peak led to erosion on ploughed land prepared for spring cereal sowing (25). Little erosion occurred on winter cereals because by that time there was adequate crop cover. The importance of the timing of rainfall had, therefore, to be addressed.

Tab. 1.
Values alloted to rainfall events to obtain Rainfall index (RI).

Rain event	Index value
30 mm in 2 days	1
30 mm in 1 day	2
60 mm in 2 days	3
60 mm in 1 day	4

Table 2.
Rainfall and soil erosion in the monitored area, South Downs, 1982-1991.

Year	1 Sept-1 Mar maximum daily rain in area (mm)	RI*	1 Sept-1 Mar* total rain (mm)	Median soil loss $m^3\ ha^{-1}$	Total soil loss $m^3\ ha^{-1}$	Number of sites
82-83	42(D)	11	724	1.7	1816	68
83-84	22(S)	3	560	0.6	27	7
84-85	30(S)	6	580	1.1	182	25
85-86	33(O)	5	453	0.7	541	49
86-87	38(O)	4	503	0.7	211	34
87-88	63(B)	13	739	5.0	13529	97
88-89	39(B)	2	324	0.5	2	1
89-90	34(S)	8	621	1.4	940	51
90-91	51(S)	5	469	2.3	1527	43
91-92	25(S)	0	298	1.2	112	14

* based on Southover rain gauge data.
Rain gauges : D = Ditchling Road ; O = Offham ; S = Southover ;
B = Balsdean ; H = Housedean.

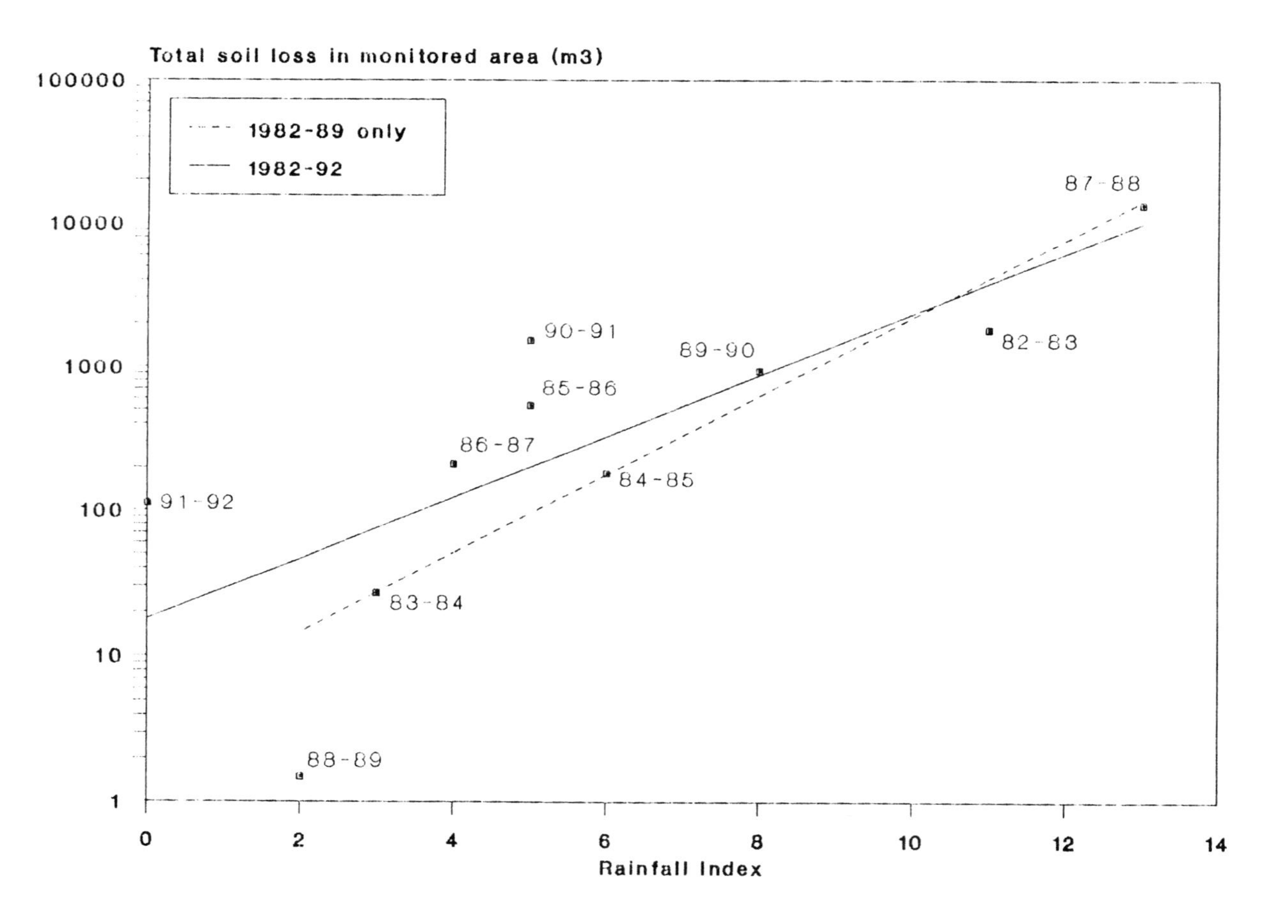

Figure 1. Erosion in the monitored area (1982-1991) and the rainfall index.

Rainfall measure	total soil loss	log_{10} (total soil loss)	median soil loss	log_{10} (median soil loss)	number of sites	log_{10} (number of sites)
total rainfall	0.241	**0.412**	0.239	0.261	**0.524**	**0.373**
max. daily rainfall	**0.585**	0.279	**<u>0.664</u>**	**0.522**	**0.513**	0.079
unweighted RI	**0.466**	**<u>0.555</u>**	**0.478**	**0.480**	**0.817**	**0.426**
RI-1	**0.470**	**0.436**	**0.505**	**0.505**	**<u>0.707</u>**	0.289
RI-2	**<u>0.599</u>**	**0.534**	**<u>0.590</u>**	**0.538**	**<u>0.833</u>**	**0.366**
RI-3	**0.508**	**0.446**	**0.415**	**0.525**	**<u>0.722</u>**	0.332
RI-4	**<u>0.553</u>**	**0.506**	**<u>0.562</u>**	**0.534**	**<u>0.798</u>**	**0.344**
RI-5	**<u>0.617</u>**	**0.441**	**<u>0.625</u>**	**<u>0.567</u>**	**<u>0.751</u>**	0.385
RI-6	**<u>0.668</u>**	**0.457**	**<u>0.632</u>**	**0.533**	**<u>0.794</u>**	0.268

BOLD is significant at the 5 % level

<u>BOLD and UNDERLINE</u> is significant at the 1 % level

Table. 3. R^2 values for rainfall measure regressions (N = 10).

METHODS

Data for the Southover rain gauge for 1982 to 1992 was examined to obtain three measures of rainfall quantity : total rainfall, maximum daily rainfall and unweighted RI for each erosion season. These values were then regressed against total soil loss, median soil loss, the number of sites at which erosion occurred, and the logarithm of each of these. The R^2 value for each regression is given in Table 3.

Six variants of the RI were also formulated and used in the analysis. The unweighted RI is unchanged from the definition given in Boardman (23), while the others assign a different weighting to each rainfall event depending on the day of its occurrence. These weightings are illustrated in Figure 2 in which time is measured from the start of the erosion season.

The first weighting, RI-1, is very simplistic ; it assumes a linear increase in crop cover (and hence an inverse decrease in erodibility) from the beginning to the end of the erosion season, for a hypothetical autumn-planted crop (Figure 2-1). Thus bare ground at the start of the erosion season is given a weighting of 1 (the same as the unweighted RI). At the end of the erosion season crop cover is assumed to reach the 30 % threshold at which the ground surface is fully protected against erosion (26). This pattern is assumed to apply equally to all land in the monitored area.

RI-2 similarly models a linear increase in crop cover for the first half of the erosion season, but overall erodibility then increases for the second half of the season (Figure 2-2). This very crudely reproduces a decline in roughness and an increase in crusting on any bare ground in the latter half of the winter.

The third and subsequent weightings include the effects of land use in the monitored area. In the years 1982-88, an average of 2.5 times as much land was devoted to winter cereal production as to spring cereals (other crops are considered to be unimportant). Therefore 28 % of the total cereal area will remain bare following autumn and winter ploughing, until the spring-sown cereal crop begins to cover the ground at the end of the erosion season. As bare ground which covers the whole of the monitored area is assigned a weighting of 1.0, bare ground which covers 28 % of the area is assigned a weighting of 0.28. (Note that this is not the same as saying that 28 % of erosion events occur in spring cereals).

RI-3 makes use of this concept ; it combines RI-1's simple winter-sown crop model with this spring cereal weighting (Figure 2-3).

The hypothetical autumn-planted crop is assumed to continue growth linearly and uniformly throughout the winter months. This is not achieved in practice : Robinson and Boardman (27) noted little winter wheat growth (and indeed some die-back of crop) during winter on the plots at their Houndean site. This is modelled in RI-4, where crop growth levels out during the coldest part of the winter, then begins again as temperatures increase (Figure 2-4). Again, 28 % of the land area devoted to spring cereals is assumed.

Robinson and Boardman (27) however found a different growth pattern in winter barley at the more sheltered Housedean site. Here growth continued throughout the winter, die-back was negligible and ground cover exceeded 30 % by the end of December. This is typical of some sites on the Downs, and some years. RI-5 simulates this, with a steeper decrease in erodibility. The area devoted to spring cereals is assumed to be 28 %.

RI-6 takes another approach to estimation of crop growth.The Erosion-Productivity Impact Calculator (EPIC) model was used to simulate winter cereal growth at a site within the monitored area (28). Values for above-ground biomass during the erosion season were computed by the model. These were used as a surrogate for crop cover, and inverted to give an RI weighting for winter cereal (Figure 2-6). Assumptions for spring cereal area were as above.

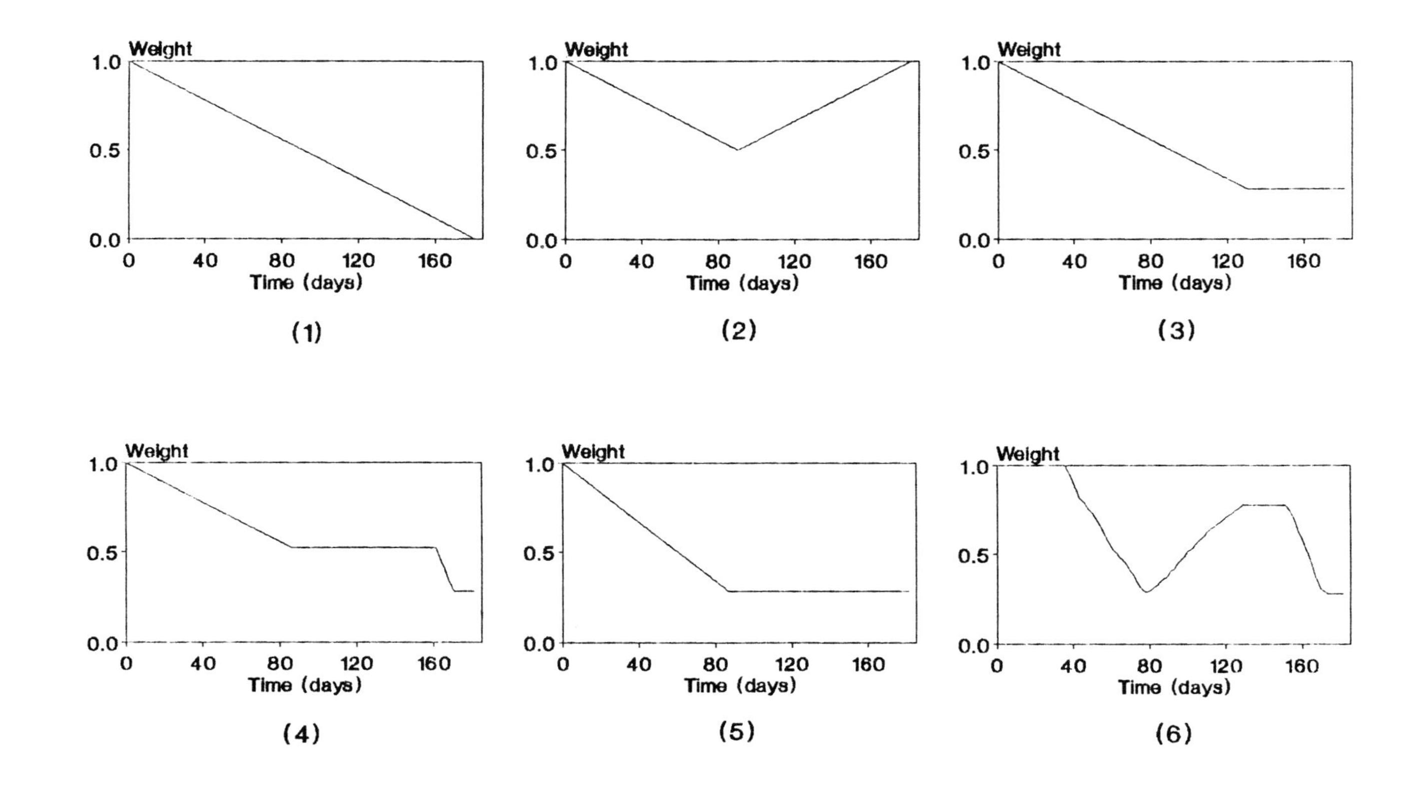

Figure 2. Weightings applied to the rainfall index (see text for explanation).

RESULTS

Every rainfall measure used - apart from maximum daily rainfall - proved better at predicting the number of sites at which erosion occurred than at predicting either total soil loss or median soil loss. The highest R^2 scores were 0.833 by RI-2 and 0.817 by the unweighted RI.

Maximum daily rainfall produced the highest R^2 value for median soil loss with an R^2 of 0.664, followed by RI-6 with 0.632 and RI-5 with 0.625. The same RI variants also performed well in terms of predicting total soil loss.

DISCUSSION

Compared to more physically-based models such as EPIC or WEPP (Water Erosion Prediction Project), this black-box modelling approach is extremely simplistic. The weighted RI method assumes crop growth to be the only factor affecting areal erodibility. There are a large number of assumptions implicit in the method. In particular the methods all neglect changes in soil type within the area considered, also local variations in tillage methods and timing of sowing the crop. The latter is particularly important in explaining differences in erosion between fields. In this simplified approach many factors are averaged both in a spatial and temporal sense.

But the method has the great benefit of simplicity. The high R^2 value for the unweighted RI and number of sites indicates that a clear linear relationship exists between these quantities in the study area. However, none of the rainfall measures tested were particularly successful in predicting soil loss amount, which would have been a more practically useful achievement. Maximum daily rainfall figures are readily available, and proved to be slightly better than any of the more sophisticated measures at predicting median soil loss.

The inclusion of the crop growth weightings seemed to improve the predictive ability of the RI here, although this may have been merely fortuitous (cf. the high R^2 value for number of sites using RI-2).

All relationships gave a better fit in their linear forms rather than the log-linear which was found to be best in Boardman (23) and Boardman et al. (24). At first glance it might appear that this is due to the decrease in influence of the single year of very high erosion rates 1987-88, as the dataset increases in size. However, further analysis does not support this hypothesis : removal of this point from the dataset does not improve the R^2 values obtained. Nonetheless it may be that the relationship is linear only up to a point ; the sharp increase in erosion in very wet winters may indicate that some discontinuity has been crossed.The crossing of an RI threshold value between 11 and 13 gives rise to a sevenfold increase in total erosion and a threefold increase in median rate of threefold (26).

There are particular problems in the use of the RI approach to predicting amounts of erosion in low rainfall years (cf. 1988-89 and 1991-92). This may be because small amounts of erosion which may be missed in a field survey can make substantial differences to the position of the point with respect to the regression line. Similarly, factors which are inadequately dealt with by our averaging approach may mean that a field erodes or does not erode as a result of rainfall close to the threshold value. Although, overall, rainfall intensity seems to be of less importance than amount, it may explain some of the 'noise' in the data. For example, small amounts of erosion in 1991-92 with an RI of zero. These effects are of less importance in a large dataset than in years with very low numbers of sites of erosion.

The fact that RI-2 performed reasonable well in terms of predicting number of sites and total soil loss (Table 3), suggests that our assumption that it is overly simplistic may not be correct. A decline followed by an increase in erodibility may adequately reflect the average soil surface conditions controlled by crop growth and roughness/crusting.

The greater ability of most rainfall measures to predict number of sites rather than soil loss amount may be in part due to the fact that the number of sites where erosion occurred can be more accurately estimated than can erosion amount. However, there is a very strong correlation between the log of the number of sites and the log of total soil loss (R^2 value of 0.921).

CONCLUSION

There are several reasons why an unsophisticated approach such as that proposed here, which gives reasonably reliable estimates of erosion, is of greater practical value than more ambitious modelling approaches. One is that for particular sites or areas, rainfall variability is not reflected in the available data. There are similar problems with obtaining accurate topographic, soil and farm management data. An approach which is conceptually straightforward, based on readily available data and requires little data processing has much to recommend it. We have attempted to show that it can be utilised to give acceptable estimates of erosion. We may be accused of accepting the second-best but would argue, with Voltaire, that, 'the best is the enemy of the good'.

ACKNOWLEDGMENTS

We thank Dr. R. Evans for valuable comments on a draft of this paper ; also the National Rivers Authority kindly provided rainfall data.

REFERENCES

1. Wischmeier, W.H. and Smith, D.D. (1978) : Predicting Rainfall Erosion Losses - a Guide to Conservation Planning. USDA Agricultural Handbook 537.
2. Pihan (1979) : Risques climatiques d'erosion hydrique des sols en France. In: Vogt, H. and Vogt, Th. (Eds.), Proceedings Seminar Agricultural Soil Erosion in Temperate non-Mediterranean Climates, University of Strasbourg, 13-18.
3. Evans, R. (1990) : Water erosion in British farmers' - some causes, impacts, predictions. Progress in Physical geography 14, 199-219.
4. De Ploey, J. (1979) : A consistency index and the prediction of surface crusting on Belgian loamy soils. In : Vogt, H. & Vogt, Th. (Eds.), Proceedings Seminar Agricultural Soil Erosion in Temperate non-Mediterranean Climates, University of Strasbourg, 133-137.
5. Fullen, M.A. (1985) : Compaction, hydrological processes and soil erosion on loamy sands in east Shropshire, England. Soil and Tillage Research 6, 17-29.
6. De Ploey, J. (1989) : Erosional systems and perspectives for erosion control in European loess areas. Soil Technology 1, 93-102.
7. Evans, R. and Cook, S. (1986) : Soil erosion in Britain. SEESOIL 3, 28-58.

8. Auzet, A.V., Boiffin, J., Papy, F., Maucorps, J. & Ouvry, J.F. (1990) : An approach to the assessment of erosion forms and erosion risk on agricultural land in the Northern Paris Basin. In : Boardman, J., Foster, I.D.L. & Dearing, J.A. (Eds.), Soil Erosion on Agricultural Land, Wiley, Chichester, 383-400.
9. De Ploey, J. (1990) : Threshold conditions for thalweg gullying with special reference to loess areas. Catena Supplement 17, 147-151.
10. Papy, F. and Boiffin, J. (1989) : The use of farming systems for the control of runoff and erosion. Soil Technology 1, 29-38.
11. Govers, G. & Poesen, J. (1988) : Assessment of the interrill and rill contributions to total soil loss from an upland field plot. Geomorphology 1, 343-354.
12. Boardman, J. (1988) : Severe erosion on agricultural land in East Sussex, UK, October 1987. Soil Technology 1, 333-348.
13. Evans, R. and Nortcliff, S. (1978) : Soil erosion in north Norfolk. Journal of Agricultural Science, Cambridge 90, 185-192.
14. Evans, R. and Morgan, R.P.C. (1974) : Water erosion of arable land. Area 6, 221-225.
15. Boardman, J. (1983) : Soil erosion at Albourne, West Sussex, England. Applied Geography 3, 317-329.
16. Reed, A.H. (1979) : Accelerated erosion of arable soils in the United Kingdom by rainfall and run-off. Outlook on Agriculture 10, 41-48.
17. Evans, R. (1981) : Assessments of soil erosion and peat wastage for parts of East Anglia, England. A field visit. In : Morgan, R.P.C. (Ed.), Soil Conservation : Problems and Prospects, Wiley, Chichester, 521-530.
18. Frost, C.A. and Speirs, R.B. (1984) : Water erosion of soils in south-east Scotland- a case study. Research and Development in Agriculture 1, 145-152.
19. Evans, R. (1980) : Characteristics of water-eroded fields in lowland England. In : De Boodt, M. and Gabriels, D. (Eds.), Assessment of Erosion, Wiley, Chichester, 77-87.
20. Harris, T.M. and Boardman, J. (1990) : A rule-based Expert System approach to predicting waterborne soil erosion. In : Boardman, J., Foster, I.D.L. & Dearing, J.A. (Eds.) Soil Erosion on Agricultural Land, John Wiley, Chichester, 401-412.
21. Boardman, J. and Favis-Mortlock, D.T. (1992) : Soil erosion and sediment loading of watercourses. SEESOIL 7, 5-29.
22. Jarvis, M.G., Allen, R.H., Fordham, S.J., Hazelden, J., Moffat, A.J. and Sturdy, R.G. (1984) : Soils and their use in South East England. Soil Survey of England and Wales Bulletin, 15.
23. Boardman, J. (1990) : Soil erosion on the South Downs : a review. In : Boardman, J., Foster, I.D.L. and Dearing, J.A. (Eds.), Soil Erosion on Agricultural Land, Wiley, Chichester, 87-105.
24. Boardman, J., Evans, R., Favis-Mortlock, D.T., and Harris, T.M. (1990) : Climate change and soil erosion on agricultural land in England and Wales. Land Degradation and Rehabilitation 2, 95-106.
25. Boardman, J. (1991) : Land use, rainfall and erosion risk on the South Downs. Soil Use and Management 7, 34-38.
26. Boardman, J. (1992) : The sensitivity of Downland arable land to erosion by water. In : Thomas, D.S.G. and Allison, R.J. (Eds.), Environmental Sensitivity, Wiley, Chichester, in press.
27. Robinson, D.A., and Boardman, J. (1988) : Cultivation practice, sowing season and soil erosion on the South Downs, England : a preliminary study. Journal Agricultural Science, Cambridge 110, 169-177.

28. Favis-Mortlock, D.T., Evans, R., Boardman, J. and Harris, T.M. (1991) : Climate change, winter wheat yield and soil erosion on the English South Downs. Agricultural Systems 37, 415-433.

Farm Land Erosion: In Temperate Plains Environment and Hills
S. Wicherek (Editor)

Effect of cultivation techniques on the hydrodynamic and mechanical behaviour of the "Lauragais-Terreforts"

D. Boudjemline [a], E. Roose [b] and F. Lelong [c].

[a] Département d'hydraulique, Centre universitaire de Bejaia, Algérie

[b] Centre ORSTOM, BP 5045, 34032, Montpellier Cédex, France

[c] Centre des sciences de la terre, Université de Bourgogne, 6 bd. Gabriel, 21000 Dijon, France

SUMMARY

The effect of various seed bed preparations on hydrodynamic and mechanical parameters was studied experimentally by simulated rainfall on clayey-loamy soils Lauragais, which had developed on chalky molasse on a slope greater than 12 %.

The level of soaked up rain ("pluie d'imbibition", Pi) is closely connected to the inital degree of saturation of the soil ; it varies from 10 to 24 mm when the soil is humid and increases when the soil is dry (14 to 47 mm). But when the soil surface is crusted the Pi is less and seems to be no longer controlled by the degree of water in the soil. Treatments with fine structural elements show the fewest possibilities of infiltration during the soaking up stage ; thickening the seed bed does not appear to have any effect on Pi. The double packing down of the soil poses a real obstacle to infiltration.

The susceptibility to runoff is all the greater as the percentage of clods less than 0.5 cm is high. Runoff is reduced when the soil is dry and increases sharply with the intensity of the rain and superficial crusting.

The intensity of infiltration permanent regime (FN) seems better on the treatments with fairly large structural elements (from 6 to 17 mm. h^{-1}).
The thickening of the seed bed did not improve FN (from 5 to 11 mm h^{-1}). On the other hand, the double thickening of the soil is a real brake on infiltration (1 to 5 mm h^{-1}). The intensity of infiltration permanent regime drops when the intensity of the rain increases.

The average solid load is very variable, probably on account of the many "lachages" of micro-structures. It is high for the treatments with fine structural elements if the soil is dry (24 to 27 g L^{-1}) and especially if the soil is humid (19 to 41 g. L^{-1}) The average solid load is less on treatments with fairly large structures (1 to 2 g. L^{-1}). It increases sharply with the slope and appears to be independent of rainfall intensity. Treatments with fine structural elements are characterized by high solid outflows (200 to 300 g.m^2 h^{-1}). They increase sharply with rainfall intensity and the slope and drop with a low level of soil saturation.

RESUME

L'influence de diverses préparations de lit de semence sur les paramètres hydrodynamiques et mécaniques a été étudiée expérimentalement par simulation de pluie sur les sols argilo-limoneux du Lauragais, à pente > 12 % et développés sur molasse calcaire.

La hauteur de pluie d'imbibition Pi est très liée à l'état d'humectation initial du sol ; elle varie de 10 à 24 mm lorsque le sol est humide et augmente lorsque le sol est sec (14 à 47 mm). Mais lorsque la surface du sol est encroûtée, Pi est plus faible et ne semble plus être contrôlé par l'état hydrique du sol. Les traitements à éléments structuraux fins montrent les plus faibles possibilités d'infiltration pendant la phase d'imbibition alors qu'un épaississement du lit de semence ne semble avoir aucune influence sur Pi. Le double tassement du sol représente un véritable obstacle à l'infiltration.

La susceptibilité au ruissellement semble d'autant plus prononcée que le % de mottes < 0,5 cm est élevé. Le ruissellement est réduit lorsque le sol est sec et augmente nettement avec l'intensité de la pluie et l'encroûtement superficiel.

L'intensité d'infiltration en régime permanent (FN) semble meilleure sur les traitements à éléments structuraux relativement grossiers (6 à 17 mm h^{-1}).

L'épaississement du lit de semence n'a pas amélioré FN (5 à 11 mm h^{-1}). Par contre, le double tassement du sol est un véritable frein à l'infiltration (1 à 5 mm h^{-1}).

L'intensité d'infiltration en régime permanent diminue quand l'intensité de la pluie augmente.

La charge solide moyenne est très variable à cause probablement des nombreux lachages de micro-structures ; elle est élevée sur les traitements à éléments structuraux fins (24 à 27 g. L^{-1}) si le sol est sec et surtout (19 à 41 g. L^{-1}) si le sol est humide.

La charge solide moyenne est réduite sur les traitements à éléments structuraux relativement grossiers (1 à 2 g. L^{-1}) ; elle augmente nettement avec la pente et semble indépendante de l'intensité de la pluie. Les traitements à éléments structuraux fins se caractérisent par des débits solides très élevés (200 à 300 $g.m^{-2}$ h^{-1}) ; ils augmentent nettement avec l'intensité de la pluie et la pente et diminuent avec le déficit de saturation du sol.

1. INTRODUCTION

An inquiry held next to Lauragais farmers has shown a certain number of characters relative to erosion in this region of France (1). The plots affected by erosion are often big ones (from 5 to 10 ha) having been subjected to developments such as the suppression of the slope. They are sloping (from 15 % to 30 %) and lenghthy enough (100 m) and according to the farmers, they are well prepared in spring, ie the seed-bed was relatively thin. The most sensitive months are April, May and June, ie during spring rainstorms. The cases of erosion are very evident : destroyed seedlings, packed ditchs. Following this inquiry, it has been decided to undertake a study using rainfall simulator for delimiting the parameters relative to the release of the surface runoff and its sediment load on the one hand, and to compare the effect of the different cultivation techniques on the other hand. Previous works (2 - 3 - 4) have really shown the role of the soil surface and the mechanical and structural characteristics of horizon surface on the susceptibility of the soil to runoff and erosion.

2. FIELDS AND METHODS

The Lauragais occupies the eastern part of the Haute-Garonne department, near Toulouse (Fig. 1). It is a multivarious farming region where predominate cereal cultivations : they are characterized by the rotations of winter farming (wheat, barley, colza) and summer farming (maize, sunflower...). The soils are brown-limestone, clayey-loamy and sandy. They are referred to as "TERREFORTS" (local appelation which evokes heavy clayey-limestone soils).

The average altitude is lower than 300 m and the drainage density is important (3 km/km^2). Slopes cartography shows that 45 % of slopes are upper than 18 % .

The Haut-Garonnais Lauragais climate is intermediate between the Mediterranean and the oceanic tendencies .

The field experimentation was held at the scale of 1 m^2 experimental plots using an ORSTOM (Fig. 2) type rainfall simulator which is well described by the way (5), and able to provoke rainfalls with an intensity varying between 30 and 150 mm/h. A guard ring whose the surface depends on the jet balancement angle, surrounds the testing plots.

Two series of tests have been made :

- A first series on 13 to 20 % slop, aimed to show :
  * The influence of an over-flowing water table or a soil surface state fastned to seed-bed preparation,
  * The role of the packing down of the soil by the machine wheel,
  *The influence of the seed-bed thickness,
  *The role of the clods size of the seed-beds,

- A second series of tests on 2 to 5 % slope aims to study the effect of intensity rainfall on hydrodynamic and mechanical parameters of the soil.

The experimental field situated in Narbons, has been ploughed at 30 cm depth after the sunflower harvest. The plough has been resumed in spring to obtain the following seed-beds :

N1 and N5 : normal plough + seed-bed at average clods.
N2 : plough packed down once + seed-bed at relatively big clods.
N3 : plough packed down twice + seed-bed at relatively big clods.
N4 : like N1 and N5 but with a thicker seed-bed.
N7 : normal plough + seed-bed with very fine clods + big clods addition.
N8 : normal clough + seed-bed with very fine clods (roll).
N2b : plough packed down once + seed-bed with very fine clods (rake).

The thickness of the seed-bed is about 8 cm for all the treatments except the N 4 one which has a more thick seed-bed (about 15 cm).

From study of the climate,which has shown danger of spring rainfalls, it has been decided to simulate rainfalls of 40 mm/h, of the annual frequency, for the first series of tests. For the second series, we have simulated three levels of rainfalls intensity (40,50 and 80 mm/h). The results of each test are represented by a discharge hydrograph and a turbidity hydrograph defining Pi, ti, Fn, Rx... (Fig. 3).

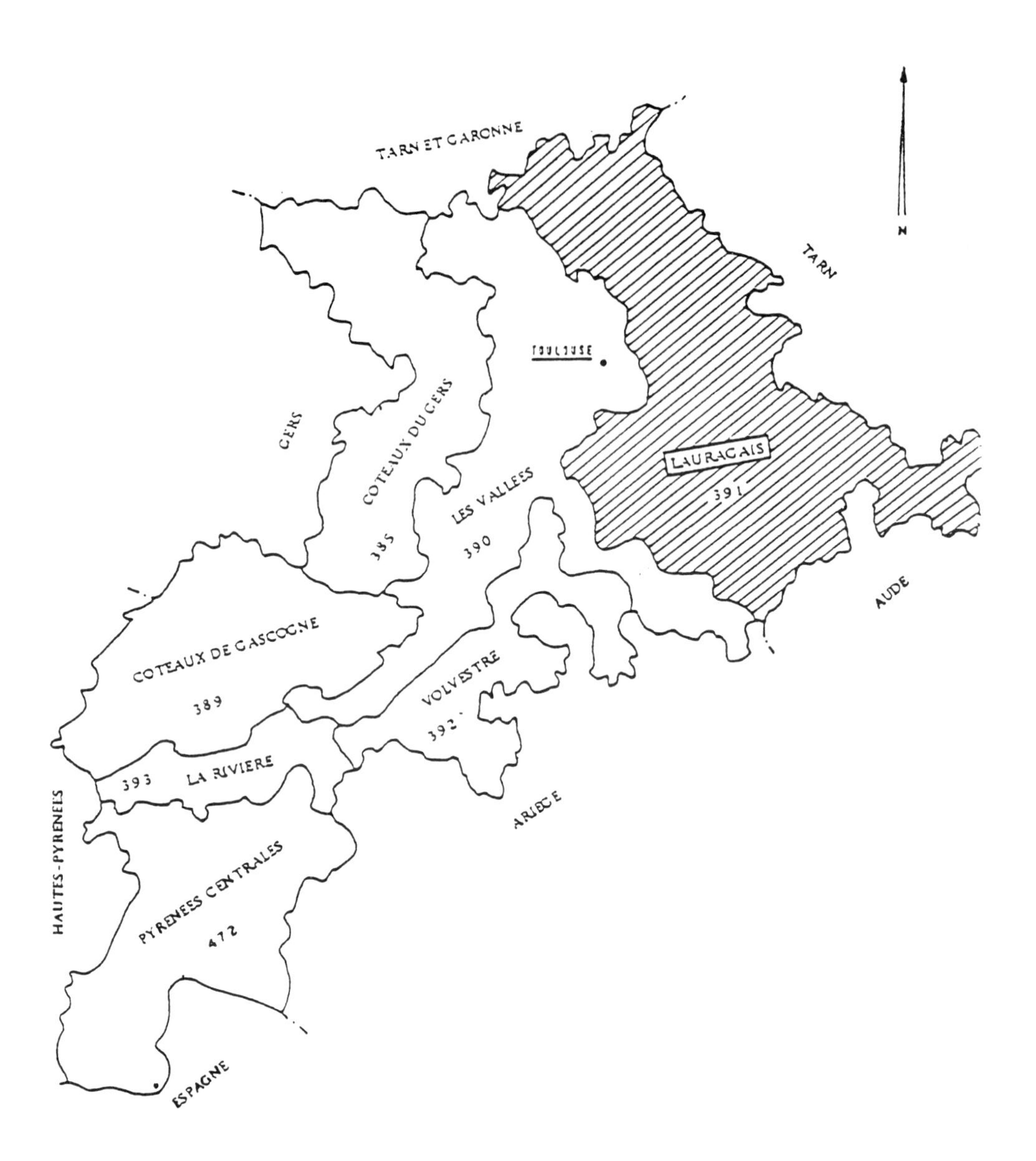

Figure 1. Situation of the Lauragais in compare with the other farming regions of Haute-Garonne.

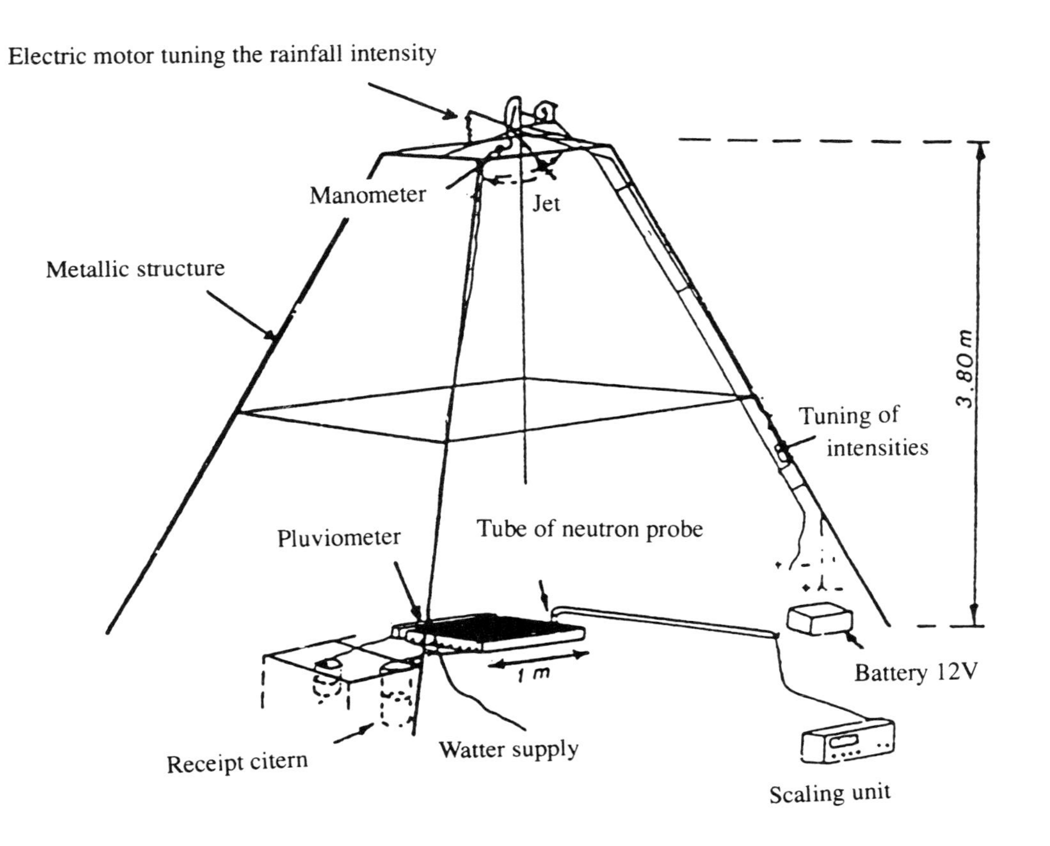

Figure 2. Schema of the rainfall simulator, ORSTOM type.

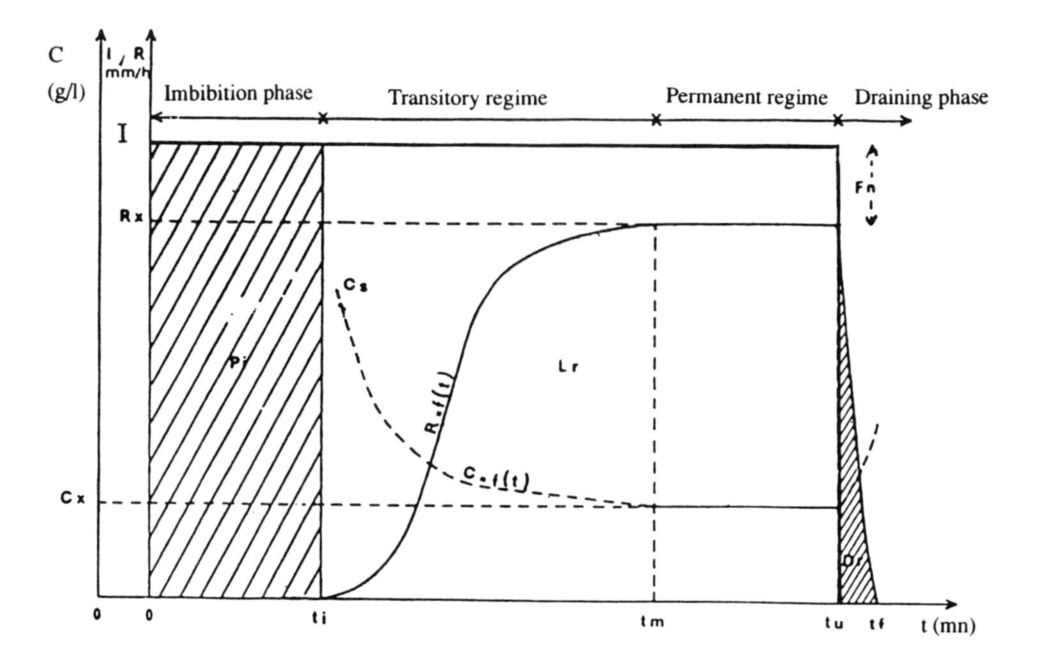

Figure 3. Theorical hydrograh of runoff and turbidigraph under constant rainfall intensity.

3. RESULTS

3.1 The level of soaked up rain,Pi.

The level of soaked up rain is closely connected to the initial degree of soil saturation when the soil surface is not degraded. The relation obtained between Pi and the descriptive factors and variables of surface states :

$Pi = 45.5 - 0.3\,D + 0.2\,i - 1.1\,S + 1.4\,C.$

D : rate of clods with a diameter inferior to 0.5 cm.

i : soil saturation deficit.

S : slope of field.

C : rate of culture residue.

Even if it explains only 60 % of the variations of Pi, this relation shows that in order to delay the runoff, it is desirable to avoid the seed-bed from crumbling, to ameliorate the soil water movement and to cover it with culture residue.

3.2 The runoff

The results of runoff (Tab. 1) allow to distinguish two phenomena :

a/ The packing down by the machine wheels : the "terreforts" of Lauragais are not much sensitive to the packing down ; were needed two passages tractor wheels to have a clear reaction vis-a-vis the runoff. The Proctor test confirms that the soils of Lauragais are not very compressible.

Table 1
Experimental parameters of the relation $\sum Lr = A+B (\sum Pu)$

Plots code	A	B	r	n	Depth of runoff observed for a 40 mm rainfall	Depth of runoff calculed for a 40 mm rainfall
N1-1	- 6,30	0,12	0,99	8	0,0	0,0
N1-2	- 13,90	0,38	0,97	21	1,7	1,6
N1-3	- 14,60	0,60	0,98	20	8,5	9,4
N1-4	- 8,77	0,71	0,98	18	19,7	19,6
N2-1	- 10,88	0,80	0,99	11	21,5	21,1
N2-2	- 9,05	0,64	0,99	11	17,5	16,5
N2-3	- 10,60	0,55	0,99	18	10,5	11,4
N2-4	- 17,80	0,80	0,99	11	14,5	14,2
N3-1	- 4,90	0,82	0,99	28	27,5	27,9
N3-2	- 4,40	0,85	0,99	20	30,0	29,6
N3-3	- 5,48	0,75	0,99	25	24,6	24,5
N3-4	- 10,10	0,88	0,99	15	25,3	25,0
N4-1	- 10,20	0,46	0,98	21	7,0	8,2
N4-2	- 13,90	0,68	0,99	25	12,3	13,3
N4-3	- 13,50	0,61	0,98	24	10,0	10,9
N4-4	- 20,40	0.54	0,98	38	1,8	1,5
N5-1	- 21,40	0,57	0,98	19	1,7	1,4
N5-2	- 10,30	0,67	0,99	13	16,6	16,5
N5-3	- 12,50	0,49	0,99	26	6,0	7,4
N7-1	- 10,20	0,58	0,99	28	12,1	13,0
N7-2	- 13,88	0,62	0,99	29	10,0	10,9
N7-3	- 11,40	0,58	0,99	27	10,5	11,7
N7-4	- 10,70	0,50	0,99	19	8,8	9,3
N8-1	- 10,50	0,56	0,99	26	10,8	11,9
N8-2	- 9,50	0,59	0,99	26	13,6	14,4
N8-3	- 12,20	0,62	0,99	29	12,0	12,6
N8-4	- 9,10	0,73	0,99	29	19,8	20,1
N2b-1	- 29,40	0,58	0,99	28	0,0	0,0
N2b-2	- 16,10	0,60	0,98	20	6,7	7,9
N2b-3	- 13,00	0,78	0,99	16	18,0	18,2
N2b-4	- 9,50	0,52	0,99	19	10,6	11,3

r : Correlation coefficient
n : Size
A and B : Experimental parameters

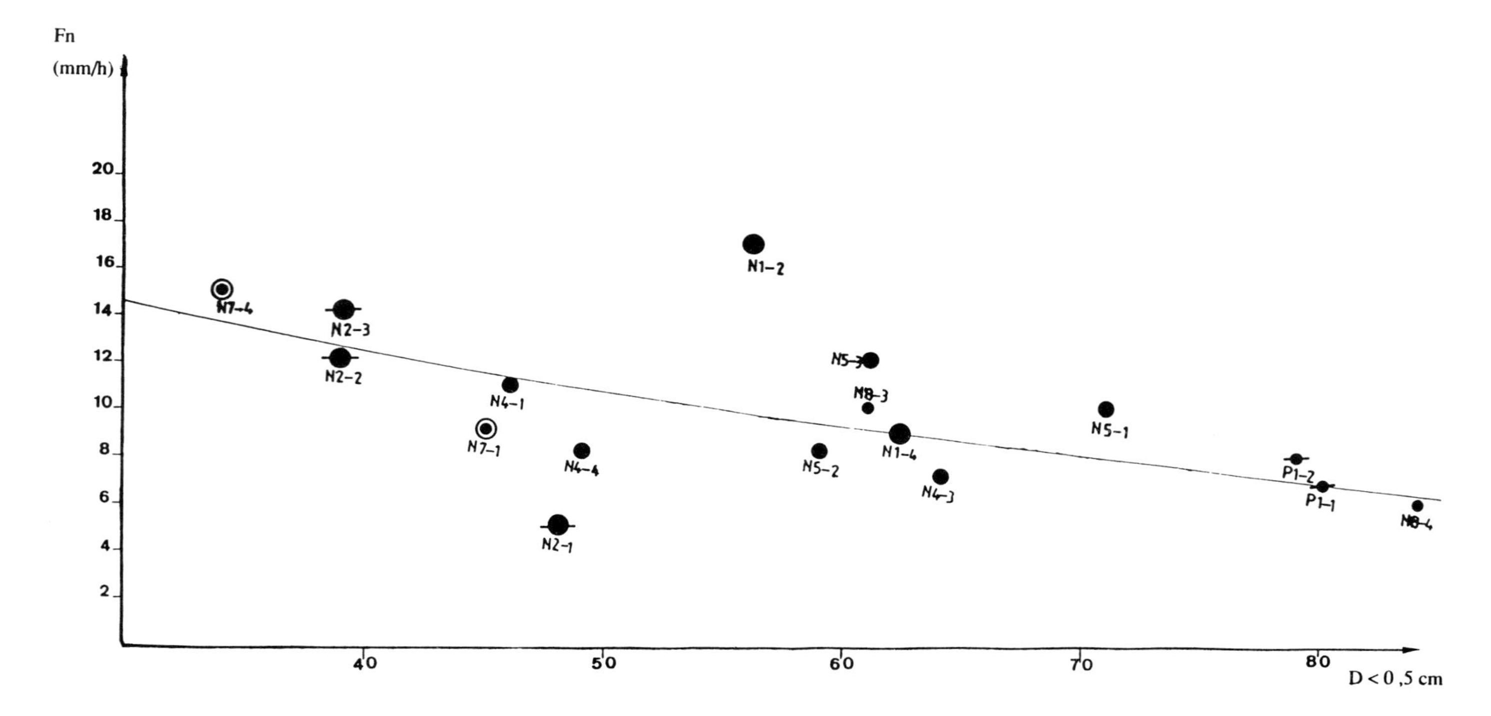

Figure 4. Evolution of Fn with the rate of clods inferior to 0,5 cm.

b/ The susceptibility of the soil to runoff (represented by the experimental parameter B of the relation Lr = f (Lr Pu), with : Lr = cumulated depth of runoff and Lr Pu = cumulated depth of rainfall,increases with the rate of clods with a diameter inferior to 0.5 cm ; it decreases slighthy when the index of roughness increases.

c/ In the various slopes studied (from 13 to 20 %), the slope does not have a clear and a significant effect on the susceptibility to runoff of the different treatments. The slope is probably not the 221explanatory factor to runoff, at least at the scale of 1 m plots.

3.3 The infiltration

We will be interested particulary in the intensity of a minimal infiltration with a permanent regime, Fn. Fn seems to be influenced by the percentage of clods less than 0.5 cm (Fig. 4) as well as by the index of roughness IR (Fig. 5), at least for plots without an intense packing down (treatment N3).

3.4 Influence of rainfall intensity on hydrodynamical parameters

The influence of the rainfall intensity on hydrodynamical parameters (Pi,Rx and Fn) has been studied on weak slope plots (from 2 to 5 %) with identical surface state (Tab. 2), with ta = time separating two successive storms. Table 2 shows that despite the disparity of the results, Pi and Fn tend to decrease when the rainfall intensity increases.

Table 2
Variation of hydrodynamics parameters function of rain intensity

Plots	I mm/h	Slope %	ta h	Pi mm	Kr 40 %	Fn mm/h
P1 to P3			1	12-16	25	
P4	40	2-5	24	24	2	4-24
P6			114	38	0	
P7 to P8			1	10-14	20-37	
	50	2-3				9-13P9
P9			144	25	5	
P12 to P13			1	9	22-53	
P10	80	4-5	16	13	19	4-6
P11			120	16	28	

3.5 The exportation of sediments

At the scale of 1 m plots, we measure only the sheet erosion which results from the detachability of particles by the "splash effect" of rainfalls ones which, in their turn, are driven by the runoff. We consider two variables :

* The average concentration,Cm (g/l) of solid particles in the runoff volume.
* The sediment discharge (g/m /h), exported from the plot.

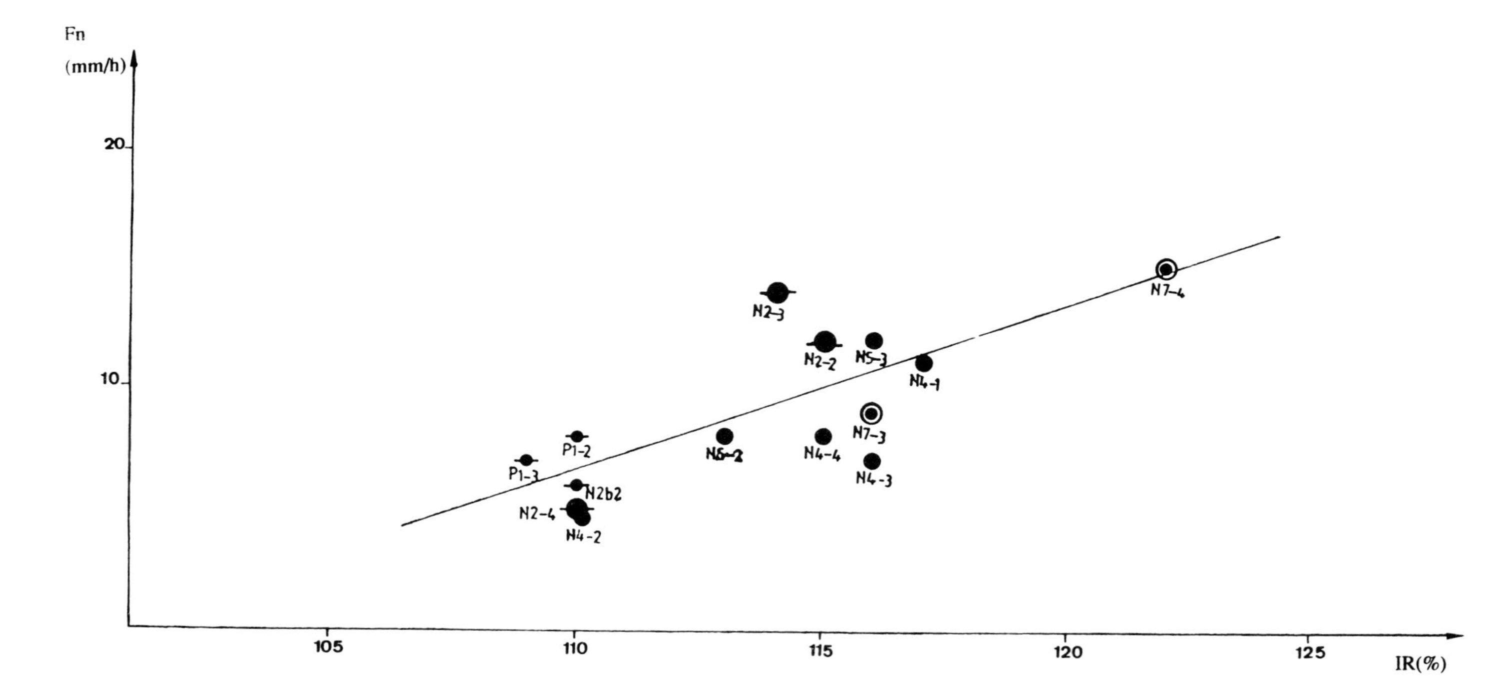

Figure 5. Evolution of Fn with the index of roughness.

3.5.1 The average concentrations,Cm,

The average concentrations increase sharply with the rate of clods less than 0.5 cm (Fig. 6) ; they also increase with the soil slope (Tab. 3) and are four to five times higher when we move from 2 % to 15 % slope soil.

Table 3
Slope influence on mean solid charge and solid debit for 40 mm rainfall during one hour

Pente %	Solid charge C_{40} g/l	Solid debit Qs_{40} $g/m^2/h$
2	0,0-2,9	0,0-29
15	0,0-13,6	0,0-175

3.5.2 The sediment discharges

The sediment discharges are three to five times higher when we move from 15 % to 2 % slope plots (Tab. 3). Then it tends to increase with very fine seed-beds (Tab. 4) and increase very quickly with the intensity of storms.

Table 4
Treatments influence on runoff, solid charge and solid debit for a 40 mm rainfall during one hour (15% slope)

Treatment	Kr_{40} %	C_{40} g/l	Qs_{40} $g/m^2/h$
Mean (N1,N4,N5)	0-30	0-8	0-75
Mean,packed down once	27-36	5-13	70-220
Mean,packed down twice	60-75	2-9	48-257
Fine + big clods addition	22-30	7-19	90-170
Fine + roll	27-50	2-20	23-280
Fine + rake	0-45	0-14	0-175

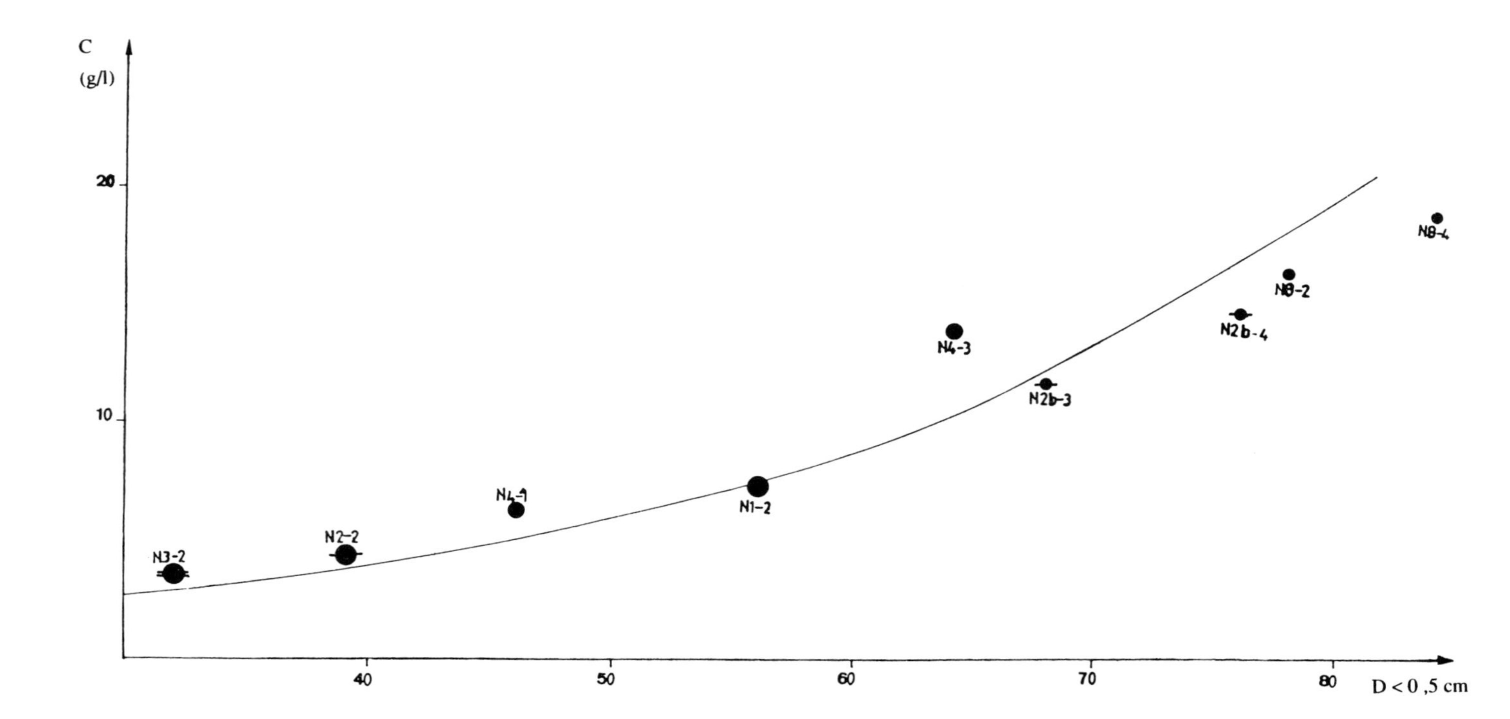

Figure 6. Evolution of the turbidity with the rate of clods at diameter inferior to 0,5 cm.

4. DISCUSSIONS

4.1 Hydrodynamical aspects

The previous results show clearly the role of surface state and size clods of seed-beds (from 5 to 15 cm), on the susceptibility to runoff. The runoff is more precocious (Pi decreases) and more intense (Rx increases) when the percentage of fine clods (diameter inferior to 0.5 cm) increases and when the index of roughness decreases. On the other hand, the slope soil does not seem to have a clear effect on the runoff intensity. Authors notice effectively various answers (positive or negative effects of the slope soil on Rx) according to the studied cases :

For Zingg (6) ; (7 - 8 - 9 - 10), the runoff increases with the slope. For the others, (11 - 12), it rather tends to decrease with the slope increase. Roose interprets this fact as the result of the partial removal of the figle packed down by the sheet erosion when the slope increases.

The intensity of storms influences both the soaked up rain duration and the intensity of runoff. The decreasing of Pi and the increasing of Rx when the rainfall intensity (I) increases, justify the big increase of the runoff coefficient of storms, with the same increasing intensity duration.

Although its dispersed, the relation Fn = f (I) is decreasing, as it has been observed elsewhere (13) on bare and deteriorated soil plots, liable subjected to a packing down and or to the formation of surface crust under the influence of rainfalls energy. The soil capping phenomenon has been analyzed by Boiffin (14). It unwinds in several phases : The desintegration of soil aggregates gives first structural crusts during the soak up rain phase by the padding of micro-depressions, then it gives a stratified sedimentation crusts in the puddles the crusts whose thickness depends on the mobilized soil mass. This phenomenon which provokes an important reduction of the structural porosity, thus of the hydraulic conductivity of soil surface, is then controlled by the desintegration of the aggregates, the detachability of particles and by the redistribution (training and redeposit) of these ones in the topographic micro-depressions. It is logical that the more rainfall intensity increases, the more kinetics energy necessary for desagregating the soil and provoking the closing of this one, increases as Moldenhauer and Long (15) observe it. In this way, we can explain the existence of the relation Fn = f (I) decreasing on soils with surface crusts, when we simulate storms with increasing intensity (13). We call back that according to Lafforgue (16), the increasing relations Fn = f (I) correspond to rough surfaces with heterogeneous permeability, and constant relations Fn = f (I), to smooth surfaces with homogeneous permeability.

4.2 Mechanical aspects : detachment and particles transport

The previous results show that the refinement of the seed-bed favours the detachability of the particles, thus, the release of these ones redistributed near at hand or exported. The exportation increases with the slope, as observed by Wischmeier and Smith (17) and Hudson (18), for the latter increases the speed of runoff flow and its competence of transport, even though this speed remains weak for 1m slope length (from 2 to 6 cm/s) according to Bryan (19) and Valentin (20), at the same time that it decreases the roughness efficiency (21) ; moreover it increases with rainfall intensity, for it acts on both rainfall (22) and on runoff discharge, thus on transport capacity (18). In short,the rainfall intensity acts on the detachment and the release of the particles, whereas the runoff intensity acts on their transport and exportation. The leveling up of the soil and the formation of surface crusts lead a partial or a total closing of the structural porosity, thus to a relative impermeabilization of soil surface (in addition to a certain mechanical stabilization of superficial film).

So, the evolution of the hydrodynamical properties noticed at the scale of 1m plots during the tests appears to be more of the consequence than of the cause of the variations of the mechanical properties of the soil. But at the scale of flank, where the concentration of the sheet flow releases the formation of channels and gullies, it is the clipping of the soil by runoff flow which is the principal cause of erosion where as the sediment transports related to these incisings are only the consequence.

4.3 Order of greatness of sheet erosion

The sediment discharges measured in the lower part of a stream of plots are rather tests of detachability at the scale of 1m than field erosion assessments, for the cumulative effect of slopes and the important spacial heterogeneity of soil are not taken into consideration. However, if we extrapolate at the scale of the hectare, the values obtained on plots (Tables 3, 4 and Tab. 5) vary with a factor 1 to 20 according to the soil slope, the soil treatment and the storms intensity. They are of 0.2 to 2 t/ha order on a thin slope (2 %) for a 40 mm/h storm intensity during one hour (storm with annual probability occurrence in the Lauragais). They are from 2 to 5 times more important for 15 % slope and they increase quickly for more important rainfall intensities. Knowing that the acceptable load sediment doorstep is from 1 to 2 t/ha/year, we understand that the optimization of the cultivation pratices (preparation of seed-beds, supply of culture residue and the ploughing calendar) is an essential aspect for the management of these soils.

Table 5
Influence of rainfall intensity on runoff and solid debits (40 mm rainfall with different intensities

I mm/h	Kr 40 %	Qs 40 $g/m^2/h$
40	0-26	0-29
50	5-37	4-35
80	19-53	15-209

5. CONCLUSION

The susceptibility to runoff and erosion of the soils of Lauragais seems to depend heavily on the soil surface state and notably on the preparation of seed-beds. The refinement of these ones favours the speed of desintegration of soil surface and the release of the particles which are redistributed under the form of a superficial film, not very permeable. The degradation of the structural properties of this film seems to control the decrease of the infiltrability, the increase of sheet runoff and transport capacity of this one.

For reducing the loss soils caused by this form of erosion, we can have to :

- reduce "splash effect" of rainfall : supply of culture residue.

avoid of refinement of the seed-bed, for it is the fine structural elements which are sensitive to the desintegration and "splash effect".

- favour the soil water movement of soils by avoiding as much as possible the packing back caused by the repeted passages of the cultivation machines.

REFERENCES

1. Etchanchu, D., N'Diaye, M. (1984) : Etude pédologique des côteaux du Lauragais. Chambre d'agriculture. Toulouse, 43 p.
2. Trevisan, D. (1986) : Comportement hydrique et suscéptibilité à l'érosion de sols limoneux cultivés : Etude expérimentale au champ sous pluies simulées. Thèse 3ème cycle, Univ. Orléans, 244 p.
3. Raheliarisoa, M.A. (1986) : Influence des techniques culturales sur le comportement hydrodynamique et la suscéptibilité à l'érosion de sols limoneux et sableux. Thèse 3ème cycle. Univ. Orléans,186 p.
4. Boudjemline, D. (1987) : Suscéptibilité au ruissellement et aux transports solides de sols à texture contrastée-Etude expérimentale au champ sous pluies simulées. Thèse 3ème cycle, Univ. Orléans, 266 p.
5. Asseline, J.,Valentin, C. (1978) : Construction et mise au point d'un infiltromètre à aspersion. Cah-ORSTOM. Hydrol. 15, 4:321-350.
6. Zingg, A.W. (1940) : Degree and length of land slope as it affects soil loss and runoff. Agr. Eng., 21:59-64.
7. Horton, R.E. (1940) : An approach toward a physical interpretation of infiltration capacity. Soil. Sci. Soc. Amer. Proc. 5:399-417.
8. Meyer, L.D., MC Cune, D.C. (1958) : Rainfall simulator for runoff plot. Agr. Eng. 39,10:644-648.
9. Albergel, J., Casenave, A., Valentin, C. (1985) : Modélisation du ruissellement en zone Soudano-Sahelienne. Simulation de pluie et cartographie des états de surface. Journées hydrologiques de l'ORSTOM. Montpellier 17-18 sept. 15 p.
10. Viani; J.P. (1986) : Contribution à l'étude expérimentale de l'érosion hydrique. Thèse Doct. es-Sci. Tech. Ecole. Polytechnique. Fédérale de Lausanne, 221 p.
11. Roose, E. (1973) : Dix-sept années de mesures expérimentales de l'érosion et du ruissellement sur un sol ferralitique sableux de basse Côte-d'Ivoire : contribution à l'étude de l'érosion hydrique en milieu inter-tropical. Thèse Doct. Ing. Fac. Sci. Abidjan.
12. Poesen, J. (1984) : The influence of slope angle on infiltration rate and Hortonian overland flow volume. CATENA.
13. Lelong, F., Darthout, R., Trevisan, D. (1992) : Susceptibilité au ruissellement et à l'érosion en nappe de divers types texturaux de sols cultivés ou non cultivés du territoire Français : Expérimentation au champ sous pluies simulées. (in press)
14. Boiffin, J. (1984) : La dégradation structurale des couches superficielles du sol sous l'action des pluies. Thèse Doct. Ing. INAPG, 297 p.
15. Moldenhauer, W.C., Long, D.C. (1964) : Influence of rainfall energy on soil and infiltration rates. Soil. Sci. Soc. Amer. Proc. 28, 6:813-817.
16. Lafforgue, A. (1977) : Inventaire et examen des processus élémentaires de ruissellement et d'infiltration sur parcelles. Cah-ORSTOM. Hydrol. 14, 4:299-344.

17. Wischmeier, W.H., Smith, D.D. (1960) : An universal soil loss estimating equation to guide conservation farm planning. Proc. of the VII th Int. Congr. Soil. Sci. MADISON., 1:418-425.
18. Hudson, N.W. (1973) : Soil conservation. BATSFORD. London, 320 p.
19. Bryan, R.B. (1976) : Considerations on soil erodibility indices and sheetwash. CATENA, 3:99-111.
20. Valentin, C (1981) : Organisations pelliculaires superficielles de quelques sols de région sub-désertique (Agadez-Niger) : dynamique de formation et conséquence sur l'économie en eau. Thèse 3ème cycle, Univ. Paris VII, 230 p.
21. Roose, E., Asseline, J. (1978) : Mesures des phénomènes d'érosion sous pluies simulées aux cases d'érosion d'Adiopodoumé. Cah-ORSTOM. Pédo. 16, 1:43-72.
Roose, E. (1980) : Dynamique actuelle des sols ferralitiques et ferrugineux tropicaux de l'Afrique occidentale. ORSTOM Sér. Trav et Doc. n° 130.
22. Fournier, F. (1967) : La recherche en érosion et conservation des sols dans le continent Africain. 12, 1:5-53.

Farm Land Erosion: In Temperate Plains Environment and Hills
S. Wicherek (Editor)

On-site and Off-site Damages by Erosion in Landscapes of East Germany

M. Frielinghaus and R. Schmidt

ZALF Müncheberg, 0-1278 Müncheberg, Germany.

SUMMARY

Off-site damages by soil and nutrient inputs into waters, hollows, and wetland biotopes are increasingly in addition to the long-known damages on crop areas caused by erosion. Up to 11 per cent of the calculated N input per year and 58 per cent P input per year into running waters are attributed to erosion.

First results have been achieved about on-site and off-site damages in the 169 km^2 catchment area of the Lake Uckersee in the East - German young moraine area.

The Lower Lake Uckersee drains via the Prenzlau weir into Ucker River discharging into the Stettiner Haff (the lagoon opening into the Bay of Pommerania) after having flown 63 km through a catchment area of 2.420 square kilometers.

The off-site damages have become evident by the fact that the water quality of the Lower Lake Uckersee decreased by one class and has been classified into class 3 now (eutrophic waters). The total load of the lake was in 1990 : 5,118 kg orthophosphate, 9,978 kg total phosphorous, 189,197 kg inorganic nitrogen.

RESUME

A côté de dégâts "in situ" il existe des risques associés qui affectent les eaux et augmentent les dangers d'entrophisation. Pour 1 tonne de sol : 3,1 à 23 kg de carbone, 0,3 à 1,3 kg d'azote, jusqu'à 0,5 kg de phosphore soluble et l'équivalent en potassium sont exportés. Dans le cas particulier des sols sur moraines la perte de particules fines < 0,0063 mm est aussi importante pour la fertilité des sols que le transport de sédiment intervenant lors des orages et des phénomènes éoliens. Il est donc nécessaire d'envisager la protection des sols dans un schéma global d'aménagement.

Concernant les dégâts "in situ" les premiers résultats ont été obtenus dans le bassin de 169 km^2 du Lake Uchersee dans une moraine récente de l'Allemagne orientale.

Le Lower Lake Uchersee s'écoule via Prenzlan dans Ucher River, se déchargeant dans le Stettiner Haff (lagon s'ouvrant dans la Baie de Pouréranie). Les dégâts sont devenus évidents au niveau qualité de l'eau du Lower Lake Uchersee qui a été déclassé de 1 à 3 maintenant (eaux eutrophiques). En 1990, la charge totale du Lac était : 5,118 kg d'orthophosphate, 9,978 kg de phosphore, 189,197 kg d'azote.

INTRODUCTION

Recently increasing off-site damages have been found out in connection with soil and nutrient displacements caused by wind and water erosions. Such damages occur outside the areas immediately affected by erosion. These loads resulting from dragouts from field-ecosystems and inputs into neighbouring ecosystems like waters, hollows, wetland biotopes etc. are some of the diffuse sources leading to eutrophisation of flowing waters and lakes. Thus 11 per cent of the calculated N input per year and 58 per cent of the P input per year into flowing waters of the Elbe-River catchment area are attributed to erosion (1) (Fig. 1). These data appear to be very high and have not been proved until now. No doubt, however, the path cultivated area - neighbouring flowing waters - river - sea plays an essential part in the material displacement. Because of the many factors causing and influencing erosion and their site-related different weightings the risk assessments should be made for individual landscapes preceded by comprehensive analyses and a clearing-up of the process. Subsequently first results are shown achieved in a small catchment area of the Lake Uckersee near Prenzlau (Land Brandenburg) affected by water erosion. The further need for investigation is indicated (Fig. 2).

The catchment area of the Upper Ucker-River is determined by the two lakes Uckersee as well as by the Lake Großer Potzlowsee (Fig. 3). The valley of the Ucker River is in a former Tertiary channel and the result of the glacial advances during the Weichsel Glacial Period.

The basal moraine originating from the Weichsel-II Glacial Period of the Pommeranian Stage is the dominating parent rock of soil formation in the flat-undulating to undulating young-moraine area surrounding the lakes. The upper and the youngest glacial till layer is between 1 and 24 m thick. The whole complex of quaternary layers can be between 18 and 80 m thick and lies on tertiary sands and chalks. There are also differences in the level of the terrain above mean sea level. Periglacial impacts on the surface-layer relief lead to linear slopes of average relief intensity and slope length of 200 to 300 m. Dells and slightly low flat crests cut through the slope areas crosswise and lengthwise. In the past the from of the relief was the reason for the considerable soil displacements by water erosion and for the development of typical erosion and deposition sites with different soil stratifications and many small waters with and without outlets (Fig. 3 ; aerial view).

Inhomogeneous fields (1) where light-coloured high slopes and round tops (erosion areas) alternate with dark wet hollows and troughs (deposition areas) are associated with moraine lakes (2), running waters (3), and small lakes flown through (4). The exemplary catchment area drains via the main ditch (5) into the Lower Lake Uckersee (6). The Lake Uckersee is connected with the Baltic Sea via the Ucker River.

All materials leaving the catchment area as solutes contribute to the pollution of the Baltic Sea, since the biodegradation is very small in the Ucker River.

Let us go back to the initial subject of material displacement by erosion.

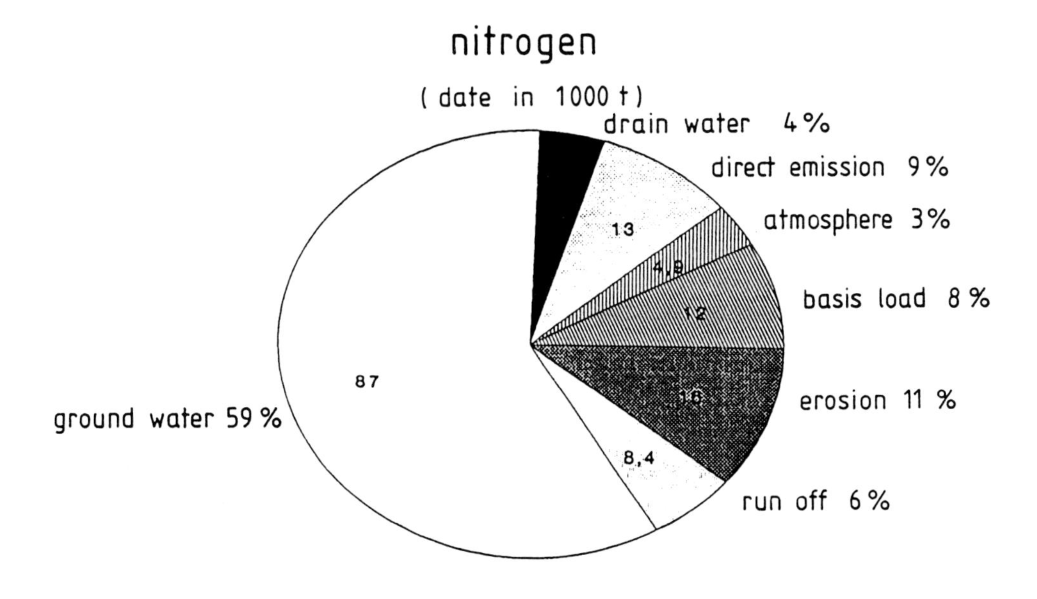

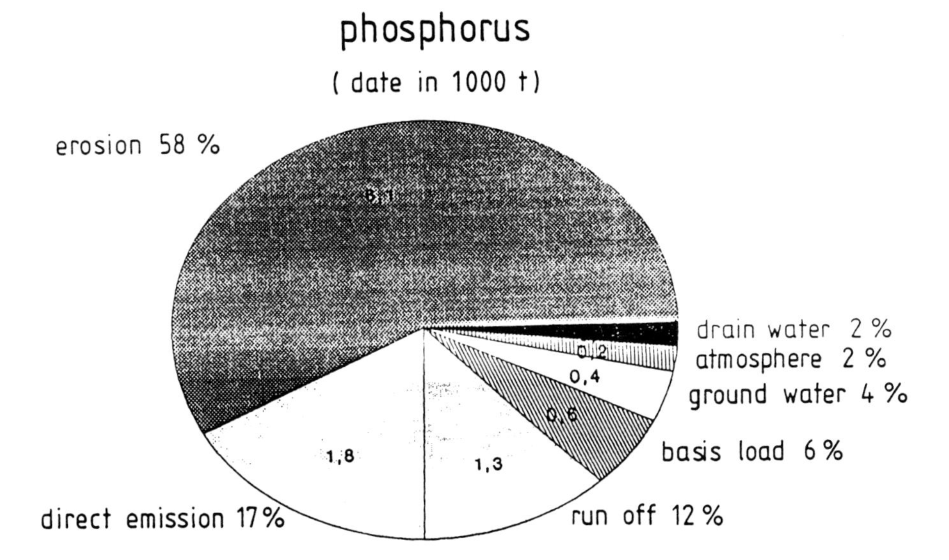

Figure 1. Calculated N- and P- emission in water bodies of the catchment Elbe 1988/1989 (date for 1 year) (Nolte and Werner, 1990).

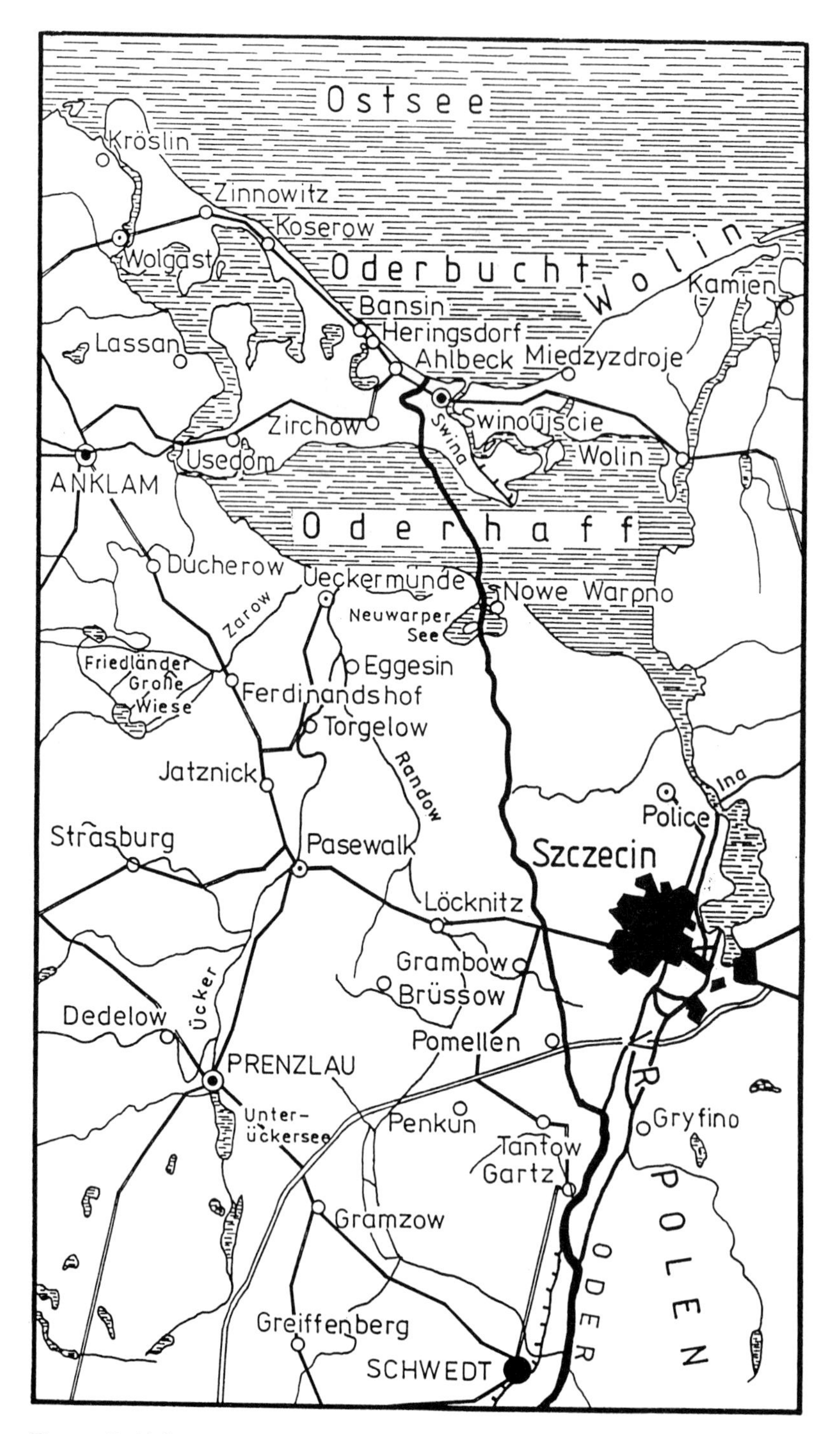

Figure 2. Uckersea catchment.

Figure 3. Aerial view of the inhomogeneous fields surrounding the lake.

RESULTS

The following changes enhancing erosion have occurred by the intensive agricultural management of areas :

- increasing differentiation between erosion and deposition areas by erosion ;
- changes in hydrological properties by reduction of the storage capacity, increase of surface runoff, and waterlogging in hollows ;
- changes of the soil-chemical environment by soil erosion down to the glacial till and eutriphication in the area of accumulation ;
- increasing contrast of important soil properties determining soil fertility, such as content of humus, pore volume, water capacity, exchange capacity, biological activity, buffer and filter capacities.

As a result of the continuous soil erosion eroded soils poor in humus in the eroding areas alternate with colluvial soils on the deposition area having a thick layer of humus. Totally different growth relations are reflected by the different development of plants.

Conventional soil cultivation characterized by concentrated and specialized plant production has lead to a further increase in water erosion on areas having been strongly heterogeneous already.

Soil erosion by precipitation measured in the catchment area of the Ucker River varies between 0.3 and 170 t . ha-1 . a-1. This great variability results from the range of factors causing and influencing erosion like relief intensity, different soil erodibility on small areas, different plant covers.

The site-specific inhomogeneity is enhanced by the effects of intensive crop production. The mechanical loading of soil has the strongest influence. Rains with intensities of more than $P > 7.5$ mm or $I_{30} > 5$ mm. h-1 have already lead to measurable soil displacement, if there were deficiencies related to soil management. On an average precipitations of such intensities were measured 14 to 22 times per year from 1961 to 1975 in Northern Germany ; the heavy rains of P more than 10 mm and I_{30} more than 10 mm occurred 7 to 11 times per year in the same period.

Normally soil erosion started in distinct wheel tracks or wheel tracks covered by soil after surface cultivation or on clogged areas on top of slopes. In 1986 and 1987 the amounts of soil translocated into the wheel tracks were significantly higher than in the loose parts of the fields measured simultaneously.

The differences are related to both the loss of carbon and the loss of nutrients and occurred less with heavy rains but with precipitations of small amounts or small intensities. On an average 3.1 kg Ct, 0.3 kg Nt, and 0.3 kg Pt were measured in 1 t of soil. The highest concentrations measured, however, were 23 kg Ct, 1.8 kg Nt, and 1.8 kg Pt per t of soil. It has not been sufficiently investigated what percentage of the load leaves the fields and reaches neighbouring waters. 10 per cent of the displaced soil material is assumed to be the diffuse material input by erosion in the following estimation, although higher dragouts are supposed to happen according to literature ([1]). The surface runoff water carried 5 ... 10 ppm N, 20 ... 30 ppm P, 100 ... 200 ppm K, and 200 ... 300 ppm Ca, with about 30 per cent being regarded as loss of the cropland ecosystem.

The off-site damages have become evident by the fact that the water quality of the Lower Lake Uckersee decreased by one class and has been classified into class 3 now (eutrophic waters). The total load of the lake was in 1990 : 5,118 kg orthophosphate, 9,978 kg total phosphorous, 189,197 kg inorganic nitrogen.

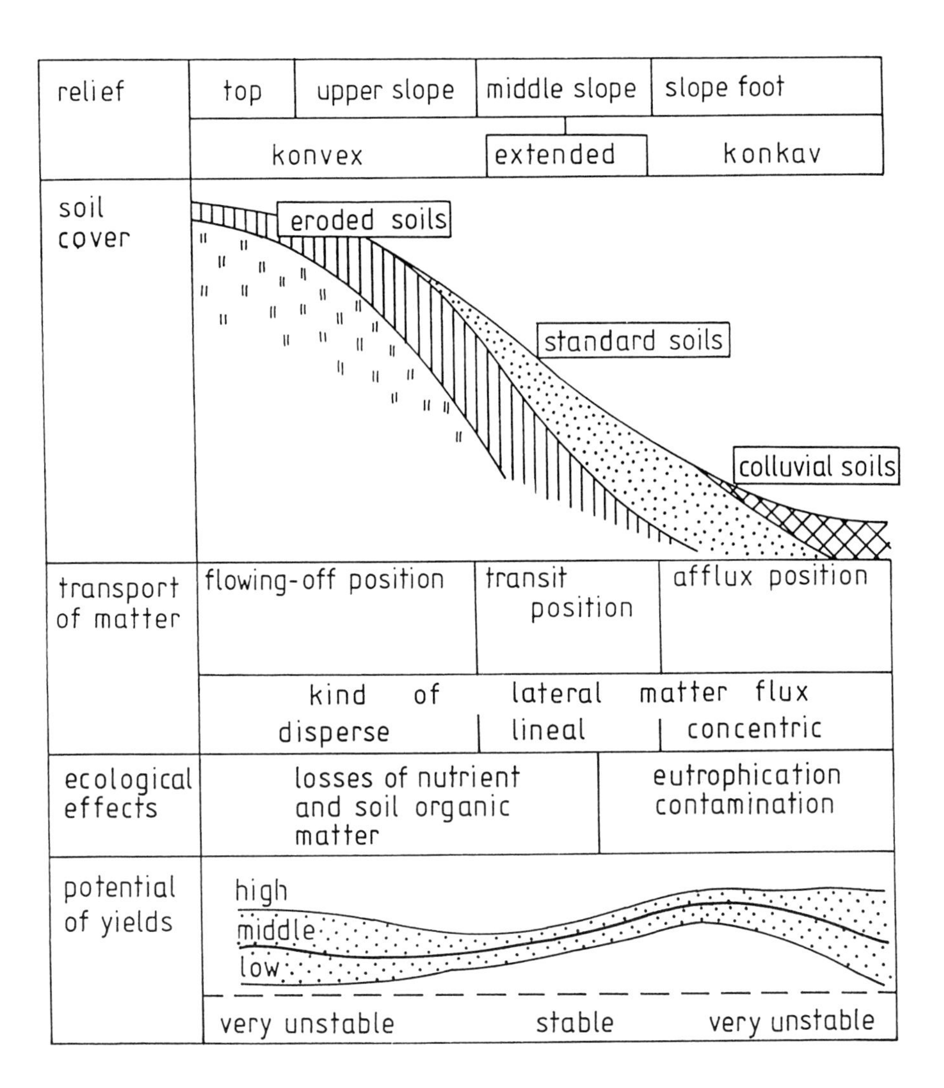

Figure 4. Type of catena for wurmian moraine areas with parts of eroded and colluvial soils and kind of matter transport.

The ratio Pt : Nin changed from 1 : 7 (1974) to 1 : 20 (1990). The N-load of 16.7 g . m^{-2} . a^{-1} suggests the water-quality class 4 (strongly eutrophic), whereas the orthophosphate load of 0.45 g . m^{-2}. a^{-1} is in the medium range.

What is the part of the diffuse material input in water pollution ?

POSITION	TOP	UPPER SLOPE	MIDDLE SLOPE	SLOPE FOOT	MIDDLE SLOPE	UPPER SLOPE	TOP
SOIL	CALCIC LUVISOL	CALCARIC REGOSOL		GLEYIC LUVISOL		CALCIC LUVISOL	CALCARIC REGOSOL
PARAMETERS	→	→	→	←	←	←	←
clay (%)	15,5 →	16,9 →	18,1 →	10,8	← 16,5	← 16,2	← 19,7
humus (%)	1,47 →	1,53 →	1,61 →	1,55	← 1,58	← 1,45	← 1,49
pH-value	6,5 →	7,1 →	7,2 →	6,5	← 5,7	← 6,1	← 7,1
T-value (mval/100 g)	9,4 →	9,9 →	8,0 →	7,4	← 7,9	← 7,3	← 10,6
P (mg/100 g)	8,7 →	4,0 →	4,7 →	7,6	← 7,2	← 6,1	← 2,7
K (mg/100 g)	6,4 →	5,7 →	6,7 →	5,4	← 10,7	← 9,3	← 7,7
Mg (mg/100 g)	6,9 →	3,3 →	4,1 →	4,8	← 7,0	← 7,4	← 3,0
ρ_d (g/cm3), 10 - 20 cm	1,67 →	1,60 →	1,65 →	1,67	← 1,54	← 1,74	-
ρ_d (g/cm3), 30 - 40 cm	1,82 →	1,80 →	1,81 →	1,77	← 1,89	← 1,87	-
ρ_d (g/cm3), 40 - 60 cm	1,67	-	1,81 →	1,71	-	1,79	-

Figure 5. Soil parameters and gradients of a catena (Schmidt, 1991).

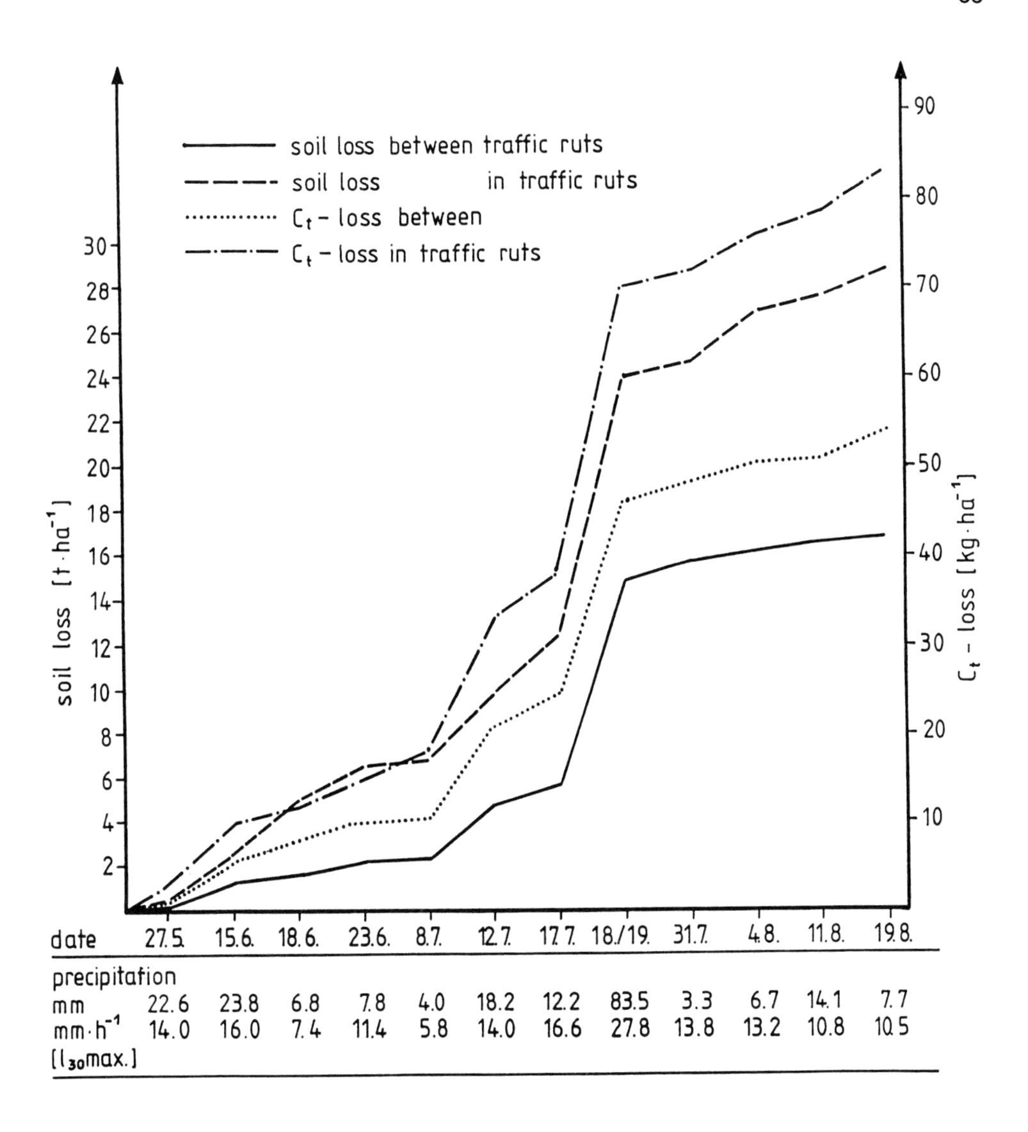

Figure 6. The summary of soil and carbon losses 1987 (May - August) in and between traffic ruts.

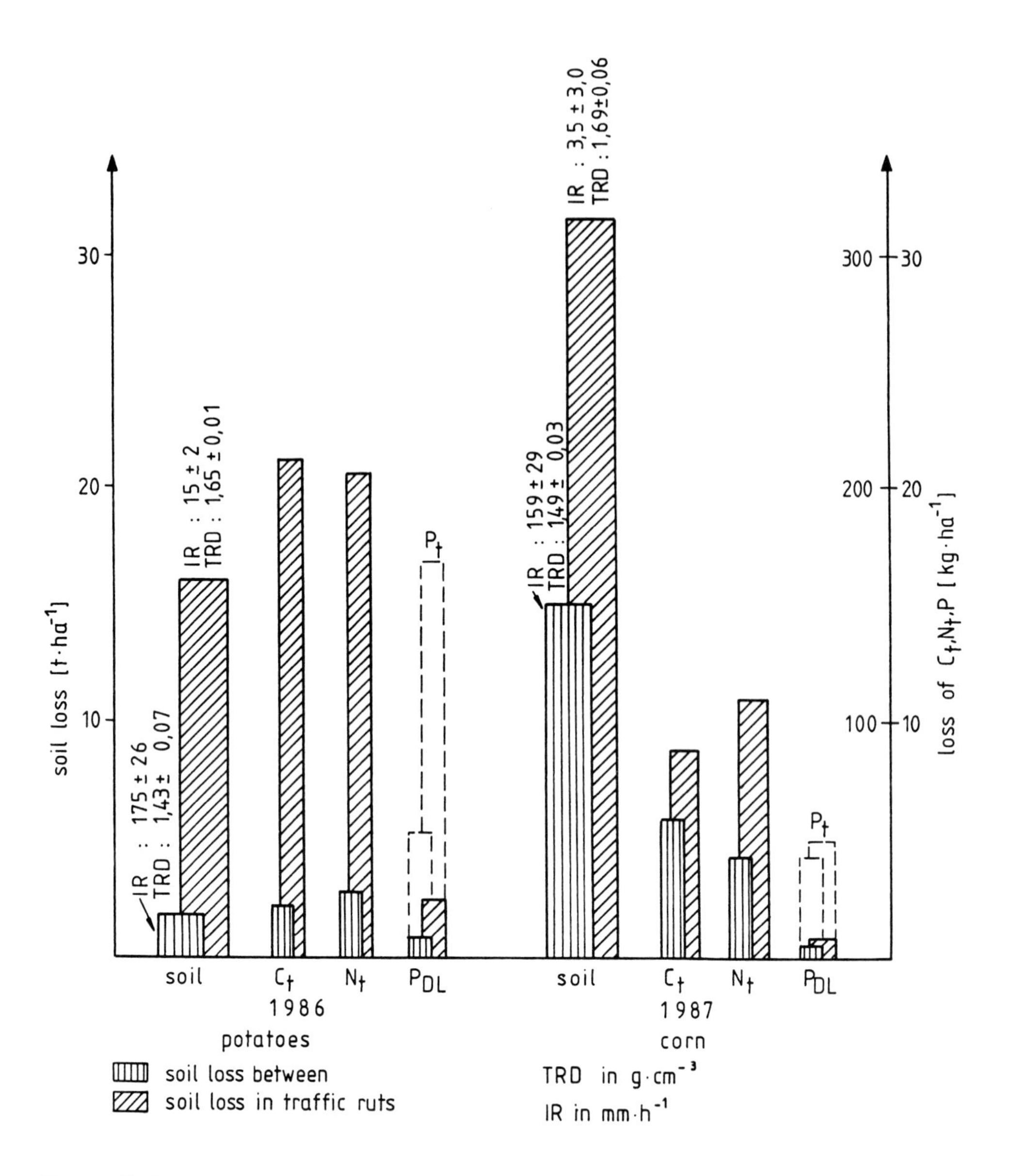

Figure 7. Total soil and nutrient losses in and between traffic ruts 1986 and 1987 (May - August) (Precipitation 1986 : 304 mm ; 1987 : 304 mm ; average of 30 years : 282 mm ; IR : infiltration rate, TRD : bulk density).

There are no exact measuring results on quantities and chemical bonds of the substances emitted by the cultivated area via the path flowing water - lake in the young moraine area. There are two ways to estimate the input : 1. estimation of the N and P inputs by erosion as percentage of the total diffuse material input (I) according to Nolte und Werner (1). 2. estimation of the N and P inputs based on erosion measurements on cultivated areas according to Frielinghaus (2). (Fig. 8)

I : N WITH SEDIMENT: 11 % P WITH SEDIMENT: 58 %) OF TOTAL
N WITH RUN OFF : 6 % P WITH RUN OFF : 12 %) POLLUTION

II: VALUES FROM EXPERIMENTS IN PLOTS

	I (NOLTE u.a. 1991) $t \cdot a^{-1}$	II (FRIELINGHAUS u.a. 1991) $t \cdot a^{-1}$
N-POLLUTION		
SEDIMENT	21,0	2,0 ... 8,8 t
RUN OFF	11,5	3,0 ... 6,0 t
P-POLLUTION		
SEDIMENT	6,0	2,0 ... 12,1 t
RUN OFF	1,2	6,0 ... 9,0 t

Figure 8. Surface run-off, soil loss and nutrient pollution from the catchment area (169 km^2 area, 80 % plant production, 7 mm flowing off in the catchment).

DISCUSSION AND CONCLUSIONS

The large range of variation in nutrient dragouts with a soil erosion of 5 t. ha^{-1} . a^{-1} indicates how difficult it is to estimate the risk in this field. The inhomogeneity caused by management, i.e. wheel tracks, soil-damaging compactions, plant cover, becomes the determining factor already when the erodibility by precipitation (R27 ... 46 KJ . mm^{-2} - h^{-1}) is moderate and the soil erodibility factor K = 0.32 +/- 103 has a variance up to 176 per cent (3 - 4).

Caused by the great heterogeneity of the soil cover the erosion process cannot be compared with the displacement processes on soil characterized by uniform substratum, e.g. in loess areas. This special aspect of the site conditions complicates the prediction of erosion by models internationally used to estimate the risk of different sites and climates (5 - 6 - 7 - 8 - 9).

The experimental examination of individual factors inducing and influencing erosion and demanded for all regions is inevitable for the North-German Lowlands before a model is adapted.

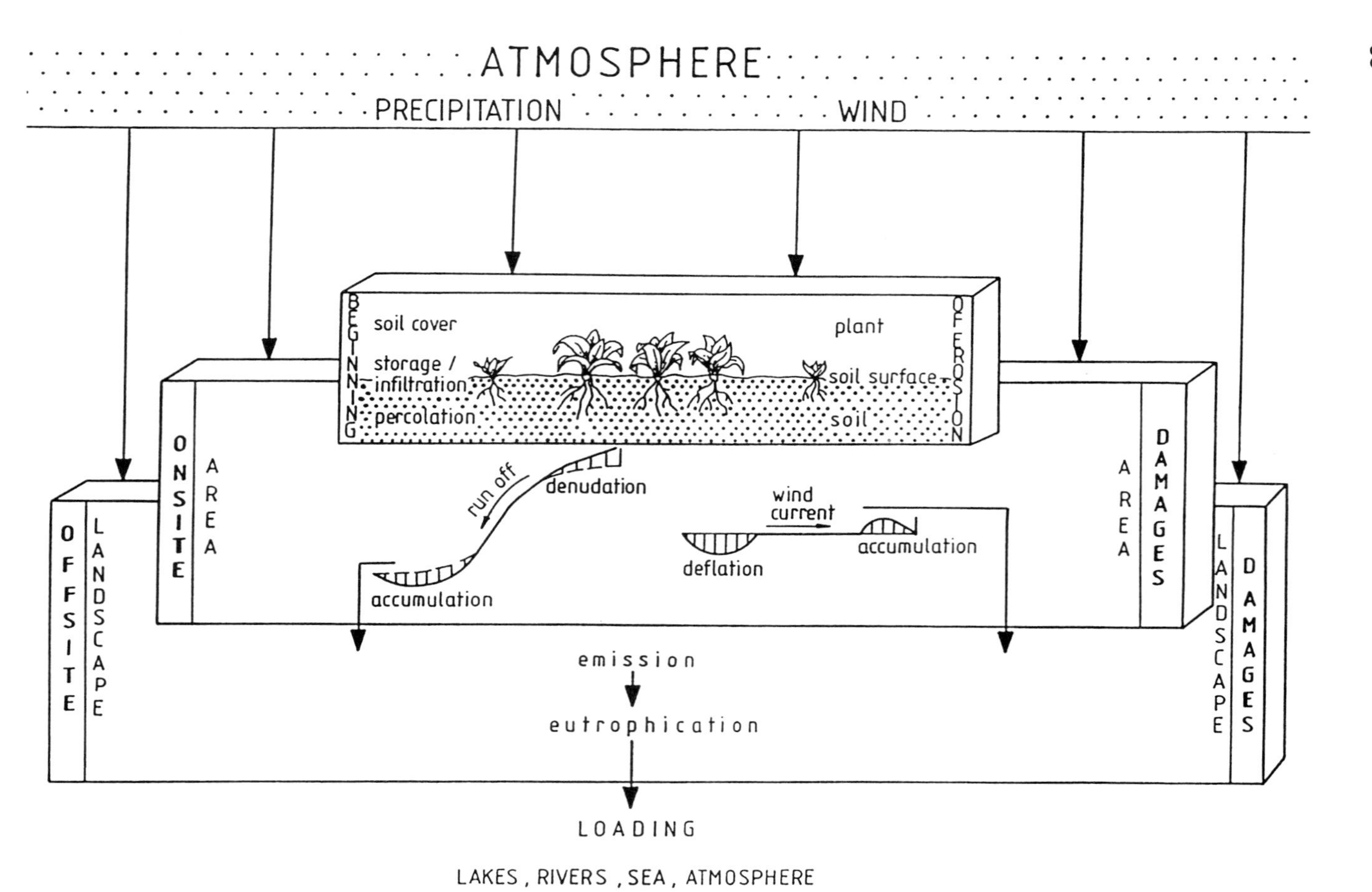

Figure 9. On site and off-site damages in the landscape.

The off-site damages by water erosion in landscapes with many waters gain significance and have to be analysed more exactly in order to take adequate protective measures against them. One precondition is to apply punctual investigations to the areas by means of validated models. The material regime on sloped areas, accumulation and sedimentation, the immediate and delayed dragout of nutrients and pollutants require further investigations and provide inputs for the risk-estimating models.

REFERENCES

1. Nolte, C., Werner, W., 1991 : Stickstoff- und Phosphateintrag über diffuse Quellen in Fließ-gewässer des Elbeeinzugsgebietes im Bereich der ehemaligen DDR. Umweltforschungsplan des BMUNR, Wasserwirtschaft, Forschungs-vorhaben Wasser 102 04 382.
2. Frielinghaus, M., Petelkau, H., Sshmidt, R., 1991 : Wassererosion im norddeutschen Jungmoränengebiet. Z. f. Kulturtechnik und Landentwicklung 33, 22-33 (1992).
3. Deumlich, D., 1987 : Untersuchungen zur Bodenerodierbarkeit auf Jungmoränenstandorten. - Arch. Acker-Pflanzenbau Bodenkd. Berlin 31, 7 - 14.
4. Deumlich, D., Gödicke, K., 1989 : Untersuchungen zu Schwellenwerten erosionsauslösender Niederschläge im Jungmoränengebiet der DDR.-Arch. Acker-Pflanzenbau Bodenkd. Berlin 33, 11, 709 - 716.
5. Wischmeier, W. H., Smith, D. D., 1978 : Predicting rainfall erosion - a guide to conservation planning. - USDA, Agric. Handbook Nr. 537.
6. Schwertmann, U., Vogl, W., Kainz, M, 1990 : Bodenerosion durch Wasser: Vorhersage des Abtrags und Bewertung von Gegenmaßnahmen. - Stuttgart, Ulmer
7. Diez, T., 1985 : Grundlage und Entwurf einer Erosionsgefährdungs-karte von Bayern. - Mitt. Dt. Bodenkd. Ges. 43, 833 - 840.
8. Kainz, M., 1985 : Verlauf des relativen Bodenabtrages unter Mais.-Mitt. Dt. Bodenkd. Ges. 43, II, 865 - 866.
9. Auerswald, K., Kainz, M., Vogl, M., 1986 : Vergleich der Erosionsgefährdung durch Maisfruchtfolgen (C-Faktor). - Bayer. landw. Jahrb. 63, 3 - 8.

Farm Land Erosion: In Temperate Plains Environment and Hills
S. Wicherek (Editor)
1993 Elsevier Science Publishers B.V.

Water input / output and soil erosion on a cultivated watershed

A. Kertesz [a], D. Loczy [a] and G.Y. Varga [b]

[a] Geographical Research Institute, Hungarian Academy of Sciences, P.O. Box Budapest H/I388, Hungary.

[b] Research Centre for Water Resources Development, Kvassay-zsilip, 1 Budapest, Hungary.

SUMMARY

The amount of water available for runoff and, therefore, for soil erosion, depends on the water budget of the catchment. In the Lake Balaton Region, Hungary, sediment and nutrient loads reaching the lake from small watersheds are studied in German/Hungarian cooperation. This paper presents attempts to quantify the water budget of a subcatchment and to find correlations between various factors of soil erosion.

RESUME

Ce travail présente les résultats des mesures effectuées durant les 10 dernières années par l'Institut de Recherche Géographique, dans le cadre de différentes stations expérimentales (Lac Balaton, Hongrie). Ces résultats sont comparés avec ceux déjà connus et une estimation globale de l'érosion des terres de culture est fournie ; elle montre des différences régionales et met en évidence le rôle des facteurs contrôlant l'érosion. Plusieurs méthodes permettant d'évaluer l'érosion des terres sont examinées de façon détaillée et critique.

Les conclusions suggèrent divers modes de mise en valeur qui préviendraient l'érosion et présentent des propositions de protection contre l'érosion des sols.

1. INTRODUCTION

Soil erosion studies are of great importance not only for surveying mere soil loss, but also for gaining information about the sites where the removed soil accumulates together with its most common transport medium, ie. water. (For agricultural crops, the availability of water is even more important at short term than the intact or eroded nature of the soil profile.) Redeposited soil masses may contain chemicals (surplus amounts of fertilizers, pesticides or herbicides), which may cause environmental pollution.

Water budgets show how the water input is redistributed among runoff, evaporation and infiltration in the catchment and thus supplement soil erosion data. Considering water and soil together, we may arrive at a better understanding of geomorphological processes as well as at a better design of soil conservation measures and farming practices.

In Hungary, erosion endangers a major natural asset, fertile soils (1). More than a third of all agricultural land (2.3 million hectares) is affected by water erosion (l3.2 per cent only slightly, l3.6 per cent average erosion and 8.5 per cent severely eroded) and l.5 million hectares by wind erosion (2).

Although geographers, hydrologists, soil and agricultural scientists have long been active in erosion studied worldwide, relatively few papers link soil erosion to the water cycle of a catchment (3 - 4). The present paper intends to evaluate the water budget of a small catchment, the sources of water and the precipitation/runoff correlations from the viewpoint of soil erosion.

2. THE STUDY AREA

The Lake Balaton system (5,775 km^2) is made up of three subcatchments with different physical conditions (Fig 1)

1. The largest catchment by far (2,622 km^2) belongs to the Zala river and became part of the Balaton drainage system, through a river capture.
2. Along the southern shore, large elongated watersheds with low relative relief are typical.
3. The northern unit comprises a series of small catchments of diverse geology, soil and land use.

Most of the non-point pollution of the lake comes from agricultural areas.

The area selected for a closer study, the watershed of the Orvényes Séd stream (Fig. 1), lies on the northern subcatchment. Soil erosion studies have been performed her for four years in cooperation between the German Research Foundation and the Geographical Research Institute, Hungarian Academy of Sciences. Among their objectives is the estimation of soil loss for each section of the watershed, using their Universal Soil Loss Equation.

On the northern shore, this is the only watershed where relatively natural conditions prevail, ie. no major construction, artificial drainage or point like pollution of industrial origin. Lying outside the main recreation zone, the area is relatively sparsely inhabited and used predominantly by agri- and viticulture as well as forestry.

The watershed has an area of 24.4 km^2 with its highest elevation 416 m above sea level and an outlet at l04 m altitude (max. relief 311 m). The main water way, the Orvényes Séd is 8.1 km long.

Geologically the area is built up of (karstic) Triassic limestones, dolomites and marls and their regolith (calcareous fragments embedded in clay matrix) as well as a mantle of unconsolidated deposits (loess, slope loess). On loess and marl the soils are either rendzinas or medium to severely eroded brown forest soils. In lowerlying areas meadow soils are found. On hilltops the parent rock is often exposed.

The watershed is divided into three sections : the upper parts belong to the Pécsely Basin (1), and through a forested gorge (2) the stream reaches the Lake Balaton Plain (3) (Fig. 2).

3. WATER BUDGET

The water budget was calculated from the formula $\Delta W = P - E - R$, where ΔW is change in water reserves, P is precipitation, E is evaporation and R is runoff.

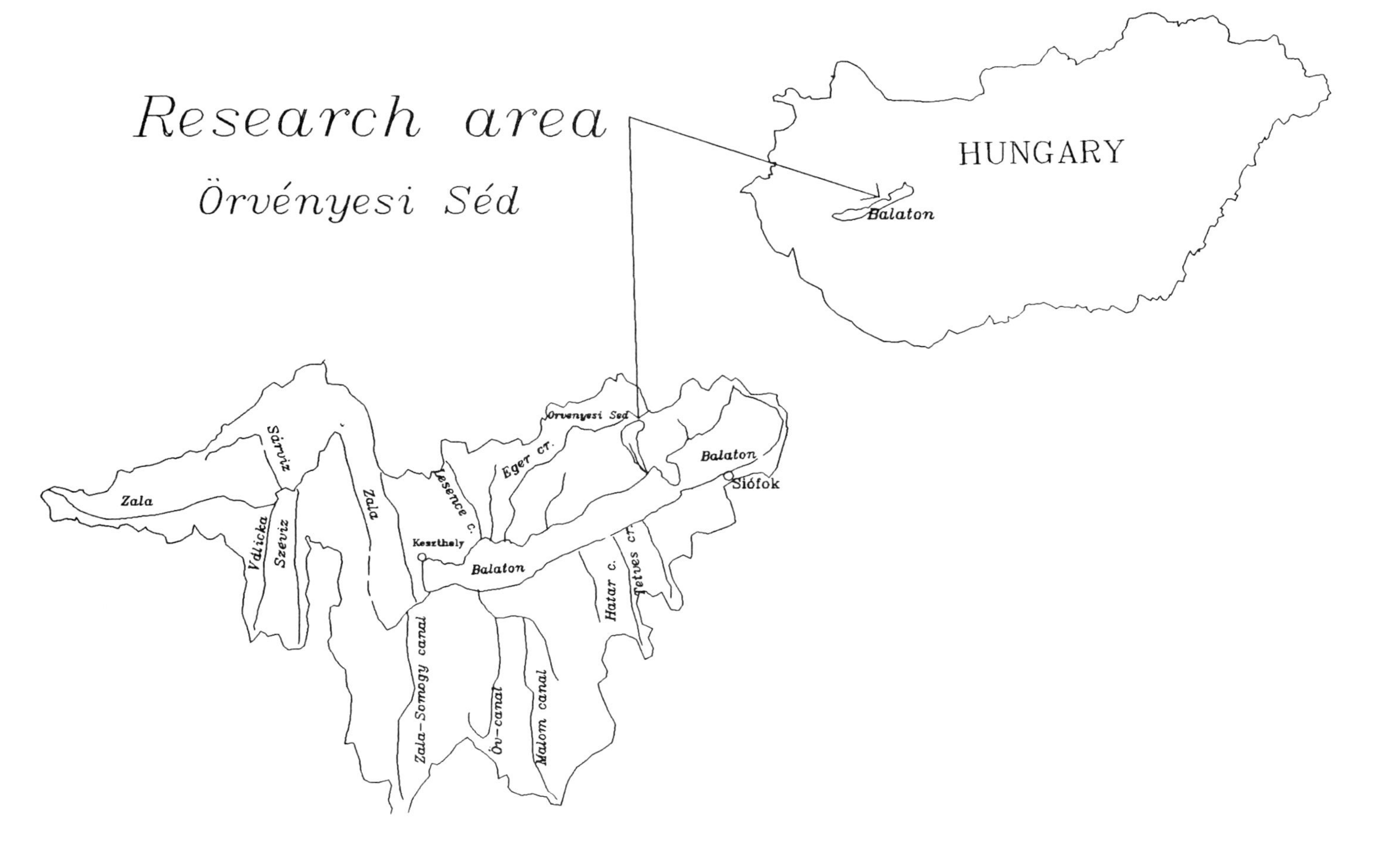

Figure 1. Research area.

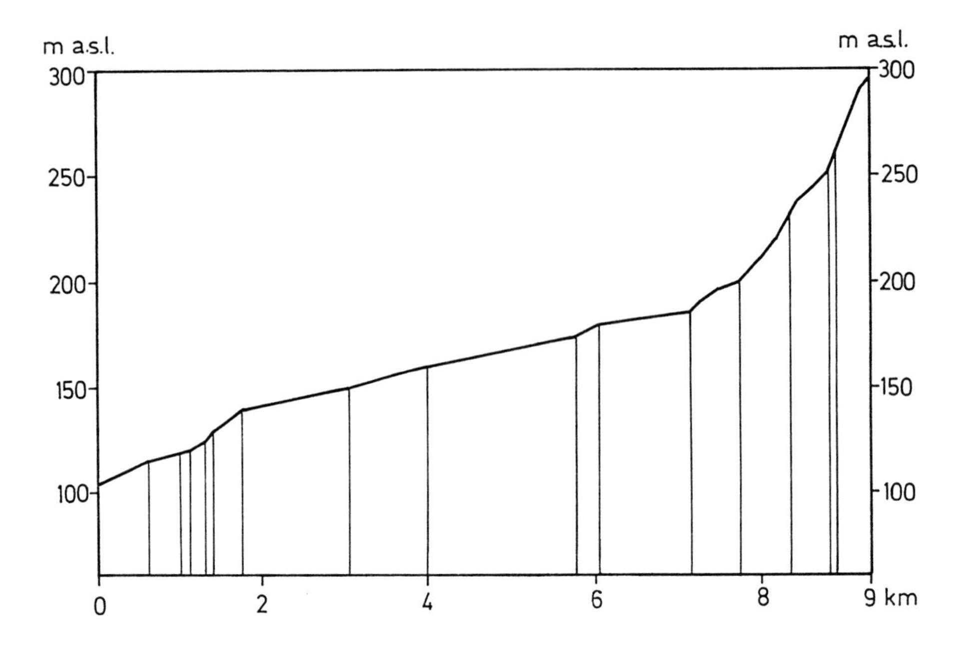

Figure 2. Longitudinal profile of Orvényesi Séd creek.

Although, according to Meteorological Survey data, long-term mean annual precipitation is 657 mm ; during the period studied (l975-l990) annual precipitation was inferior to this most years. The average for these l5 years is 592 mm, 10 per cent below the longterm average.

Runoff data were supplied by the gauging station installed at the stream outlet (watermill of Orvényes). From the continuous record the Research Centre for Water Resources Development (VITUKI) computed monthly mean discharges (Table 1).

In the changes of water reserves (Table 2) negative deviations from the long-term averages are also observed. The drop in values of evaporation from the long-term average of 522 mm correlates with the l0 % loss mentioned for precipitation. This dry spell is also apparent in records for other parts of the country. Even greater extremities are observed in the last three years investigated (l988-l990), when annual precipitation was only 495 mm, mean evaporation 427 mm/ca. 25 per cent reduced compared to the respective long-term values. The l5-year mean runoff is ll7 mm/year, while the average for l988-l990 is only 66 mm/year-44 per cent lower (!).

Even if increased water intake for irrigation is considered, hydrological observations show an exagerated response to changes in meteorological factors.

Table 1.
Mean monthly discharge of Orvényes Séd stream at Orvényes / m^3 / s /

Year/Month	XI	XII	I	II	III	IV
1975/76	0.088	0.086	0.107	0.109	0.122	0.130
1976/77	0.119	0.203	0.167	0.263	0.243	0.262
1977/78	0.084	0.079	0.070	0.080	0.070	0.030
1978/79	0.050	0.050	0.071	0.088	0.071	0.075
1979/80	0.060	0.061	0.061	0.101	0.080	0.082
1980/81	0.102	0.129	0.071	0.068	0.078	0.064
1981/82	0.029	0.040	0.082	0.058	0.087	0.070
1982/83	0.072	0.122	0.118	0.118	0.205	0.170
1983/84	0.074	0.069	0.059	0.078	0.070	0.073
1984/85	0.093	0.097	0.080	0.083	0.161	0.130
1985/86	0.105	0.151	0.171	0.179	0.300	0.298
1986/87	0.066	0.078	0.073	0.111	0.095	0.141
1987/88	0.092	0.106	0.098	0.122	0.153	0.133
1988/89	0.041	0.045	0.039	0.044	0.047	0.050
1989/90	0.026	0.020	0.024	0.021	0.023	0.029

Year/Month	V	VI	VII	VIII	IX	X
1975/76	0.120	0.125	0.119	0.096	0.085	0.103
1976/77	0.181	0.172	0.140	0.116	0.068	0.068
1977/78	0.080	0.090	0.100	0.080	0.050	0.050
1978/79	0.073	0.073	0.072	0.084	0.065	0.040
1979/80	0.045	0.078	0.091	0.060	0.069	0.072
1980/81	0.071	0.070	0.073	0.033	0.042	0.031
1981/82	0.071	0.071	0.083	0.068	0.049	0.052
1982/83	0.144	0.108	0.076	0.059	0.060	0.063
1983/84	0.088	0.065	0.058	0.067	0.074	0.080
1984/85	0.178	0.166	0.106	0.103	0.113	0.067
1985/86	0.242	0.197	0.134	0.102	0.076	0.060
1986/87	0.131	0.122	0.091	0.091	0.079	0.075
1987/88	0.112	0.081	0.045	0.037	0.042	0.041
1988/89	0.046	0.039	0.026	0.024	0.031	0.027
1989/90	0.028	0.033	0.023	0.015	0.021	0.020

Table 2.
Water budgets for the Orvényes Séd watershed (mm)

Hydrological year	1975/76	1976/77	1977/78	1978/7919	1979/80
Precipitation	521	674	609	550	643
Evaporation	490	661	630	557	584
Runoff	-42	220	101	88	94
Balance	- 111	-207	-122	-95	-35
Hydrological year	**1980/81**	**1981/82**	**1982/83**	**1983/84**	**1984/85**
Precipitation	525	687	566	704	616
Evaporation	493	553	421	566	540
Runoff	92	84	-44	94	150
Balance	-60	+50	+1	-44	-74
Hydrological year	**1985/86**	**1986/87**	**1987/88**	**1988/89**	**1989/90**
Precipitation	664	594	486	487	513
Evaporation	519	547	344	455	482
Runoff	219	126	-17	48	32
Balance	-74	-79	+25	-16	-1

4. PRECIPITATION/RUNOFF CORRELATIONS

In order to estimate the water effective in soil erosion, we have to know the relationship between the water input (precipitation) and output (sediment-laden inflow into the lake). With regard to seasonal differences, the data series were divided into spring (March to May), summer (June to August), autumn (September to November) and winter (December to February) sets.

The linear regression analysis provided a correlation coefficient (Table 3) which indicates *no direct correlation* between precipitation and runoff for the period investigated.

In this case, runoff must have additional sources, in our hypothesis, the water reserves of karst aquifers.

Table 3.
Correlation coefficients between precipitation and runoff

season	correlation coefficient
winter	0,365103
spring	0,154285
summer	0,107838
autumn	0,283304

5. KARST SPRINGS AND THE WATER BUDGET OF THE WATERSHED

Table 4 shows yields of the two most important karst springs of the area (Zàdor and Vàszoly springs) for the last six hydrological years. (The other springs on the watershed supplied negligible amounts of water in this dry period.) Although the yields are based on one measurement per month, the data allow some conclusions to be drawn.

Comparing spring yields with runoff data, it becomes clear that a considerable portion (ca. 50 per cent) of stream discharge is of subsurface origin. Subsurface catchments are mostly delimited with much uncertainty and do not overlap with surface ones. An analysis of the broader region would reveal the relationships ; this type of analysis was not possible.

We had decided on investigating the correlation between the precipitation and the reduced discharge (Q_r) for the six years, using the formula

$Q_r = 1 - \frac{Q_f}{Q}$, where Q_f is mean annual spring yield and Q is total discharge.

This time, the received correlation (c = 0.24) shows *no relationship* between precipitation and reduced runoff.

Table 4.
Joint yields of Zàdor and Vàszoly springs /m³/sec/

Year/Month	XI	XII	I	II	III	IV
1984/85	0,009	-	-	-	0,018	0,028
1985/86	0,0017	0,018	0,052	0,039*	-	0,092
1986/87	0,014*	0,030	-	0,024*	0,028	0,034
1987/88	0,034	0,033	0,025	0,024	0,035	0,046
1988/89	0,025	0,020	0,030	0,025	0,022	0,019
1989/90	0,018*	0,015*	0,012	0,011*	0,014	0,010*

Year/Month	V	VI	VII	VIII	IX	X	annual average
1984/85	0,029	0,032	00,30	0,021	0,017	-	0,023
1985/86	0,099	0,098	0,063	0,056	0,056	0,049	0,060
1986/87	0,041	0,043	0,050	0,040	0,037	0,034	0,037
1987/88	0,039	0,034	0,031	0,037	0,030	0,041	0,034
1988/89	0,018	0,005*	0,019	0,017	0,012*	0,013*	0,022
1989/90	0,011	0,008*	-	0,004*	0,008*	0,009*	0,011

* yields of only one spring ; not included in annual average

6. TRANSPORT OF SUSPENDED LOAD

The concentration of sediment transported in suspension generally depends on discharge. For the Orvényes séd *no correlation* was found between discharge and suspended load. The low correlation coefficient (0.31) obtained from 180 pairs of data also underlines that a considerable part of discharge does not derive from surface runoff.

7. FLOOD WAVES AND FLOOD-GENERATING RAINFALLS

From the time series of discharge 32 flood waves had been selected, their water mass and the precipitation of the previous 20 days was calculated. The correlation between flood-generating precipitation and the water mass of flood waves was 0.33, while with total previous precipitation (including the flood-generating precipitation) was only 0.02. This means that *no correlation* exists between these parameters.

8. ESTIMATED SOIL EROSION

Since the German-Hungarian joint research project is not yet completed, the data by Dezseny (5) although not founded on field measurements, but of the same magnitude as ours - are used in the estimation (Table 5). The data point to severe soil losses from the watershed. Large-scale sheet-wash in spring is evident in the field and also confirmed by soil analyses.

Table 5.
Soil loss (tons/hectare) from the Orvényes Séd watershed by land use and slope categories, computed using USLE, from VIZITERV (Water Constructions Planning Enterprise) data (after Dezsény 1982)

Land use	Slope categories (per cent)				
	<5	5-12	12-17	17-25	25<
Arable					
Upper section	3,3	12,6	27,3	50,4	-
Lower section	4,0	15,4	33,4	61,6	-
Vineyards and orchards					
Upper section	17,8	68,3	148,2	273,2	409,8
Lower section	15,0	57,5	124,8	230	345,0
Meadow and pasture					
Upper section			1.6*		
Lower section			2.5*		

*undifferentiated by slope categories

Along with the eroded soil, suspended load and washed-off chemicals reach the stream (Table 6). Unfortunately, the table fails to show the impact of storm runoffs, responsible for extreme loads, exceeding the average values 5-10 times over.

Some of the chemicals found in the water of the Orvényes Séd may naturally come from karst water (this may explain the high Ca^{++} concentration), but evidence is also found for overfertilization and improper manure treatment.

Table 6.
Load of Orvènyes Séd stream (VITUKI observations, averages for the years I970-I980)

Water discharge $m^3 sec^{-1}$	Total mg/l	phosphorus tons/year	Total nitrogen mg/l	tons/year	Total suspended load mg/l	tons/year
0.090	0,I32	04	6,365	I8,I	I8,2I3	I8,2

9. CONCLUSIONS

Since much of the discharge of the Orvényes Séd comes from a subsurface catchment difficult to delimit, the water budget compiled for the surface catchment cannot reflect the true conditions of runoff and of water reserves. This explains why no correlation was found between precipitation and runoff, between discharge and total suspended load and between the water transport of flood waves and flood-generating rainfall.

At the same time, soil erosion at a considerable rate-although spatially variable as a function of local erosion controls is observed on the watershed. Given the basin character of the upper (largest) portion of the catchment, it is assumed that the soil masses removed from slopes accumulate on the basin floor. Notwithstanding, the load of the stream with sediment and chemicals is considerable. This indicates that a substantial amount of eroded soil *does* reach the stream, and indirectly Lake Balaton.

The following conditions are instrumental for further research into runoff and soil erosion :

I. A more comprehensive hydrological study of the broader environs is necessary and the investigation of water budget should be repeated in the light of its findings.
2. More frequent measurements of karst spring yields and an evaluation of the significance of other karst springs in the water budget of the watershed.
3. An automatic gauging of flood waves on the stream.
4. An estimation of transpiration and water consumption by vegetation (6) on the basis of land use mapping, in order to develop our hydrological water budget into a geoecological one.
5. An estimation of water retention capacities of soils (soil sediment accumulations) and unconsolidated cover deposits (loess) from soil survey data.

REFERENCES

1. Kertesz, A., Loczy, D. and Olah, I. (1990) : Soil conservation Policy and Practice for Croplands in Hungary. In : Boardman J. and al. (eds) : Soil Erosion on Agricultural Land. John Wiley, Chichester, 605-6I9.
2. Stéfanovits, P. and Vàrallyay, G.Y. (1992) : State and Management of Soil Erosion in Hungary. Soil Erosion and Remediation Workshop US - Central and Eastern European Agro/Environmental Program, Budapest, April 27-May I 1992, Proceedings, Rissac, Budapest 79-95.

3. Bryan, R.B., Dakshinamurti, C. and Boswas, T.D. (l962) : Soil erosion and infiltration as a function of rainfall. In : Comm. de l'érosion continentale, Coll. de Bari, Publ. 59 de l'Assoc. Int. d. Hydrol. Sci., Gent, 133-147.
4. Kirwald, E. (1969) : Wasserhaushalt und Einzugsgebiet, Essen.
5. Dezsény, Z. (1982) : A Balaton részizgyüjtöinek összehasonlito vizsgàlata az erozioveszélyesztetettség alapjàn (A comparative study of erosion hazard for the subcatchments of Lake Balaton). Agrokémia es Talajtan 31, 405-421.
6. Sponagel, H. (l980) : Zur Bestimmung der realen Evapotranspiration landwirtschaftlicher Kulturpflanzen. Geologisches Jahrbuch Hannover, F9, 3-87.

Farm Land Erosion: In Temperate Plains Environment and Hills
S. Wicherek (Editor)

Characteristics of runoff generating rains on bare loess soil in South-Limbourg (The Netherlands)

F.J.P.M. Kwaad

Laboratory of Physical Geography and Soil Science, University of Amsterdam, Nieuwe Prinsengracht 130, 1018 VZ Amsterdam, The Netherlands.

SUMMARY

In 1985 a plot study was started in Dutch South-Limbourg to evaluate the effects of various cropping systems of fodder maize on runoff, erosion and crop yield. Also permanently bare soil was included in the study. In the course of the period of measurement of three years and four months, 73 run-off events were registered on bare soil, 35 in summer and 38 in winter. In this paper the following characteristics of the runoff generating rains are presented : amount, duration, average intensity and max. 5-minute intensity. From an analysis of the collected data it appeared, that the modal runoff generating rain during the period of study was characterised by a rainfall amount of 2-4 mm, a duration of 30-45 minutes, an average intensity of 2-3 mm/hour and a max. 5 minute intensity of 5-10 mm/hour. These are not high values. It is important to note, that runoff generation was not restricted to high intensity summer thunder storms. Runoff was also generated by low intensity winter rainfall. This can be interpreted to mean that Horton overland flow was not the sole or even the dominant form of overland flow at the site of study. Saturation or storage controlled overland flow may have been just as important. This must be given due consideration in devising soil conservation measures, which traditionally aim at increasing the *infiltration capacity* in stead of the *moisture storage capacity* of the soil. Also, in many soil erosion models Hortonian overland flow is postulated or tacitly implied. There seems to be room for erosion models based on soil storage excess overland flow.

RESUME

En 1985, nous avons commencé à étudier des parcelles afin d'envisager les effets de divers systèmes de cultures du maïs sur le ruissellement, l'érosion et le rendement.

Des sols nus ont également été inclus dans notre étude. Sur l'une des parcelles nues a été installé un système de mesures "HS" équipé d'un enregistreur d'eau automatique. Au cours des 3 ans et 4 mois pendant lesquels les mesures ont été effectuées, 73 événements de ruissellement ont été enregistrés sur ces parcelles - 35 pendant l'été et 38 pendant l'hiver. Les précipitations, facteurs de ruissellement, ont été enregistrées à l'aide d'un pluviographe mécanique à siphon de type Lambrecht (taux de transfert sur papier : 1 mm / 3 minutes). Nous présentons ici les caractéristiques des orages générateurs de ruissellement : quantité, durée, intensité moyenne et intensité maximum à 5 minutes. L'analyse des données ainsi recueillies fait apparaître que les orages générateurs de ruissellement se caractérisent par une quantité de pluie de 2 à 4 mm, une durée de 30 à 45 minutes, une intensité moyenne de 2 à 3 mm / heure et une intensité maximum à 5 minutes de 5 à 10 mm / heure. Ces valeurs ne sont pas élevées.

Fullen et Reed ont publié des valeurs comparables en 1986, avec un seuil d'intensité de 2 mm / heure pour un début d'érosion du sol sur des sols à croûte de battance en Angleterre. Il est aussi important de noter que le ruissellement n'est en aucun cas généré uniquement par des orages d'été de haute intensité : il est également produit par des pluies d'hiver de faible intensité. Ceci signifie que le modèle proposé par Horton est insuffisant. Il faut tenir compte d'un tel fait dans l'élaboration de mesures de conservation du sol, mesures qui traditionnellement visent à augmenter la *capacité d'infiltration* du sol au détriment de la capacité de *rétention du sol.* Il semble qu'il y ait place pour des modèles fondés sur un excès de rétention.

1. INTRODUCTION

In hilly Dutch South-Limbourg, rainfall induced accelerated soil erosion of loess soils with associated off-site effects (flooding, sedimentation) occurs (1 - 2 - 3). In 1985 a plot study was started to evaluate the effects of various conservation cropping systems of fodder maize on runoff, soil loss and crop yields (4 - 5 - 6 - 7 - 8). As part of this study, rainfall measurements were carried out. Objectives of the rainfall measurements were : (a) to assess rainfall conditions which give rise to overland flow and erosion on loess soils in South-Limbourg, and (b) to determine the value of the EI_{30} -index or rainfall erosivity index as part of the procedure to establish the K-factor or soil erodibility index from soil loss measurements on permanently bare fallow plots (9). In this paper results regarding the first objective are presented. Not very many data are available for western Europe on rainfall characteristics related to soil erosion. For Belgium data are given by Bollinne (10), Bollinne et al. (11) and Sinzot et al. (12). Data for England were summarised by Evans (13). From these data it appears that rainfall that is excluded from the calculation of the R-factor value of the Universal Soil Loss Equation in the USA (9), contributes to accelerated soil erosion in West-Europe.

2. MATERIALS AND METHODS

In 1985, twelve plots were laid out in a randomized block design with three replications per treatment (14). The plots were located near the village of Wijnandsrade (Fig. 1) on a uniform 6 % slope. Plot length and plot width were 22.00 and 1.80 m resp. (15). Runoff was collected and stored in three storage tanks with divisors per plot (16). Runoff volume and sediment concentration of the runoff was determined with four weekly intervals. For budgetary reasons more frequent visits of the experimental site were not possible. Fodder maize was grown continuously on the plots during 1986, 1987, 1988 and 1989. Cropping systems of fodder maize are described in detail by Geelen (4) and Kwaad (6 - 7). Systematic measurements of runoff and soil loss were carried out from May 1987 until March 1990, both in summer and winter. Precipitation was measured with a standard raingauge and a mechanical recording raingauge with a paper transfer rate of 1 mm per 3 minutes (Lambrecht, type 1509).

Figure 1. Location of Experimental Farm Wijnandsrade in South-Limbourg.

Besides nine plots with three replications of three maize cropping systems, three permanently bare fallow plots were laid out, one in each block. The fallow plots were tilled (ploughed and harrowed) once a year, at the same time as the conventional maize cropping system (around May 1st). One of the fallow plots was equipped with an HS-flume (17) and a Munro water level recorder (type IH 89 with vertical rotating drum). The data obtained with the water level recorder were used to sort out the runoff generating rain storms from the pluviograph recordings. The resolution of the water level recorder (one drum rotation in 28 days, or a paper transfer rate of 1 mm per 96 minutes) did not allow determination of the runoff volume of single runoff events, but only approximate timing of the occurrence of runoff peaks. Due to the low resolution of the stage recording and the fact that soil loss was not determined on a single event basis, but as a four-weekly total, it was not possible to establish the contribution of single storms to total runoff and soil loss. The available data were used to analyse the characteristics of the runoff generating rain storms.

Soils of the experimental site were imperfectly drained, truncated gleyic luvisols with less than 0.80 m colluvium, overlying the (lower part of the) original argillic horizon. Parent material was decalcified loess. A (perched) water table was present within 1.00 m of the soil surface during part of the year.

Average yearly rainfall amount of the site is 750 mm with rain in all seasons. High intensity rains only occur in the summer half of the year ; 30-minute intensity that is exceeded once a year is 24 $mm.h^{-1}$ (18). Average yearly rainfall erosivity (R-factor), as defined by Wischmeier and Smith (9), is 75 (metric units) according to Bergsma (19), resp. 60 (metric units) according to Bollinne et al. (20). These values are relatively low, compared to large parts of the USA (9).

3. RESULTS

Total yearly rainfall amounts of the years 1987, 1988 and 1989 at the site of study were 873.2 mm, 780.4 mm and 695.8 mm resp. These amounts are not far from normal (750 mm). Therefore, the period of study is considered as fairly representative of the rainfall regime of the region. From Fig. 2 and 3 it appears that, on the permanently bare plots, soil losses were higher in summer than in winter, whereas runoff amounts and runoff percentages were (much) higher in winter than in summer.

3.1 Frequency and seasonality of runoff generating rains

During the 40 months period of observation, 73 runoff events were recorded with the HS-flume : 9 in 1986, 17 in 1987, 27 in 1988 and 20 in 1989, or an average number of 22 runoff events per year on permanently bare soil. In fact, a few more events have occurred, which were not recorded due to malfunctioning of the equipment, mostly by sediment clogging the connecting tube between the flume and the stilling well of the stage recorder. Of the total number of runoff events 35 (48%) occurred in summer and 38 (52 %) in winter. Summer is taken here as the period from sowing to harvest of maize, approximately May 1st to November 1st.

3.2 Storm rainfall amount

From Tab. 1 it appears that runoff was generated by rains with rainfall amounts of 1.4 - 33.3 mm. Summer runoff was generated by rains of 2.1 - 33.3 mm and winter runoff by rains of 1.4 - 22.3. Average rainfall amount of all runoff generating storms was 7.6 mm (Tab. 2). Average amounts of summer and winter storms were about equal, 7.9 and 7.4 mm resp. Rainfall amount of the modal rain was 2-4 mm for both summer and winter rains (Fig. 4). In Fig. 5 rainfall amount of all runoff generating rains is compared with the long term frequency distribution of all "rains" occurring in the Netherlands (21). It appears that 12 % of all "rains" fall in the modal class (2-4 mm) of the runoff generating rains.

3.3 Storm duration

Duration of all runoff generating rains ranged from 6 minutes to 22 hours (table 1). Duration of runoff generating summer storms was 6 - 420 minutes and that of winter storms 13.5 - 1320 minutes. Average duration of all runoff generating storms was 3 hours 7 minutes (table 2). Average summer storm duration was about 2 hours and winter storm duration 4 hours. Modal storm duration was 30-45 minutes for all runoff generating storms, with a second modal duration of 1-15 minutes for the summer storms (Fig. 6).

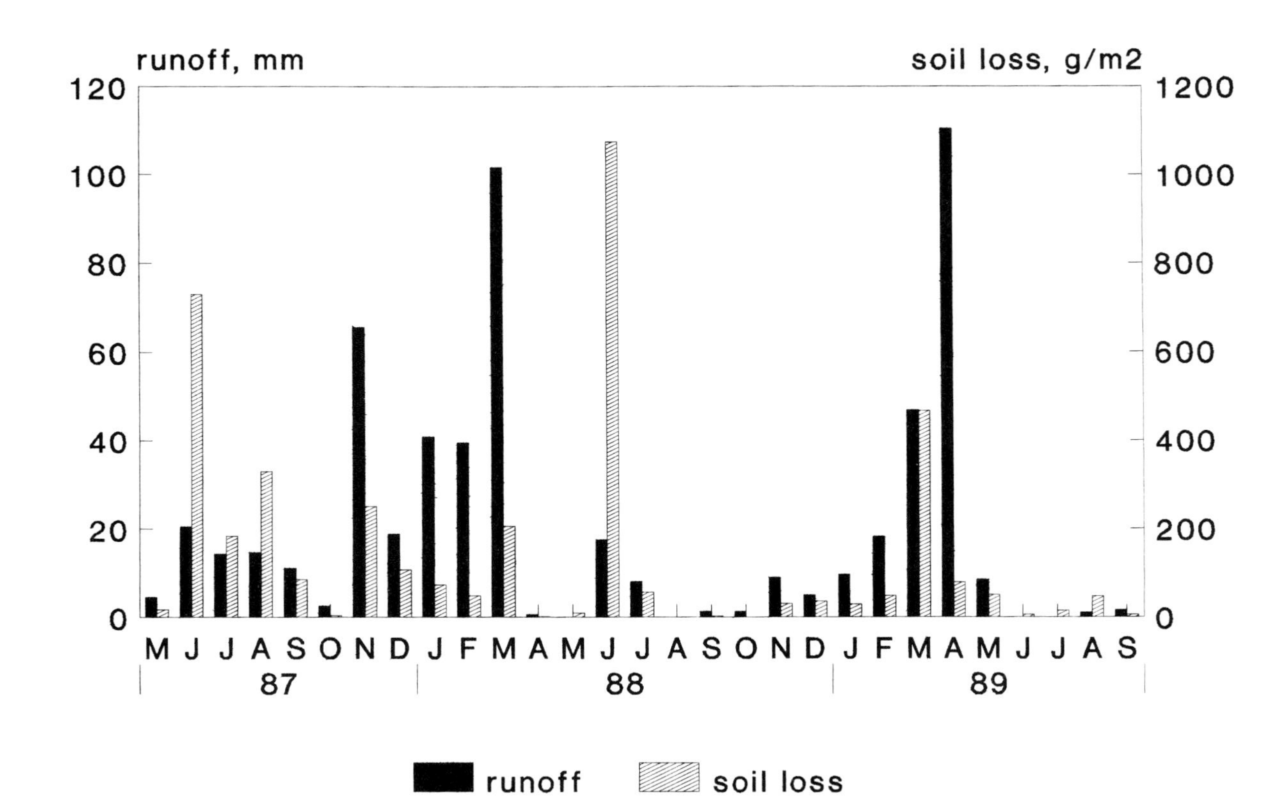

Figure 2. Monthly runoff and soil loss of permanently bare soil (average of three fallow plots).

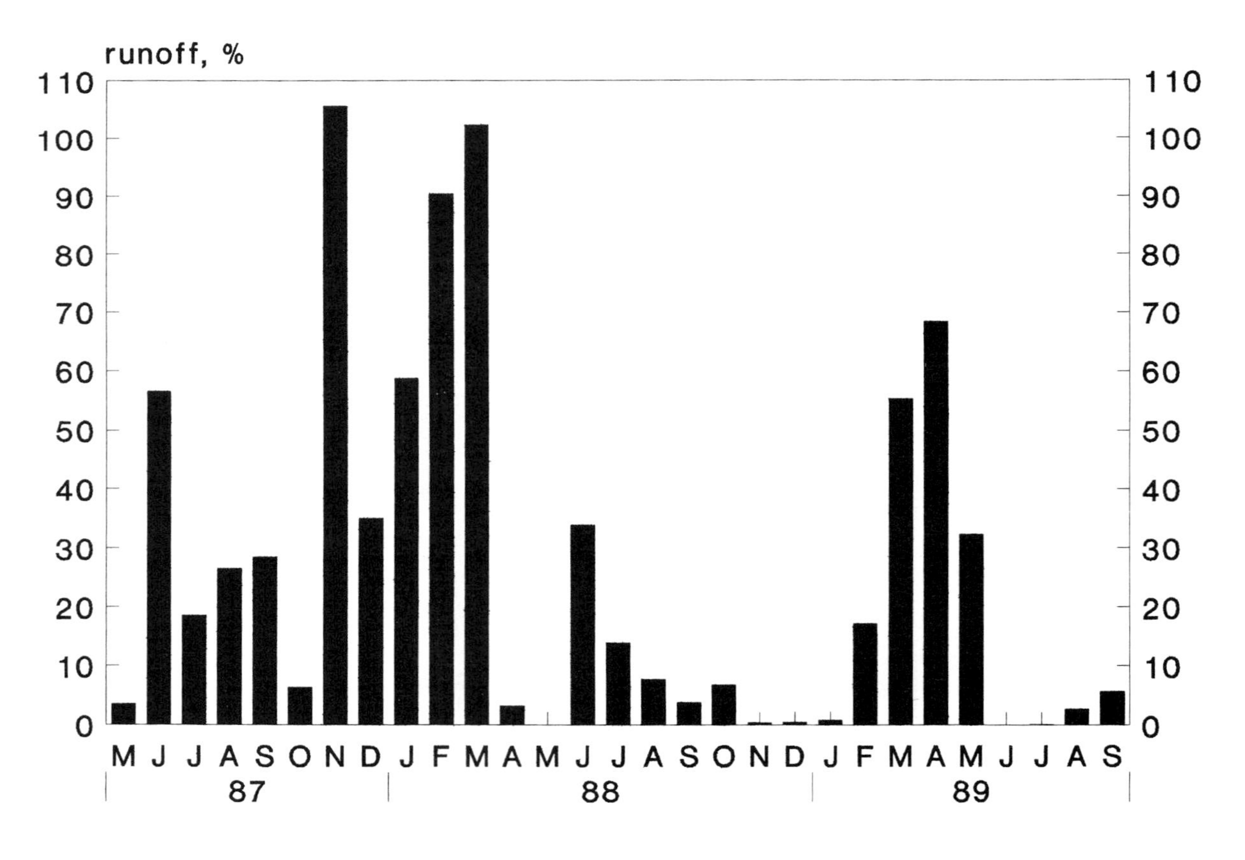

Figure 3. Monthly runoff percentages of fallow plot, equipped with HS-flume.

Table 1.
Characteristics of runoff generating rains.

date	rainfall amount, mm	duration, mm	average int., mm/h	max. 5'-int., mm/h
05/07/86	18.3	420	2.6	
06/07/86	4.8			
07/07/86	3.9			
08/07/86	4.2	6	42.0	42.0
04/08/86	6.7	10.0	40.2	40.2
23/08/86	4.8	36.0	8.0	20.0
03/09/86	6.5	157.0	2.5	16.8
15/12/86	5.3			
18/12/86	5.6			
27/02/87	16.4			
18/07/87	5.0	33.0	9.1	24.0
21/07/87	10.3	45.0	13.7	45.0
30/07/87	2.1	11.0	11.5	11.5
31/07/87	3.3	225.0	0.9	2.5
04/08/87	5.0			
23/08/87	10.0	27.0	22.0	48.0
23/08/87	7.9	10.5	45.0	54.0
04/09/87	12.1	144.0	5.0	20.0
22/09/87	8.8	73.0	7.2	32.6
22/09/87	10.0	102.0	5.9	43.2
24/09/87	4.4	87.0	3.0	21.0
12/11/87	12.4	438.0	1.7	33.0
16/11/87	4.0	60.0	4.0	13.3
16/11/87	3.8	34.5	6.6	25.0
19/11/87	11.6	225.0	3.1	13.2
20/11/87	8.6	174.0	3.0	15.6
02/01/88	4.2	108.0	2.3	13.2
04/01/88	3.0	24.0	7.5	40.0
12/03/88	21.7	1320.0	1.0	6.0
13/03/88	16.8	768.0	1.3	9.0
15/03/88	5.1	210.0	1.5	10.0
16/03/88	3.9	282.0	0.8	4.5
21/03/88	6.8	411.0	1.0	3.0
23/03/88	12.4	216.0	3.4	25.0
23/03/88	2.5	78.0	1.9	3.0
25/03/88	9.1	213.0	2.6	6.3
25/03/88	2.9	39.0	4.5	9.3
25/03/88	2.7	72.0	2.3	18.0
25/03/88	2.6	22.5	6.9	15.6
26/03/88	4.2	150.0	1.7	8.4
26/03/88	3.8	108.0	2.1	12.0
21/06/88	33.3	46.5	43.0	94.0
05/07/88	14.5	354.0	2.5	7.6
05/07/88	3.9	153.0	1.5	6.4
14/07/88	5.0	90.0	3.3	12.8
16/07/88	9.8	367.0	1.6	8.0
16/07/88	3.3	72.0	2.8	15.0
23/07/88	9.5	63.0	9.0	38.0
26/07/88	3.3	126.0	1.6	8.0
28/09/88	5.1	57.0	5.4	12.4
06/10/88	6.9	354.0	1.2	2.5
06/10/88	8.6	189.0	2.7	39.0
10/10/88	3.8	24.0	9.5	27.0
15/02/89	5.7			

Table 1. (suite)
Characteristics of runoff generating rains.

08/03/89	17.9	828.0	1.3	5.0
15/03/89	3.1	177.0	1.0	2.4
15/03/89	1.6	45.0	2.1	10.0
16/03/89	4.8	13.5	21.3	49.2
16/03/89	4.1	90.0	2.7	12.0
22/03/89	2.1	33.0	3.8	10.8
24/03/89	6.7	396.0	1.0	2.0
24/03/89	14.1	108.0	7.8	31.2
12/04/89	6.6	462.0	0.9	7.0
13/04/89	22.3	489.0	2.7	9.1
19/04/89	2.8	37.5	4.5	10.0
19/04/89	1.4	16.5	5.1	10.0
21/04/89	16.5	663.0	1.5	15.0
25/04/89	2.1	97.5	1.3	9.6
11/05/89	18.1	231.0	4.7	42.0
30/07/89	6.2	21.0	17.7	52.0
07/08/89	4.0			
27/08/89	3.2	33.0	5.8	10.8
14/09/89	9.7			17.0

Table 2.
Average values of characteristics of runoff generating rains.

	amount, mm	duration, min.	intensity, mm/hour	max.5'-int., mm/hour
all rains	7.6 ± 5.9	187.1 ± 232.3	7.0 ± 10.2	20.0 ± 17.1
summer rains	7.9 ± 5.9	118.9 ± 117.7	11.0 ± 13.3	27.1 ± 19.9
winter rains	7.4 ± 5.8	247.3 ± 285.7	3.4 ± 3.7	13.7 ± 10.7

3.4 Average storm intensity

Table 1 shows that average intensity of the runoff generating rains ranged from 0.8 to 45.0 mm/hour. Average intensity was 0.9-45.0 mm/hour for the individual summer storms and 0.8-21.3 mm/hour for the individual winter storms. Collective average intensity of the runoff generating summer storms was 11.0 mm/hour and that of the winter storms 3.4 mm/hour (table 2). Modal average intensity of all storms was 2-3 mm/hour, with a second modal class of 1-2 mm/hour for the winter storms (Fig. 7).

3.5 Max. 5 min. storm intensity

Max. 5'-intensity of all runoff generating rains ranged from 2.0 to 94.0 mm/hour (table 1). Summer range was 2.5-94.0 mm/hour and winter range 2.0-49.2 mm/hour. Average max. 5'-intensity of the runoff generating summer storms was twice that of the winter storms, 27.1 and 13.7 mm/hour resp. (table 2). Modal max. 5'-intensity was 10-15 and 40-45 mm/hour for the summer storms and 5-10 mm/hour for the winter storms (Fig. 8).

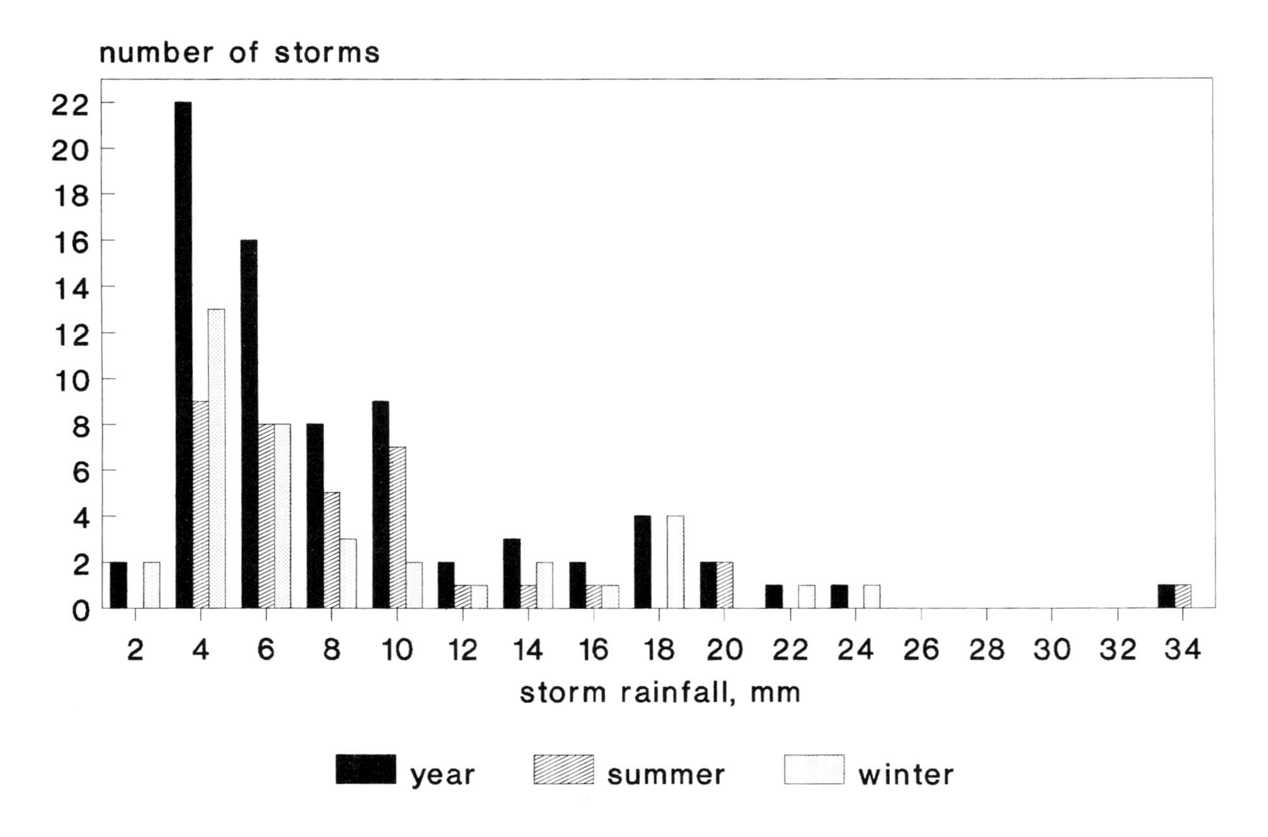

Figure 4. Frequency of runoff generating rains by rainfall amount.

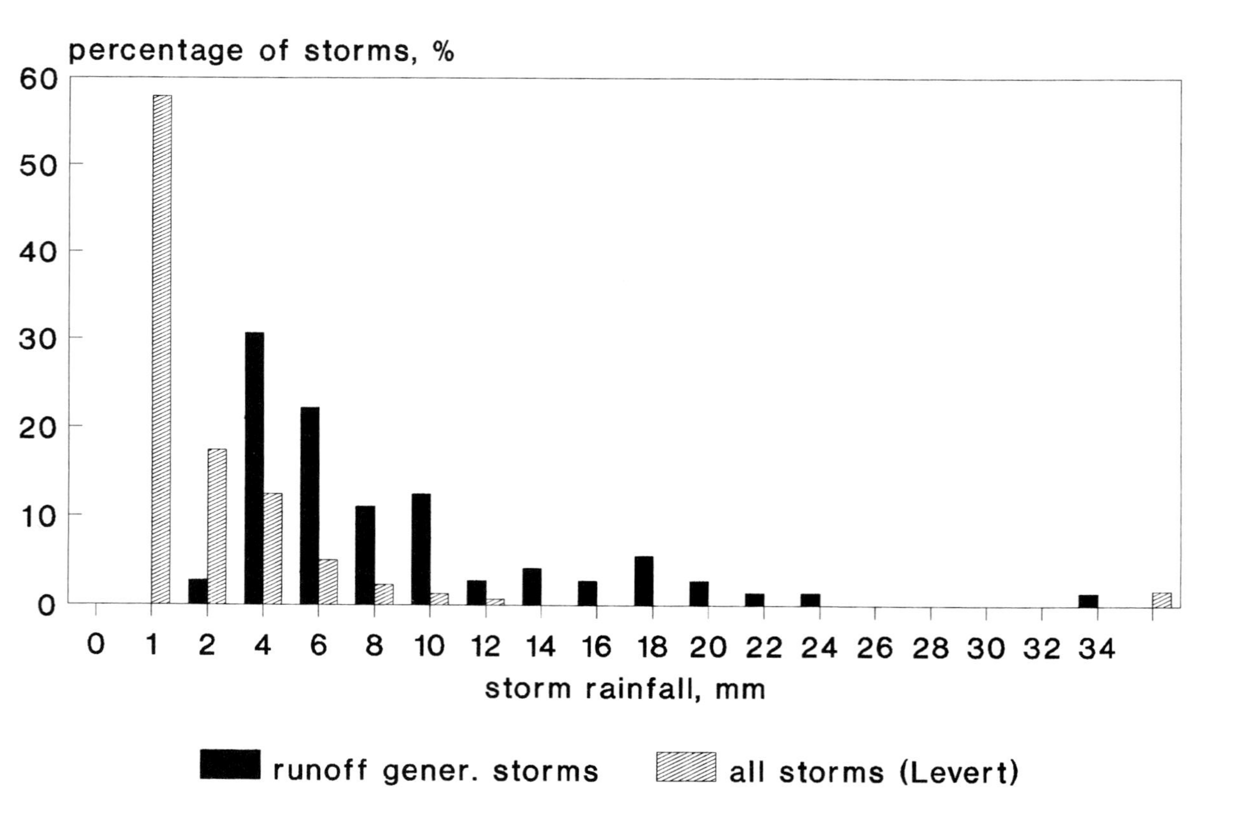

Figure 5. Frequency of runoff generating rains by amount, as compared to long term frequency of all rains in the Netherlands (21).

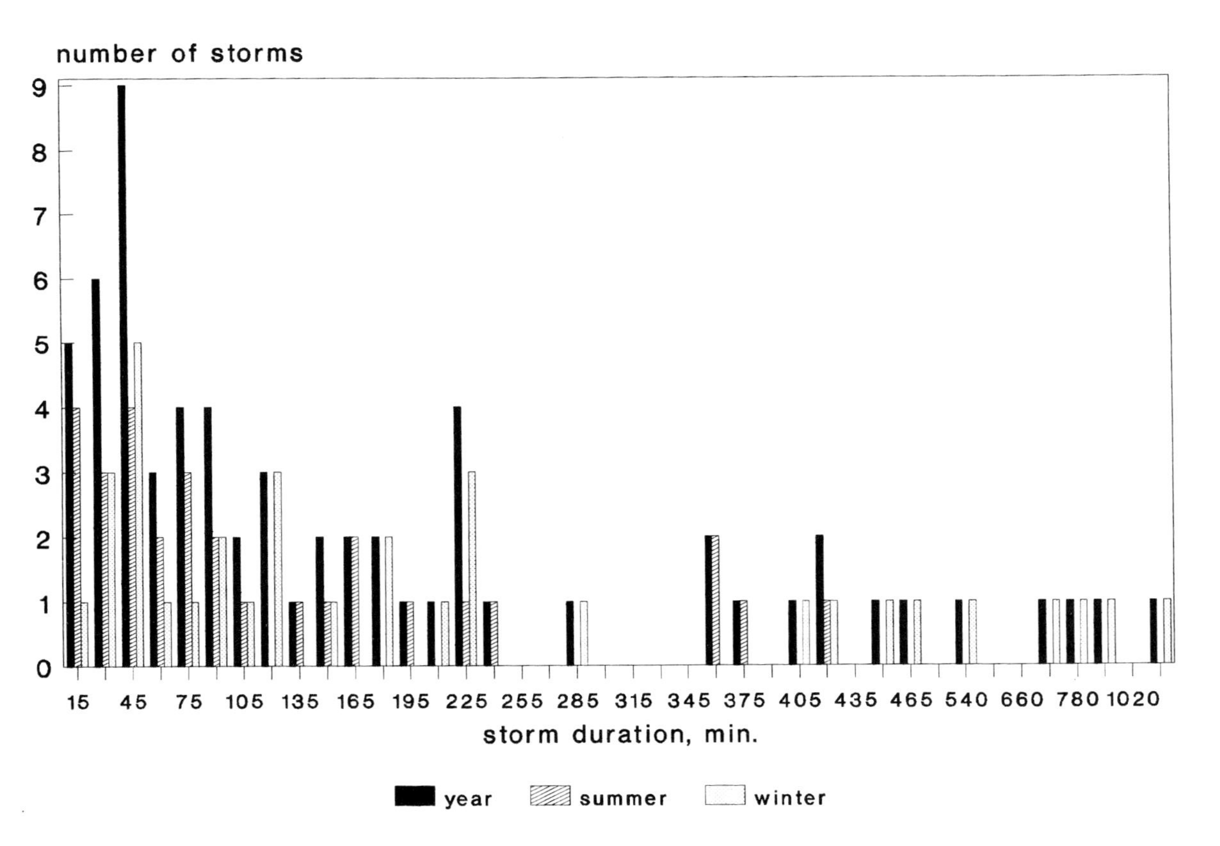

Figure 6. Frequency of runoff generating rains by duration of rain.

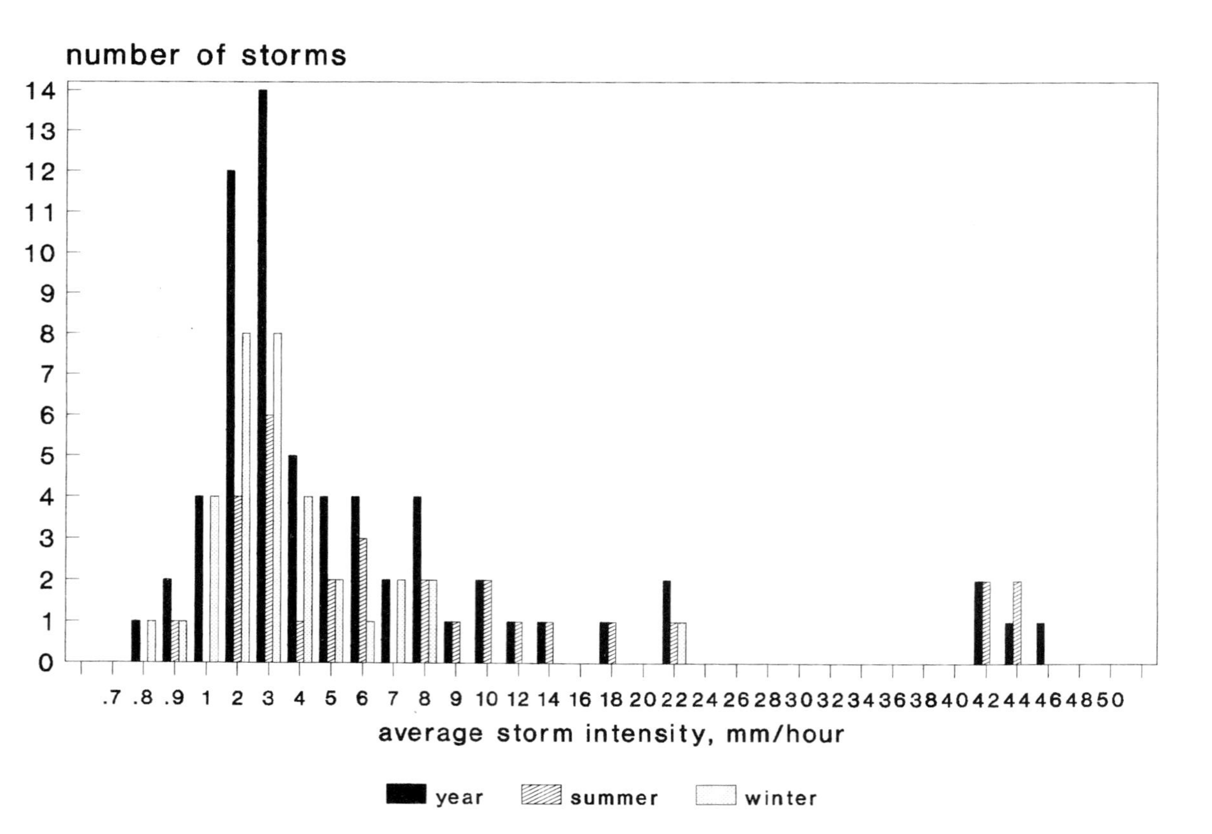

Figure 7. Frequency of runoff generating rains by average intensity.

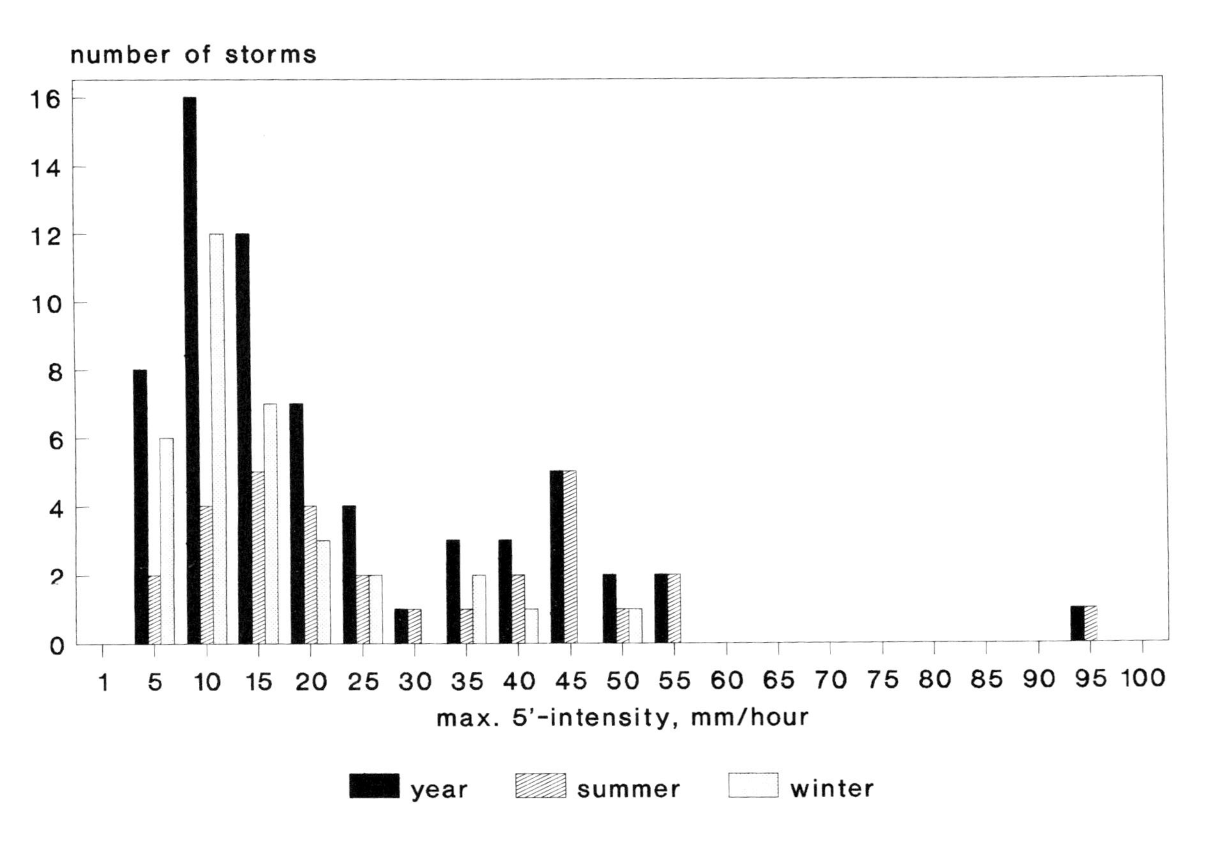

Figure 8. Frequency of runoff generating rains by max. 5 min. intensity.

4. DISCUSSION

The rather high number of 73 runoff events in 40 months, as inferred from the HS-flume recordings, corresponds with the fact that runoff and erosion has occurred on the permanently bare plots during 31 (or 58.5 %) of the total number of 53 four-weekly intervals of measurement (4 - 22 - 23 - 24). A comparable frequency of potentially erosive rains was found by Fullen and Reed (25). They observed erosion on 50 occasions during an 18 month plot study in Shropshire (England) with 125 field visits. Unexpected was the finding that about half of the runoff events occurred in winter. Before the start of the plot study it was supposed that only summer storms on slaked loess soils would cause overland flow and erosion. The fact that 16 (51.6 %) of the 31 four-weekly periods with runoff and erosion occurred in summer and 15 (48.4 %) in winter, fully confirms the runoff event recording. Evidently, not only summer but also winter conditions were conducive to runoff generation (see also fig. 3). Summer and winter regimes of runoff generation and soil erosion on the bare experimental plots have been discussed in some detail by Kwaad (5). Different modes of runoff generation are envisaged for the experimental site in summer and winter, viz. Hortonian overland flow dominating in summer, and saturation overland flow or storage controlled overland flow (26), with a perched water table reaching the soil surface, coming into play in winter and early spring.

Rainfall amount of only 11 (15 %) of the 73 runoff generating rains was more than 12.7 mm, the lower limit for inclusion of a storm in the R-factor of the USLE (9). Therefore, the results of this plot study seem to support the recommendation of Sinzot et al. (12) to set a lower limit for the inclusion of a storm in the calculation of an erosivity index for West-Europe. They recommend a rainfall amount of 8 mm as lower limit. This would include 25 (34 %) of the 73 runoff generating rains of this study. Fullen and Reed (25) found that 3.25 mm of rain can be sufficient to initiate plot erosion. This would include 62 (85 %) of the 73 runoff generating rains of the present study.

Duration of runoff generating winter rains was on average twice that of the runoff generating summer rains. No literature data are available to verify this. The difference fits in with the idea of Hortonian overland flow dominating in summer and saturation overland flow being important in winter. Summer runoff was caused by higher intensity rainfall than winter runoff. Average and max. 5'-intensity of the summer rains was 3 resp. 2 times higher than that of the winter rains. This also is in agreement with Horton overland flow generation in summer. Fullen and Reed (25) found that an intensity of 2.08 mm/hour is necessary for soil erosion to begin. This would include 54 (74 %) of the 73 rains of the present study.

5. CONCLUSIONS

The results of the runoff and rainfall recording on the experimental site indicate that : (a) overland flow generating events occur quite frequently, 22 per year in the period of study, (b) the most frequent (or modal) runoff event during the period of study was characterized by a rainfall amount of 2-4 mm, a duration of 30-45 min., an average intensity of 2-3 mm/hour and a max. 5'-intensity of 5-10 mm/hour, (c) overland flow events occur just as frequently in winter as in summer, (d) seasonal characteristics of the runoff generating rains are in agreement with the hypothesis of a dominance of Horton overland flow in summer and of saturation overland flow in winter.

Because of the different modes of runoff generation that can occur, soil conservation measures should not be directed exclusively at the key variable of the Horton model of overland flow generation, viz. the infiltration capacity of the soil surface, but also at the key variables of the saturation model of overland flow generation, viz. the presence of an impeding layer in the soil and the moisture storage capacity of the zone above this layer.

ACKNOWLEDGEMENTS

Financial support was received from the Foundation Experimental Farm Wijnandsrade, the Landscape Engineering Service at Roermond and the Provincial Government of Limbourg. The plots were located on property of the Experimental Farm Wijnandsrade. Ing. P. Geelen, agricultural research officer at the Experimental Farm, is thanked for stimulating discussions regarding the design of the experiment, and Mr. A. Kerckhoffs, general manager of the Experimental Farm, for carrying out the agricultural operations on the plots.

REFERENCES

1. Provinciale Staten van Limburg (1987) : Streekplan Zuid-Limburg (Algehele Herziening). Maastricht.
2. Schouten, C.J., Rang, M.C. and Huigen, P.M.J. (1985) : Erosie en wateroverlast in Zuid-Limburg. Landschap, 2, 118-132.
3. Schouten, C.J. and Rang, M. (1987) : Bodemerosie in Zuid-Limburg. Natuur en Milieu, Volume 11, pp. 9-13.
4. Geelen, P. (1987) : Bestrijding van watererosie. Jaarverslag over 1986 van de Proefboerderij Wijnandsrade, pp. 83-85.
5. Kwaad, F.J.P.M. (1991) : Summer and winter regimes of runoff generation and soil erosion on cultivated loess soils (The Netherlands). Earth Surface Processes and Landforms, 16, pp. 653-662.
6. Kwaad, F.J.P.M., (1992a) : Cropping systems of fodder maize to reduce erosion of cultivated loess soils in South-Limbourg (The Netherlands). Proceedings First International ESSC Congres, Silsoe, in press.
7. Kwaad, F.J.P.M., (1992b) : A delivery ratio to characterize soil erosion events. Proceedings First International ESSC Congres, Silsoe, in press.
8. Kwaad, F.J.P.M. and Van Mulligen, E.J. (1991) : Cropping system effects of maize on infiltration, runoff and erosion on loess soils in South-Limbourg (The Netherlands) : a comparison of two rainfall events. Soil Technology, 4, pp. 281-295.
9. Wischmeier, W.H. and Smith, D.D. (1978) : Predicting rainfall erosion losses. A guide to conservation planning. U.S. Department of Agriculture, Agriculture Handbook n° 537.
10. Bollinne, A. (1985) : Adjusting the universal soil loss equation for use in western Europe. In : S.A. El-Swaify, W.C. Moldenhauer and A. Lo (editors), Soil Erosion and Conservation. Soil Conservation Society of America, Ankeny, pp. 206-213.
11. Bollinne, A., Florins, P., Hecq, Ph., Homerin, D., Renard, V. and Wolfs, J.L. (1984) : Etude de l'énergie des pluies en climat tempéré océanique d'Europe Atlantique. Zeitschrift für Geomorphologie, Supplement Band 49, pp. 27-35.

12. Sinzot, A., Bollinne, A., Laurant, A., Erpicum, M. and Pissart, A., (1989) : A contribution to the development of an erosivity index adapted to the prediction of erosion in Belgium. Earth Surface Processes and Landforms, 14, pp. 509-515.
13. Evans, R., (1990) : Water erosion in British farmers' fields - some causes, impacts, predictions. Progress in Physical Geography, 14, pp. 199-219.
14. Quenouille, M.H., (1953) : The design and analysis of experiment. Griffin, London, 356 pp.
15. Mutchler, C.K., Murphree, C.E. and Mc Gregor, K.C., (1988) : Laboratory and field plots for soil erosion studies. In : R. Lal (editor), Soil erosion research methods. Soil and Water Conservation Society, Ankeny, pp. 9-36.
16. Brakensiek, D.L., Osborn, H.B. and Rawls, W.J. (editors) (1979) : Field ma¡nual for reasearch in agricultural hydrology. U.S. Department of Agriculture, Agriculture Handbook n° 224.
17. Bos, M.G. (editor) (1976) : Discharge measurement structures. Publication n° 20, ILRI, Wageningen.
18. Buishand, T.A. and Velds, C.A. (1980) : Neerslag en verdamping. K.N.M.I., de Bilt.
19. Bergsma, E. (1980) : Provisional rain-erosivity map of the Netherlands. In : M. de Boodt and D. Gabriels (editors), Assessment of erosion. Wiley, Chichester, pp. 121-126.
20. Bollinne, A., Laurant, A. and Boon, W. (1979) : L'érosivité des précipitations à Florennes. Révision de la carte des isohyètes et de la carte d'érosivité de la Belgique. Bulletin Société Géographique de Liège, 15, pp. 77-99.
21. Levert, C., (1954) : Regens. Een statistische studie. Mededelingen en Verhandelingen KNMI, nr. 102-62, Staatsdrukkerij, Den Haag, 246 pp.
22. Geelen, P. and Kwaad, F.J.P.M. (1988) : Bestrijding van watererosie. Jaarverslag over 1987 van de Proefboerderij Wijnandsrade, pp. 95-99.
23. Geelen, P. and Kwaad, F.J.P.M. (1989) : De bestrijding van watererosie bij de continuteelt snijmais. Jaarverslag over 1988 van de Proefboerderij Wijnandsrade, pp. 143-150.
24. Geelen, P. and Kwaad, F.J.P.M. (1990) : De bestrijding van watererosie bij de continuteelt snijmais. Jaarverslag over 1989 van de Proefboerderij Wijnandsrade, pp. 133-140.
25. Fullen, M.A. and Reed, A.H., (1986) : Rainfall, runoff and erosion on bare arable soils in East-Shropshire, England. Earth Surface Processes and Landforms, 11, pp. 413-425.
26. Kirkby, M.L., (1978) : Implications for sediment transport. In : M.J. Kirkby (editor), Hillslope Hydrology, Wiley, Chichester, pp. 325-363.

Farm Land Erosion: In Temperate Plains Environment and Hills
S. Wicherek (Editor)
1993 Elsevier Science Publishers B.V.

Assessment of soil erodibility : the relationship between soil properties, erosion processes and susceptibility to erosion

Y. Le Bissonnais [a], M. J. Singer [b] and J.M. Bradford [c]

[a] SESCPF, INRA - Centre d'Orléans, 45160 Ardon, France

[b] Land, Air and Water Resources, Univ. of California, Davis, CA, USA

[c] National Soil Erosion Research Laboratory, Purdue Univ., West Lafayette, IN, USA

SUMMARY

By means of literature data, the relationship between soil properties and erosion is analysed. The processes of soil erosion by water, and the main soil characteristics influencing erosion, are reviewed.

Particular interest is focused on the erosion phenomena occurring in cultivated areas, where runoff is increased by surface crusting. The outline of a typology of erosion situations is presented, which is based on the dominant erosion processes. Because of the complexity of erosion processes, soil erodibility should be assessed within the context of this typology. Aggregate breakdown play a significant role in erosion and a standardized method for measuring this parameter is needed.

RESUME

L'objet de cette communication est d'établir une typologie des relations entre processus d'érosion et propriétés du sol, fondée sur des recherches récentes dans le domaine des mécanismes de l'érosion. Cette typologie comprend la détermination de facteurs d'érodabilité précis dans toute situation.

On a déjà beaucoup travaillé sur cette approche, notamment en séparant érosion linéaire en rigoles et érosion entre les rigoles ou en étudiant la dynamique de la croûte, mais beaucoup reste encore à faire avant de parvenir à un modèle général déterministe de l'érosion du sol.

INTRODUCTION

Soil erosion results from the interaction of many factors, the most important of which are rainfall characteristics, geomorphology, land use, and soil properties. Soil erodibility has been defined as the inherent tendency for soils to erode at different rates that are solely due to the differences in the soils themselves. Since soil erodibility is thought to be a function of chemical and physical properties, numerous studies have focused on statistical relations between soil properties and erodibility. The most important works were the field studies reported as the Universal Soil Loss Equation and the erodibility nomograph (1). Other studies have predicted field behaviour through simple laboratory tests, such as the aggregate stability test (2 - 3 - 4).

Both approaches generally allow for isolation of the particular soil properties influencing erosion within the study. Unfortunately, situations or group of soils are found that do not fit either of these approaches, and in some cases the results can be contradictory (5). This may be due to the interaction with other factors that were not taken into account in standardized measurements and tests.

Because of the inherent complexity and the dynamic character of soil erosion, it is difficult to establish an universal soil-erodibility equation. An alternative to the empirical and statistical approaches is the deterministic approach (6).

As a contribution to this objective, we will try in this paper : (i) to identify and understand the processes of erosion ; (ii) to determine the main properties that influence each of these processes ; (iii) to identify situations in which a process is predominant, i.e. to establish a typology of soil-erosion situations that will help in the assessment of soil erodibility.

1. EROSION PROCESSES

Soil erosion results from the action of water on soil which : (i) reduces the infiltration rate and decreases roughness by aggregate breakdown, thus increasing the risk of runoff; and (ii) detaches and transports particles by raindrop impact and runoff.

In cultivated soils, runoff and erosion most commonly result from surface crusting, which is closely related to aggregate breakdown. The other causes of erosion mainly correspond to long-duration and/or high-intensity rainfall. In this paper, we will focus on the case of erosion resulting from the structural degradation of cultivated soils.

1.1 Crusting and runoff

According to the works of Boiffin (5), Farres (7), Loch (8) and Le Bissonnais (9), a well-structured soil surface exposed to successive rainfall is affected by a series of processes that can lead to the formation of a seal, which becomes a crust after drying (10). It is possible to distinguish between three groups of processes : (i) disaggregation, producing microaggregates or primary soil particles from initial structural units ; (ii) displacement of particles or larger units at the soil surface; and (iii) reorganization of these materials into new, denser and more continuous structural units.

1.1.1 Disaggregation

The disaggregation, or breakdown by water results from various physico-chemical and physical mechanisms, which can occur separately or simultaneously during a rainfall event. Four main mechanisms can be identified :

(i) *Physico-chemical dispersion* results from the reduction of the attraction forces between colloidal particles while wetting (11). It is related to the exchangeable sodium percentage (ESP), and generally induces very fast crusting, low infiltrability, and very high particle mobility in water because of the small size of elementary particles.

(ii) S*laking*, i.e. the disaggregation by compression of entrapped-air during wetting is a function of the agggregate porosity and moisture content (4), as well as of the

rate of wetting and the shear strength of wet aggregates (12). Slaking occurs when dry aggregates are immersed in water or rapidly wetted.
(iii) *Microcracking* results either from moderate slaking (partially saturated aggregates), or from differential swelling during wetting (soils with a high clay content).
(iv) *Breakdown by raindrop impact* usually occurs in addition to the other mechanisms, when rainfall intensity and the kinetic energy of individual raindrops are sufficient. The size distribution of the particles resulting from these various mechanisms greatly vary (4 - 8).

1.1.2 Displacement of particles

To induce crust formation and/or erosion, particles or aggregates detached through breakdown must be displaced. We can distinguish two mechanisms of displacement :
(i) *Raindrop impacts* can remove particles as the same time as breakdown takes place. They can also transport the particles previously detached by other mechanisms, so breakdown products may be displaced several times during a rainfall event. Raindrop impacts are the main process of particle displacement under rainfall before runoff start.
(ii) *Runoff and flow turbulence* can detach particles, or transport particles previously detached. While detachment by flow depends on the shear strength of the soil, the transportation of particles is a function of particle size, shape and density. Thus, the shear velocity required to detach particles is much higher than the flow velocity needed to transport small particles produced by aggregate breakdown (13).

Many attempts were made to measure the detachment of particles at the soil surface (splash), in order to characterize soil erodibility (14 - 15), but unfortunately the relation is not so simple. The splash amount varies during crust development and the variation is not related to the erosion rate (7). Furthermore, the size of breakdown products is a much more important factor than their amount, both for crust properties and the availability of particles for erosion by overland flow (8 - 16).

1.1.3 Reorganization

Disaggregation and particle displacement are associated with reorganization processes, which mainly depend on the nature and size of the detached units and on the moisture and rainfall conditions. The reorganization processes are :
(i) *Vertical translocation,* or washing-in, of detached units by the infiltrated water, generally leading to clogging of the pores within the first millimeters of soil.
(ii) *Compaction of the surface layer* by the rearrangement and coalescing of soil particles under the impact of raindrops, which reduces the pore size and porosity, and therefore the hydraulic conductivity. The raindrops are responsible for surface compaction, by increasing the plasticity of the material when wetting, by their mechanical impacts, and by inducing suction below the crust.
(iii) *Sedimentation* of particles in microdepressions occurs after a period of puddling. It forms a continuous layer that commonly consists of several microbeds resulting from the sorting and differential deposition of particles in puddles (5).

All reorganization processes lead to a decrease in infiltration capacity of the soil surface. However, other processes may have an opposite effect : sheet erosion removes particles from the surface, thus limiting crust development (17), and cracking of the surface crust during dry periods enhances the infiltration capacity (18 - 19).

1.1.4 Combination of the crusting processes

When combining the crusting processes, two main morphological types of crust can be distinguished. *Structural crusts* are formed by an in-situ reorganization with very limited particle displacements, and without sorting and sedimentation. They result from the gradual packing and coalescing of soil units which are mainly produced by slaking and microcracking. *Depositional or sedimentary crusts* result from particle displacement and sorting under puddling conditions. The breakdown mechanisms involved are mainly dispersion and raindrop impacts. Bresson and Boiffin (20) showed that these two crust morphologies correspond to two successive stages in a general crusting development pattern.

1.2 Generation of rills and gullies

Rill incision begins as soon as hydraulic tractive forces over come a resistance threshold for the soil surface (13). This hydraulic threshold depends on mechanical properties and roughness of the soil. The mechanical resistance of a topsoil can be expressed as the shear strength of soil, which usually is measured with a cone penetrometer or a vane shear tester, and is greatly influenced by both moisture and structure for a given soil. However, the resistance basically depends on soil texture and binding components (iron oxides and organic polymers). The total roughness can be divided into grain roughness and soil-surface roughness (21). A good relationship between the apparent cohesion measured by torvane and the effective shear velocity of the flow, which is related to grain shear stress, was found by Rauws and Govers (13).

Major spatial variations of the soil resistance to rill initiation can be caused by topographic parameters, or can be induced by human activities (compaction, wheel tracks), or rainfall (crusting, moisture variations). Soil resistance to rill erosion also varies greatly throughout the seasons.

Rill and gully formation is closely related to runoff generation, for which reason it is commonly associated with crust formation that increases the potential runoff. However, crusting and rilling areas are not necessarily located in the same part of the landscape. The spatial interactions between areas where runoff generates and those where runoff concentrates must be considered within spatial units of hydrological functioning, i.e. elementary catchment areas.

2. SOIL PROPERTIES AFFECTING EROSION PROCESSES

The primary soil properties affecting the aggregate stability and erosion have been relatively well known since the early works by Yoder (2), Hénin (22) and Ellison (23). However, results that contradict the findings by these authors are sometimes reported as well. We will review and discuss the most recent results concerning the effects that soil texture, clay mineralogy, organic matter, cations, Fe and Al oxides, and $CaCO_3$ have on erosion.

2.1 Soil texture

It is generally found that, as the silt or silt and fine-sand fractions increase and the clay content decreases, erodibility increases (24 - 4 - 25). This is due to three reasons : (i) the aggregating and bonding effect of clay ; (ii) the transportability of fine and non-aggregated particles, i.e. silt ; and (iii) the detachability of sand and silt particles. However, with a large range of soils, Pierson and Mulla (26), as well as Le Bissonnais and Singer (27), did not find a significant correlation between clay content and erosion. Furthermore, mineralogical and chemical characteristics may modify the effect of clay minerals on aggregation and stability.

For sandy soils in Africa, Obi et al. (28) found that runoff and erosion were best predicted by the sand percentage with a negative relationship. It was also shown that rock fragments on surface may either decrease or increase crusting and erosion, depending on their shape and position (29).

2.2 Clay mineralogy

Clay mineralogy clearly influences erosion processes, but the effect is difficult to assess because soils usually contain a mixture of various clay minerals. Using pure clay minerals, Emerson (30) showed that swelling clays like montmorillonite are less subjected to slaking than kaolinite or illite, because the pressure that is developed by entrapped air is released by swelling ; however, fissuring of montmorillonite can occur. The behaviour of soils differs from that of a pure clay. Smectitic clays should be more efficient than other clays as aggregating particles because of their large specific surface and high physico-chemical interaction capacity (31). Trott and Singer (32) showed that soils with kaolinite were less eroded than soils with montmorillonite, which is probably due to iron-kaolinite interactions that produce very strong cohesive forces when compared with the clay-clay interactions for smectite.

2.3 Organic matter

Organic matter is one of the most important and best known stabilizing agents in soils (24). There has been considerable experimentation and discussion on the mechanisms of the effect that organic fractions have on soil. Ekwue (33), studying the effect of the form of organic matter, found that grass treatment reduced erosion by increasing aggregate stability, while peat acted as a mulch. Recently, Robertson et al. (34) showed that aggregate stability may increase with an increase in the heavy fraction of carbohydrates (polysaccharides), although the total organic content was not modified. Therefore, the total organic matter, which is usually measured, probably is not the best factor for predicting aggregate stability or erosion. The microbial effect on aggregate stability is due to the presence of micro-organisms (fungal hyphae, bacterial mucilage) as well as to that of their by-products that have a chemical and physico-chemical action. Among these by-products, polysaccharides are considered to be very efficient. The effect of polymers is highly variable depending on molecular weight, the charge of the polymer, and the electrolyte concentration (35 - 36 - 37).

For very high contents of organic matter, the induced hydrophobic effect may become a problem, and can inverse the relationship between organic matter content and aggregate stability or erosion. In addition, the effect of organic matter canvary over time, because of differences in the decomposition rates of organic matter (3) and because of seasonal or yearly climatic variations (38).

2.4 Sodium and other cations

The nature and amount of exchangeable cations can influence erosion through their effect on clay dispersion/flocculation processes, which are closely related to clay mineralogy. A well known effect is that of an increase in the exchangeable sodium percentage (ESP), which causes a decrease of the infiltration rate and subsequent erosion. Physico-chemical dispersion results from the reduction of the attractive forces between colloidal particles during wetting (11).

Even though almost all studies found a positive correlation between ESP, and dispersion, crust formation and erosion, the effect greatly differs between soils. Stability or dispersion depends on the size and valence of cations. The increase of electrolyte concentration reduces the dispersion effect of Na, and chemical dispersion may be prevented by spreading phosphogypsum, or another readily available electrolyte source, over the soil surface (39). The effect of sodium was not observed in tropical soils, which have high content of iron oxides due to the cementing action of iron and the non swelling of kaolinite. Furthermore, the type of cations influences the adsorption of polymers that will help in the bridging of soil particles.

2.5 Fe and Al oxydes

Most workers have found a positive correlation between sesquioxides and soil-structure stability (40 - 32 - 7). Different theories explain the mechanisms of the sesquioxides effect:

(i) Iron and aluminium in solution act as flocculants, preventing dispersion in a similar manner as calcium does in other soils. Because of the pH of the soil solution, it is probable that this process concerns aluminium more than iron.

(ii) Sesquioxides play a role in the interactions between clay particles and organic polymers. In this case, iron and organic matter together create a suitable condition for stable aggregation. Le Bissonnais and Singer (27) showed that soils with more than 3 % organic C and 2.5 % CBD extractable Fe + Al do not induce infiltration decrease or erosion, which support this theory.

(iii) Sesquioxides can precipitate as a gel on clay surfaces.

Different iron and aluminium behaviours can be observed depending on several factors such as pH, clay minerals, climate, organic matter, and the composition of the soil solution. For this reason, the various forms of free iron and aluminium oxides can play varying aggregating roles. The two basic types of iron in soil are iron present in primary minerals and free iron ; the latter includes crystalline and amorphous iron oxides, and organically associated iron. The total free iron, extracted by citrate-bicarbonate-dithionite (CBD), is correlated with erodibility (10, 32).

2.6 $CaCO_3$

The effect of $CaCO_3$ on aggregate stability and infiltration or erosion is not well studied. Ben-Hur et al. (41) observed no effect of $CaCO_3$ on the infiltration rate under simulated rainfall. However Merzouk and Black (42) in a study of Maroccan soils found that, among other parameters, active $CaCO_3$ was a good indicator of soil erodibility. Castro and Logan (43), found an ambivalent effect of liming, which is a classical soil improvement method : they observed a short-term negative effect on the soil structure and a long-term positive effect on the reduction of erosion. In theory, from a chemical point of view, calcareous soil should be favourable to high aggregate stability and good infiltration because of the presence of Ca cations.

However, the effect depends probably on the size distribution of $CaCO_3$ particles and on the clay content.

3. TYPOLOGY OF SOIL EROSION AND ASSESSMENT OF SOIL ERODIBILITY

Water erosion consists of the detachment and displacement of soil particles at the earth's surface by water. The main cause of water erosion is surface runoff. It is possible to distinguish between two main types of erosion situations : no or little change in water-infiltration capacity, and a progressive decrease in infiltration capacity.

For the first type of erosion, the soil structure does not greatly change during rainfall. Runoff is the result of particular rainfall conditions such as a very high intensity that exceeds infiltrability, whatever the soil properties, and/or very long rainfall events resulting in the saturation of the whole soil profile, and/or a permanent low percolation capacity of the profile due to specific physical conditions of the soil or subsoil (compacted subsurface soil layers, bedrock, heavy clay, etc). This situation may occur in heavily vegetated non-arable land during long rainstorms. Generally, the vegetation and land use greatly influence erosion,in which case the soil characteristics are less important. For this type of erosion it is clear that aggregate stability is not a good indicator of soil erodibility, because aggregates themselves may be transported. Finally, it is possible to distinguish between various situations within this erosion type, depending on geographical parameters, geology and climate.

For the second type of erosion, the infiltration capacity decreases after rainfall events because of surface sealing and crusting processes (16 - 9), and infiltration may be reduced to 1 mm/h (5). This situation corresponds to most of the cultivated land which generally is medium textured. Silty and loamy soils are the most prone to sealing, and in addition are more easily transported by runoff. However, this effect is complicated by the fact that crusting may increase the shear strength of soil and its resistance to rill formation. Water erosion in cultivated areas is separated into two components i.e. rill and interrill erosion, which commonly are closely associated in the field. Sealing may affect both interrill and rill erosion. It increases runoff, and therefore splash, transport and detachment capacity, i.e. rill initiation. For the crusting-related erosion situations, aggregate stability certainly is a good indicator for erodibility. In fact, it measures the susceptibility of soil to aggregate breakdown, which is the main process responsible for crusting. However, as stressed by Farres (7), the magnitude of erosion is not just controlled by when breakdown takes place. Equally important are the characteristics of the detached particles (8 - 44). In fact, the size of the detached particles determines both the physical properties of the crust (infiltration) and the transportability of particles.

This type of erosion corresponds to various situations. We can distinguish between pure interrill erosion, erosion with parallel rills, ephemeral gully erosion, and all possible combinations between these elementary forms.

CONCLUSIONS

Soil erosion is inherently highly complex and dynamic, which implies that a single universal soil-erodibility equation probably will never be established. We have tried to clarify the relationships between soil properties and erosion processes, in order to propose a typology of erosion situations. The criteria for this typology are related to the erosion processes. A first level makes the distinction between tilled soils, which are subject to human and rainfall induced structural modification, and non-tilled soils, that commonly are covered by vegetation, or are subject to extreme climatic conditions. A second level should take into account the processes within each main erosion type : sealing and crusting, rill and gully formation, etc. Such a typology will need much more work before it is fully established. However, this will be necessary to arrive at the possibility of making objective and really accurate assessment of soil erodibility.

Aggregate breakdown is the key process and the main cause of both crusting (which decreases surface infiltrability) and the production of small particles (which are then eroded). This means that it should be taken into account in all assessments of soil erodibility. However, standardized methods for the analysis of the aggregate stability and crusting, and their relationships with erosion, are not yet available (45).

Although the effects of many of the primary soil properties on aggregate breakdown and erosion are well established from experimental studies, the relationships do not necessarily give a mechanistic explanation of soil behaviour. Furthermore, these relationship generally are difficult to apply to the whole range of soils and situations, because the processes can differ.

REFERENCES

1. Wischmeier W.H., Johnson C.B. and Cross B.V. (1971) : A soil erodibility nomograph for farmland and construction sites. J. Soil Water Conserv., 20, 150-152.
2. Yoder R. E. (1936) : A direct method of aggregate analysis of soils and a study of the physical nature of erosion losses. J. Am. Soc. Agron., 28, 337-351.
3. Monnier G. (1965) : Action des matières organiques sur la stabilité structurale des sols. Thèse, Univ. of Paris.
4. Le Bissonnais Y. (1988) : Comportement d'agrégats terreux soumis à l'action de l'eau : analyse des mécanismes de désagrégation. Agronomie, 8, 87-96.
5. Boiffin J., (1984) : La dégradation structurale des couches superficielles du sol sous l'action des pluies. Thèse de docteur-ingénieur. INA-PG, 320 p.
6. Bradford J.M. and Huang C. (1991) : Mechanisms of crust formation : physical components. Proc. Int. Symp. on soil crusting. Athens, Georgia.
7. Farres P.J. (1987) : The dynamics of rainsplash erosion and the role of soil aggregate stability. Catena, 14, 119-130.
8. Loch R. J. (1989) : Aggregate breakdown under rain : its measurement and interpretation. PhD Thesis, Univ. of New England, Austr.
9. Le Bissonnais Y. (1990) : Experimental study and modelling of soil surface crusting processes. In : Catena supplement 17 : Soil Erosion : Experiments and models. R.B. Bryan (Ed), 13-28.
10. Römkens M.J.M., C.B. Roth and D. W. Nelson (1977) : Erodibility of selected clay subsoils in relation to physical and chemical properties. Soil Sci. Soc. Am. J., 41, 954-960.
11. Sumner M. E. (1991) : The electrical double layer and clay dispersion. Proc.Int. Symp. on Soil Crusting. Athens, Georgia.
12. Nearing M. A. and Bradford J. M. (1985) : Single waterdrop splash detachment and mechanical properties of soils. Soil Sci. Soc. Am. J., 49, 547-552.
13. Rauws G. and Govers G. (1988) : Hydraulic and soil mechanic aspects of rill generation on agricultural soils. J. of Soil Sci., 39, 111-124.

14. Poesen, J. and Govers, G. (1986) : A field-scale study of surface sealing and compaction on loam and sandy soils. 2 : Impact of soil surface sealing and compaction on water erosion processes. In : Caillebaut, Gabriels, De Boodt (eds) : Assessment of soil surface sealing and crusting, Ghent, Belgium.
15. Truman C.C. and Bradford J. M. (1990) : Effect of antecedent soil moisture on splash detachment under simulated rainfall. Soil Science, 150, 787-798.
16. Moore D.C. and M. J. Singer (1990) : Crust formation effect on soil erosion processes. Soil Sci. Soc. Am. J., 54, 1117-1123.
17. Valentin, C. (1981) : Organisations pelliculaires superficielles de quelques sols de région subdésertique (AGADEZ. République du NIGER) Dynamique de formation et conséquences sur l'économie en l'eau. Thèse Paris. 213 p.
18. Hardy, N., Shainberg, I., Gal, M. and Keren, R. (1983) : The effect of water quality and storm sequence upon infiltration rate and crust formation. J. Soil Sci., 34, 665-676.
19. Levy G., Shainberg I. and Morin J. (1986) : Factors affecting the stability of soil crusts in subsequent storms. Soil Sci. Soc. Am. J., 50, 196-201.
20. Bresson, L.M. and Boiffin, J. (1990) : Morphological characterization of soil crust development stages on an experimental field. Geoderma, 47, 301-325.
21. Rauws G. (1988) : Laboratory experiments on resistance to overland flow due to composite roughness. J. of Hydrol., 103, 37-52.
22. Hénin S. (1938) : Etude Physico-chimique de la stabilité structurale des terres. Thèse Paris, 70 p.
23. Ellison W.D. (1945) : Some effects of raindrops and surface flow on soil erosion and infiltration. Trans. Am. Geophys. Union, 24, 452-459.
24. Wischmeier W.H. and L. V. Mannering (1969) : Relation of soil properties to its erodibility. Soil Sci. Soc. Am. Proc., 33, 131-137.
25. Gollany H.T., Schumacher T.E., Evenson P.D., Lindstrom M. J. and Lemme G.D. (1991) : Aggregate stability of an eroded and desurfaced typic argiustoll. Soil Sci. Soc. Am. J., 55, 811-816.
26. Pierson F. B. and Mulla D.J. (1990) : Aggregate stability in the Palouse region of Washington : effect of landscape position. Soil Sci. Soc. Am. J., 54, 1407-1412.
27. Le Bissonnais Y. and Singer M. J. (1992) : Seal formation, runoff and interrill erosion from 17 California soils. Soil Sci. Soc. Am. J., 56, in press.
28. Obi M. E., Salako F. K. and Lal R. (1989) : Relative susceptibility of some southeastern Nigeria soils to erosion. Catena, 16, 215-225.
29. Poesen J. (1986) : Surface sealing as influenced by slope angle and position of simulated stones in the top layer of loose sediments. Earth Surface Processes and Landforms, 11, 1-10.
30. Emerson W.W. , (1964) : The slaking of soil crumbs as influenced by clay mineral composition. Austr. J. Soil Res., 2, 211-217.
31 Young R.A. and Mutchler C. K. (1977) : Erodibility of some Minnesota soils. J. Soil Water Conserv., 32, 180-182.
32. Trott K. E. and M. J. Singer (1983) : Relative erodibility of 20 Californian range and forest soils. Soil Sci. Soc. Am. J., 47, 753-749.
33. Ekwue E.I. (1990) : Effect of organic matter on splash detachment and the process involved. Earth Surface Processes and Landforms. 15, 175-181.
34. Robertson E.B., Sarig S. and Firestone M. K. (1991) : Cover crop management of polysacchardes-mediated aggregation in an orchard soil. Soil Sci. Soc. Am. J., 55, 734-739.
35. Chenu C. (1985) : Etude Physico-chimique des interactions argiles-polysaccharides neutres. Thèse Univ. Paris VII. 185 p.

36. Shainberg I., Warrington D. and Rengasamy P. (1990) : Effect of PAM and gypsum application on rain infiltration and runoff. Soil Sci., 149, 301-307.
37. Le Souder C., Le Bissonnais Y. and Robert M. (1991) : Influence of a mineral conditioner on the mechanisms of disaggregation and sealing of a soil surface. Soil Science, 152, 395-402.
38. Alberts E.E., Laflen J.M., Spomer R.G. (1987) : Between year variation in soil erodibility determined by rainfall simulation. Trans. of the ASAE, 30, 982-987.
39. Kazman Z., Shainberg I. and Gal M. (1983) : Effect of low level of exchangeable sodium and applied phosphogypsum on the infiltration rate of various soils. Soil Science, 135, 184-192.
40. El-Swaify S.A. and Dangler E. W. (1977) : Erodibility of selected tropical soils in relation to structural and hydrologic parameters. In : Soil Erosion : Prediction and Control. Spec. Pub. 21 Soil Conserv. Soc. Am., p 105-114.
41. Ben-Hur M.I., Shainberg I., Bakker D. and Keren R. (1985) : Effect of soil texture and CaCO3 content on water infiltration in crusted soils as related to water salinity. Irrig. Sci., 6, 181-194.
42. Merzouk, A. and Blake, G.R. (1991) : Indices for the estimation of interrill erodibility of Moroccan soils. Catena, 18, 537-550.
43. Castro C.F. and Logan (1991) : Liming effect on soil stability and erodibility. Soil Sci. Soc. Am. J., 55, 1407-1413.
44. Le Bissonnais Y., Bruand A. and Jamagne M. (1989) : Laboratory experimental study of soil crusting : relation between aggregates breakdown and crust structure. Catena, 16, 377-392.
45. Bryan R.B., Govers G., Poesen J. (1989) : The concept of soil erodibility and some problems of assessment and application. Catena, 16, 393-412.

Farm Land Erosion: In Temperate Plains Environment and Hills
S. Wicherek (Editor)

The effects of tillage system and annual crop residue on rill morphology

G. F. McIsaac and J. K. Mitchell

Department of Agricultural Engineering, University of Illinois at Urbana-Champaign
1304 West Pennsylvania Avenue, Urbana, Illinois 61801 USA

SUMMARY

Rainfall and overland flow were applied to two soils and three tillage treatments following soybeans in a corn-soybean rotation. Changes in soil surface elevations were measured with a pin type rillmeter. Equations were developed to describe mean rill width as a function of cross sectional area. Estimates of rill width were calculated for various conditions of discharge. The relationship between rill width and discharge were compared to equations developed by Elliot (1) and Gilley et al. (2). The equations proposed by Elliot (1) accounted for 94 % of the variation in rill widths estimated from field plot data. However, this model tended to under predict rill widths observed for the no-till treatments by 20 to 40 %.

1. INTRODUCTION

Soil erosion modelling in recent years has attempted to characterize runoff and erosion processes, such as rill hydraulics, soil detachment and sediment transport. The influence of tillage, management and cropping practices on rill morphology has been identified as a topic needing more research (3, p. 236). Equations reported by Van Liew & Saxton (4) indicate as much as 50 % greater wetted perimeter per unit of rill cross sectional area for soils with incorporated crop residues. Brown (5) measured rill geometry and erosion from a soil with four levels of incorporated corn residue, immediately after and 30 days after incorporation. For most flow rates tested, the soil eroded beyond the tillage depth to a less erodible layer. In general, incorporated residue did not have a statistically significant affect on parameters describing rill hydraulic radius, wetted perimeter or top width as functions of cross sectional area. However, rills in the freshly tilled, high residue treatment tended to be narrower, perhaps because residue reduced the rate of rill widening.

Elliot (2) and Gilley et al. (2) reported empirically derived parameters for the following equation relating rill width to rill discharge :

$$Wt = C\,Q^m \qquad [1]$$

where : Wt is width of water surface (m),
Q is discharge (m^3/s), and
C and m are constants.

Elliot (1) analyzed widths of rills in 18 fallow soils with slopes ranging from 3.3 to 8.9 % which had been "deep tilled" by various means and lightly disked 3 to 12 months prior and ridged immediately prior to rainfall simulation experiments. Elliot (1) proposed that the exponent, m, of equation [1] should be 0.3 and the coefficient, C, calculated according to the equation :

$$C = 0.757 - 4.75 \times 10^{-6} (Si - 30)^3 + 274 (Tw)^{-3} + 2.56/(So - 0.06) \quad [2]$$

where : Si is soil silt content (%),
Tw is water temperature (°C), and
So is soil slope (%).

Gilley et al. (2) measured widths of rills formed in 10 fallow soils with slopes ranging from 3.8 to 9.8 % which had been moldboard plowed 3 to 12 months prior to rainfall simulation. Rather than having pre-formed ridges, the plots were disked and raked smooth immediately prior to rainfall. Sixty two percent of the variation in rill widths were explained by equation [1] with a coefficient of 1.13 and an exponent of 0.303. These parameters are used in the current version of the Water Erosion Prediction Project model (6).

The objectives of this study were to 1) compare rill widths as functions of cross sectional flow area in different soil and tillage treatments and 2) compare rill widths observed in these soils and tillage treatments to widths estimated by the equations of Elliot (1) and Gilley et al. (2).

2. PROCEDURES

2.1 The Soils

In 1982, field experiments were conducted in east central Illinois, south of Champaign, on a Catlin silt loam soil (fine silty, mesic typic Arguidolls) at slopes ranging from 2 to 5 percent. Field experiments were also conducted in 1985 at the Northwestern Illinois Agricultural Research and Demonstration Center near Monmouth, Illinois, in a Tama silt loam soil at slopes ranging from 6 to 13 percent. Both soils have 3 to 4 % organic matter, 67 to 75 % silt, moderate permeability (15-50 mm/hr) and soybeans had been grown in the soils during the year prior to the experiments. The Tama soil tended to have a slightly greater clay content (20 to 25%) than the Catlin soil (15 to 20 %).

2.2 The Tillage Treatments

The tillage systems analyzed in this study were implemented up-and-down the slope on blocks of land approximately 50 m long by 15 m wide. The tillage systems studied included chisel plowing, moldboard plowing and no-till.

For the chisel plow treatment, the land was chisel plowed approximately 24 cm deep in the fall following soybean harvest. In the spring, secondary tillage consisted of two disk passes and one field cultivation for the Catlin soil, and two field cultivations for the Tama soil.

The moldboard plow treatment was identical to the chisel plow treatment, except a moldboard plow was substituted for a chisel plow. This treatment was only implemented on the Tama soil.

For the no-till treatment, corn was planted without pre-plant tillage and the only other soil disturbance was that caused by injecting anhydrous ammonia in the spring prior to planting corn. The previous soybean crop was also grown without fall or spring tillage. Weeds were controlled with chemical herbicides.

For all tillage treatments, corn was planted 0-7 days prior to rainfall simulation. Average plot slope, residue cover prior to the rainfall simulation experiments and soil silt content for the tillage treatments appear in Tab. 1.

Table 1.
Average slope and soil residue cover for the soil and tillage treatments studied.

Soil	Tillage	Number of plots	Slope	Residue	Silt
			------------------(%)---	-------------------	-------------------
Catlin	Chisel	4	2.2 c *	6 c	67 a
Catlin	No-till	2	2.4 c	16 b	71 a
Tama	Moldboard	4	9.7 a	2 d	70 a
Tama	Chisel	4	7.6 b	3 cd	72 a
Tama	No-till	2	10.0 a	54 a	70 a

* Values within a column followed by any identical letters are not statistically different at the 5 % level of significance according to Scheffe's test.

2.3 Rainfall Simulation

Rainfall simulations were conducted using a rotating boom rainfall simulator of the type described by Swanson (7). Simulated rain was applied at approximately 64 mm/hr to 2 plots, each 3 m wide by 11 m long, simultaneously. The initial simulated rainfall event (run 1) was conducted until a constant rate of runoff was observed for ten minutes. During rainfall simulations, runoff rate was measured and sediment concentration samples were collected every three to five minutes. During steady runoff conditions, colored dye was injected into the runoff and the rate of advance of the leading edge of the dye was measured over the middle third of the plot.

At the conclusion of run 1, the rainfall simulator was stopped for at least one hour. During this pause, various data were collected. The second rainfall event (run 2) was conducted for thirty minutes. After a 15 to 30 minute pause for data collection, an additional 30 minute run (run 3) was conducted. After run 3, approximately 0.6, 1.3 and 2.0 L/s of clear water was applied uniformly across the top of the plot to simulate runoff from longer slope lengths while rainfall was applied (runs 4, 5 and 6, respectively). For two plots of each treatment on the Tama soil, additional events were simulated by applying 2.0, 1.3 and 0.6 L/s clear water across the top of the plots without rainfall (runs 7, 8 and 9, respectively).

A pin-type rillmeter was used to measure elevations of the soil surface before and after rainfall simulations. The rillmeter spanned the width of the plot and rested on steel rails which were anchored to the ground by 15 cm long spikes. Rill meter measurements were taken at three locations across the plots : 3, 5.5 and 8 meters from the up-hill boarder. The pins of the rillmeter were spaced 25.4 mm apart. Black and white photographs were taken of the rillmeter to record the soil surface elevation measurements.

2.4. Analyses

The black and white photographic negatives of the soil profile meter were projected onto a translucent screen and the soil surface elevations were digitized by hand. The pin elevations were corrected for image distortion introduced by the angle of the camera and slide projector lenses. For a particular cross section, the average, absolute deviation between the digitized elevation and the actual elevation was 1.25 mm.

Straight lines between pin elevations were used to approximate the soil surface. To quantitatively characterize relationships describing flow top width as a function of cross sectional area, a series of water surface elevations were assumed at 1.25 mm intervals starting from the lowest elevation on a soil cross section. For each assumed water surface elevation, top width and cross sectional area were calculated for individual rills up to a cross sectional area of 2,000 mm^2 for the chisel and moldboard plow treatments and 3,500 mm^2 for the no-till treatment. These upper limits approximated the maximum estimated rill cross sectional area which occurred for a tillage treatment.

Non-linear regression analysis was used to estimate parameters for an equation describing top width as a function of cross sectional area for each tillage treatment and rainfall simulator run according to the model :

$$Wt = a\,(Ar)^b \qquad [3]$$

where : Wt is rill top width (mm),
Ar is rill cross sectional area (mm^2), and
a and b are parameters estimated by regression.

For each tillage treatment and rainfall simulator run, a range of average rill widths was estimated by substituting estimates of rill cross sectional flow area per rill into regression equation [3] with parameter estimates determined for the corresponding run and treatment. The average cross sectional flow area per rill was estimated by dividing the discharge in the middle of the plot by the average flow velocity and the estimated number of rills. Average flow velocity was estimated to be 0.74 times the measured rate of dye advance (2). To approximate the runoff rate in the center of the plot (the location of the runoff velocity measurement), the average of the discharge measured at the end of the plot and the inflow at the top of the plot was calculated.

The number of rills per plot was estimated by two methods : 1) counting the local minima in the elevation cross sections measured with the rillmeter and 2) inspecting photographic slides of the plots. The average number of rills per plot by each method and 95 % confidence limits were calculated. Lesser and greater estimates of cross sectional flow area and discharge in an average rill for each tillage treatment and run were calculated by dividing the total cross sectional flow area and average discharge in the center of the plot by the 95 % upper and lower confidence limits of the number of rills. A greater and lesser estimate of mean rill width was calculated by substituting the greater and lesser estimate of rill cross sectional area into equation [3] with the corresponding parameters for tillage treatment and rainfall simulator run.

Estimates of rill width were also calculated using equations of Elliot (1) and Gilley et al. (2) and compared to width estimates calculated from the rillmeter data using the procedures described above.

3. RESULTS

3.1. Soil Surface Cross Sections

Fig. 1 illustrates a typical plot cross section with rill incision readily apparent after run 6 on the chisel plow treatment for the Tama soil. Rill incision was also readily apparent for the moldboard plow and no-till treatments in the Tama soil. For the Catlin soil, rill incision was not readily apparent but there appeared to be some widening of concentrated flow areas.

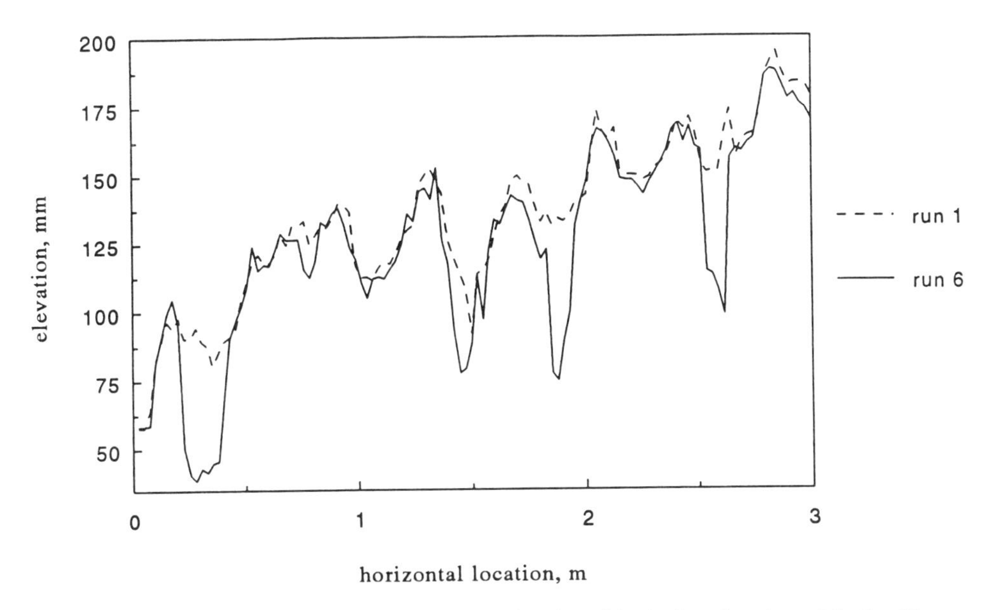

Figure 1. Typical soil surface cross section for the chisel plow treatment in the Tama soil after runs 1 and 6.

3.2. The Number of Rills

For the Tama soil, where rill incision was readily apparent, the number of rills estimated by examining the photographic slides was not statistically different from that determined from examining the rillmeter graphs (Tab. 2). The average numbers of rills per plot in the chisel and moldboard plow treatments were statistically greater than the number of rills in the no-till treatment.

For the Catlin soil, where rill incision was less apparent, concentrated flow areas were difficult to identify. The numbers of rills determined by the two methods were statistically different. By analyzing photographic slides of the plots, it appeared that the chisel plow treatment had significantly greater number of rills than the no-till treatment. However, the nombers of rills for the two tillage treatments was not statistically different when estimated by analyzing cross sections developed from rillmeter data.

For the Tama soil before rainfall simulation, the estimated rill widths as a function of cross sectional flow area were similar for all treatments (Fig. 2). Mean rill width for a given cross sectional flow area tended to decrease with cumulative rainfall simulation. This narrowing of rills would appear to be consistent with the rill incision observed in Figure 1. The one exception to this trend was for the chisel plow treatment, where mean rill width per unit of cross sectional area appeared to increase during the first two rainfall simulator runs.

Table 2.
Average number of rills estimated for each soil and tillage treatment by viewing photographic slides of the plots and by visually analyzing the rillmeter measurements.

		Number of Rills					
Soil	Tillage	from photo slides			from rillmeter		
		mean	LCL1/	UCL2/	mean	LCL	UCL
Catlin	Chisel	7.3	6.9	7.6	6.4	6.1	6.7
Catlin	No-till	4.7	4.1	5.2	6.8	6.0	7.6
Tama	Moldboard	7.0	6.7	7.4	7.2	6.7	7.8
Tama	Chisel	6.4	6.0	6.8	6.4	5.9	6.9
Tama	No-till	4.1	3.3	4.9	5.5	4.7	4.7

1/ LCL = Lower 95 % confidence limit.
2/ UCL = Upper 95 % confidence limit.

3.3. Rill Flow Width as a Function of Cross Sectional Area

The parameters describing rill width as a function of cross sectional area are presented in Tab. 3. Since the two equation parameters vary simultaneously, the implications of statistical differences in these parameters are difficult to interpret without graphical presentation.

For the Catlin soil, mean rill width for a given cross sectional flow area appeared to increase with cumulative rainfall simulation. This suggests that concentrated flow areas tended to widen during the rainfall simulation procedure. It appeared that more widening occurred in the chisel plow treatment than for the no-till treatment.

The 95 % confidence intervals for expected mean values of rill width for a given cross sectional area ranged from 4 to 25 mm. In general, a difference of 20 mm between two expected means would be statistically significant at the 95 % level of probability. The 95 % confidence intervals for an individual value of rill width for a given cross sectional flow area were between 100 and 200 mm.

3.4. Estimates of Mean Rill Width and Discharge

The equation of Elliot (1) appeared to predict the rill widths estimated from field observations more accurately than did the equation of Gilley et al. (2). Linear regression analysis indicated that Elliot's equation accounted for 94 % of the variation in rill width estimated from the field data, with the regression coefficient not statistically different than 1 and the intercept not different than 0 at the 5 % level of significance. Alternatively, the equation of Gilley et al. (2) accounted for 51 percent of the variation in rill width estimated from the field data, but the regression coefficient was 0.33 and statistically less than 1.0.

For the Catlin soil, the average ratio of rill width estimated from the field data to that estimated by Elliot's equation was 1.0 and 1.2 for chisel plow and no-till, respectively (Table 4). The equation of Gilley et al. (2) tended to underestimate rill widths by more than a factor of 2. Since the equation of Gilley et al. (2) was developed from observed widths of incised rills, and since rill widening, rather than rill incision appeared to occur more commonly on the Catlin soil, it is not surprising that this equation under predicted the actual flow width.

Table 3.
Parameters describing rill flow surface width as a function of rill cross sectional area according to the equation Wt = a(Ar)b, where Wt is the surface width in mm and Ar is the cross sectional area in square mm.

Run	Parameter Estimates for Wt = a(Ar)b a	b	r^2	n
	Catlin Soil, Chisel plow treatment			
Before	3.76 b1/	0.504 a	0.86	1086
1	9.61 a	0.415 b	0.81	532
2,3	10.2 a	0.424 b	0.77	842
4,5,6	9.52 a	0.442 b	0.80	1238
	Catlin soil, No-till treatment			
Before	4.85 b	0.466 a	0.82	615
1,2,3	9.31 a	0.416 b	0.73	1050
4,5,6	9.11 a	0.434 ab	0.71	845
	Tama Soil, Moldboard Plow treatment			
Before	2.83 b	0.543 a	0.79	1254
Run 1	6.32 a	0.429 b	0.68	817
Run 2	6.46 a	0.404 b	0.68	886
Run 6	7.62 a	0.380 bc	0.70	463
Run 9	8.54 a	0.325 c	0.59	498
	Tama Soil, Chisel Plow treatment			
Before	2.97 b	0.542 a	0.84	1187
Run 1,2	7.60 a	0.434 b	0.77	1262
Run 6,9	7.83 a	0.388 b	0.64	806
	Tama soil, No-till treatment			
Before	3.92 a	0.497 a	0.81	683
Run 1,2	4.40 a	0.484 a	0.77	1013
Run 9	5.28 a	0.430 b	0.70	696

1/ Parameter estimates within a column, for a particular soil and tillage treatment are not statistically different at the 95 % level of probability if followed by any identical letters.

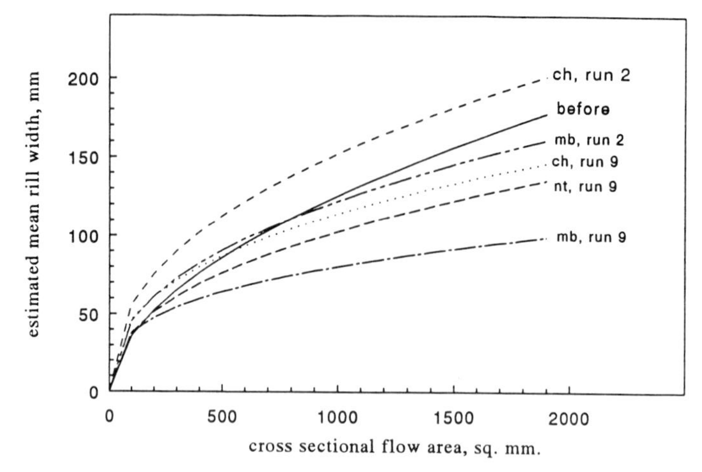

Figure 2 : Estimated mean rill widths as a function of cross sectional flow area at various stages of the rainfall simulation procedure on the Tama soil : "ch," "mb" and "nt" refer to the chisel, moldboard plow and no-till treatments, respectively ; and "before", refers to the soil condition prior to rainfall simulation.

Table 4.
Average ratio of rill width estimated from field plot data to rill width estimated by the equations of Elliot (1) and Gilley et al. (2).

		Ratio rill widths estimated from field plots to that estimated by	
		Elliot (1)	Gilley et al. (2)
Soil	Tillage	Wt =C Q0.3	Wt= 1.13*Q0.303
Catlin	Chisel	1.0	2.2
Catlin	No-till	1.2	2.6
Tama	Moldboard	1.0	1.1
Tama	Chisel	1.0	1.4
Tama	No-till	1.4	1.6

For the Tama soil, the average ratios of rill width estimated from field observations to that predicted by Elliot's equation were 1.0, 1.0 and 1.4 for the moldboard, chisel plow and no-till, respectively. The ratio of rill widths estimated from the equation of Gilley et al. (2) to that estimated from field data were 1.1, 1.4 and 1.6 for the moldboard plow, chisel plow and notill, respectively.The moldboard and chisel plow treatments in the Tama soil are more representative of the conditions under which the equation of Gilley et al. (2) was developed and it is also the conditions in which this equation most accurately estimated rill widths. The apparent robustness of Elliot's equation in this study is probably related to the fact that it attempts to account for the effects of slope steepness.

There was a tendency for Elliot's equation to underestimate mean rill widths for the no-till treatments by 20 to 40 %. Furthermore, for a given soil, Gilley's equation underestimated mean rill widths in the no-till to a greater degree than it underestimated rill widths for the other tillage treatments. This suggests that there was a an influence of tillage treatment which was not captured by either model.

4. SUMMARY AND CONCLUSIONS

Rainfall and overland flow were applied to two soils and three tillage treatments following soybeans in a corn-soybean rotation. Runoff rate and velocity were measured and cross sectional flow area was calculated. Changes in soil surface elevations were measured with a pin type rillmeter. Equations were developed to describe mean rill width as a function of cross sectional area. Estimates of rill width were calculated for various conditions of discharge. The relationship between rill width and discharge were compared to equations developed by Elliot (1) and Gilley et al. (2).

The relationship between rill width and discharge did not appear to be greatly affected by the estimated number of rills. Slope steepness (or soil type) appeared to have a large affect on the rill width. For a given discharge, rills for the Tama soil at 9 % slope tended to be narrower than for the Catlin soil at 2.5 % slope. The equations proposed by Elliot (1) accounted for 94 % of the variation in rill widths estimated from field plot data. However, this model tended to under predict rill widths observed for the no-till treatments on both soil types by 20 to 40 %.

Mathematical Symbols used

a is a coefficient estimated by regression
Ar is rill cross sectional area (mm^2)
b is an exponent estimated by regression
C is a coefficient estimated by regression
LCL is the lower 95 % confidence limit
m is an exponent estimated by regression
n is the number of observations
Q is discharge (m^3/s)
r2 is the coefficient of determination
Si is soil silt content (%)
So is soil slope (%)
Tw is water temperature (°C)
UCL is the upper 95 % confidence limit
Wt is width of water surface (m or mm)

REFERENCES

1. Elliot, W. J. (1988) : Relationship between rill width and flow rates for 1987 WEPP cropland soils. Unpublished USDA-ARS memo obtained from W.J. Elliot, Department of Agricultural Engineering, Ohio State University, Columbus, Ohio, USA, 6 p.

2. Gilley, J.E., Kottwitz, E.R. and Simanton, J.R. (1990) : Hydraulic Characteristics of Rills. Transactions of the American Society of Agricultural Engineers, 33, 1900-1906.

3. Moore, I.D. and Foster, G.R. (1990) : Hydraulics and overland flow. In : M.G. Anderson. and T.P. Burt. (eds), Process Studies in Hillslope Hydrology, John Wiley and Sons, Ltd, Sussex, England, 215-254.

4. Van Liew, M.W. and Saxton, K.E. (1983) : Slope steepness and incorporated residue effects on rill erosion. Transactions of the American Society of Agricultural Engineers, 26, 1738-1743.
5. Brown, L.C. (1988) : Effects of incorporated crop residue on rill erosion. Unpublished PhD Thesis, Purdue University, West Lafayette, Indiana. 197 p.
6. Flanagan, D. C. (1991) : Personal communication. USDA-ARS, National Soil Erosion Research Laboratory, West Lafayette, Indiana, USA.
7. Swanson, N.P. (1965) : Rotating-boom rainfall simulator. Transactions of the American Society of Agricultural Engineers ; 8, 71-72.

Farm Land Erosion: In Temperate Plains Environment and Hills
S. Wicherek (Editor)
1993 Elsevier Science Publishers B.V.

On the Question of Aggregate Stability of Soils

O. Nestroy

Department of Engineering Geology and Applied Mineralogy, Graz University of Technology, Rechbauerstrasse 12, A-8010 Graz, Austria.

SUMMARY

For numerical determination of the structural stability of topsoils and subsoils (5 and 20 cm depth), four sites used for different agricultural purposes (slope planosol, non calcareous regosol, typical planosol, gleyed silicate clastic sediment cambisol) in Eastern Styria were investigated.

The samples taken from depths of 5 and 20 cm at four dates (April 30, June 26, September 9 and October 24 1991) show in their evaluations clear seasonal variations, with stability generally decreasing in spring and increasing in summer. After the harvest, structural stability was again observed to decrease.

This cycle may be attributed the structure-improving effect of vegetation during the growth period and the negative impact of the use of tractors and farming machines and of ploughing.

Further improvement of the method and more extensive studies are to corroborate these first results.

The Universal Soil Loss Equation - USLE - according to Wischmeier and Smith, which tries to ascertain soil erosion, is well known. Some partial factors of this formula, such as Length (L) and gradient of the slope (S) can be determined very precisely, whereas others are more or less estimated values, for example the assessment of the cover and cultivation factor (c).

This is why we are often uneasy when we apply this formula.

Thus a test series was started to determine the structural stability of soils in Styria, Austria. All test sites are located within the same geographical region, only the way of agricultural use differs.

I would like to give a first report on this study, which has been carried out together with Mr. A. Rechberger.

The following sites were examined :

A : Gleisdorf, Unterlassnitz. Slope planolsol of loam, non-calcareous , farmland and grassland of poor quality.
Crop rotation : Silo maize (1986-1987), grain maize (1988), silo maize (1989), barley (1990), silo maize (1991).

B : Gleisdorf, Frösaugraben. Non-calcareous regosol, poor-quality farmland.
Crop rotation : 55% silo and grain maize, 45% cereals.

C : Gleisdorf, Hofstätten. Planosol of loam, medium-quality farmland.
Crop rotation : Grain maize (1986-1988), barley (1989), grain maize (1990 and 1991).

D : Wildon, Empersdorf. Gleyed clastic sediment cambisol, non-calcareous, of sandy sediments, high-quality farmland.
Crop rotation : Pumpkin (1986), grain maize (1987 to 1989), spelt (1990), grain maize (1991), always with intercropping.

The examinations for structural stability were performed by way of the turbidity method, which is briefly decribed below ; 0. 1 g of air-dried soil sieved to 1-2 and 2-4 mm resp. are placed in tubes (6 x 1.7 cm), mixed with 10 ml of de-ionized water and shaken at 15 turns per minute. After a settling time of 2 minutes, 3 ml of the suspension are put into a cuvette and absorption is measured at 620 nm wave length. The procedure is repeated four times.

The samples for this examinations were taken every two months (during the vegetation period) from a depth of 5 and 20 cm respectively. The exact dates were April 30, June 26, September 5, and October 24.

The following results can now be presented (Fig.1) :

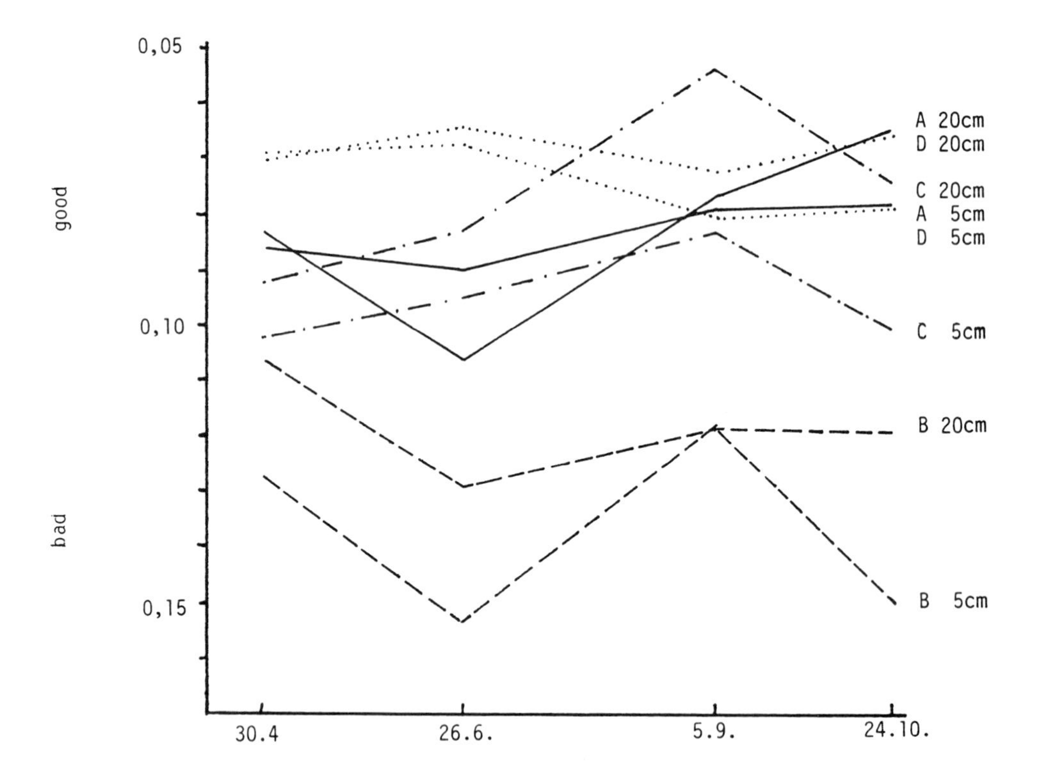

Figure 1. Changes in structured stability during the year (1991)
A - D : Sites
Stability according to the turbidity method.

1. The values are better at a 20 cm depth than at 5 cm depth.
2. Seasonal variations have a greater stability in summer, a slightly reduced one in fall, and a very low one in fall.
It can be said that vegetation has a structure-stabilizing effect.
3. There are very pronounced differences among the four sites described :
A : The values from sampling date 1 to sampling date 2 deteriorated due to tillage, and improved at date 3 due to the development of soil life.
B : Worst site due to anaerobic conditions.
C : This site improved from date 1 until date 3 due to shading of soil life, and then deteriorated because of harvesting and the spreading of liquid manure.
D : Generally favorable situation due to non-ploughing management and mixed cultivation.
4. We observed that the intensified use of tractors and annual ploughing weakens the structural stability of soils.
5. We believe that we have found a practicable method of determining structural stability.

Any criticism and suggestions will be appreciated.

Farm Land Erosion: In Temperate Plains Environment and Hills
S. Wicherek (Editor)

The role of test plot measurements in a long-term soil. Erosion research project in Switzerland.

D. Schaub and V. Prasuhn

Forschungsgruppe Bodenerosion, Geographisches Institut, Universität Basel, Spalenring 145, CH-4055, Basel, Switzerland

SUMMARY

The main goal of the project is the estimation of the spatial and temporal variability of erosion and the influence of changes in land use. These date can be used to check the applicability of computer simulation models as decisions-making tools for soil conservation planning. The approach includes measurements on groups of test plots (with areal extensions between 2.5 - 60 m^2) within each watershed, which represent average soil and slope conditions. At least one plot per site is kept in permanent fallow. Long-term soil losses from the test plots were in the upper range of middle-European measurements (13-22 t/ha/yr), higher than soil loss rates assessed by field throughs, damage mapping or sediment discharge for regularly cultivated arable land in the watersheds, which fell into the lower range of middle-European date (0.3 - 5.0 t/ha/yr). Therefore, the results of the test plots in permanent fallow were considered as quantitative indicators of the maximum potential erosion risk for an area and used for direct comparisons between individual watersheds.

RESUME

L'objectif principal du projet étant d'estimer les variations à la fois spatiales et temporelles de l'érosion et l'influence des changements dans l'exploitation des terres. On a utilisé les données ainsi recueillies pour vérifier si les modèles de simulation par ordinateur pouvaient servir dans la prise de décisions pour la conservation. Une telle approche comprend des mesures sur des groupes pour des parcelles d'expérimentation (d'une superficie allant de 2,5 à 60 m^2) situées à l'intérieur de bassins et représentant les conditions moyennes de sol et de pente. Une parcelle par site au moins est maintenue de manière permanente en jachère. Les pertes de sol à long terme dans ces parcelles d'expérimentation se situent parmi les plus importantes de l'Europe centrale (13-22 t/ha/an). Les résultats concernant les parcelles d'expérimentation maintenues de manière permanente en jachère, ainsi considérés comme des indicateurs quantitatifs du risque potentiel maximum d'érosion d'une région, ont été utilisés pour établir des comparaisons directes entre différents bassins.

1. INTRODUCTION

Bounded runoff plots (test plots) are probably the most common tool in soil erosion research. Due to known boundary conditions (area, slope steepness and length, soil type) and a certain standardization of layout and instrumentation (e.g. 1), they are considered to give the most reliable data on soil loss. Thus, test plots were of great importance for the soil erosion research programme of the Department of Geography, University of Basel, which started in 1975. A number of these experimental stations in landscape-ecologically different test areas in Switzerland have been operated for more than 10 years or are still in use (Tab. 1).

However, there are some inherent sources of error involved with runoff plots (specified e.g. by Hudson 2) and two of them were supposed to be of particular significance in our case :

- The fact that the plot is a closed system, always representing the typical upslope position without input of runoff from above. Due to the special hydraulic conditions (limited runoff generation and flow velocity), this is the part of the slope least endangered by erosion .
- The influence of edge effects along the bounds of the plots, which could be greater in our case, since in general the plots are smaller than the 22.1 x 1.8 m USLE standard field plot (Tab. 1), due to the difficulties of finding appropriate sites in a country like Switzerland, with its irregular topography and high density of population.

To circumvent these disadvantages, alternative field methods such as sediment throughs or damage mapping were equally considered already at the start of the project (3). Detailed experimental and instrumental set-ups of the test plots as well as the other methods have already been laid down (recently by Prasuhn 4 - 5). This paper presents the most important results of the long-term test plot measurements and discusses their applicability compared to the results of the afore-mentioned alternative techniques.

2. RESULTS

2.1 Areas and techniques

At least one plot per site was kept in permanent fallow and plane surface (no inserting of drills etc.). Twice a year, the plot was manually plowed and smoothed, corresponding to corn and winter wheat cultivation in the surrounding test area. This was considered as worst-case management, so that the results of these plots were deemed to be quantitative indicators of the maximum rill and interrill erosion risk for an area. Even though the results were obtained as point measurements, they represent real areal values, because the test plots represent the average soil and slope conditions of a test region. However, this also means that several variables are introduced (e.g. diverse slope steepnesses) and differences between test areas cannot always be directly attributed to single factors.

2.2 Long-term soil losses on test plots

Long-term soil losses from the test plots (Fig. 1, Tab. 2) were in the upper range of Middle-European measurements (6 - 7), higher than soil loss rates assessed by field throughs, damage mapping or sediment discharge for regularly cultivated arable land in the watersheds, which fell into the lower range of Middle - European data (< 5 t/ha/year ; 8 - 9). A possible explanation of this apparent contradiction is that the potential soil erosion risk (i.e the effect of natural factors influencing erosion) in Switzerland is generally high, but the - compared to the other Western -European countries - lesser degree of industrialization of Swiss agriculture with more diverse crop rotations seems to keep soil loss at a lower level. A further intensification of agriculture however might be leading to a serious aggravation of the situation.

Table 1
Chart of investigation areas and test sites. Note the different measurement periods and the different sizes of the plots. See Prashun 1992 for the exact locations

Area	High Rhine Valley		Swiss Jura Plateau		Napf Highland	
Site Code	T1	T2	T30	T50	T300	T350
Investigation Period	since 1975	1975-1984	since 1978	since 1983	1980-1982	1980-1982
Plot Size [m] (Width x Lenght)	1x10 1x10 1x10	1x10 1x10 1x10	2x10 2x10 1x20	3x20	1x10 1x10 1x10	1x10 1x10
Slope [%]	14	13	17	21	31	29
Soil	Gravel and Alluvial Loess Eutric Cambisol	Loess Eutric Cambisol	Loam Slope Deposits Stagno-Gleyic Cambisol	Loam Deposits on Slopes with Lime-stone Debric Calcic Cambisol	Sand Eutric Cambisol	Sand Eutric Cambisol
Soil Texture (A_h)	silty loam	loamy silt	loamy clay	loamy clay	loamy sand	silty-loamy sand
Erodibility(K-Factor) [kg $hN^{-1}m^{-2}$]	0.026	0.052	0.022	0.020	0.024	0.033

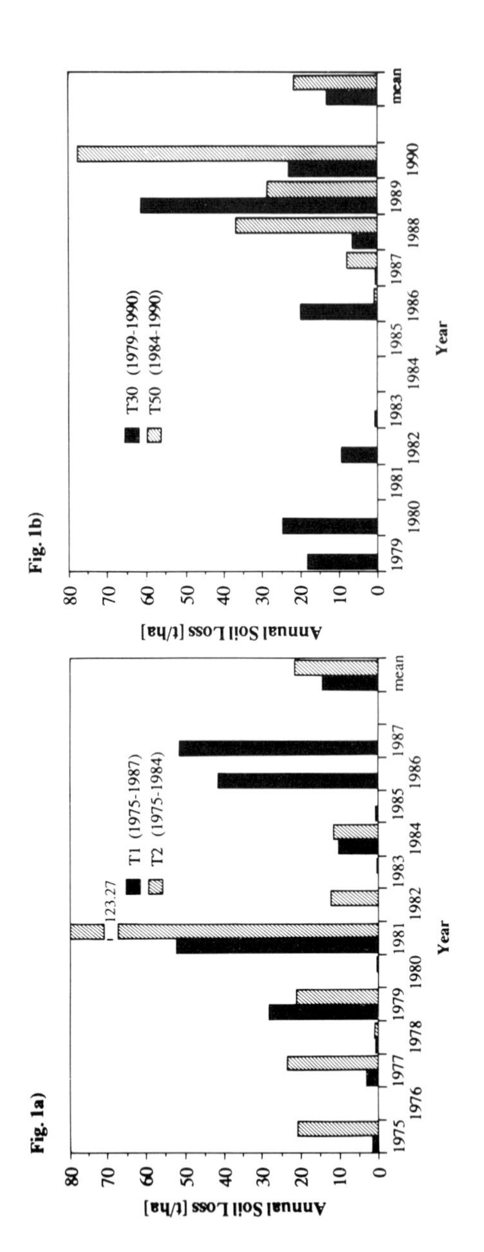

Figure 1. Long-term soil losses from the test plots T1, T2, T30 and T50 (permanent fallow, plane surface).

Figure 1 Shows the annual soil loss amounts from different test plots and the long-term mean soil losses thereby calculated. There are great differences between the particular annual values, mainly due to climatic variabilities. Even 10-year mean values can be distinctively altered by two consecutive wet and erosive years. This emphasizes the importance of long-term measurements and indicates that our previous results can only be regarded as provisional orders of magnitude. Presumably reliable means will not be reached before 20 years of monitoring. However, keeping up a consistent measurement activity for such a long period requires a rather expensive organization due to problems of test plot maintenance.

Even in long-term data series, the distribution of individual values exhibits a marked positive skewness with few large amounts of soil loss and many small losses (8 - 9). More than 50 % of single events amount to only 5 % of the total soil loss. The importance of extreme events is illustrated by the following numbers : On T30 the major one of the totally 31 erosion events which occurred during a period of 12 years produced 37 % of all the eroded material, the five major ones added up to 80 %. For T1 the corresponding figures are 42 % for the major two of 65 events in 13 years and 60 % for the major five respectively. The ranges for T2 and T50 are similar.

The seasonal distribution of soil loss is outlined in Fig. 2. Hence, erosion almost exclusively (ca. 95 %) occurs during the hydrologic summer half-year with a maximum in July in the High-Rhine region and in August in the Jura. These months are characterized by numerous thunderstorms with high erosivity. Nevertheless, linear regressions between rainfall parameters (rainfall amount, EI-30-index, I30) and soil loss or runoff quantities in the Jura region showed only very weak relations or none at all. This can be explained by frequent hailstorms not sufficiently covered by the conventional computation of erosivity as well as by variable soil moisture. High-erosive storms on dry soils with desiccation cracks cause less soil loss than an otherwise inconsiderable rainfall with unfavourable erodibility conditions.

Different measuring units to characterize soil erosion dynamics are listed in Tab. 2. The effects of particular erosion process factors differ from one region to another. A rather plausible way to express this is the sediment concentration (soil loss per unit of runoff). In the Napf Highland much more runoff was measured, the melting of snow contributing considerably to this fact. Thus runoff often occurs when the soil surface is not especially susceptible to erosion due to the lack of concurrent rainstorms of high intensity causing detachment of soil particles. Hence the sediment concentration remains quite low. Loess soils take an intermediate position in this respect. Clay-rich soils on the other hand showed the smallest amount of runoff, but a sediment concentration twice as high as on the High Rhine plots. The higher top soil infiltration capacity in the Jura area, due to the greater aggregate stability of the clay soils and the occurence of desiccation cracks in dry summer periods (10), leads to a higher threshold for runoff generation. However, the greater length and steepness of the slope substantially increase the transport capacity, and yet larger, stable soil aggregates, i.e. a greater quantity of soil can be eroded. Thus the regional differences in erosivity, erodibility and characteristic topography are equalized to a similar value of potential erosion risk in all three compared landscapes.

2.3 Limitations of test plot measurements

Detailed investigations have shown that the applicability of the above-mentioned indicators is limited. Generally test plots are too small to represent

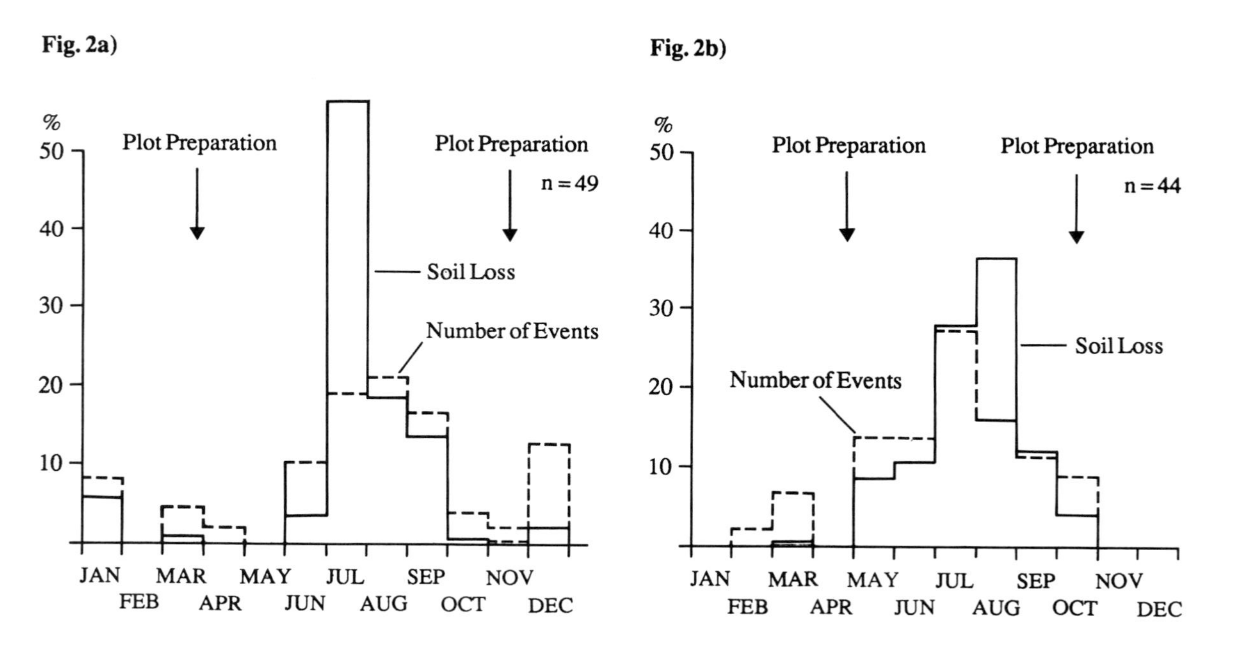

Figure 2. Seasonal distribution of erosion events and soil loss on the test plots T2 and T50.

Table 2
Characteristic values of the erosion measurements on test plots T1, T2, T30, T50, T300, T350. As already mentioned, investigation period and plot dimensions may differ. Data from Rohrer 1985, Schaub 1989 and Prashun 1991

Area	High Rhine Valley		Swiss Jura Plateau		Napf Highland	
Site Code	T1	T2	T30	T50	T300	T350
Mean Annual Precipitation [mm]	1000	1025	1047	1000	1454	1473
R-Factor [N h^{-1}]	89	89	115	105	139	138
Mean Annual Soil Loss [t ha^{-1}]	14.59	21.50	13.39	21.60	8.58	22.29
Mean Annual Runoff [l m^{-2}]	16.42	27.13	7.59	11.43	36.04	54.99
Sediment Concentration [g l^{-1}]	88.8	79.2	176.4	189.0	23.8	40.5
Runoff in [%] of Total Precipitation	1.9	2.9	0.7	1.1	2.5	3.7
Runoff in [%] of Erosive Rainfalls	11.9	23.4	9.5	7.5	11.1	14.0

anything other than the typical upslope situation already mentioned with homogeneous, regular relief conditions. Thus, they can mainly be considered as interrill areas and, confirming this assumption, there were hardly any (small) rills observed on our test plots. Other slope forms (e.g. slope depressions) and other erosion forms (talwegs as well as erosion due to exfiltrating subsurface flow, wheeling or furrows) cannot be represented although they contribute up to 86 % of the total soil loss in the regularly cultivated areas of the Jura catchments.

As suspected, edge effects had a considerable influence. To elucidate this, runoff and soil loss on the 3 meter wide plot T50 were collected in 3 separate gutters. Luff and lee effects on the 15 cm high metal edges of the plot lead either to more or to less rainfall and rainsplash in the fringe area of the plot. Moreover the anastomosic overland flow is hampered in this zone by the sheet metal. This straightens the runoff thereby increasing flow velocity and transport capacity. Thus, soil and water loss was always significantly smaller in the middle of the plot (Tab. 3).

Table 3
Total runoff and soil loss of the 30 main erosion events from different segments of test plot T50 (3 x 20 m). Each one of the three gutters separately covered a 1 x 20 m plot strip (T50/1, T50/2, T50/3). In the middle of the plot (T50/2) significantly less water and sediment was collected

Plot strip	T50/1	T50/2	T50/3
Soil Loss [kg] [%]	15.07 50	5.34 18	9.55 32
Runoff [l] [%]	67.58 44	34.63 23	50.93 33
Sediment Concentration [g l^{-1}]	223	154	188

Changing soil properties due to permanent fallow and soil loss affect erodibility in course of time. The constant absence of vegetation causes a fast mineralization and leaching out of nitrogen as nitrate and other nutrients on test plots after their installation. Gradually this entails a decrease of organic matter and aggregate stability. The high initial organic contents of the plot top soils in the Jura area were reduced to about half this value, a distinctly lower level than on the adjacent cultivated land (Tab. 4).

Table 4
Gradual decrease of the organic content in the parental and erosion material on T30 and T50. The organic content in the A horizons of the regularly cultivated arable land in the watersheds, as comparison, lies between 2.6 and 3.8 % (9)

Plot T30		Plot T50	
organic content [%], top soil			
1978 1988 1990	4.2 2.3 2.3	1983 1989 1990	4.8 2.3 2.5
organic content [%], erosion material			
1988 1989 1990	2.5 2.6 2.5	1987 1988 1989	3.5 2.9 2.7

Related to this drop of organic matter is the decline of aggregate stability (11). Compared to samples from regularly cultivated arable land in the watersheds, top soils of test plots showed a lower aggregate stability (Tab. 5). This enhances the formation of puddles and crusts, reduces the infiltration capacity, and forces the generation of runoff and soil loss.

Table 5
Aggregate stability of top soil samples from test plots in the Jura test area (T30, T50) compared to adjacent regularly cultivated fields. Aggregate stability was examined under artificial rainfall (11)

	water stable aggregales > 1 mm [%]	
	air-dried samples	pre-wetted samples
test plot T30 test plot T50	55 47	80 63
adjacent fields (average value)	74	90

The deterioration of the soil structure, the absence of vegetation and the thereby induced alteration of microclimate, water and nutrient regime in the top soil layer as well as the use of herbicides (to keep the plot in permanent fallow) affect soil fauna in a negative way. The abundance of earthworms on an examined runoff plot in the Jura Plateau with 32 individuals/m^2 as well as the biomass of 15g fresh mass/m^2 was considerably below the respective numbers for regularly cultivated fields in this test area (with an average of 166 individuals/m^2 and 162g fresh mass/m^2). This causes a further decrease of infiltration capacity (reduction of biogenic macropores) and aggregate stability (due to the decrease of stable earthworm-casts on the soil surface).

2.4 Influence of plot management

Plot management, comprising methods and timing of soil cultivation, preparation of the soil surface, and control of crusts, rills and weeds, has a significant influence on erosion. It is virtually impossible to create uniform initial conditions for all sites. Measurements on adjacent, identically prepared and therefore seemingly uniform plots showed a considerable variance of the results (Tab. 6, upper section). This variance can only be explained with invisible differences in micro-topography and surface roughness.

Table 6
Runoff and soil loss on uniformly prepared plots with (middle and lower section) or without (upper section) insertion of drills. The three sections are not directly comparable because they do not represent identical investigation periods

Site T1, all plots in permanent fallow, no drills, 1 year			
plot	number of events	runoff [l m^{-2}]	soil loss [kg m^{-2}]
T1/1	4	19.04	1.96
T1/2	4	21.75	2.10
T1/3	4	16.99	1.46
Site T1, all plots in permanent fallow, 4 years			
T1/1 no drills	21	77.85	8.15
T1/2 contour drills	14	44.10	5.89
T1/3 down slope drills	15	46.09	5.33
Site T2, all plots in permanent fallow, 5 years			
T2/1 no drills	27	178.42	17.82
T2/2 contour drills	20	51.13	8.53
T2/3 down slope drills	18	58.48	10.59

Small deviations in the preparation of the test plots, like the insertion of drills to simulate a fresh seedbed, decreased soil losses to a surprisingly great extent (Tab. 6, lower section). In particular this applies not only to contour drills but also to the up-and-down-slope culture. Despite the smallness of these rills, they act on the plot surface as linear structures of considerably higher roughness and thus higher infiltration capacity. Due to the anastomosic form of the overland flow on the plot, especially in the early phase of an erosion event, runoff and removed soil particles are trapped in downslope drills too. This causes an increase of the triggering limit for runoff generation and thus reduces the number of soil loss events.

3. CONCLUSION : TEST PLOTS VS. ALTERNATIVE FIELD METHODS

Measurements on test plots are thus less suitable for predictions of areal soil loss. They do, however, enable the elucidation of processes responsible for erosion, especially the evaluation of erodibility and erosivity parameters, under the constriction that sufficient attention is paid to the inherent methodical problems. These sometimes "hidden" problems (e.g. those connected with plot surface preparation) considerably reduced the comparability of the results between the different Swiss test regions. It must be emphasized that this inference is certainly not of general validity. In those parts of the world where a single factor dominates the physiognomy of soil erosion (like the high rainfall erosivity in the tropics and sub-tropics), this preponderance may also reduce the importance of edge effects as well as the influence of slope position, changing soil properties and plot preparation, and thus lead to dependable outcomes on bounded standard runoff plots.

For assessing actual erosion losses in temperate plains environments and hills, however, it appears as though field techniques such as damage mapping and sediment trapping on regularly cultivated fields are more useful and reliable than test plot measurements. The use of field throughs (100 - 300 cm wide ; a further development of the Gerlach-through, 12) can be regarded as measuring soil loss on open test plots. The disadvantage of a possible uncertainty about the catchment area for this method is offset by avoiding edge effects, the flexibility of operating a larger number of sample sites, and the fact that soil erosion is monitored under real agricultural conditions (see also Roels, 13). Thus field throughs enable a process based estimation of the spatial and temporal variability of erosion and the influence of changes in land use, the latter being a key process for increasing the danger of soil erosion in Central Europe over the past 40 years.

In regions where interrill erosion is prevailed by rill erosion, separate estimation of the latter can be carried out best by event-related damage mapping. All visible forms of erosion are included and classified in a quantitative way through precise measurement of the length, mean depth and width of the rills. Furthermore, for each erosion system a form is filled out which contains detailed supplementary informations (14). This allows not only a survey of the areal distribution of erosion but also the recognition of the dominant process factors (15). Combined with large-scale geomorphological mapping these data can be used for erosion hazard assessment (16) or as decisions-making tool for soil conservation planning as it was successfully shown by Herweg (17) in a research area in Tuscany (Italy).

ACKNOWLEDGEMENT

The authors wish to express their cordial thanks to the Schweizerischer Nationalfonds zur Förderung der wissenschaftlichen Forschung (SNF) for having supported this project since 1978, and to Mrs. Suzanne de Roche Pfeffinger for brushing up the English.

REFERENCES

1. Mutchler, C.K., Murphree, C.E. & K.C. Mc Gregor (1988) : Laboratory and field plots for soil erosion studies.LAL, R. (Ed.): Soil erosion research methods, Ankeny I0, 9-36.
2. Hudson, N.W. (1957) : The design of field experiments on soil erosion. J. Agric. Engng. Res. 2, 56-65.
3. Leser, H. (1980) : Soil erosion measurements on arable land in north-west Switzerland. Geography in Switzerland, Bern, 9-14.
4. Prasuhn, V. (1992) : A geoecological approach to soil erosion in Switzerland. - Tato, K. & H. Hurni (Eds.: Proc. 6th ISCO Conference Addis Abeba 1989 : Soil conservation for survival, Vol. 2, Bern, Addis Abeba (in print).
5. Unterseher (1992) : Der landschafts- und der geoökologische Ansatz der Bodenerosions-/Bodenschutzforschung und ihre Anwendung im Landschaftsmanagement. Forschungsstelle Bodenerosion, Universität Trier (in print).
6. Dikau, R. (1986) : Experimentelle Untersuchungen zu Oberfläche-nabfluss und Bodenabtrag von Messparzellen und landwirtschaftlichen Nutzflächen- Heidelberger Geogr. Arbeiten, H. 81,
7. Jung, L., & R. Brechtel (1980) : Messungen von Oberflächenabfluss und Bodenabtrag auf verschiedenen Böden der Bundesrepublik Deutschland. - Schr.reihe d. Dtsch. Verb. f. Wasserwirtschaft und Kulturbau (DVWK), H. 48, Hamburg, 139 p.
8. Schaub, D. (1989) : Die Bodenerosion im Lössgebiet des Hochrheintales (Möhliner Feld/Schweiz) als Faktor des Landschaftshaushaltes und der Landwirtschaft. - Physiogeographica, Basler Beiträge zur Physiogeo-graphie, Bd. 13, Basel, 228 p
9. Prasuhn, V. (1991) : Bodenerosionsformen und -prozesse auf tonreichen Böden des Basler Tafeljura (Raum Anwil, BL) und ihre Auswirkungen auf den Landschaftshaushalt. Physiogeographica, Basler Beiträge zur Physio-geographie, Bd. 16, Basel, 372 p.
10. Prasuhn, V. & D. Schaub (1991) : The different erosion dynamics of loess and clay soils and the consequences for soil erosion control. - Z. Geomorph. N.F., Suppl.-Bd. 83, 127-134.
11. Prasuhn, V. (1989) : Aggregatstabilitätsmessungen unter künstlicher Beregnung. Bull. Bodenkundliche Ges. d. Schweiz 13, 137-142.
12. Seiler, W. (1980) : Messeinrichtungen zur quantitativen Bestimmung des Geoökofaktors Bodenerosion in der topologischen Dimension auf Ackerflächen im Schweizer Jura. Catena 7 (1980), 233-250.
13. Roels, J.M. (1985) : Estimation of soil loss at a regional scale based on plot measurements - some critical considerations. Earth Surface Processes and Landforms 10, 587-595.
14. Rohr, W., Mosimann, T., Bono, R., Rüttimann, M. & V. Prasuhn (1990) : Kartieranleitung zur Aufnahme von Bodenerosionsformen und -schäden auf Ackerflächen. - Materialien z. Physiogeographie, H.14, Basel, 56 p.
15. Mosimann, T., Maillard, A., Musy, A., Neyroud, J.-A., Rüttimann, M., Weisskopf, P., Ammon, H.-U. & A. Crole-Rees (1991) : Erosionsbe-kämpfung in

Ackerbaugebieten. Ein Leitfaden für die Bodenerhaltung - Themenbericht des Nationalen Forschungsprogrammes "Nutzung des Bodens in der Schweiz", Liebefeld-Bern, 187 S.
16. Crole-Rees, A., Baril, P. & D.Schaub (1990) : Cartographie des risques d'érosion : Une approche multidisciplinaire. Soil Technology 3, 351-366. Heidelberg, 195 p.
17. Herweg, K. (1988) : The applicability of large-scale geomorphological mapping to erosion control and soil conservation in a research area in Tuscany. Z. Geomorph. N.F., Suppl.-Bd. 68, 175-187.
18. Rohrer, J. (1985) : Quantitative Bestimmung der Bodenerosion unter Berücksichtigung des Zusammenhanges Erosion-Nährstoff-Abfluss im oberen Langete-Einzugsgebiet. - Physiogeographica, Basler Beiträge zur Physiogeographie, Bd.6, Basel, 242 p.

Farm Land Erosion: In Temperate Plains Environment and Hills
S. Wicherek (Editor)

Assessment of factors affecting ephemeral gully erosion in cultivated catchments of the Belgian Loam Belt

K. Vandaele

Laboratory for Experimental Geomorphology, Katholieke Universiteit te Leuven, Redingenstraat 16bis, B-3000 Leuven, Belgium.

ABSTRACT

A field survey was carried out from October 1989 to October 1991 in the Loam Belt of Central Belgium in order to assess ephemeral gully erosion and its contribution to total soil loss. The mean annual ephemeral gully erosion ranges from 0.85 to 1.15 m^3/ha/y. However, due to the fact that ephemeral gully erosion is strongly controlled by high-intensity events, it shows a high temporal and spatial variability. The relative importance of ephemeral gully erosion appears to be strongly influenced by the season. The mean ratio of ephemeral gully erosion to rill erosion equals 1.6 during low-intensity high-frequency rainfall events in winter and 0.4 during high-intensity low-frequency rainfall events in summer.

In addition ephemeral gully length, estimated on aerial photographs, is related to influencing factors. Between-years variations are explained by rainfall regime and landuse. The years with high-intensity rainfall events in late spring or early summer show the highest ephemeral gully densities. During high-intensity low-frequency rainfall events between-catchment variations in ephemeral gully length are dominantly controlled by the area under spring-sown crops. No significant correlations were found for the years with low-intensity high-frequency rainfall events in winter.

RESUME

Les études sur les processus et les taux d'érosion dans la région loessique de Belgique étaient menées sur de petites parcelles ou sur des champs en pente expérimentaux, mais également par l'utilisation de traceurs au moyen de sondages. Pour la plupart, ces études ne couvrent qu'une période de temps limitée et/ou ne s'intéressent qu'à une unité spatiale très restreinte. Ainsi dans le cadre d'un bassin-versant, les facteurs contrôlant l'intensité des processus d'érosion et leurs variations spatiale et temporelle, demeurent encore très mal connus.

Une étude portant sur trois bassins versants cultivés (50-200 ha) situés dans le Centre de la Belgique a été conduite d'octobre 1989 à octobre 1991 dans le but d'analyser globalement le système d'érosion à l'intérieur de ces bassins. Diverses questions se posent qui méritent une réponse : Où et quand l'érosion se produit-t-elle ? Peut-on évaluer l'importance des divers processus d'érosion ? Quels sont les facteurs qui interviennent et dans quelle mesure ces facteurs expliquent-ils les variations spatiales et temporelles ?

Cette étude inclut l'analyse des caractéristiques morphologiques, pédologiques et agricoles des bassins-versants, ainsi que la cartographie des processus d'érosion et des volumes déplacés. L'installation d'un pluviomètre automatique dans chaque bassin fournit des informations détaillées.

1. INTRODUCTION

The term ephemeral gully is used to describe the linear erosion forms that are larger than rills and occur where overland flow concentrates in the landscape, i.e. either in natural drainage ways as in (or along) linear landscape elements e.g. parcel borders, field roads, plough furrows, etc. Ephemeral gullies are temporary features, usually removed by tillage and reoccurring in the same place (1 - 2). Ephemeral gully erosion is an important source of sediment that is frequently being overlooked and not accounted for in soil erosion studies. Estimates of soil erosion in cultivated fields and/or catchments made by most soil loss equations do not include the soil loss due to concentrated flow in topographically defined flow paths (3 - 1 - 4 - 5 - 6). The recognition of ephemeral gully erosion as a separate erosion class is relatively recent. However there is field evidence that ephemeral gullies are responsible for significant erosion (7 - 8 - 1 - 9 - 10 - 11 - 2 - 12 - 13 - 14 - 15 - 16 - 17 - 18 - 19 - 20 - 21). On the other hand, little is known about the factors controlling ephemeral gully erosion and its relative contribution to sediment production in agricultural catchments has not been assessed. The objective of this paper is to present the results of a study that was designed to meet the following objectives.

a) Assessment of ephemeral gully contribution to total soil loss for two cultivated catchments in the Belgian Loam Belt.
b) To relate ephemeral gully length, estimated on aerial photographs, to infuencing factors.

2. STUDY AREA AND PROCEDURES

The field study was carried out in the Loam Belt of Central Belgium, between Brussels and Leuven. The land in this region is dominantly used for crops, of which the most important are winter wheat, barley, potatoes, sugar beets, maize and chicory. The area is characterized by a semi-continuous loess cover and the topsoils mostly have a very high silt content (70 - 80 %), a clay content between 10 - 20 % and a sand content below 15 %. Field observations were carried out in two catchments. Some morphological characteristics of these catchments are given in Tab. 1. Both catchments were visited after eacht important rainfall event. Data on both rill and ephemeral gully erosion were rather rapidly obtained, with an acceptable precision, by volumetric measurements. In this study ephemeral gullies are defined as channels with (1) a size larger then a square foot (22), (2) occurring along topographically defined flow paths and (3) temporary features, usually removed by tillage and reoccurring in the same place (1 - 2 - 16). Observations were carried out from october 1989 until december 1991.

In order to cover a larger study area, aerial photographs from 1963, 1969, 1971, 1981 and 1986 were analysed (Tab. 2). From these aerial photographs it was possible to detect lineair erosion features (ephemeral gullies) in natural drainage ways. The ephemeral gully erosion pattern was copied from the photographs to a topographic map (1/25000). Total length of the ephemeral gullies was quantified on the topographical map using a curvimeter. Morphological characteristics (Tab. 3) were obtained from topographical maps. Land use, i.e. percentage of area with winter crops or with spring-sown crops was measured on the photographs.

Table 1.
Some characteristics of the studied catchments in Central Belgium.

Location	Hammeveld catchment (Leefdaal)	Ganspoel catchment (Huldenberg)
catchment area (ha)	50	120
% of area with slopes ≥ 10 %	10	30
drainage density (km/km^2)	45	65

Table 2.
Date and scale of aerial photographs and number of catchments included in study.

date of aerial photograph	scale	number of catchments included in study
29/07/1963	1/16000	11
09/06/1969	1/18500	27
02/05/1971	1/15000	24
15/04/1981	1/21000	27
25/06/1986	1/21000	30

3. RESULTS

3.a. Field Observations

Data on major rainfall events and soil losses due to ephemeral gully erosion from october 1989 to october 1991 are presented in table 4.

Total soil loss due to ephemeral gully erosion over the study period amounted to 80 m^3 for the Hammeveld and 292 m^3 for the Ganspoel catchment. Of this amount about 62 % was produced during two events in early summer in the Hammeveld catchment. These two events had a recurrence period of 5 and 10 years (23 - 24). In the Ganspoel catchment about 35 % of the soil loss due to ephemeral gully erosion was produced by one summer storm event with a recurrence interval of 30 years (23 - 24). The mean annual ephemeral gully erosion rate equals 0.85 and 1.15 m^3/ha/y (Tab. 5).

Table 3.
Morphological characteristics of study catchments (aerial photographs).

	Catchment area(1) (km²)	Maximum height difference(2) (m)	Length of catchment(3) (km)	Relief-ratio(4) (km/km)	Total length of drainage lines(5) (km) (2/3)	Drainage density (6) (km/km²) (5/1)
Mean	0.35	24.69	0.83	0.03	1.18	3.89
Std. Dev.	0.26	6.82	0.39	0.02	0.67	1.10
Minimum	0.03	12.50	0.16	0.02	0.16	1.90
Maximum	1.12	37.50	2.13	0.08	2.13	5.88

Maximum height difference = difference between highest and lowest points in catchment.
Length of catchment = horizontal distance along longest dimension of catchment parallel to principal drainage line.

Table 4.
Rainfall, maximum rainfall intensities and soil loss due to ephemeral gully erosion for some major erosion events from october 1989 to october 1991.

Hammeveld

Date	Total rainfall (mm)	Maximum rainfall intensity during 15 minutes (mm/h)	Soil loss due to ephemeral gully erosion (m³)	% of total soil loss
27/06/1990	26.0	46.0	23.4	29
03/01/1991	11.8	5.6	11.1	14
08/07/1991	23.2	36.8	25.8	32
Total for whole period			80.0	100

Ganspoel

Date	Total rainfall (mm)	Maximum rainfall intensity during 15 minutes (mm/h)	Soil loss due to ephemeral gully erosion (m³)	% of total soil loss
28/10/1990	32.0	27.5	25.5	9
11/12/1990	13.5	-	17.4	6
03/01/1991	10.0	6.0	17.8	6
10/01/1991	13.0	16.0	79.4	27
08/07/1991	35.0	66.0	100.0	34
Total for whole period			292.0	100

Table 5.
Ephemeral gully erosion rates.

	Mean annual ephemeral gully erosion rate (m^3/ha/y).		
	1989/90$^{(*)}$	1990/91$^{(**)}$	1989-1991
Hammeveld	0.60	1.10	0.85
Ganspoel	0.10	2.20	1.15

(*) measurements are from October 1989 to October 1990.
(**) measurements are from October 1990 to October 1991.

	Seasonal ephemeral gully erosion rate (m^3/ha).			
	1989/90		1990/91	
	winter	summer	winter	summer
Hammeveld	0.1	0.5	0.6	0.5
Ganspoel	0.1	0.0	1.4	0.8

However, due to the fact that ephemeral gully erosion is strongly controlled by high-intensity events, it shows a high temporal and spatial variability. For the Hammeveld catchment there is a twofold increase in annual ephemeral gully erosion rate from 1989/90 to 1990/91. The Ganspoel catchment however showed a 20 fold increase. Little ephemeral gully erosion occurred during the winter of 1989-90 in both catchments. During the summer of 1990 we measured, due primarily to the absence of extreme storms, no ephemeral gully erosion in the Ganspoel catchment while a high-intensity low-frequency rainstorm caused a lot of ephemeral gully erosion in the Hammeveld catchment. The relative importance of ephemeral gully erosion in these catchments can be assessed from the ratio between soil loss due to ephemeral gully erosion and soil loss due to rill erosion (Tab. 6).

Mean annual ephemeral gully erosion equals 70 to 75 % of the mean annual rill erosion. However this ratio appears to be strongly influenced by the season. During low-intensity high-frequency rainfall events in winter the ratio ranges between 1.5 and 1.7 while during high-intensity low-frequency summer storms the ratio is lower than one and ranges between 0.3 and 0.5.The ratio appears to increase with decreasing recurrence period of the storm events. Ephemeral gully erosion can thus be considered as an important source of soil loss in agricultural catchments in the study area. This corresponds findings of other studies. However, their results should be used with great care because they are partly based on methods with a high degree of uncertainty. Volumes of soil loss due to ephemeral gully erosion on tilled agricultural land were quantified using large scale aerial photographs (9 - 14). Sheet-rill erosion on the hillslopes is estimated using the Universal Soil Loss Equation. Both studies indicated that ephemeral gully erosion can account for more than 30 % of the total soil loss from agricultural catchments. The SCS reported (cited by Laflen et al.,8 - 2) that the ratios of ephemeral gully
Table 6.

Table 6.
Ratio between ephemeral gully erosion and rill erosion.(*)

	Mean annual ratio (October 1989 - October 1991)
Hammeveld	0.70
Ganspoel	0.75

	Seasonal ratio	
	winter	summer
Hammeveld	1.58	0.52
Ganspoel	1.66	0.35

(*) ratio = ephemeral gully erosion / rill erosion.

erosion to sheet-rill erosion are quite similar and equaled 25 % for in Iowa with different soils and slopes. For these data the sheet-rill erosion rate is estimated using the Universal Soil Loss Equation. Alabama sites show a much higher contribution of ephemeral gully erosion (2) primarily due to a higher runoff coefficient (25 - 26). The results of Spomer and Hjelmfelt (10) indicate that ephemeral gully erosion is responsible for 25 to 50 % of the average measured long term sheet-rill erosion. Results from Grissinger & Murphey (15) indicate that erosion by ephemeral gully incision amounted to about 60 % of the total annual soil loss. Spomer & Hjelmfelt and Grissinger & Murphey used the sediment yield at the outlet of the catchment to estimate the sheet-rill erosion rate. Ephemeral gully erosion was measured during field surveys and with aerial photographs. For three sites in northern Mississippi, the ratio between ephemeral gully erosion and sheet-rill erosion ranged from 0.13 to 0.70 for one crop year (11). The only other researcher who did volumetric measurements of both rill and ephemeral gully erosion was Auzet et al. (21). Auzet et al (21) found that ephemeral gully erosion during winter equals more or less 80 % of the soil loss due to rill erosion in agricultural catchments in northern France. These results indicate that soil loss due to ephemeral gully erosion is an important and sometimes even dominant source of sediment in cultivated catchments. The failure to include this sediment source in soil loss equations and models can lead to significant underestimates of soil erosion's severity.

3.b. Aerial photographs

3.b.1. Between-year variations in ephemeral gully erosion

Between-year variation in mean ephemeral gully incision density (total ephemeral gully incision length / total catchment area) are significantly controlled by the year of occurrence. Mean ephemeral gully incision percentages for the years 1963 and 1986 are significantly higher from the other years (Tab. 7). About 30 % of the variation in mean annual ephemeral gully incision density is explained by the time factors. In effect rainfall regime and landuse are the variables changing with time.

Table 7.
Ephemeral gully incision density and analysis of the effect of time factor (year) on ephemeral gully erosion.

Mean ephemeral gully incision density. (km/km^2)	Year	Means with the same letter are not significant different.
2.88	1963	A
2.76	1986	A
1.28	1981	B
0.98	1969	B
0.68	1971	B

Analysis of variance was carried out using the SAS-procedure GLM (SAS Institute,1988).
Differences between means were tested on significance level ($P<0.001$) using the Bonferroni T-test.

3.b.2. Between-catchment variations in ephemeral gully erosion.
Ephemeral gullies result from concentrated flow erosion. Sediment detachment and removal is assumed to be a function of flow intensity (27 - 12 - 4 - 15). Therefore, one would expect the length of the incision of the ephemeral gullies to be related to runoff discharge and to thalweg slope. In a first approach catchment area has been used in this study as a surrogate for the volume of runoff involved in ephemeral gully incision. This is based on the assumption that in landscapes where Hortonian overland flow dominates, runoff volume increased proportional to catchment area. This is an acceptable hypothesis for this study considering the fact that ephemeral gully erosion is mainly caused by short, high-intensity rainfall events. The relief ratio of the catchment (defined as the ratio between maximum difference in height of the catchment and length from watershed divide to outlet via main drainage line) was a good representative for the mean slope of drainage lines and ephemeral gully length as a measure for ephemeral gully erosion. For the years 1963 and 1986 we found that the length of the ephemeral gullies was strongly related to the catchment area. However, the runoff productions is not equally distributed over the whole catchment area. Therefore the correlation is improved when the combination of catchment area and the fraction of the catchment under spring-sown crops, which is believed to be a measure for the potential runoff contributing area in late spring or early summer, is used. No significant correlations were found between these measures of flow intensity and ephemeral gully length for the years 1969, 1971 and 1981. In a second approach a non-linear regression is carried out, fitting the results to the equation. EGL = a AREA b ; EGL = ephemeral gully length (m) ; AREA = catchment area (ha) or potential runoff contributing area (ha) (catchment area*fraction of area under spring-sown crops) a, b = parameters

Using the Non Linear regression SAS-procedure (1988) we can calculate the parameters and the coefficients of determination (Tab. 8). For the years 1963 and 1986 again the best results are obtained with the potential runoff contributing area. The area exponent (b) is almost identical for both years equals circa 0.5 No significant correlations were found for the years 1969, 1971 and 1981.

Table 8.
Parameters of the model and coefficient of determination.

AREA = catchment area.

	a	b	R^2
1963	1.56	0.67	0.79
1986	1.55	0.59	0.55

AREA = potential runoff contributing area.
(catchment area * % of area under spring-sown crops)

	a	b	R^2
1963	1.65	0.48	0.96
1986	2.17	0.52	0.54

The coefficient of determination is calculated as : 1 - (residual variance/corrected total variance).
EGL = total length of ephemeral gullies.

4. DISCUSSION

For 1963 and 1986 the observed ephemeral gully erosion can be surely associated with extreme rainfall events. From Tab. 9 it can be seen that these extreme rainfall events are situated in late spring, early summer for 1963 and 1986. During these short-duration rainstorms the peak intensity can reach several tens of mm per hour.

On the other hand the resistance of loamy material is extremely dependent on the initial moisture content. Due to the quick wetting of the dry top layer during the low-frequency high-intensity summer rainfall events there is a pronounced reduction in strength of the soil top layer which causes a higher erodibility of that layer (28 - 29 - 30). The results obtained from the years with high-intensity rainfall events indicate that ephemeral gully erosion during low-frequency high-intensity rainstorms is dominantly controlled by flow intensity. The catchment area, used as a surrogate for flow discharge, appears to be control incision length. The results indicate that the slope gradient of the main drainage line has no or little influence on the length of incision. This is rather surprising while flow erosivity is generally considered to be strongly dependent on slope. Upslope contributing area and

Table 9.
Date of storm events.

date of aerial photographs	storm event	
	date	rainfall (mm)
29/07/1963	12-13/06/1963	94.0
09/06/1969	-	-
02/05/1971	-	-
15/04/1981	-	-
25/06/1986	6-7/06/1986	37.3

- : No extreme storms.

slope energy are often used in the present models and equations predicting the location and amount of ephemeral gully erosion (1 - 12 - 4 - 31 - 20 - 21). Assuming the cross-sectional area of the ephemeral gully to be more or less contstant, our results indicate that the ephemeral gully erosion on or by unit area basis, decreases with catchment area. This may be due to the negative relation between catchment area and slope (Tab. 10). On the other hand, the relative importance of ephemeral gully erosion in the two study catchments appears to be strongly influenced by the season. Due to the occurrence of high-intensity rainfall events and the higher erodibility of the soil top layer (29 - 30) in summer, runoff coming down from the hillslopes reaches almost transport capacity. As a result the concentrated water in the drainage lines have a low degree of erosivity compared with the erosion that occurs on the hillslopes.

Table 10.
Correlations between catchments characteristics.

	Catchment area	Relief ratio	Drainage density	Maximum height difference	Total length drainage lines
Catchment area					
Reliefratio	-0.63***				
Drainage density	-0.67***	0.53**			
Maximum height difference	0.57**	-0.26	-0.27		
Total length drainage lines	0.86***	-0.65***	-0.36	0.53**	
Length of the catchment	0.87***	-0.75***	-0.57**	0.66***	0.73***

Pearson correlation coefficients.
Level of statistical significance: * $P<0.05$, ** $P<0.01$, *** $P<0.001$.
Number of observations, n = 30.

5. CONCLUSIONS

This paper presented some results to illustrate the importance of ephemeral gully erosion and to identify some influencing factors. Soil losses due to ephemeral gully erosion amounted to 80 and 292 m^3 over a two year period, of which most was produced during a few rainfall events. These amounts equalled 70 to 75 % of the mean annual rill erosion in these study catchments. The ratio of ephemeral gully erosion to rill erosion is higher in winter than in summer. The failure to include soil erosion due to concentrated flow can lead to significant underestimates of soil erosion's severity. During high-intensity low-frequency rainfall events in spring or early summer the length of the ephemeral gullies is dominantly controlled by the area of the catchment under spring-sown crops. There is however little or no influence of the slope energy of the thalweg. This is partly due to a high degree of correlation between catchment area and slope energy of the thalweg.

ACKNOWLEDGEMENT

Many thanks are expressed to Prof. Govers and Prof. Poesen for their support and constructive comments they made during the various stages of this work.

6. REFERENCES

1. Foster, G.R., (1986) : Understanding Ephemeral Gully Erosion. In : National Research Council, Board on Agriculture. Soil Conservation : Assessing the National Research Inventory, National Academy Press, Washington, DC, 2 : 90-118.
2. Laflen, J.M., Watson, D.A. and Franti, T.G., (1986) : Ephemeral gully erosion. Proceedings of the Fourth Federal Interagency Sedimentation Conference, March 24-27, at Las Vegas, Nevada, pp. 3.29-3.37.
3. Beasley, D.B., Huggings, L.F. and Monke, E.J., (1980) : ANSWERS : A model for watershed planning. Transactions of the ASAE, 23 (4) : 938-944.
4. Merkel, W.H., Woodward, D.E. and Clarke, C.D., (1988) : Ephemeral Gully Erosion Model (EGEM). In : Modelling agricultural, forest and rangeland hydrology. Proceedings of the 1988 international symposium, 12-13 December 1988, 315-323.
5. Borah, D.K., (1989) : Sediment discharge model for small watersheds. Transactions of the ASAE, 32 (3) : 874-880.
6. Bingner, R.L., Murphree, C.E. and Mutchler, C.K., (1989) : Comparison of sediment yield models on watersheds in Mississippi. Transactions of the ASAE, 32 (2) : 529-534.
7. Thorne, C.R., Grissinger, E.H. and Murphey, J.B., (1984) : Field study of ephemeral cropland gullies in Northern Mississippi. American Society of Agricultural Engineers, Paper n° 84-2550, 15 pp.
8. Laflen, J.M., Franti, T.G. and Watson, D.A., (1985) : Effect of tillage systems on concentrated flow erosion. In: Pla, I.S. (Editor), Soil Conservation and Productivity. Proceedings of the Fourth Internation Conference on Soil Conservation, 3-8 November 1985, at Maracay, Venezuela, 2 : 798-809.

9. Thomas, A.W., Welch, R. and Jordan, T.R., (1986) : Quantifying concentrated-flow erosion on cropland with aerial photogrammetry. Journal of soil and water conservation, 4 : 249-255.
10. Spomer, R.G. and Hjelmfelt, A.T., (1985) : Concentrated flow erosion on conventional and conservation tilled watersheds. American Society of Agricultural Engineers, Paper no. 85-2050, St. Joseph, Michigan.
12. Thorne, C.R., Zevenbergen, L.W., Grissinger, E.H. and Murphey, J.B., (1986) : Ephemeral gullies as sources of sediment. Proceedings of the Fourth Federal Interagency Sedimentation Conference, Las Vegas, Nevada, 1: 3.152-3.161.
13. Evans, R. and Cook, S., (1987) : Soil erosion in Britain. In : C.P. Burnham and J.I. Pitman (Editors), Soil Erosion. SEESOIL, 3 : 28-59.
11. Forsythe, P., Parkman, J.S. and Ulmer, R., (1986) : Evaluation of concentrated flow erosion estimation procedures, Mississippi, 1986. Report to the SNTC, SCS, Fort Worth, Texas, 29 pp.
14. Thomas, A.W. and Welch, R., (1988) : Measurement of ephemeral gully erosion. Transactions of the ASAE, 31 (6) : 1723-1728.
15. Grissinger, E.H. and Murphey J.B., (1989) : Ephemeral gully erosion in the loess uplands, Goodwin Creek Watershed Northern Mississippi, USA. Proceedings of the 4 th. International Symposium on River Sedimentation, 5-9 June 1989, at Beijing, China, 51-58.
16. Poesen, J., (1989) : Conditions for gully formation in the Belgian Loam Belt and some ways to control them. Soil Technology Series, 1 : 39-52.
17. De Ploey, J., (1989) : Erosional systems and perspectives for erosion control in European loess areas. Soil Technology Series, 1: 93-102.
18. De Ploey, J., (1990) : Threshold conditions for thalweg gullying with special reference to loess areas. Catena supplement, 17 : 147-151.
19. Poesen, J. and Govers, G., (1990) : Gully erosion in the Loam Belt of Belgium : typology and control measures. In : J. Boardman, I.D.L. Foster and J.A. Dearing (Editors), Soil Erosion on Agricultural Land. John Wiley and Sons Ltd., pp. 513-530.
20. Papy, F. and Douyer, C., (1991) : Influence des états de surface du territoire agricole sur le déclenchement des inondations catastrophiques. Agronomie, 11: 201-215.
21. Auzet, A.V., Boiffin, J., Papy, F., Ludwig, B. and Maucorps, J., (1992) : Rill erosion as a function of the characteristics of cultivated catchments in the North of France. Catena, 19.
22. Hauge, C., (1977) : Soil erosion definitions. California Geology, 30 : 202-203.
23. Laurant, A., (1976) : Nouvelles recherches sur les intensités maximums de précipitations à Uccle. Courbes d'intensité-durée-fréquence. Annales des Travaux Publics de Belgique, n° 4, 9 pp.
24. Demareé, G., (1985) : Intensity-duration-frequency relationships of point precipitation at Uccle ; Reference period 1934-1983. K.M.I. Publikaties, Reeks A, 116, 52 pp.
25. Murphree, C.E. and Mutchler, C.K., (1981) : Sediment yield from a flatland watershed. Transactions of the ASAE, 24 : 966-969.
26. McGregor, G.R. and Greer, J.D., (1982) : Erosion control with no-till and reduced till corn for silage and grain. Transactions American Society of Agricultural Engineers, 25 : 154-159.

27. Foster, G.R. and Lane, L.J., (1983) : Erosion by concentrated flow in farm fields. In : R.M. Li, P.F. Lagasse and L. Simons and Associates (Editors). Proceedings of the D.B. Simons Symposium on Erosion and Sedimentation, Colorado State University, Fort Collins, pp. 9.65-9.82.
28. Rauws, G. and Auzet, V., (1989) : Laboratory experiments on the effects of simulated tractor wheelings on linear soil erosion. Soil and Tillage Research, 13 : 75-81.
29. Govers, G., Everaert, W., Poesen, J., Rauws, G., De Ploey, J. and Lautridou, J.P., (1990) : A long-flume study of the dynamic factors affecting the resistance of a loamy soil to concentrated flow erosion. Earth Surface Processes and Landforms, 15 : 313-328.
30. Govers, G., (1991) : Time dependency of runoff velocity and erosion : the effect of the initial soil moisture profile. Earth Surface Processes and Landforms, 16 : 713-729.
31. Thorne, C.R. and Zevenbergen, L.W., (1990) : Prediction of ephemeral gully erosion on cropland in the South-eastern United States. In : J. Boardman, I.D.L. Foster and J.A. Dearing (Editors), Soil Erosion on Agricultural Land. John Wiley & Sons Ltd., pp 447-460.
32. Moore, I.D., Burch, G.J. and Mackenzie, D.H., (1988) : Topografic effects on the distribution of surface soil water and the location of ephemeral gullies. Transactions of the ASAE, 31 (4) : 1098-1107.

Farm Land Erosion: In Temperate Plains Environment and Hills
S. Wicherek (Editor)

Impact of agriculture on soil degradation : modelisation at the watershed scale for a spatial management and development

S. Wicherek
In collaboration with G. Chêne and M. Mekharchi

Centre de Biogéographie-Ecologie, U.R.A. 1514 C.N.R.S., Ecole Normale Supérieure de Fontenay - Saint-Cloud, Le Parc, 92 211 Saint-Cloud, France

SUMMARY

With the aid of experimental equipment installed on the downward part of the main drain of the Arbre à Robert watershed, surveys have been taken over three years' time. This follow-up carried out under real conditions and synchronous timing, provides information of transiting flow in the basin as well as data on modes of rainfall distribution. The upper basin receives an average of 30 % more rainfall than the lower part.

The watershed studied spreads over 28 ha, with a low slope (2 to 6 %) and fits into the larger Erlon basin ; it can be considered as representative of the Eastern Parisian Basin, large-scale agricultural area.

The main parameters commanding runoff and land degradation modes are discussed and a hierarchical study is made.

Runoff and degradation occur with variable intensity in function of rainy periods and rainfall intensity and duration. Spring and Summer rains are the most efficient, those in Autumn and Winter are less efficient.

The rainfall sequences recorded show that one of the triggering limits for runoff and major erosion corresponds to a minimal rainfall intensity of 30 to 40 mm/hour in 10 minutes' time. A minimum of 12,5 mm of precipitation in 1 hour constitutes another limit (these values decrease strongly after rainy periods). 10-15 minutes are necessary if the water is to concentrate in the drains (talwegs). Nevertheless, the role of existing vegetal cover in the watershed should be taken into account, for this role can drastically modify the results obtained, in particular concerning the mass of mobilized material which can vary considerably.

Damage recorded affects the physical quality of the soil, its finest components are taken charge of, and among these, organic material are in considerable quantity. On the other hand, mobilization of larger particles is always limited. Solid debit could reach 13 % of the entire mobilized group.

Methods and techniques of soil utilization, and occupation of the soil also intervene in accelerating or slowing down the phenomenon of runoff and degradation, however, the effect of these agricultural methods is less than one might imagine.

RESUME

Grâce à un équipement expérimental installé à l'aval du talweg principal du bassin versant de l'Arbre Robert, des relevés ont été effectués durant trois ans. Ce suivi réalisé en conditions réelles en temps instantané et synchrone, fournit des informations sur les flux transitant dans le bassin ainsi que des données sur les modalités de répartition des précipitations. L'aval du bassin reçoit en moyenne 30% de pluies de plus que ne reçoit l'amont.

Le bassin versant étudié s'étend sur 28 ha, à pente faible (2 à 6 %) il est emboîté dans le bassin plus vaste d'Erlon ; il peut être considéré comme représentatif de l'est du bassin parisien, domaine de la grande culture.

Les principaux paramètres commandant les modalités du ruissellement et de la dégradation des terres sont discutés et hiérarchisés.

Ruissellement et dégradation se manifestent avec des intensités variables en fonction des périodes pluvieuses de l'intensité et de la durée des pluies ; celles de printemps et d'été sont les plus efficaces, celles d'automne et d'hiver le sont moins.

Les séquences de pluie enregistrées montrent que l'un des seuils de déclenchement du ruissellement et de l'érosion importante correspond à une intensité de pluie minimale de 30 à 40 mm/heure en 10 mn. Un minimum de 12,5 mm de précipitations en 1 heure constitue un autre seuil (ces valeurs diminuent fortement après des périodes pluvieuses), 10 à 15 mn sont nécessaires pour que l'averse s'accompagne de la concentration des eaux dans les talwegs. Néanmoins le rôle du couvert végétal existant alors dans le bassin versant demande à être pris en compte, ce rôle peut modifier les résultats obtenus de façon importante, en particulier en ce qui concerne la masse de matériaux mobilisés qui peut varier de façon considérable.

Les dégâts enregistrés affectent la qualité physique des sols, les composantes les plus fines sont prises en charge, parmi elles, les matières organiques occupent une place non négligeable. Par contre la mobilisation des particules plus grossières demeure toujours limitée. Les débits solides peuvent atteindre 13 % de l'ensemble mobilisé.

Les méthodes et techniques d'utilisation des sols et l'occupation de ceux-ci, interviennent également en accélérant ou en ralentissant le phénomène de ruissellement et de dégradation, cependant l'effet de ces façons culturales est moindre qu'on n'aurait pu l'imaginer.

INTRODUCTION : EQUIPPING A WATERSHED

The Erlon community, situated 25 km north of Laon, 140 km away from Paris, covers an area of 1200 ha of which 800 (fertile land, loess) belong to property farmers and are used for large-scale agriculture (Fig. 1).

This community, like most in the region, was replotted in 1971. As of this date, there was an accelerated degradation of lands by run off. Following numerous floods affecting Erlon and its infrastructures, the population became aware and began to want to find solutions. Profitability of farms is certainly threatened and cost of reparations is high for the community budget.

The "Pays de Serre" association, created a few years ago, started to play a quite important role in financing research and work in the community.

Currently numerous organizations work on the problem : the Chamber of Agriculture and the Agronomical Station of the Aisne as well as our Research Center Biogéographie - Ecologie, ENS/CNRS.

Work on erosion of farm lands are carried out on the level of *the elementary watershed* or of *a complex of watersheds.* At this level formation and concentration of runoff can be studied with the most precision : surveys of rills (localization and dimension) measures of solid debit at the foot of the slope, characterization of transported sediment and localization of colluvium.

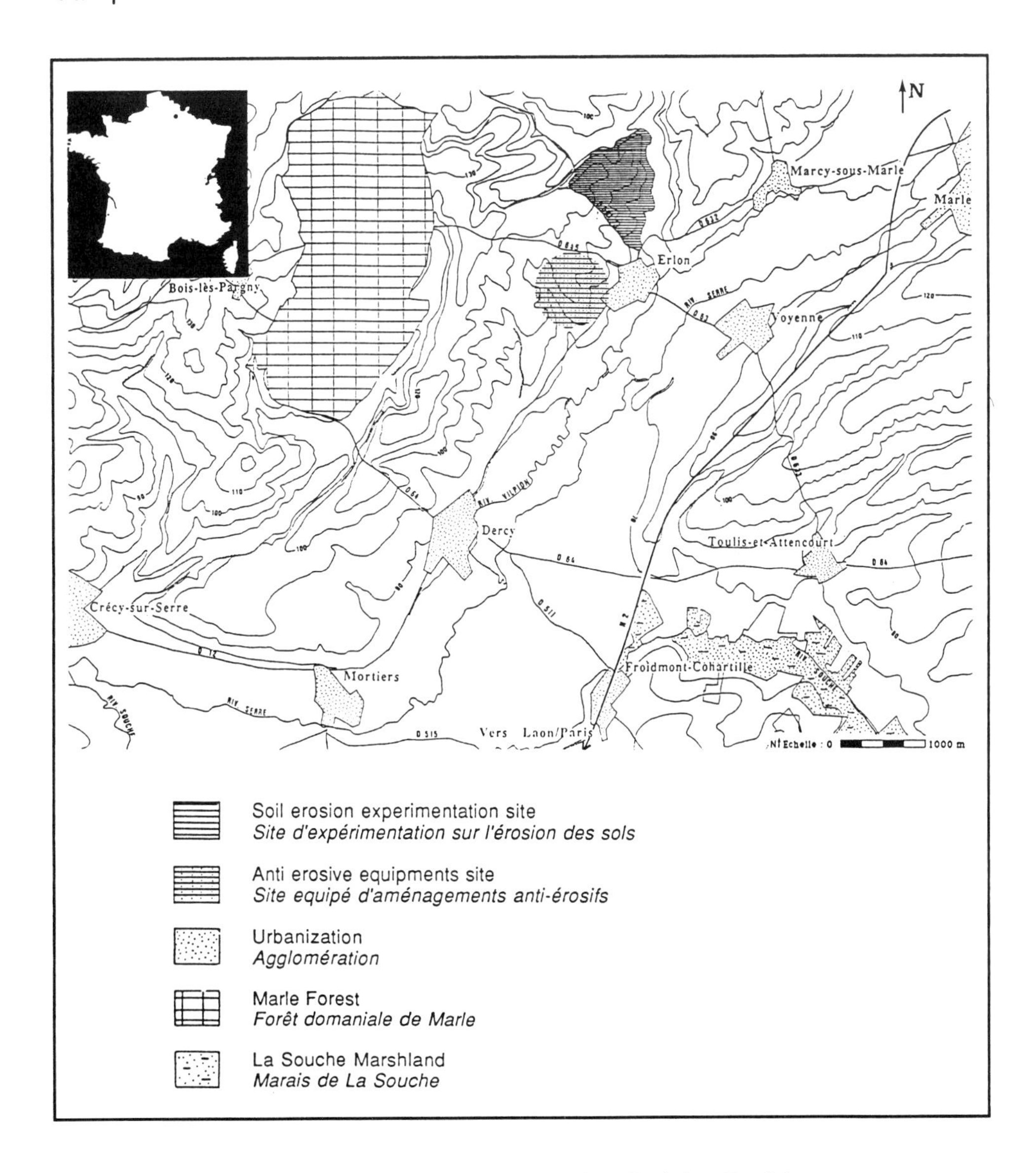

Figure 1. Experimentation site of Erlon (Laonnois - Parisian Basin).

The farmed basin called "Arbre Robert" (28 ha) belongs to the larger Erlon watershed was equipped by our laboratory and with the scientific support of the

INRA (Laon) in order to forsee movement and rate of flow (water, dissolved elements, gross transported material...) (1), (Fig. 2).

This experimentation furnished flow information under real conditions. Follow-ups enable us to give special attention to interrelationships between agricultural value of plots, techniques used, circulation of liquid and solid debits, and perceptible damage in the watershed.

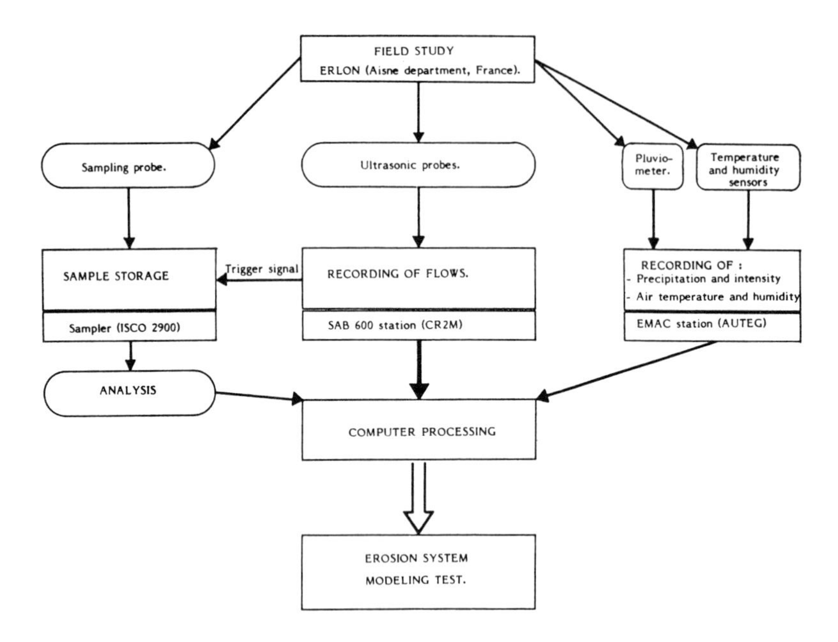

Figure 2. Operating principle of the experimental system. Reprinted from : S. Wicherek, Soil Technology, a cooperating Journal of Catena, June 1991, p. 102 (1).

Soil : Such an approach imples that watershed detail reliefs and spatial, vertical and lateral differentiations of superficial formations and soil be taken into account. A detailled soil cartography, at 1/5000 th was carried out by J. Maucorps (2) and 5 pedological ditches were dug in order to characterize the soil . The most complete soil profile was described close to the crest line dominating the basin ; two loessic loam deposits, decarbonated, surmount loamy colluvium, which are lying on brown-red loessic loam, in which a leached or brown-leached paleosoil, truncated ; chalk is not quite 5.20 meters thick. On the southern slope, a single loamy area, the most ancient, was identified, lying on loamy colluvium and a brown-red paleosoil. In the dale axe, close to the measurement equipment, there are more than 2 meters of loamy-pebbly colluvium, and 3.75 meters close to the B.V.E. exurgence. On the northern slope, a last profile shows a brown superficial limestone soil, developed in decarbonation clay, juxtaposed with anthropic rendzina.

The soil map shows the effect of erosion, which leaves very little loam, poor in clay, at the surface: on the dale head, and the southern slope, clayey loam is largely represented, except for the upper part of the northern slope where ancient alluvium, and more or less altered clay outcrop.

DISCUSSION AND RESULTS

Follow-up and gathering of data were effected over three years' time, from July 1989 to June 1992, on a permanent and synchronous timing (Fig. 2). Results obtained by triggering these processes have, for the most part, confirmed studies carried out at the measurement area scale over the past twenty years (3 - 4).

At the measurement area level we have been able to take constant and precise measurements under maximal homogeneous conditions (soil, climate...). These concerned mostly climatic parameters, the evolution of physical, hydric and mechanical characteristics of the soil, evaluation of the surface state, runoff and sediment loss. This scale of measurement is the only one which can truly be considered for continuous study, for example, the evolution of surface state, formation and extent of runoff during rainfalls, splash problems and eolian transportation (5), influence of vegetal cover, and thus its importance in understanding the mechanisms and verification of simulation protocol in a laboratory setting. However, the measurement station does not take the diversity of slopes, of watersheds and complexes of watersheds nor situational diversity into account. This level is necessary but not sufficient, it explains why the catchment area as a spatial and fundamental unit for environmental studies is important. Furthermore this area is constantly used for agriculture, which adds more importance and thus this approach is necessary for more global types of research, for example, typology of high-risk erosion zones (6).

Distribution of rainfalls on a watershed

We have observed that for a temperate oceanic climate, total pluviometry is relatively weak. Distribution of trimestrial pluviometry varies from 1 to 3 (example from 7/1/90 to 9/30/90, 85 mm, and from 10/1/92 to 10/31/92, 220 mm). It is quite clear that in its totality, the downhill slopes recieve 29 % more rainfall than the upper slopes, with a rainfall intensity which is often higher (Tab. 1). Maximal precipitation occurred *in January and February and March* and the difference in distribution over the catchment area can reach 45 %.

This difference is 42 % for *the October-November-December period*, but these rains, usually weak in intensity and long-lasting, are not erosion triggering rains. If we add vegetal cover - such as winter wheat, green fertilizers or straw - this is the ideal situation and period for limiting damage due to erosion in spite of a beginning runoff phenomenon, without solid charge. On the other hand, if soil is left bare (except for recent cultivation), this phenomenom becomes dangerous, and a system of planting in strips of a few dozen or few hundred meters in width (example: bare soil, wheat, green fertilizer, etc) becomes quasi-obligatory in order to limit damages over this period that is recommended to farmers.

The July through September period has an almost negative pluviometry, and distribution of rain over uphill and downhill slopes does not exceed 20 %. These precipitations are generally short in duration (a few minutes) and of heavy intensity sometimes exceeding 60 mm/hour in 10 minutes' time. The soil in place (loess) not being the most absorbant soil (at best, its capacity for absorption is 25-30 % in relation to soil volume), it is nearly impossible for such a quantity of water to be absorbed over such a short time, not to mention the crusting or hardened soil layer that forms. These three parameters, either taken separately or together, will diminish permeability. It is precisely this excess water that will then provoke damage, directly acting on loss of natural fertility. It is clear that distribution and intensity of rains and the relationship with vegetal cover are at the center of understanding the erosion phenomenon.

Table 1
Precipitation : totals and distribution over the Erlon watershed

Periods	*Upstream* (in mm)	*Downstream* (in mm)	*Downstream / Upstream*
from 01-07-89 to 30-09-89	125,4	136,5	1,09
from 01-10-89 to 31-12-89	127,1	174,0	1,37
from 01-01-90 to 31-03-90	135,3	186,5	1,38
from 01-04-90 to 30-06-90	143,5	163,5	1,14
from 01-07-90 to 30-09-90	85,7	107,5	1,25
from 01-10-90 to 31-12-90	156,0	222,0	1,42
from 01-01-91 to 31-03-91	109,8	156,5	1,43
from 01-04-91 to 30-06-91	92,8	119,0	1,28
from 01-07-91 to 30-09-91	115,0	130,0	1,13
from 01-10-91 to 31-12-91	119,3	165,0	1,38
from 01-01-92 to 31-03-92	101,4	147,0	1,45
from 01-04-92 to 30-06-92	196,2	235,5	1,20
from 01-07-89 to 30-06-92	1507,5	1943,0	1,29

During *the April to June period*, rain is abundant over a relatively long duration and in spite of maximal intensity, it rarely exceeds 60 mm/h in 10 minutes, provoking severe erosion (Tab. 2) (Photo 1). This period also corresponds to extensive farming (for ex. sowing, phyto-sanitary treatments), so they provoke a settling of soils barely covered with vegetation, which favors lower resistance of the epiderm of the soil to strong rain. This period should be considered as the most favorable to triggering runoff.

To conclude

Consideration of these factors, on both a quantitative and qualitative level in establishing overall evaluations of degradation is essential ; not taking them into consideration, often introduces errors in hydric erosion evaluation. Thus the real problems which must be handled are water surplus above mentioned and its distribution, in order to propose better soil management.

Table 2
Comparison between strong rain, some of which generate erosion - Erlon watershed

Dates	*Rain duration*	*Total precipit.* (mm)	*Maximum intensity* for 10 mn (mm/hr)	for 1hr (mm/hr)	*Runoff* total (l)	maxi inst. discharge (l/mn)	*Solid matter* total (kg)	runoff maxi %	*Nitrates* (mg/l)
08-08-89	0h22	5,5	44,0	8,4	0	0	0	0	-
24-09-89	0h34	12,5	36,0	12,2	275	6	7	3	-
01-11-89	3h36	12,5	25,5	8,8	0	0	0	0	-
13-12-89	0h08	2,5	12,5	3,1	0	0	0	0	-
14-02-90	11h00	17,0	4,5	1,9	1170	5	-	-	11
15-02-90	5h00	15,5	5,1	4,4	8500	43	-	-	11
27-02-90	4h23	27,5	16,5	12,7	12800	110	58	0,7	28
22-04-90	1h34	9,5	24,0	10,3	200	10	-	-	-
08-05-90	1h36	19,0	51,0	18,5	130000	3000	9500	11	-
27-06-90	0h11	6,5	12,6	4,8	2500	600	-	-	-
30-06-90	0h08	2,0	6,6	1,7	550	65	-	-	-
15-08-90	3h46	19,0	19,0	7,5	0	0	0	0	-
28-10-90	11h34	23,0	4,8	3,2	120000	340	730	0,7	12
25-11-90	8h50	8,5	3,3	2,1	72000	450	114	0,2	6
21-03-91	3h00	14,0	6,6	6,0	0	0	0	0	-
23-03-91	2h45	11,5	12,0	6,8	0	0	0	0	-
27-06-91	0h06	2,0	9,4	2,0	0	0	0	0	-
06-07-91	2h04	15,5	60,0	16,1	3000	300	283	13	100
07-07-91	0h08	2,5	12,7	3,9	475	50	37	10	27
24-07-91	0h23	7,0	34,7	6,9	0	0	0	0	-
08-08-91	1h00	5,0	20,0	5,0	50	5	0	0	-
25-09-91	0h33	6,5	22,5	6,5	50	<1	0	0	-
11-11-91	5h30	11,5	3,8	3,5	450	2	0	0	-
12-11-91	1h23	7,0	31,7	5,9	500	2	0	0	-
18-11-91	0h21	5,5	35,0	5,5	200	4	0	0	-
22-03-92	6h20	12,0	4,2	3,2	200	1,5	0	0	-
17-04-92	3h00	4,0	3,2	1,9	20	0,3	0	0	-
26-04-92	1h06	4,5	12,8	4,4	0	0	0	0	-
09-05-92	0h21	3,0	11,3	3,0	0	0	0	0	-
28-05-92	0h42	12,5	43,0	12,5	200	25	4	2,5	41
28-05-92	1h20	5,5	9,2	4,1	250	16	10	4	57
30-05-92	0h11	5,0	28,5	5,0	11000	900	300	3	57
31-05-92	0h05	2,0	12,0	2,0	600	60	14	2,7	58
31-05-92	0h11	12,5	72,0	12,5	400000	5000	40000	13	22
02-06-92	3h10	16,5	18,0	8,5	160000	2000	14000	12	13

Photo 1. Intensive erosion. Several dozen tons of soils washed away on the Erlon watershed, bumper state of vegetal cover (corn). S. Wicherek, May 1990.

Rainfall intensity and duration.

The most important elements influencing runoff are intensity and duration of rainfall. For example, at the end of June, only 12.6 mm/h in 10 min. with total precipitation of 6.3 mm in 11 min. sufficed to trigger minor runoff with no solid charge. Three days later, a 2 mm in 8 min. rainfall, intensity 6.6 mm/h in 10 min again provoked runoff, in spite of vegetal cover over 50 % of the surface (Tab. 2). The triggering threshold for erosion was as of 2.5 mm rainfall in 8 min. with an intensity of 12.7 mm/h in 10 min. when the precedent periods were also slightly rainy and the soil was wet (for example see table 2, 5/28/92 - 6/2/92). In this context, erosion is high, solid debit reaches 13 % of liquid debit volume, including 0,5 % organic matter. It is certain that 12,5 mm of rain in 34 min. with maximal intensity of 36 mm/h in 10 min. triggers erosive processes ; roughly corresponding to the September-October period when the soil left bare after harvesting, has not yet been tilled, except for land left in straw ; this is actually a good protection, and soil should be left this way as long as possible. The most critical erosive phase starts as of 60 mm/h in 10 min. with minimal rainfall of 15 mm in 1 hour.

To better visualize these ideas and to be more explicit, over two different periods we have presented correlations between total rainfall, rain intensity, liquid and solid debits and their totals :

- Spring-Summer, full vegetation (Fig. 3 and 4)
- Autumn - after harvest (Fig. 5), (Tab. 3)

During the first period (Fig. 3), rainfall intensity and duration are sufficient to trigger runoff followed by strong erosion, in spite of good vegetal cover (Tab. 4). One observes a gap between the start of rainfall and that of runoff - about 12 minutes or so - proving that a certain length of the catchment area is functionning.

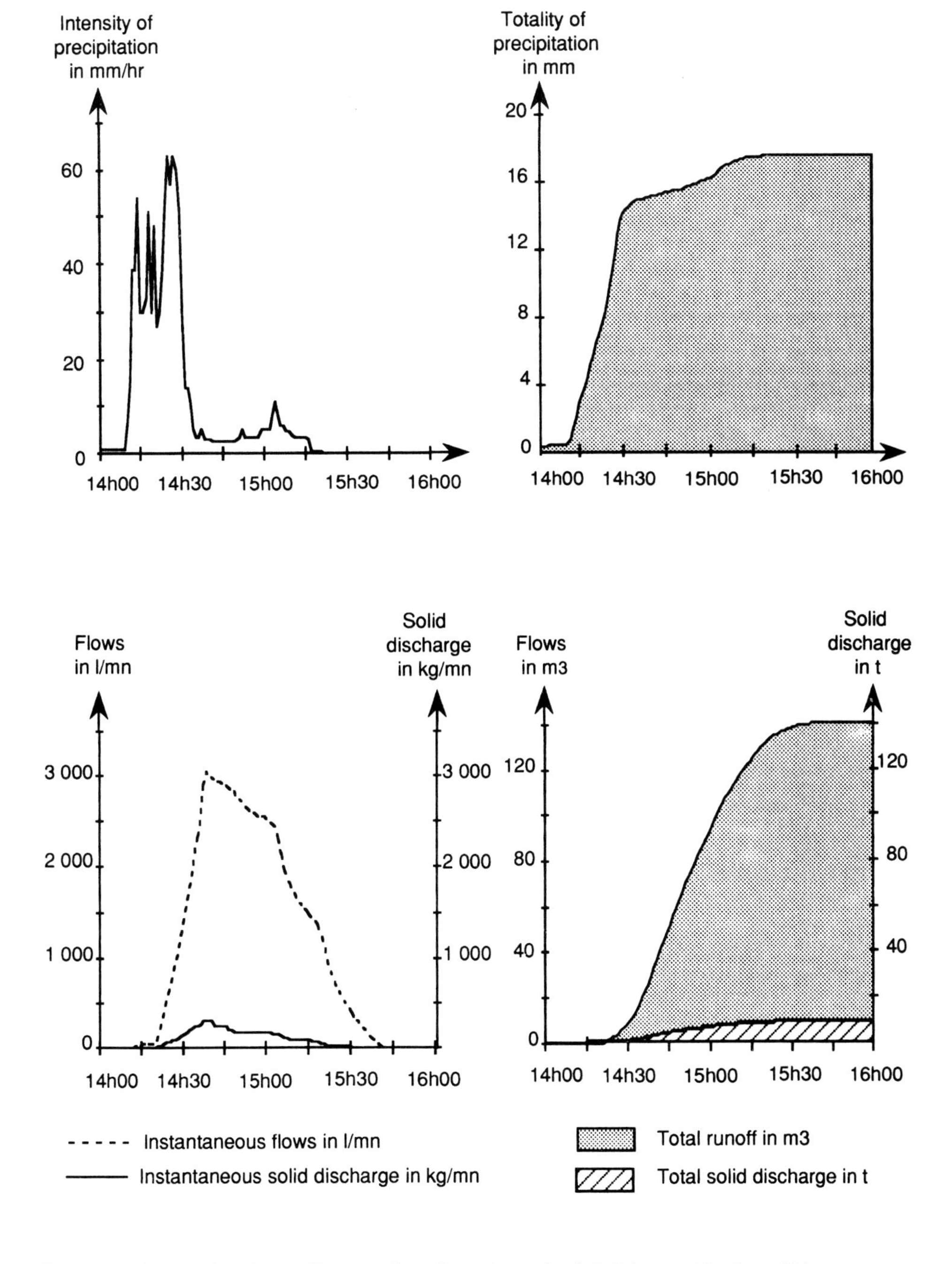

Figure 3. Example of runoff recording (function of rainfall intensities) on Erlon watershed (P.B.), the 5/8/90.

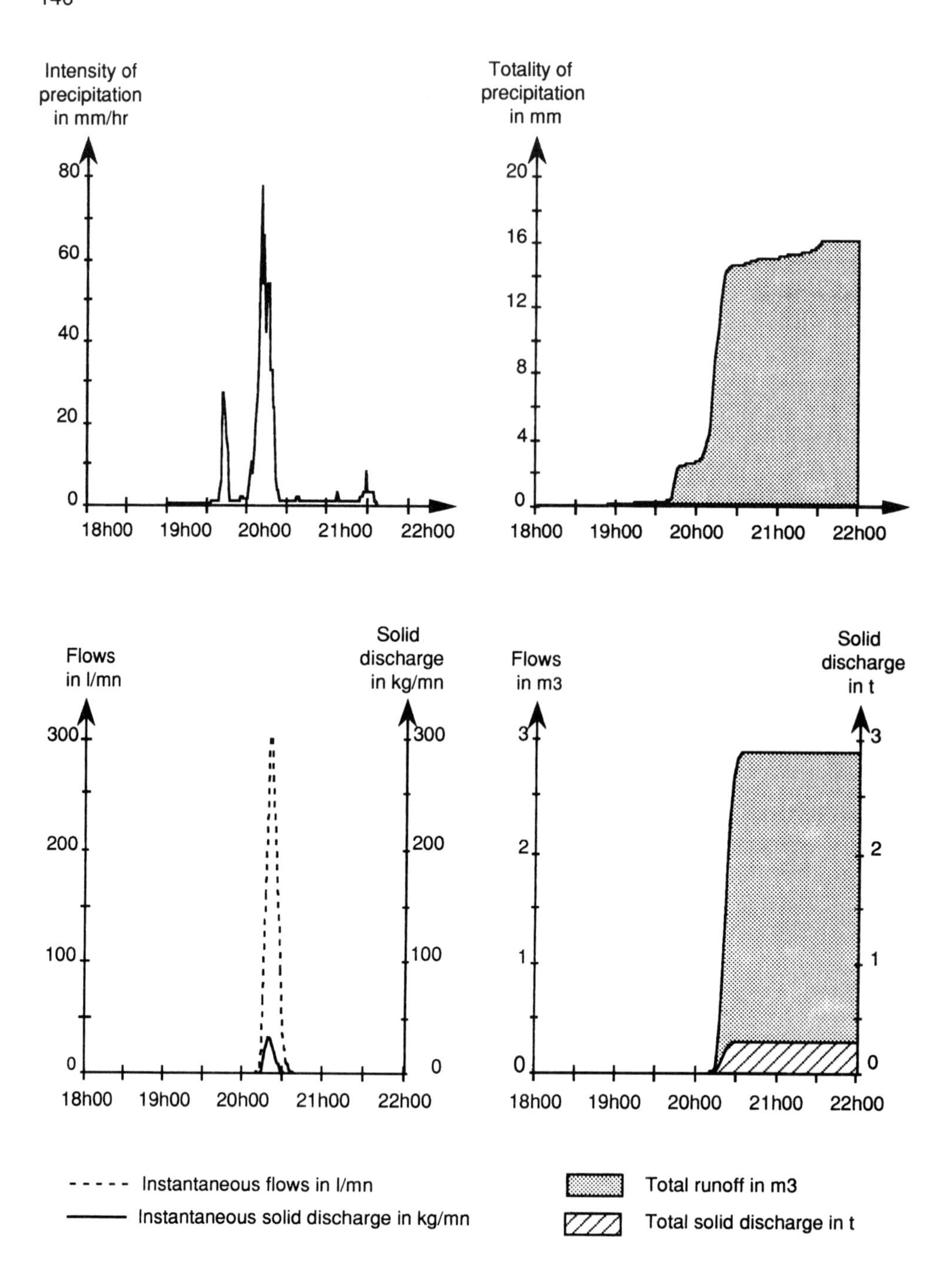

Figure 4. Example of runoff recording (function of rainfall intensities) on Erlon watershed (P.B.), the 07/06/91.

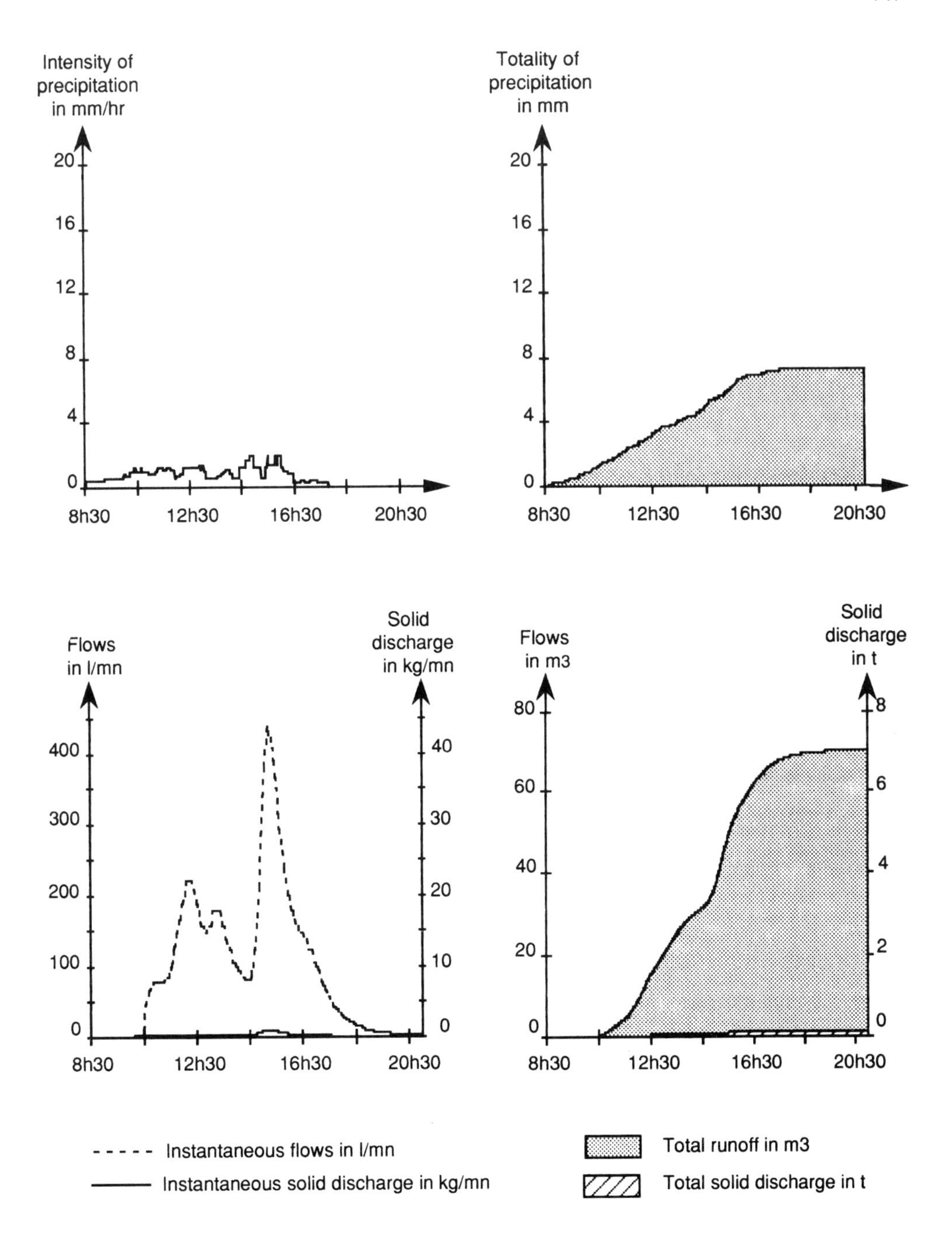

Figure 5. Example of runoff recording (function of rainfall intensities) on Erlon watershed (P.B.), the 11/25/90

Table 3
Consequences of cultural practices and techniques on erosion

Dates	*Rain duration*	*Total precipitation* during X previous days (mm)			*Precipitation*			*Runoff*		*Solid matter*	*Nitrates*
		X=8	X=4	X=1	total (mm)	maximum for 10mn (mm/hr)	maximum for 1hr (mm/hr)	total (l)	maximum discharge (l/mn)	total (kg)	(mg/l)
Before tillage											
28-10-90	11h34	24,5	24,0	4,0	23,0	4,8	3,2	120000	340	730	12
29-10-90	2h07	51,0	50,0	29,0	3,0	2,1	1,5	23000	285	-	10
29-10-90	1h09	54,0	53,0	32,0	2,0	5,3	1,8	10000	150	-	10
31-10-90	0h55	57,0	37,0	0,5	3,0	5,1	2,7	32000	90	18	7
13-11-90	1h45	8,5	8,0	1,5	2,0	3,5	1,5	6000	130	6	7
13-11-90	3h00	10,5	10,0	3,5	1,0	0,5	0,5	2000	20	-	7
14-11-90	6h10	12,0	10,5	4,0	6,0	3,6	1,4	22000	150	16	7
18-11-90	0h35	22,0	11,5	2,5	2,5	6,0	2,2	7000	200	112	7
18-11-90	1h26	24,5	14,0	5,0	4,0	12,8	3,6	27000	770	-	6
19-11-90	0h38	32,5	15,5	11,0	1,5	2,1	1,2	17000	440	-	6
20-11-90	1h27	30,0	17,5	6,5	2,0	4,0	1,4	6000	80	218	6
20-11-90	4h36	32,0	19,5	8,5	5,5	5,1	2,6	43000	500	-	6
24-11-90	0h14	27,5	10,0	2,0	2,5	10,7	2,4	13400	400	47	6
25-11-90	8h50	28,5	5,0	4,5	8,5	3,3	2,1	72000	450	114	6
After tillage											
25-12-90	0h32	20,5	13,5	0,0	2,0	5,6	1,4	0	0	0	-
25-12-90	0h40	22,5	15,5	2,0	2,0	5,1	1,9	0	0	0	-
26-12-90	2h45	26,0	7,0	6,0	6,0	3,7	2,9	0	0	0	-
27-12-90	0h25	35,0	14,0	9,0	1,5	5,6	1,4	0	0	0	-
29-12-90	0h32	35,5	22,0	9,5	1,5	3,8	1,2	0	0	0	-
29-12-90	3h00	37,0	23,5	11,0	6,5	6,7	3,5	0	0	0	-
31-12-90	1h37	30,0	16,0	0,0	2,5	1,4	1,3	0	0	0	-
31-12-90	2h33	32,5	18,5	2,5	7,5	11,6	5,3	0	0	0	-
02-01-91	2h16	42,5	25,0	2,0	4,5	3,5	2,1	0	0	0	-
02-01-91	2h40	47,0	29,5	6,5	3,5	1,9	1,6	0	0	0	-
03-01-91	5h44	48,5	23,5	11,0	7,5	16,5	3,9	0	0	0	-
03-01-91	6h13	56,0	31,0	9,5	9,0	3,8	3,1	0	0	0	-
10-01-91	1h48	47,0	13,0	11,0	5,5	5,5	3,9	0	0	0	-
11-01-91	3h17	41,0	16,5	11,5	7,5	6,4	2,8	0	0	0	-

Table 4
Relationship between vegetal cover and erosion. Ex. Erlon watershed (Parisian Basin)

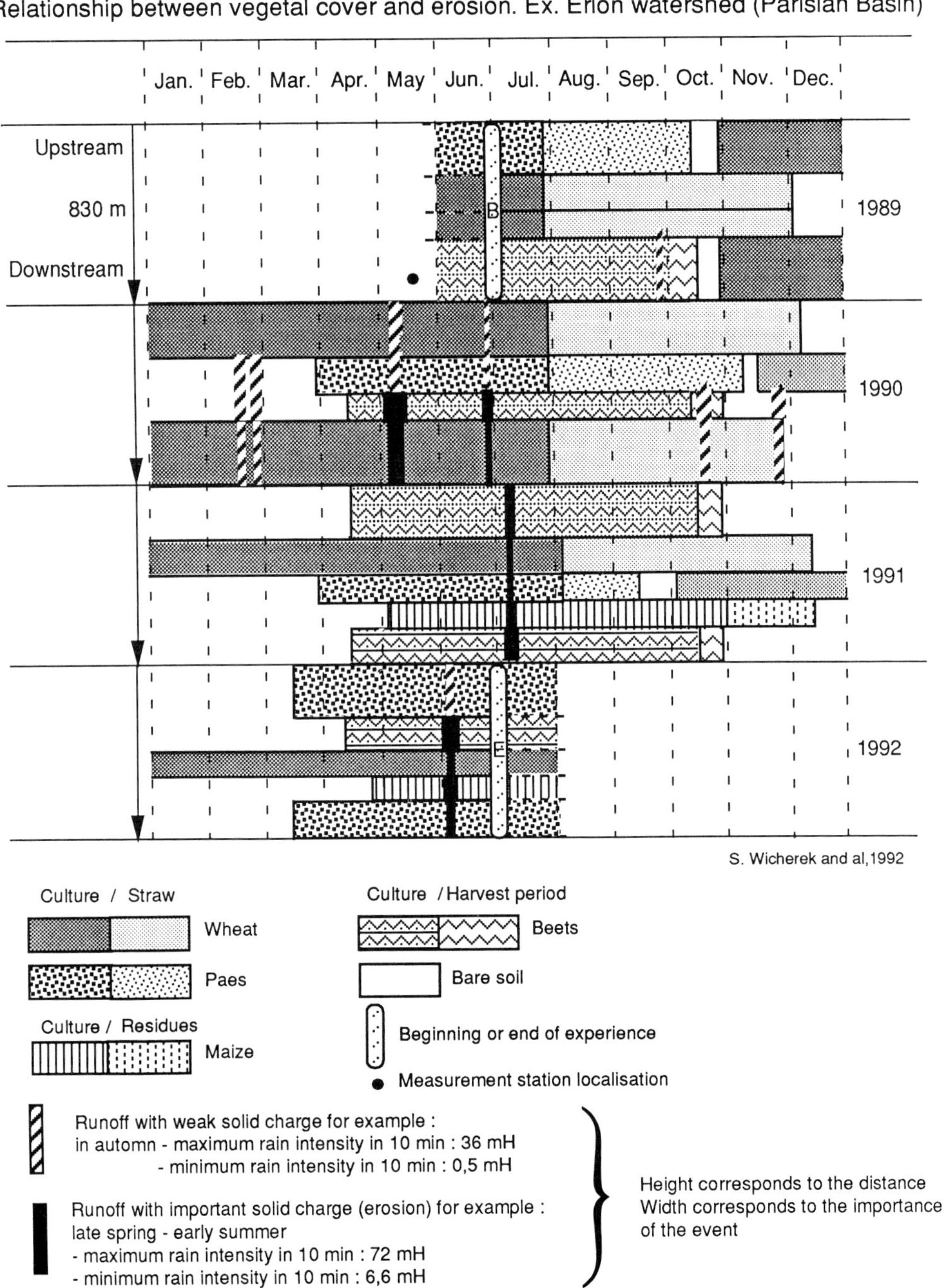

The plots are arranged depending on the contour lines (recommended method)
on the graph the width stripes correspond to the real widths. Their lengths are from 400 to 500 m

Other experiments with polystyrene capsules - used as a tracer - of different colors, spread over various transects on the entire catchment area, confirmed this runoff activity. However, figure 4 shows the need for a certain duration of rainfall in order to accelerate erosion.

In the second period (Fig. 5) rainfall intensity was quite weak but duration and totals were high and moreover, the soil left bare after the beet harvest was highly saturated with water. All conditions are united to trigger runoff with little suspended material. The time needed to trigger is higher than that of preceding cases because of the uneven surface of the soil.

To conclude

During the experimental period, on dry soil, we have not noticed runoff when the rainfall intensity is weak. However, there was also the starting up of runoff with solid charge provoking real damage if rain intensity exceeded 30 mm/h in 10 min. with a minima of 12.5 mm in 1 hour. In these conditions solid debit is less than 1 % of the total transiting volume in the catchment area. If rain attains 16-17 mm solid debit will take up 3 % and when total rainfall approaches 19 mm, solid debit is over 10 %. Thus the importance of water retention of the soil (which is low for this type of soil) and minimal pluviometry to saturate them. Only 2 mm/h is enough, in 10 min. with total precipitation of 3 mm in 2 hours, to generate erosion. We have seen that erosion is a threshold factor, increase in intensity and duration of rain will, firstly, saturate the soil and then trigger runoff then erosion, vegetal cover and certain farming practices can reduce or accelerate this phenomenon.

Vegetal cover and agricultural practices

The experimental site is cultivated with high performance equipment, machines pass roughly 15 times per year (tilling, sowing, treatments, supplementing N, P, and K). Weight of these machines directly affects the compaction and infiltration capacity of the soil, especially for these types of soils, and when the level reaches 30-35 cm deep, corresponding to tillage pan, hindering infiltration.

It is clear that most soil degradation results from :

- Runoff especially in sheets, very damaging, eroding particularly fine elements, organic matters and chemical fertilizers ; it is often favored by crusting. At contrary rills play a virtually role by washing away gross material in lower quantity.

- Subsuperficial runoff (7) is generated at the tillage pan level, because of farming practices, carrying off natural fertilizers and chemical nitrates. This occurs in solifluction phenomena even, displacing the arable layer almost entirely, which can correspond to a few cm or even a few meters over the year (Photo 2). At this level, something must be done, and preventive measures taken, such as breaking up the tillage pan to facilitate infiltration.

Vegetal cover also plays an important role in soil degradation (4 - 8). The majority of soils examined (Tab. 4) remained bare over 6 months' time, because of new practices adhering to market demands, with accelerated erosion, especially for corn and beets. These damages occur over two periods : sowing time in April-May, beginning of lenghty growth, with no protection of the soil by foliage, and the harvest period, end of September - early October. These periods correspond to the maximum damages done, without taking into account the loss of soil carried off when beets are harvested (about 20 tons/hectare). To limit this damage, tilling must be done as early as possible (Tab. 3).

Photo 2. Plough pans, subsurface runoff and their impact on soil degradation (corn field). Faucoucourt B.P. R. Boy, January 1991.

To conclude

We have seen different periods of runoff over the year but real danger only in three periods (Tab. 2).

Winter (late February - early March), Autumn (late September - mid November) : soil is quite bare, rainfall is long in duration but low intensity, sufficiently strong to provoke runoff but not enough to cause erosion.

Periods that are truly dangerous are those of sowing of beets and corn in early May and late June to early July when rainfall is minimal but of strong intensity, the water's moving force is such that solid debit is into movement, beet vegetal cover, corn or potatoes do not hinder erosion ; only wheat (especially winter wheat) serves to protect the entire vegetative phase, this is also true for straw and sweet pea crops.

These different thresholds of soil stability, in function of rainfall intensity are presented in figure 6 where one can remark the relation between vegetal cover and erosion.

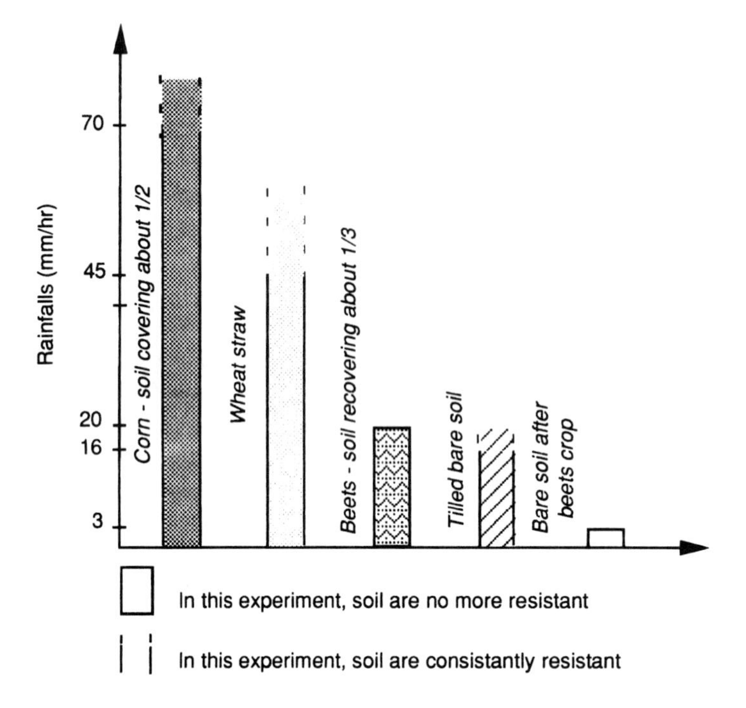

Figure 6. Different thresholds of agricultural land stability in function of precipitation intensity.

GENERAL CONCLUSION

Changes in agrarian structure, techniques of soil use, increase in productivity over the past 25 years by specialized techniques, increasing fertilizers, pesticides and mechanization... have led to a progressive structural and natural fertility degradation, particularly to a decrease in organic matter (decrease of 0.10% over three years for our case), in clay, and fine loam carried off by runoff, without mentioning the consequences of the biological life within the soil. One can evaluate the quantity of moved soil as hundreds of tons on the total watershed, with more of 40 tons issuing from it for a 3 years period. Direct effects of farming practices leading to compaction from weight and numerous passages create a serious problem, a top crust favoring hypodermic runoff, not quite visible but as dangerous as sheet erosion. For this reason it is indispensable to break this tillage pan for example by installing a claw behind every plough share in order to facilitate infiltration. Chemical degradation is another issue, for instance increase in nitrates. We have measured these in transiting flow, with concentrations exceeding the EEC norm, at 50mg/litre for drinking water. This degradation affects water quality, rivers and water tables but the soil determines water quality and must be protected first. Infiltration must be encouraged in spite of an inevitable surplus water which will provoke erosion in such a loessic soil. This water quantity is equally depending directly on intensity and duration of rainfall, or 30 mm/h in 10 min. and 12,5 mm in 1 hour. This flow can be modified by using certain crops at different phenological stages of the year thus playing the role of a damper in order to decrease the force of runoff. This could also be assured by straw, plants residues and a recent tillage. Liquid debit can be charged up to 13% in solid matter, particularly during early May to early July. This period is crucial due to rain intensity often exceeding 60mm/h in 10 min., and it is almost indispensable that cynetic energy of raindrops (splash) be decreased through use of a good vegetal cover or by plants residues reducing the phenomenon without stopping it altogether. At this period, degradation of the soil attains its maximal threshold, followed by the harvest period for "weeded" plants as of late September.

This type of study at the catchment area level as an elementary unit of environmental studies allows us to measure total flow by correlating it with use of this space by man ; this type of study can thus have real impact and influence management of the rural space and the soil asset facilitate risks cartography and open up a new field of research concerning the relationship between environment and health.

With many thanks for F. Grégoire collaboration

REFERENCES

1. Wicherek, S. (1991) : New approach to the study of erosion in cultivated lands. (Nouvelle approche pour l'étude de l'érosion des sols cultivés). Soil Technology, Catena, 4, 2, 99-110.
2. Maucorps, J. (1992) : Carte pédologique du bassin versant d'Erlon au 1/5 000ème, p. 75, In Ouvrage collectif. Excursion guide, the Parisian Basin. In International Symposium titled "Farm lands erosion in temperate plains environments and hills", Saint-Cloud 25 - 29 May 1992. Editeur Centre de Biogéographie - Ecologie, ENS / URA 1514 CNRS, 148 p.
3. Wicherek, S. (1986) : Ruissellement, érosion sur les versants de la France des plaines et des collines. Exemple : Cessières (02 Aisne). Hommes et Terres du Nord, 4, 254-261.
4. Wicherek, S. (1988) : Les relations entre le couvert végétal et l'érosion en climat tempéré de plaines. Exemple : Cessières (Aisne-France). (Relationship between the vegetal cover and erosion under temperate climates of plains. Example of Cessières (02 Aisne - France)). Zeitschrift fûr Geomorphologie, 32, H. 3, 339-350.
5. Wicherek, S. (1989) : Impact of splash and eolian transport on erosion ; case study : North of Parisian basin, France. (Rôle du splash et du transport éolien sur l'érosion. Terrain d'étude : Nord du Bassin Parisien). Earth Surface Processes and Landforms, v. 14, 6/7, 461-467.
6. Veyret, Y. and Wicherek, S. (1992) : Réflexion sur une cartographie des risques d'érosion des terres agricoles en région tempérée de plaines et de plateaux. *B.A.G.F.*, 2, 169-176.
7. De Coninck, F., Bock, C., Grégoire, F., Maucorps, J. and Wicherek, S. (1991) : Les transferts latéraux de solutions du sol dans un système lande sur podzol - tourbière acide. (Lateral transportation of soil solutions in a heath on a podzol and bog system). Bull. Ecologie, T.22, 3-4, 27-46.
8. Wicherek, S. . (1992) : Ouvrage collectif. Excursion guide, the Parisian Basin. In International Symposium titled "Farm lands erosion in temperate plains environments and hills", Saint-Cloud 25 - 29 May 1992. Editeur Centre de Biogéographie - Ecologie, ENS / URA 1514 CNRS, 148 p.

Farm Land Erosion: In Temperate Plains Environment and Hills
S. Wicherek (Editor)

Remarks on Flow for an Evaluation of Erosion Risks (examples in the Parisian Basin)

M. Dacharry[a], Y. Veyret[b], S. Wicherek[b]

a Université Lille I, UFR Géographie, BP 36, 59655 Villeneuve d'Ascq, France.

b Centre de Biogéographie-Ecologie, ENS Fontenay St Cloud, Le Parc, 92211 Saint-Cloud, France.

SUMMARY

The watershed, a fundamental and dynamic unit, serves as the framework for studying water circulation and carried material.

Because of insufficient quantity, or dense vegetal cover and/or "absorbant" soil, water which makes its way by flow into the watershed does not always come back out again. Numerous cases exist, either the rills and ravines installed on slopes are functional, or else, the basic talwegs (drains) function at the same time ; however in both cases, if land associated with drains suffers degradation, the only transfer is that occuring at the watershed. Accumulation areas serve as relays, and are useful for study.

If water reaches certain limits, flow outside of the basin becomes a possibility, this is carried out through the talweg (drain) network. This flow, which can cause floods, is the raison behind the most spectacular damage affecting the large-scale Parisian Basin agricultural regions. Flow conditions force us to take the mophological character of watersheds into account, which translates factors of a more or less high potential for risk.

RESUME

Le bassin versant, unité élémentaire et dynamique sert de cadre pour envisager la circulation de l'eau et des matériaux déplacés.

En quantité insuffisante ou en raison d'un couvert végétal dense et/ou de sols "absorbants", l'eau qui parvient à s'écouler dans le bassin versant ne sort pas toujours de celui-ci. Plusieurs cas peuvent exister, soit seules fonctionnent les rigoles et ravines installées sur les pentes, soit, fonctionnent aussi les talwegs élémentaires mais dans les deux cas, s'il y a dégradation des terres, associée aux drains il n'y a que transfert au sein du bassin versant. Les secteurs d'accumulation constituent des relais, utiles à étudier.

Si la quantité d'eau tombée atteint certains seuils, l'écoulement de l'eau hors du bassin est possible, il s'effectue par l'intermédiaire du réseau de talwegs. Cet écoulement qui peut donner naissance à des crues est à l'origine des dégâts les plus spectaculaires affectant les régions de grandes cultures du Bassin Parisien. Les modalités de l'écoulement forcent à prendre en compte les caractères morphologiques des bassins versants qui se traduisent par des potentialités de risques plus ou moins importantes.

INTRODUCTION

The study on the 1 September 1987 flood at Brie (Petit Morin basin) (1) and work done towards establishing a cartography of erosion risk in the Parisian Basin (2) have led us to consider the catchment area as a most pertinent elementary and dynamic unit for study. The choice of the watershed became necessary in order to complete highly developped research carried out up to this time on experimental plots.

Evaluating erosion risk within the catchment area framework leads us to define the elementary catchment area and to envisage an acceptable scale. Our choice could vary between areas of a few square kilometers up to those of a few dozen square km, and depends mostly of climatic factors, these dimensions allowing centers of maximal intensity to affect the entire catchment area. Beyond proposed areas, the storm will only affect part of the catchment.

Risks recorded downhill from the catchments and which affected the inhabitants and the infrastructures of Vierzy (cf Vierzy 1988) are one of the major problems of space management in the Parisian Basin. (1-2-3-4).

While Article 1 of the 13 July 1982 Law on natural catastrophes had been applied (Vierzy...) cost was heavy for the communities. These risks resulted in concentration of water and mobilization of soil particles in natural drains; they are often assimilated to risk caused by runoff, generating rills and gullies on slopes. In fact there are two phenomena which should be considered separately. One of the distinctions made is that of Cl. Cosandey (5) between rapid flood-related runoff designating "the rapid volume of water flowing into a collection area and provoking rapid increase of debits) and runoff affecting slopes, perhaps generating rills.

1. FLOW AND DEGRADATION IN THE WATERSHED: WATER AND MATERIAL TRANSFER WITH NO EXPORTATION OUTSIDE OF THE WATERSHED AND NO EXTERIOR DAMAGE.

In function of agricultural practices (bare soil, wheeltracks in the sense of the slope, "fourrières"...) traces of gullies contributing to land degradation, appear in the watershed. Different kinds of runoff exist, which must be distinguished, as is useful to stress successive and continuous runoff.

F. Morand and S. Wicherek (6) have established the connection between splash and "film" runoff, using observations and results on experimental plots. This was done through justaposition of a multitude of fine blades of water, and should also be considered separately from runoff in more or less anastomosed streams. These different types of runoff make up the more generalized runoff which causes uprooted and salient gross elements (twigs, grains of sand), runoff in stream can turn into concentrated runoff and provoke rills. Even if flow by channels is the most spectacular kind, there are other sorts, sometimes more efficient.

The case of Jumigny catchment near Soissons, is very interesting. On a slope with mean slope reaching 8 %, gullies are rare. Most of the land loss affecting the top of the slope (20 cm over 20 years) comes from sheet erosion. Observations effected at Erlon show that on prominent plateaus, the water film that forms above a certain precipitation threshold and/or in relation to the state of the soil, is characterized by a low erosion capacity, however, this increases as does the possibility of carrying off particles of soil, as soon as the slope increases in steepness. Runoff can continue in the form of film or small streams or in veritable canals.

Concentrated runoff, which can be the result of scratches (2 cm wide and 2 cm max. thickness), of rills (10 cm wide and 25 cm deep) , of gullies (25 cm wide and 25 cm deep) and large gullies (exceeding 3 meters in width and 5 meters in depth) often acts on portions of the watershed or on the level of the field; quantities of material displaced are difficult to transpose to the level of the entire catchment area. Much of the material is not exported out of the catchment, there is simply transfer within the catchment. For this reason, even though it is useful for better understanding processes at work, data concerning gullies and transported material is insufficient for understanding erosion and its risks.

Nevertheless detection, localization, functionning of concentrated runoff is useful in understanding flow dynamics in catchment areas or in groups of catchment areas.

1. Localization of gullies

A long-term study over a few years, for example, is difficult, even impossible because most of the scratches, gullies are temporary phenomena, and can be eliminated during cultivation. We should however, note that many forms of concentrated runoff are reconstituted each year in the same places. The Vierzy basin offers numerous examples of this kind of surface water flow at the top of valleys.

Research carried out on plots has allowed Govers (7) to point out that gullies frequently appear on the convex side of the slope, at the transition point between the prominent plateaux and average-sized slopes.

Numerous authors (8 - 9) insist on these slope ruptures as being the site where rills and gullies start to form. They also insist on slope angle and length. Threshold values for triggering concentrated runoff have been fixed between 3° and 7° (10).

Multiplying slopes due to the presence of numerous talwegs existing in the watershed is a factor that favors formation of gullies. In function of the set-up of the watershed : distribution of slopes, convexity of slopes, interruptions in the slope, ledges, abundance of talwegs and associated slopes, preferential sectors allowing development of traces of concentrated runoff.

Other factors favor appearance of these traces. Absence of vegetal cover, low organic matter content and soil settling reduce non-capillary superficial porosity and in consequence, reduce infiltration. The soil must be sufficiently "coherent" if the longitudinal incision is to be maintained without caving in. Sometimes, currents or their functionning downhill is hindered by the presence of some minor obstacle in its path or even by the existance of a cultivated strip, which favors water and transported material being trapped (Erlon) (Wicherek, infra).

2. Functionning of gullies

If gullies only affect the length of a field without attaining the drainage network in the catchment area, there is only minimal displacement of soil, which takes the form of a transfer from one part of the slope to another. There will thus be deposit and accumulation at the edge of the field or deposits of earth in the cultivated or tree-covered stripo (1). This mobilized material reflects the loss at the slope or field level which undergoes more or less pronounced degradation, but this material is not taken into account in calculating flow coming downhill from the catchment.This is the aforementioned notion of relay (2). Displaced and accumulated material can undergo another mobilization, in function of how the soil is used following deposit of material. Possibilities of remobilization depending on the nature of material that has accumulated and its capacity for undergoing the effects of surface sealing deserve more close study. Although few studies have been carried out in this domaine up to now, this area is potentially of great interest; it will enable us to evaluate the ways in which material accumulates and is displaced within the catchment area.

Exportation of displaced material out of the catchment area, through rills, implies that gullies are linked into the talweg network and that all the large gullies and talwegs function. Impact of gullies on land degradation throughout the catchment area should be taken into account, notably in order to foresee proper equipment of more vulnerable sections of slopes.

3. Functionning of talwegs within the catchment area

Degradation of land in the catchment area can take place in direct relation to the functionning of talwegs as observations made at Erlon have shown.

The "Arbre Robert" catchment area, sub-catchment within the vast Erlon area, covers 28 ha. Cut into Coniacian chalk, it has mild and regular slopes from 4° to 6°, which turn into veritable glacis downhill, close to Vulpion. These slopes with convex-concave profile, are covered with loam of uneven thickness (many meters on the Eastern slopes, while the Western slopes have only very thin deposits). In this catchment area, shallow talwegs (about 10 m in drop), are found, with the largest converging towards a major confluence followed by a single largely canal closed off by recent loamy and pebbly colluvial formations (about 3.5 meters thick at the base of the talweg).

The functionning of the Arbre Robert catchment area has been followed over a few years (S. Wicherek) (cf. infra) ; the 28 October 1990 rainfall was the object of a specific study. The rainfall was a good example of a winter storm. Precipitation was of long duration (18 hours) feeble intensity (maximal recorded intensity was 4 mm/hour). 17.8 mm of water fell upstream from the watershed and 27.5 downhill on already humid soil that had little protection because the beets had already been harvested. There were 120,000 litres of water in runoff. The beginning of the runoff in the main talweg occurred 4 hours after the start of the storm. On the catchment area slopes, there was no rill or gully formation ; material carried off (730 kg for the entire basin and 5 kg/mn for the mobilized maximum) was made up essentially of fine elements and was carried off by film runoff (Fig. 1).

The downhill catchment area where the slope tapers off and where a topography of glacis appears, water and material flow remained moderate and the water spread out, depositing elements caught up along the way. A deposit of a few millimeters modified the granulometric composition of the under-lying soil and made it less permeable and more humid, which explains the development of specific vegetation.

Furthermore, when transportation of dissolved matter (notably fertilizer) contributed to the enrichment of the downhill-glacis land with water accumulating, stagnating and then slowly infiltrating, cereal crop growth was more rapid than elsewhere in the catchment area. These crops which grow too high too quickly are often recumbent during strong Spring rains.

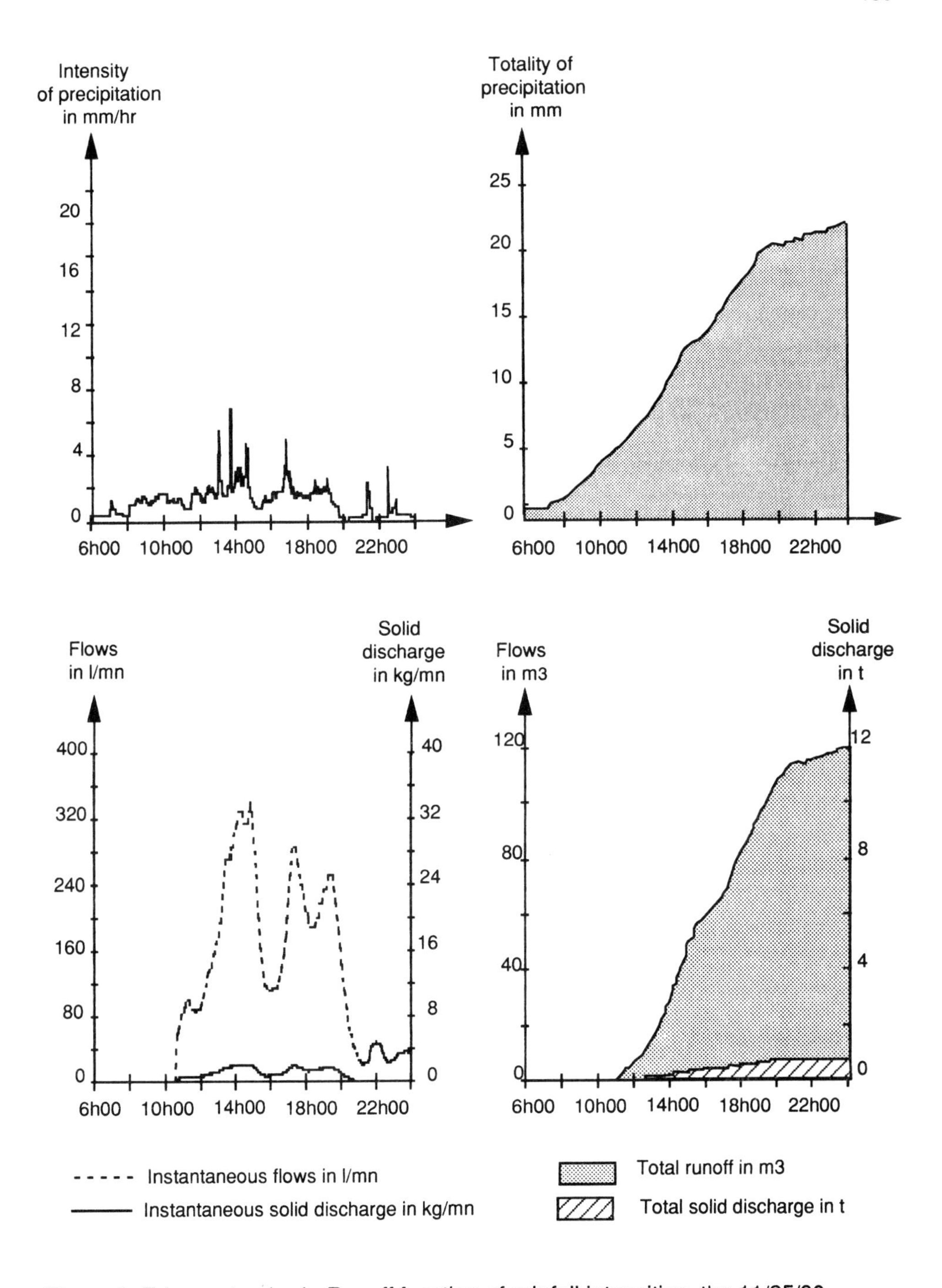

Figure 1. Erlon watershed - Runoff function of rainfall intensities, the 11/25/90.
(S. Wicherek, G. Chêne and M. Mekharchi, 1992)

This functionning which often occurs with relatively modest pluviometric totals (always below 40 mm) mobilizes non-negligeable quantities of suspended matter and products in solution, revealing the poverty of uphill soil. This degradation occurred without the presence of rills or gullies, under the combined effect of film runoff and mobilization of water in existing talwegs.

II. RAPID FLOOD RUNOFF

Catastrophic phenomena observed at Vierzy (16 May 1988) at Erlon (8 May 1990) on the Petit Morin (1 September 1987) are all the result of water concentrating in the talweg network. They can be analyzed under two different aspects : on one hand, water volume in movement largely determines their force, and on the other hand, rapidity of the hydrological response to pluviometric impulsion. The more or less soaring form of the flood hydrogram, which gives a graphic translation of the phenomenon, depends firstly on the original rainfall, its extensiveness, its volume, its intensity and its progress. But sudden increase in debit is also the result of morphological variables, catchment area characteristics, whose relative influence seems to be increased when the temporal scale is reduced.

1. Analysis of the Pluviometric Phenomenon

In both cases, recorded precipitation reached avout 40 mm corresponding to maximal local daily rainfall of 10-year frequency. M. Dacharry (1) cites the strongest daily rainfall occuring since the start of the measurements in 1962, for the Orgeval catchment area, 49.2 mm (on 14 July 1980), the reoccurrence of this kind of rainfall being re-estimated at 60 years' interval. On the Petit Morin, the 1 July 1987 flood was linked to precipitations of 75-85 mm occuring over a brief lapse of time (one hour). Such rain is considered to be a centennial value. It was marked by a particular intensity ; at Saint Germain sous Doue, close to the sector affected by the flood, 226 mm/hour during 4 minutes was recorded.

Floods affecting the catchment areas not evenly covered by vegetation allowed the threshold values for runoff and water concentration to be confirmed in the existing talweg network and justifying damage in the lower watershed. The 40 mm threshold seems to be a constant, beyond that, the flood is much more intense. Gross totals of rain fall on the Petit Morin had very high runoff coefficients. "Pulsation of gullies (brooklets) closely imitated the pulsations of water falls, on the steepest slopes, 90 % runoff was recorded". The Arbre Robert catchment area (Erlon) though well-protected by vegetation, underwent heavy runoff in the talweg network and overflow in the lower section, the water flowed over the road before the perennial exurgence. At the time of the flood, the catchment area's upper portion is covered with pea crops, covering the soil quite completely, the lower part winter wheat about 50 cm high covered the soil. A strip of beets about 150-200 wide, just planted (early April) was only 5 % covered. In this area, sheet runoff occurred and wheeltracks collected and evacuated water.

Some of the water and mobilized material was lost in the wheat and was not taken into account in the measurement taken at the observation station. Nevertheless, the talweg network evacuated 10 % of the rainfall out of the catchment area. The start of concentrated runoff in the talwegs occurred 10 minutes after the rain began to fall. For about 10 % of the barely covered surface, 4 t. of carried material were recorded in the talwegs at the bottom of the catchment area.

The relationship between the total precipitation, rain intensity and runoff has been the object of probability analyses of which two are commonly used : 1) the relationship between duration and intensity for precipitation thresholds of a certain recurrence period 2) the relationship between maximal rain intensity and the area of the storm's extent. We can thus define the probabilities of rain maxima in function of time and space, and estimate hydrological yield (in terms of total debit volume) in the same catchment area.

2. Accouting for the catchment area

Nevertheless a complimentary approach is needed to compare catchment areas and later, to do a cartography of the high-risk zones. (2). We need to pinpoint the specificity of the rain/debit relationship in different catchment areas undergoing the same inital extensive rainfall, by integrating the physiographical characteristics of these catchment areas. This will give the speed of flood concentration, which can be modified by a third group of variables linked to soil occupation. In other words, geomorphological parameters resulting from the morphostructural and morpho-dynamic history of the region are indispensable for explaining runoff. The preliminary step is therefore establishing a morphological and morphopedological map of the catchment area. Among the different criteria linked to runoff rapidity, a few morphometric aspects are of particular interest.

Low slope topographies (< 3 %) directly contribute in favoring concentration as soon as the slope increases and talwegs form. Catchments which served as reference for our studies, were characterized by a high percentage of prominent plateaus - 74.5 % for the Fonderie catchment, 72.5 at Charnesseuil, 80 % for la Vorpillière and 80 % at Vierzy (Fig. 2).

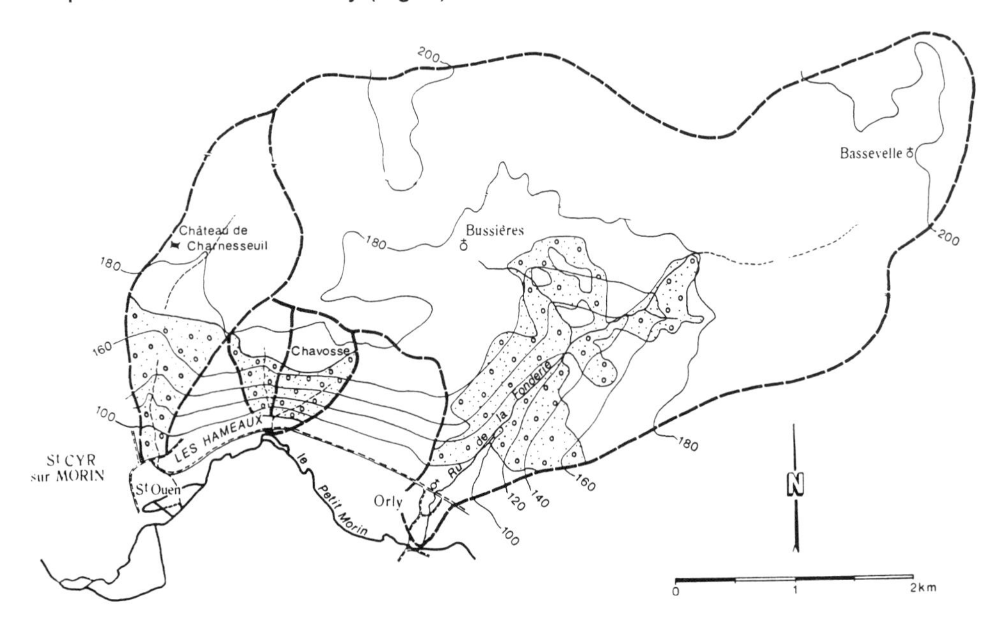

Figure 2 : Catchments map. Fonderie, Chavosse, Charnesseuil catchments from (M. Dacharry, 1988) (Dotted line : Woody zone).

The importance of these surfaces for catchment area functionning has become quite clear. Beyond the reserve threshold for plateaux, the proportion of runoff rain increases with the size of the storm following a non-linear relationship : concentration time depends upon this factor.

The shape of the catchment area also plays a role in favoring rapid or slower flow. The most compact catchment areas are characterized by a lower concentration than that on longer catchments, but the concentration and flow peak is more marked in the first case. The consequences is more intense flooding.

Density of drains and the design of their networks is quite important. "Class drains" (Strahler), very numerous, allow for rapid concentration by multiplying slopes, can favor development of gullies. At Vierzy, field work and analysis of geological and topographical maps has shown the permanent nature of the talweg network, formed by colluvial accumulation. Analysis of this colluvium of considerable thickness, often a few meters thick, shows, "inverse sedimentation", the result of stripping and transportation of original material from the plateau surface. A single drain , Class 2 or a hierarchical network of the "Reticule" type have also had different consequences of water transit and transported material. In the second case, there is greater efficiency. Confluence of two different branches of comparable importance brings maximal supply. Generally speaking, augmentation of confluences is a factor of heightened runoff. M. Dacharry showed that the Fonderie gully in a 14.75 sq. km catchment area had considerable efficiency during the rain episode of 1 September 1987, notably because of the convergence of two branches of equivalent size, before the shrinking of the talweg downward. In the case of the single drain, the exported quantity is more spread out (cf. the Charnasseuil gully), flowed water is absorbed.

Slope and profile are other factors to be taken into account. On the one hand, drop has to be considered for it characterizes the entire catchment area and justifies flow differing from that occurring in less dislevelled catchment areas. Role of length profile in the principal drain, was discussed, runoff and erosion should increase when the slope increases and when the length increase. Such conclusions have been discussed and challenged by certain authors and the best approach is to qualify these conclusions. We can, however, remark the role of slightly incised talwegs whose profile offers a convex-concave trace with the result of increasing water speed. The size and the number of such drains should be taken into account in the catchment areas.

Finally, examining the state of the catchment area at a given moment is necessary for understanding its functionning during heavy rains. The notion of dynamic drainage is fundamental, notably in the case of perennial drains (Vorpillière gully) implying that the hydrological state must be taken into account. In episodically functionning catchment areas, it is also useful to examine functionning of drains by integrating data on soil state and vegetal cover whose intercepting capacities are modified during rainfall. This approach allows different scenarios of risk appreciation to be established.

CONCLUSION

Analysis of runoff manifestations needs to take into account phenomena on different scales within the catchment area : gullies, elementary talwegs, and complex talweg networks. Relationships between these three elements are numerous. Gullies can feed into an elementary talweg and the entire network, or on the contrary, can be isolated. "Sheet " discharge and hypodermic flow differred in time also have a complex impact which should not be ignored.

The size and speed of floods, and associated risks can be linked, as are rain-related risks, to catchment basin characteristics. The Soissonnais Petit Morin Erlon catchment areas are at high risk potential due to the increase of technical equipment and utilization of the land.

Adding up local parameters and discontinuous spatial and temporal phenomena does not yield any real synthesis for each catchment area unit, because transformation under rainfall comes from a combination of various elements, as well as the relationships established between these elements. Calculating global results at the exsurgence of the catchment area is the best way of integrating these factors.

REFERENCES

1. Dacharry , M. (1988) : Averse et crue du 1er septembre 1987 en Brie (Bassin du Petit Morin) Hydro Continentale. ORSTOM, vol. 3, n° 1, p. 3-17.
2. Veyret, Y., Wichereck, S. (1992) : Réflexions sur une cartographie des risques d'érosion des terres agricoles en région de plaines et de plateaux, BAGF (2), p. 168-176.
3. Boardman, J. (1990) : Soil erosion on the South Down. A review in Boardman J. Dearint, J.A. and Foster, I.D.L., ed. Soil erosion on agricultural Land Chichester, J., Willey and Sons, Ltd, 87, 107.
4. Papy, F., Douyer, C. (1991) : Influence des états de surface du terroir agricole sur le déclenchement des inondations catastrophiques. Agronomie, p. 201-215.
5. Cosandey, Cl. (1990) : L'origine des crues dans les bassins versants élémentaires. Du ruissellement sur les versants à l'écoulement de crue. Annales de Géographie, t. 556, p. 641-659.
6. Morand, F. and Wicherek, S. (1987) - Douze parcelles de mesure d'érosion sur un versant de la France des Plaines : l'exemple de Cessières (1977-1983). (Twelve plots for erosion measurement of a slope in a lowland area (France). The example of Cessieres (1977-1983)). In Processus et mesure de l'érosion, Ed. du CNRS, 271-290.
7. Govers, G. (1991) : Rill erosion on arable land in Central Belgium rates, controls and predictability. Catena, vol. 18, p. 133-155.
8. Colborne, G.J. and Staines, S.J. (1985) : Soil erosion in South Somerset. Journal Agricultural Science. Cambridge, 104, p. 107-112.
9. Evans, R. (1990) : Water erosion in British farmer's fields some causes, impacts and predictions. Progress in physical geography n° 14, p. 199-219.
10. Wicherek, S. (1989) - Impact of splash and eolian transport on erosion ; case study : North of Parisian basin, France. (Rôle du splash et du transport éolien sur l'érosion. Terrain d'étude : Nord du Bassin Parisien, France). Proceedings of 4th Benelux Colloquium on geomorphological processes and soils, (I.G.U.-COMTAG), 24 April-2 May 1988, Amsterdam *Earth Surface Processes and Landforms*, v. 14, 6/7, 461-467.
11. Gregory, K.J. and Walling, D.E. (1973) : Drainage basin form and process. A geomorphological approach. Arnold Ed., 449 p.

Farm Land Erosion: In Temperate Plains Environment and Hills
S. Wicherek (Editor)
1993 Elsevier Science Publishers B.V.

Soil erosion under irrigation in Bulgaria

G. Dochev, M. Neshkova, R. Rafailov, V. Krasteva

N. Poushkarov Research and Development Institute of
soil Science and Agro-ecology, Sofia, Bulgaria.

SUMMARY

Irrigation of farm lands in Bulgaria is the cause of erosion affecting 50 % of the soil in North Bulgaria, and 70 % in the South. These phenomena have been studied since the 1960's, in three different sectors and under corn and soy crops. The effets of irrigation by gravity and by aspersion have been examined.

RESUME

L'irrigation des terres cultivées en Bulgarie est à l'origine d'une érosion qui affecte 50 % des sols de Bulgarie du Nord et 70 % de ceux du Sud du pays. Ces phénomènes ont été étudiés depuis les années 60, dans trois secteurs différents et sous cultures du maïs et du soja. Les effets de l'irrigation par gravité et de l'irrigation par aspersion ont été examinés.

Irrigational erosion means a series of soil destruction processes under the influence of irrigation water and the combined action of pouring rains over the irrigated areas. Most of the arable lands in Bulgaria are potentially endangered by erosion techniques. In the Northern part of the state these lands amount to 50 % of the irrigated areas and in the Southern part to about 70 %. The investigations of water-erosion processes caused by atmospheric rains date quite far back, while those connected with erosion degradation of soil under irrigation were carried out during the last 30 years. Soil erosion under irrigation has been investigated with the help of two basic methods of irrigation - gravitational and spray irrigation. Mass trials of gravitational irrigation were conducted in different areas during the period 1960-1981. The objectives of this investigation can be devided into two main groups : A. Objectives examined for determining the degree of soil erosion under gravitational irrigation in different geographical regions - . forest soils, typical chernozem, calcareus chernozem, alluvial meadow soil ; B. Objectives examined for determining the character and the behaviour of soil erosion under gravitational irrigation of the main types of soil.

Different agricultural crops such as tobacco, grain corn, sunflower, paprika and raspberries were used in the investigation depending on the region.

Identical methodology was used for all investigations. Furrow length (L) varied from 20 to 350 m, water flow (Q) from 9 to 108 l/min/furrow, slope of the terrain (I) from 2 to 18 %, wide scale of times of irrigation (T) from 1 min to 8 hours and hence wide range of realized single irrigation rates (SIR) (M). These indexes have different values in accordance with the aim of the investigation. General scheme of integrated methods for determining the character and the behaviour of soil erosion under irrigation of the main soil types in the state, requiring furrow inclination (notwithstanding the slope of the terrain) I up to 3,5 %, L from 20 to 120 m, Q from 9 to 90 l/min, T = TL and tilling of soil surface was used. Different kind of furrows were successively investigated.

The degree of erosion was studied at the end of irrigated furrows of different length. In some plots observations were carried out on 2 to 4 places along the length of the furrows at intervals of 10 to 150 meters.

Table 1.
Principle erosion indexes for objectives of a group.

Objective	Variant by:		Furrow length	Irriga-tion water-jet	SIR	Liquid runoff in % of SIR	Solid runoff
			m	l/min	m^3/dca	%	kg/dca
Cinnamonic	Irrigation	1.	30	20	16,2	5,2	22
forest soil	water-jet	2.	30	55	28,5	33,2	818
tobacco, I_{furrow} 0,6%							
Typical	Furrow	1.	250	48	111,0	31,8	2230
chernozem	length	2.	350	53	110.0	20,0	2660
corn, $I_{fur.}$ 2%							
Alluvial	Time of	T_L	6	18	5,0	71	27
meadow soil	irrigation	$T_{nec.}$	6	18	228,0	38	727
tobacco,		T_L	16	18	7,2	22	12
$I_{fur.}$ 2%		$T_{nec.}$	16	18	86,0	22	121

$T_{nec.} = T_{necessary}$

Some principle results of the investigations in A group are given in Tab. 1.

This value of eroded soil was determined as soil loss tolerance and the water-flows that determine it were specified as critical in this respect (Qcr). The meaning of Qcr could not be isolated from that of the other elements of the irrigation technique. Qcr could be determined as the water-jet, applied with gravitational irrigation and which, if omitted, at the beginning of the irrigation furrow, causes erosion at the end of the irrigated area at the rate of 20-25 Kg/dca with only one watering. It is valid with the use of furrows of middle length for the region and for the irrigated crop with slope (notwithstanding the slope of the terrain) from 0,5 to 3,5 %, duration of the irrigation T = TL and tilling of soil surface.

Investigations of the objectives of B group show that on Vertisoils Qcr is in the range of 25-90 l/min, on North-Bulgarian chernozems from 12 to 54 l/min, on Light grey forest soils from 12 to 36 l/min. Although these extreme values generally show the susceptibility of soils to erosion depending on genetic types, their wide variation does not give any possibility for applying the genetic principle of classification. For that reason the symptomatics of different soil characteristics was adopted as erosion criterion : aggregate and mechanical composition, nutrient and colloidal content. On this basis the dependances for Qcr were worked out in view of their utilization in practice when determining the erosion-safe water flows. These dependances are from first degree of the type $Y = ax + b$:

$$Y = 1{,}164\, x^1 - 22{,}2 \qquad (1)$$
$$Y = 9{,}545\, x^2 + 11{,}14 \qquad (2)$$
$$Y = 9{,}667\, x^3 + 9{,}00 \qquad (3)$$

where Y is the erosion behaviour of soils, expressed by Qcr/l/min, x^1 is the physical clay content in the plough layer in percents, x^2 is the humus content of the plough layer in percents and x^3 is the colloidal content (by Robinson) percentage. The average deviations from the straight line (in percentages) are 30, 32 and 23 % respectively. The equation (1) is valid for the values of the independent variable $x^1>30$.

For the purpose of forecasting the degree of water runoff the dependances of the percentage between the actual and the maximum water permeability of soil during irrigation and of the water flow head velocity along the furrow bed were found. These dependances are of first degree too, consequently they could be estimated by the formula :

$$Y = 0{,}228\, x^1 - 7{,}820 \qquad (4)$$
$$Y = 3{,}760\, x^2 - 0{,}906 \qquad (5)$$

where Y is the water flow percentage compared to the quantity of applied water, x^1 = Wactual/Wmax, expressed as a percentage, x^2 is the water flow velocity in m/min. Deviations from the straight lines are 30 and 15 % respectively.

Continuous investigations and the results obtained furnished the basis for developping a complete research project on soil erosion due to gravitational irrigation for the whole state. A detailed inventory of 1,150,000 ha of irrigated land was made using 20,073 soil profiles of 862 irrigated plots, selected from 24 different irrigated soil types. The weighed averages for Qcr and the coefficients of variation for every profile were determined. Outlining of areas with close values of Qcr outlined the areas with similar erosion resistance under irrigation too. An operating classification was worked out capable of assessing this erosion resistance from very low (class 1) to very high (class 7) with the help of specified interval of 10-15 l/min Qcr. The data obtained were summarized for the whole state and for the different regions on a 1 : 1,000,000 scale. The planimetring of the aforested material showed that 45 % of the irrigated land has very low, 40 % mean, 14 % good and only 1 % very high erosion resistance. There are three zones of distribution of the soil most susceptible to irrigational erosion : the extreme Southern and South-western parts of the state, the Southern slopes of Stara planina and the Danube river valley.

In addition to the above-mentioned basic investigations of erosion under gravitational irrigation, many further analyses, contributing to the more profound study of the problem, were carried out.

A series of problems like that of solid runoff distribution along the furrow length, water runoff selectivity, the effect of soil tilling on the degree of erosion, the movement of some nutrients through the surface runoff with furrow irrigation, the effect of the polymer Separan 2610 on erosion resistance of one soil type, the effectivity of piercing underground tunnels in gravitationally irrigated soil for erosion reduction, were investigated.

To determine the solid runoff quantity (S in gr/m) as a function of the average quantity of flowing water along the furrow length (q) and the time of soil wetting, the average slope of the terrain (I) and the length of the irrigated area (L) to following equation was developed :

$$S = q(I)^{1/3} = (I).e\ 0,57/0,0108.L \qquad (6)$$

where n is an experimentally determined index.

The above dependence was worked out for the conditions of gravitational irrigation of Leached meadow soils.

The investigation of erosion under spray irrigation was started in the early 1980's and have continued up to the present time. Observations were made on three types of soil - Calcareus chernozem, Leached chernozem, light grey forest soil with two staples for Bulgaria - corn and soybeans.

Spray irrigation was conducted using installations widely spread in our practice. Spray installation Maritza-45, completed with sprinklers Sila-30, was used for Grey forest soils. Corn planted on Leached chernozem and Calcareus chernozem soils was irrigated using one PDI-20M installation, completed with sprinklers Sila-30. and soybean, planted on Calcareus chernozem soil was irrigated using PDI-20K, completed with sprinklers P-50C. For the purpose of the experiment special irrigation wings depending on the parameters of each technique used were developped. Erosion indexes were studied on runoff plots of average area 30 sq.m. The liquid and solid runoff were studied in dynamics and in total. The main improvement parameters used in the experiment are given in Table 2.

Two experimental studies were conducted in the conditions of calcareus chernozem. The first one was aimed at determining the anti-erosion effectivity of runoff-retaining furrows when spraying corn in two variants. Variant 1 is conducted without the above measures and Variant 2 with them. The second study was aimed at determining the most satisfactory method of applying SIR in order to achieve minimum soil degradation. For that purpose three methods were studied m^1 - application of the whole SIR at one time (single), m^2 - application in two portions (twofold) and m^3 - application in three portions (threefold).

The effect of artificial rain intensity and the slope of the terrain on the erosion processes development under spray irrigation of corn was studied in the conditions of Leached chernozem soil. The intensity was formed in three intervals : i^1 - from 0,10 to 0,21 mm/min, i^2 - from 0,22 to 0,28 mm/min and i^3 - from 0,28 to 0,43 mm/min. Two groups of terrain inclination were used : group I^1 - from 0° to 3° and group I^2 - from 3° to 6°. A similar experiment was conducted with spray irrigation of corn on Grey forest soil in which the effect of tilling direction on soil loss reduction under spray irrigation of soybean was studied. In Variant 1 tilling was directed down the slope and in Variant 2 it was directed across it.

	Spray irrigation of corn				Spray irrigation of soybean			
	Variant	SIR	Rain intensity	Slope	Variant	SIR	Rain intensity	Slope
		m^3/dca	mm/min	degrees(°)		m^3/dca	mm/min	degrees(°)
Calcareus chernozem	a.1.	40	0,26	12	a. M_1	60	0,18	7
	b.2.	40	0,26	12	b. M_2	2x30	0,18	7
					c. M_3	3x20	0,18	7
Leached chernozem	a.I_1i_1	45	0,10 to 0,28	0,04-1,3				
	b.I_1i_3	55	0,33 to 0,43	0,4-1,3				
	c.I_2i_1	46	0,10 to 0,28	2,8-5,8				
	d.I_2i_3	56	0,33 to 0,43	2,8-5,8				
Light grey	a.i_1	53	0,16 to 0,21	1-2	a.1.i_1	48	0,12	1-2
	b.i_2	54	0,22 to 0,27	1-2	b.1.i_3	49	0,30	1-2
	c.i_3	56	0,28 to 0,31	1-2	c.2.i_1	48	0,12	1-2
					d.2.i_3	49	0,30	1-2

Table 2. Principle meliorative parameters of the spray irrigation trials.

TABLE 3. Water and soil losses under spray irrigation

Soil type	Investigated period	Variant	Spray irrigation of corn		Spray irrigation of soybean	
			Liquid runoff m^3/dca	Solid runoff kg/dca	Liquid runoff m^3/dca	Solid runoff kg/dca
Calcareus chernozem	1984-1988	a.	162,01	146,20		
		b.	58.37	73.20		
	1988-1990	a.			132.79	322.73
		b.			96.86	197.32
		c.			59,53	169.16
Leached chernozem	1986-1988	a.	0.5	1,6		
		b.	3,0	17,3		
		c.	6,1	106.6		
		d.	7,8	139.1		
Light grey forest soil	1984-1985	a.	1,00	0,05		
		b.	4,47	4,08		
		c.	24,77	50,52		
	1982-1983	a.			27.85	15,00
		b.			14,81	50,30
		c.			-	-
		d.			3,09	4,70

In table 3 the main erosion indexes (solid and liquid runoff) by variants are given.

For the assessment of erosion caused by spray irrigation we accepted the allowable losses adopted by V. Krasteva, i.e. 5% of the applied SIR for water losses and 25 Kg/dca from one single watering for soil losses. We found out that it is possible to irrigate by spraying Leached chernozem soil with very gentle slope of the terrain (I^1),with spraying intensity being 43 mm/min without any danger of erosion. Water losses amount to 5 % of SIR and soil losses to 17,3 Kg/dca. But with slope of 3-6° even less intensive spraying of 0,10-0,28 mm/min results in 13 % water losses and 4 times higher than the adopted allowable soil losses. With steeper slopes the difference between erosion losses caused by the different spraying intensities is not clearly expressed. Water and soil losses increase by about 30 % with the increase of spraying intensity.

The allowable spraying intensity amounts to 0,28 mm/min when spraying corn on gently inclined terrain of Light grey forest soil. Higher intensities of the artificial rain lead to 44 % water losses and twice as high as the allowable soil losses. The effect of the different intensity levels on increasing soil loss on this soil type is more clearly expressed. But under the same soil type conditions and planted by the standard method (along the slope) soybean water losses significantly exceed the allowable ones with soil losses being the same as for corn. The change of planting direction (across the slope) has an obvious effect on erosion loss reduction as they actually get eliminated.

When spraying corn on Calcareus chernozem soil with 12° slope, liquid runoff of the variant with presence of runoff-retaining furrows is reduced 2,8 times and the solid runoff about twice. But when spraying soybean, the application of SIR in two portions reduces the liquide runoff 1,4 times and the solid runoff 1,6 times while with application in three portions these reductions are 2,2 and 1,9 compared to the variant with one application, respectively. When the experiment was conducted on Calcareus chernozem, the utilization of both antierosion measures did not lead to reduction of soil losses in the limits of the adopted soil loss tolerance because of the steep slope of the terrain.

The assessment of erosion indexes for the three soil types and two crops studied imposes the initial conclusion that when one and the same crop is being irrigated by spraying, the soil type and the slope of the terrain play a significant role in the extent of soil erosion losses. While in using irrigation furrows, their slope is liable to shaping but when spraying installations are disposed in natural relief conditions, the effect of the slope on the erosion development is a decisive factor.

Farm Land Erosion: In Temperate Plains Environment and Hills
S. Wicherek (Editor)
1993 Elsevier Science Publishers B.V.

Long-term effects of erosion on soil degradation

Ph. Duchaufour

8, allée de la Doucerie - 78620 L'Etang la Ville, France.

Most of the papers delivered at this important symposium on soil erosion concerned short-term effects and their immediate incidence on agriculture. We should, however, remember that erosion exerces a long term effect leading to degradation - sometimes even complete disappearence of the soil. This short article will highlight certain of these degradation processes, choosing among the most serious (which are often irreversible) : these processes have been observed and described by the author or certain of his colleagues following field excursions or local research in different European countries and in Northern Africa. The most spectacular examples concern forest lands. A forest in good condition, well-developed (tree clusters) protects its soil, whereas clearing of timber has a quite brutal effect, provoking erosion and degradation of the forest soil, and dangerously threatening fertility.

We should stress that on forest land, nutrients tend to concentrate at the surface in the humus horizons, through the *bio-geochemical cycle* : tall trees absorb in-depth nutrients thus bringing them up to the surface, within litter that makes up the primary matter of humus. The efficiency of this process is evident, when soil is acid, and formed on poor material, the contrast between the humus horizon and mineral horizons becomes quite great. Rapid elimination of humus by the erosion process weakens soil, stripping it and leading to acidification and eventually modifying forest flora.

A characteristic example is furnished by the Kroumiria (Tunisia) oak forest. This example, studied by Selmi (PhD dissertation, Nancy(1) is situated on a mountain; its soil is made up of weathered material from highly acidic sandstone. The Chernozem humus horizon , 20 cm in thickness, has a pH level of 6.5; the Calcium/Aluminium ratio (exchangeable) is 50, which eliminates aluminium toxicity. This horizon contains almost all the nitrogen reserves or 6,000 kg per ha. In eroded plots, with the acid mineral horizon totally uncovered, the pH level is only 4.5, and the Ca/Al ratio falls to 0.30, engendering a high level of toxic aluminium ; at this stage, the nitrogen reserve disappears. Rich neutrophilous flora is replaced by a poorer flora, made up of acidophilous species.

Three types of degradation will be discussed in this article : 1) irreversible elimination of certain paleo-soils 2) pedological degradation observed in temperate climates 3) pedological degradation observed in arid climates.

1. IRREVERSIBLE ELIMINATION OF PALEO-SOILS

The original matter of certain soil is often *paleo-soil*, with formation dating from 100,000 years back : when such a paleo-soil is eliminated by erosion, it can not be reconstituted by human effort. Two examples will be used, from the Mediterranean region, the first concerns decarbonated hard clay (the "*terra rossa*", the result of very slow accumulation of silica residue, by pellicular weathering of the corroded limestone surface ; the second example concerns ancient river terraces with pebbles and fine earth intermixed.

In the original Mediterranean forest, which contained mostly green oaks, the layer of *terra rossa* was usually thick enough to allow root development. Unfortunately this forest was almost completely destroyed, and only rare markers with intact profiles (and humus surface horizons) remain, like the one observed in Spain at Bailen. Disappearance of the green oak forest was accompanied by massive erosion of the *terra rossa*, causing elimination of first the humus horizons, then of all mineral horizons, with subsequent formations made up of xerophilous shrubs like the Kermes oak (*Quercus coccifera*), - creating the "*garrigue*", or scrubland, in place of the forest : this can be observed in all the Mediterranean countries - including southern France, Italy, Greece, and Marocco (2).

As for the ancient terraces, the most interesting examples are those of Durance (Costière and Crau in Provence). Under well-irrigated and well-maintained prairies, the soil has been relatively well-preserved, and retains a high percentage of fine earth rich in humus and clay. However, under certain vineyards, or in orchards with wide spacing between trees, fine earth has been eliminated by selective erosion, and in place of fine earth are pebbles, so that the vines have to seek necessary nutrients from deep within the soil, instead of its surface.

2. SOIL DEGRADATION BY PEDO-GENETIC CHANGES IN TEMPERATE CLIMATE

This type of degradation can have various aspects, depending on the granulometric composition of original material (clay or loam), or depending on the structural heterogeneity between surface or deeper horizons of the soil.

1) Selective erosion seen on clayey material

Lysimetrical experiments, using a catchment area covered by peduncle oaks on Triassic Keuper marl, were carried out in 1974 and 1975 by the CNRS Pedological Center at Nancy (3). Original soil contained about 50 % partially swelling clay. Selective erosion had affected surface horizons causing lateral removal of 400 kg/ha/year of clay because selective erosion occurred during violent winter rainstorms. Certain comparisons made concerning similar material have allowed the stages of this degradation to be reconstructed in the following way :

1) "Darkening" of the soil. Only part of the clay is eliminated, forest humus remains, the structure is preserved, soil ventilation is even improved at the surface ; at this stage, no real degradation has taken place.
2) Oxydation-reduction, caused by settling and disappearance of large pores. The soil becomes asphyxiating, and humus deteriorates, with the soil turning greyish-green in color, covered with rust-colored spots.
3) Planisolization, characterized by the near-total disappearance of clay, of humus and of free iron, leaving only bleached, loamy-sandy quartz skeletal forms; deeper horizons remain unchanged and a brutal contrast in texture and structure between horizons A and B can be observed (4).

2) Formation of a perched water-table with oxydation-reduction on loessic loam (brown leached soil)

This type of degradation has been studied by Becker (5) in the Charmes forest (peduncle oak) in the Lorraine area. The initial soil, well-preserved under the forest, is a *brown leached soil*, with its illuvial B horizon situated at 50 cm in depth, is still not strongly marked by hydromorphy. The entire upper part of the profile has darkened, turning brown, the humus being a well-structured and well-ventilated "mesotrophic mull".

Within large clearings in the forest, hydromorphic degradation can be seen : it is the result of three simultaneous processes : 1) an increase in leaching of the clay causes sealing off, or waterproofing of the Illuvial horizon, 2) stripping of humus horizons by erosion, and the B Illuvial horizon moving closer to the surface, 3) destruction of the over-all structure, so there is a resulting decrease in porosity.

These three processes contribute to the formation of a perched winter water-table, now attaining the surface and causing asphyxiation ; the soil is then invaded by a more resistant vegetation (*Molinia coerulea*), which creates a new acidic humus tending to decompose more slowly ; the soil is transformed into *pseudo-gley*, inhibiting the natural regeneration of the oak trees. A recent study by Beyer, (6) in Holstein, has confirmed Becker's observations.

3) Soil degradation with a deep structure favoring damage

Various examples of this type of degradation have been observed : they all concern mobile and porous surface horizons, lying on deeper, compact and even hardened horizons, laterally orienting liquids, and favoring erosion of mobile surface horizons. Leached Fragipan soils, soils cultivated with "tillage soles" and lastly, ironpan hardened podzols can be cited here. The case of hydromorphic ironpan podzols was studied by RIGHI (7) : a very hard and thick ironpan, high in iron content, formed in zones where water had surfaced. The water had flowed very slowly, carrying dissolved ferrous iron, which progressively incrusted the ironpan when the water table surfaced : only a thin sandy horizon remained, taken over by dry heather moors (*Erica cinerea, Calluna vulgaris*). Erosion then rapidly eliminated the horizon, thus baring the hardened ironpan: all vegetation disappeared.

3. SOIL DEGRADATION IN ARID OR SEMI-ARID REGIONS

In this type of climate, two processes have been observed and described : 1) Superficial hardening of limestone incrustations, and 2) Salinization of surface horizons.

1) Hardening of limestone incrustations

This phenomenon was observed in the semi-arid zones of different Mediterranean countries, and in particular, in southern Spain and northern Africa (cf. research by Müller (8) in Algeria).

In these regions, colluvium from the bottom of slopes, on limestone outcroppings is covered with green oak forests or protective shrub formations - like *Oleo-lentiscetum* (*Olea and Pistacia leutiscus*). "Brown" soil is characterized by a mobile dark red humus horizon, reposing on a sandy chalk incrustation, which has not yet hardened, and is often made up of discontinuous ironpan with roots growing between them. When this formation has been destroyed - often by goats - the fine earth of the upper horizons is eventually eliminated. Furthermore, the incrustation surface undergoes alternate solution and cristallization, causing hardening, characterized by formation of a *zonal crust* : initial lignitiferous vegetation will then be replaced by a barren steppe made up of xerophilous graminae (ex.*Stipa*) and by sagebrush (*Artemisia herba-alba.*).

2) Salinization caused by erosion

This process is characteristic of the dry plains of Central Europe (Hungary, Rumania) and of Eastern Europe (Caspian plain) : it has been studied by Kovda and Szabolcs (9). The process is characteristic of deep soils with saline fossile deposits (or with salty levels - the "nappes salées"). Kovda and Szabolcs have determined a "*critical depth*" threshold, for saline deposits (or these "nappes salées") affecting the surface : this is the principal effect of erosion, which in carrying off surface horizons, brings the saline layer closer to the surface. Irrigation of the soil, without any simultaneous drainage, considerably worsens this phenomenon. If the salt is sodium (Na Cl, Na_2 SO_4), it will come up to the surface by capillary ascent and will cause exchangeable ion calcium to be replaced by ion sodium, which destroys the structure and makes the soil even more sensitive to erosion. Little by little, the whitish sterile saline crust moves up to the surface and all vegetation disappears, except for certain halophytic species (salsolacae).

CONCLUSION

Long-term effects of erosion are often ignored by farmers, because they are not immediately noticeable : these effects do, however, seriously threaten the future of the land. Only a few examples have been given in this short article, but we should consider that the entire planet is concerned. The following remark by Commander Cousteau is thus particularly appropriate : "We have a most urgent duty to bequeath a habitable world to our grandchildren ".

REFERENCES

1. Selmi, M. (1985) : "Différenciation des sols et fonctionnements des écosystèmes forestiers sur grès numidien de Kroumiria (Tunisia)". Thèse Nancy, 198 p.
2. Lemee, G. (1978) : Précis d'écologie végétale Masson édit. Paris, 285 p.
3. Nguyen-Kha and Paquet, H. (1975) : "Mécanisme de redistribution des minéraux argileux dans les pélosols", Bull. Sc. Geol. 28, (1), 15-28.
4. Baize, D. (1983) : "Les planosols de la Champagne humide", INRA (PhD dissertation U. of Nancy, 285 p.).
5. Becker, M. (1989) : "Relations sol-végétation en condition d'hydromorphie dans une forêt de la plaine lorraine" (PhD dissertation, U. of Nancy), 225 p.).
6. Beyer, L. (1989) : "Nutzungseinfluss auf die Stoffdynamik Schleswig Holstein Boden". Inst. f. Bodenkunde Kiel, 197 p.
7. Righi, D. and Chauvel, A. (1987) : "Podzols et Podzolisation", INRA, 228 p.
8. Muller, S. (1954) : "Beobachtungen an rezenten Kalkrindenboden im nordlichen Algerien", Zeitschr. f. Pflanzenernähr. 65, 107-117.
9. Kovda, V.A. and Szabolcs, I. (1979) : "Modelling of soil salinization and alkalinization", Agrokemia es Talajtan, 28, 208 p.
10. Duchaufour, Ph. (1991) : Abrégé de pédologie. Sol-végétation-environnement. Masson, édit. Paris, 289 p.

Farm Land Erosion: In Temperate Plains Environment and Hills
S. Wicherek (Editor)

Extent, Frequency and Rates of Rilling of Arable Land in Localities in England and Wales

R. Evans

Department of Geography, University of Cambridge, Downing Place, Cambridge CB2 3EN, United Kingdom

SUMMARY

Water erosion needs to be monitored by a field-based scheme. Such a scheme is described. Seventeen localities in England and Wales were monitored between 1982-1986. The extent, frequency and rates of erosion of the 24 landscapes (soil associations) occurring in these localities are given. Where erosion and its impacts are known to be a problem, on average more than 5 % of the arable land was affected by erosion per year. The maximum area affected by rilling per year was in Nottinghamshire where 13.9 % of the arable land eroded. Rilling was most widespread on sandy and coarse loamy land. Mostly, less than 2% of arable land eroded per year. Fields rilled on average in most localities at frequencies of 1- 3 years, and mean rates per field were 3-<5 m^3 ha^{-1} yr^{-1} in a small number of localities but were generally less than 1.5 m^3 ha^{-1} yr^{-1}. Rates of erosion were highest where soils were 'silty' or coarse loams. Over the landscape as a whole soil transported exceeded 0.1 m^3 ha^{-1} yr^{-1} in six localities which had coarse loamy, sandy or 'silty' soils, and was less than 0.01 m^3 ha^{-1} yr^{-1} in six landscapes. A field-based study such as this explains why farmers do not consider erosion much of a problem, and shows that where rilling occurred in the landscape is closely related to soil type. Erosion in valley floors was more common on heavier textured soils, on lighter soils rills were most common on slopes.

RESUME

L'érosion hydrique a souvent été mesurée sur parcelles, plus rarement sur des champs ou à l'échelle de régions. Sur parcelles soumises à des pluies simulées, les quantités de matériaux érodés ont été mesurées ; cependant il reste difficile d'extrapoler ces résultats à l'ensemble d'une région.

L'érosion hydrique dans 17 sites différents de superficie variant de 30 à 100 km en Angleterre et au Pays de Galles. Ces sites ont été sélectionnés en fonction des types de sol, et des possibilités du déclenchement de l'érosion. Un suivi par photographies aériennes a été réalisé chaque année entre 1982 et 1986. Il a permis de repérer les champs érodés. Le travail de terrain qui a suivi a confirmé les observations précédentes et a conduit à mesurer les volumes de sols érodés lors du ruissellement par rigoles.

Une discussion est présentée qui concerne l'ampleur et la fréquence de l'érosion hydrique dans les champs et dans les régions où les sites choisis se situent.

1. INTRODUCTION

To assess the problem of water erosion, its extent, frequency and rates need to be known, and how these parameters vary between different landscapes. It is important to have this information rather than to apply theory, as is generally done, for example, when using the Universal Soil Loss Equation for assessing erosion. It is not known how the Soil Loss Equation relates to what actually happens in the field. Field assessments, made within large blocks of land, rather than measurements made on plots are needed, therefore. The best way to do this is to monitor individual fields in a landscape throughout the year, as was done by Boardman (1) on the South Downs of Sussex. However, this is time-consuming and to monitor many localities needs much person-power, which funding bodies are not willing to support. Rills and deposits can often be clearly seen on aerial photographs and trial projects in the late 1970s suggested that this was a way of cutting down the time and person-power needed to monitor erosion in fields within large blocks of land.

Rill erosion was monitored in 17 localities in England and Wales between 1982 and 1986. The year refers to the harvest year of the crop. In some of these localities erosion was known to be widespread. The scheme was funded by the Ministry of Agriculture and for four years administered by the then Soil Survey of England and Wales. Rills and deposits were identified on 1:10 000 scale (1:15 000 scale of one locality) aerial photographs taken in spring or early summer and then checked in the field in September to October after the cereal harvest. Areas considered susceptible to rilling but which did not appear to have eroded were also field-checked, and fields in which erosion had occurred after the aerial photographs were taken were located during the field visit. Amounts eroded were simply measured. Rill lengths and areas of deposition were estimated by pacing or from measurements taken from the aerial photographs, and rill cross-sectional areas and depths of deposition were measured by tape. A detailed assessment of three of the five years' data has already been given (2), although it is not easily available because of problems of copyright.

Information was collected to evaluate the magnitude of the problem of water erosion, not to assess its causal factors. The date when rilling occurred often could not be ascertained with certainty. Data on crop cover and rainfall intensities at the time of erosion were not collected. Rainfall also varied within localities so it proved not possible to identify amounts critical for initiating erosion.

Land use and soil type were the main factors controlling the extent and rates of erosion. Within localities, amounts eroded could not be related to slope angle or relief as measured from 1:25 000 topographical maps. However, between localities there were correlations between median slope angle and median relief within fields and volumes eroded (3).

Field-based estimates of rill erosion appear satisfactory. Generally amounts are between half to twice the mean amount estimated by a number of workers to have eroded (4). Splash and wash erosion appear of negligible importance in Britain (3). Where data on extent and frequency of erosion have been collected from similar landscapes with similar rainfalls, again they compare well (4).

2. LOCALITIES MONITORED

The 17 localities were widely scattered, 16 in England, one in Wales (Fig. 1).

Monitored localities

1. Bedfordshire	*10.* Norfolk – Holt
2. Cumbria	*11.* Norfolk – Walsingham
3. Devon	*12.* Nottinghamshire
4. Gwent	*13.* Shropshire
5. Dorset	*14.* Somerset
6. Hampshire	*15.* Staffordshire
7. Herefordshire	*16.* Sussex – Weald
8. Isle of Wight	*17.* Sussex – Downs
9. Kent	

Figure 1. Monitored localities.

The mean area of arable land at risk of erosion in these localities ranged in size from almost 100 km² in Bedfordshire and adjacent counties to just under 30 km² in the Isle of Wight, Sussex Downs, Shropshire and near Holt in Norfolk (Fig. 2).

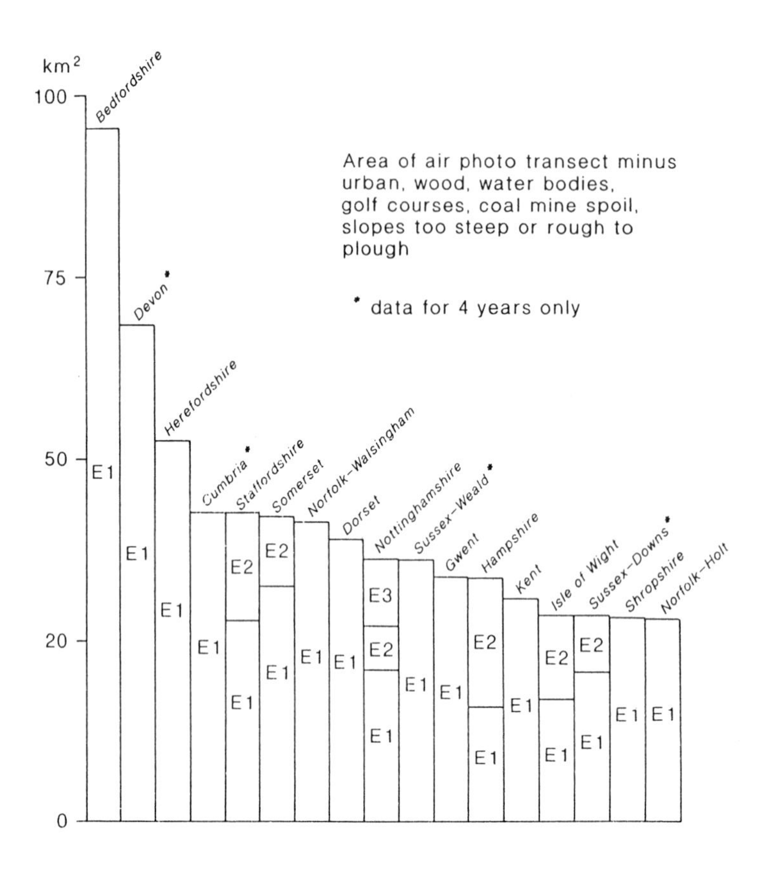

Figure 2. Mean areas of arable land in the monitored transects and landscapes (E1-E3) within transects.

Six of the localities covered more than one type of landscape. Landscapes have a related pattern of relief and soils and in turn these, with climate, control land use. Erodibility varied accordingly. Some localities were not photographed every year. Cumbria, Devon and Sussex Weald were photographed on four occasions, and Sussex Downs on three. However, data was provided for the Sussex Downs for one of these unphotographed years by John Boardman, as the aerial photograph transect fell within the area monitored by him (1).

3. SOILS AND LAND USE OF THE MONITORED LOCALITIES

Two pieces of information are needed to explain much of what follows - (i) the dominant texture of the topsoils in the eroded fields (5), and (ii) the land use of the landscapes. In three localities (Nottinghamshire, Staffordshire, and Shropshire) soils of eroded fields were dominantly sandy (Fig. 3) and in two (Isle of Wight and in Norfolk near Walsingham) dominantly coarse loamy (sandy loams). In Norfolk near Holt many eroded fields had coarse loamy or sandy textures. In Hampshire, Kent, Somerset and the Sussex Downs soils contained much silt, although the textures ranged from coarse silty (silt loam) through fine silty (silty clay loam) to fine loamy (clay loam). However, on the Sussex Downs and in Kent much of the silt was derived from windblown material, whilst in Somerset and Hampshire the silt was derived from weathering of the bedrock and these soils also incorporated much fine sand. The latter soils are best described as 'silty', for the silt fraction was probably mostly comprised of the coarser part (0.02-0.06 mm). In other landscapes (Sussex Weald, Herefordshire) where soils contained high silt, often of the finer silt fractions, the fine silty (silty clay loam) soils generally contained more clay. In Cumbria, Devon and Gwent soils in eroded fields were often fine loamy and mostly clayey in Dorset and Bedfordshire.

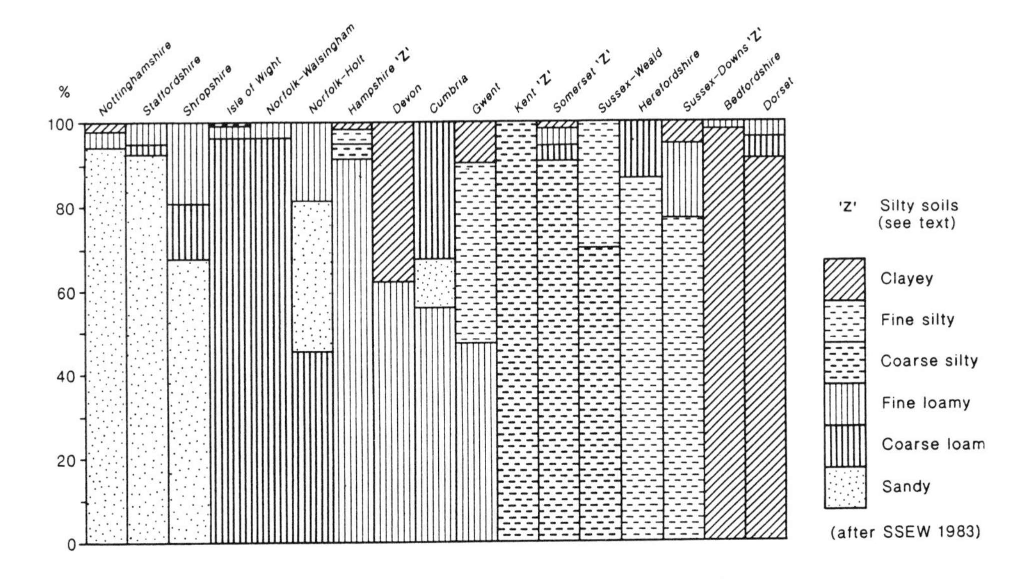

Figure 3. Topsoil textures of eroded fields (after SSEW 1983).

Erosion occurred dominantly in fields sown to winter cereals in 11 localities (Fig. 4), and in 8 of these more than 50 % of all erosion occurred in cereal fields drilled in autumn. Elsewhere, erosion occurred most often in sugar beet fields or in fields sown to vegetables such as lettuce, onions, and parsnips ; these crops, especially in Kent, were often irrigated. Some fields drilled to spring cereals eroded in most (15) localities whilst reseeded grass leys and potato fields commonly eroded in 10 localities. In general, the widest range of crops was grown on the more easily worked sandy, especially, and 'silty' soils.

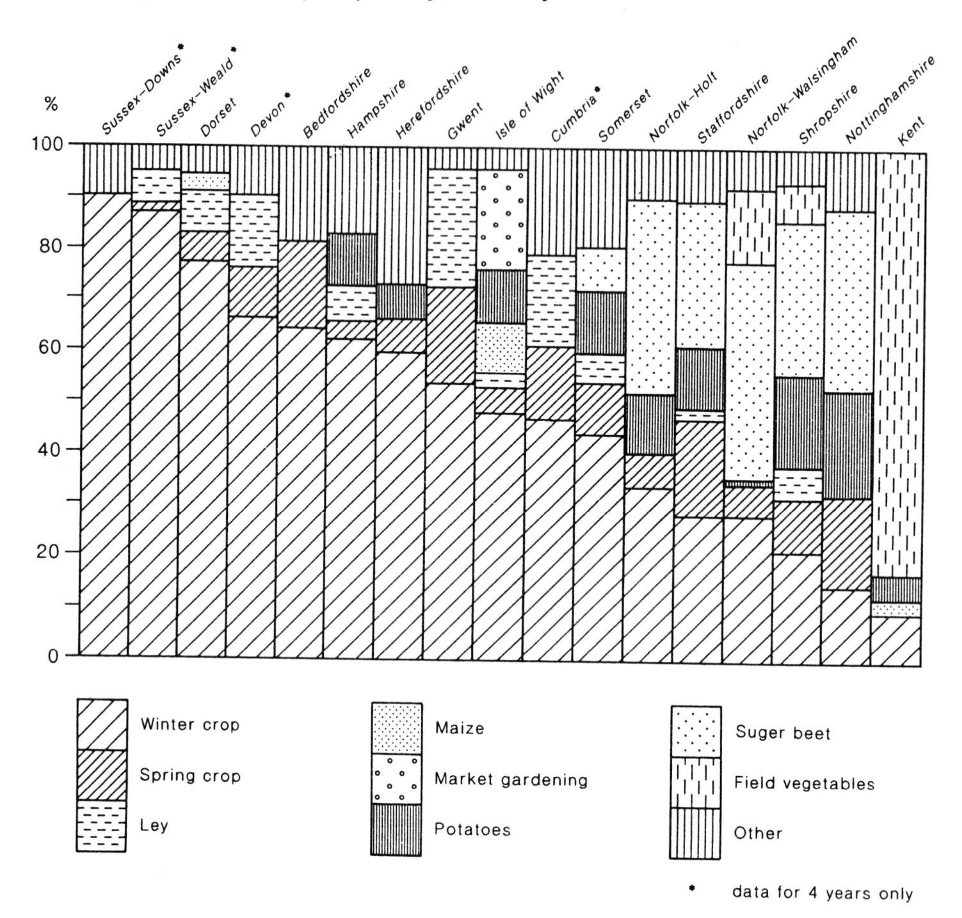

Figure 4. Crop types in eroded fields.

4. EXTENT OF RILLING

The maximum extent of eroded land was found on the sandlands of Nottinghamshire (Fig. 5) where, on average over the 5 years, 13.9 % of the arable land was affected by erosion, ranging from 1.5 % in 1982 to 24.0 % in 1983. In other sandy and coarse loamy landscapes erosion was also extensive, from a mean of 10.2 % per year in the Isle of Wight (range 2.4-19.7 %) to 8.3 % (2.5-14.1%) in Norfolk near Holt. Next came the 'silty' soils of the Sussex Downs (7.7 %, 3.6-12.6 %), although these figures were for four years only, and Hampshire (6.4 %,

1.6-19.8 %). Erosion was least extensive on clays and finer textured soils, being less than 3 %, and often about 2 % of arable land per year, although occasionally 1% or less. The range of values was also much less on the finer textured soils, for example, from 0.9-4.5 % in Dorset and 0.0-0.5 % in Somerset.

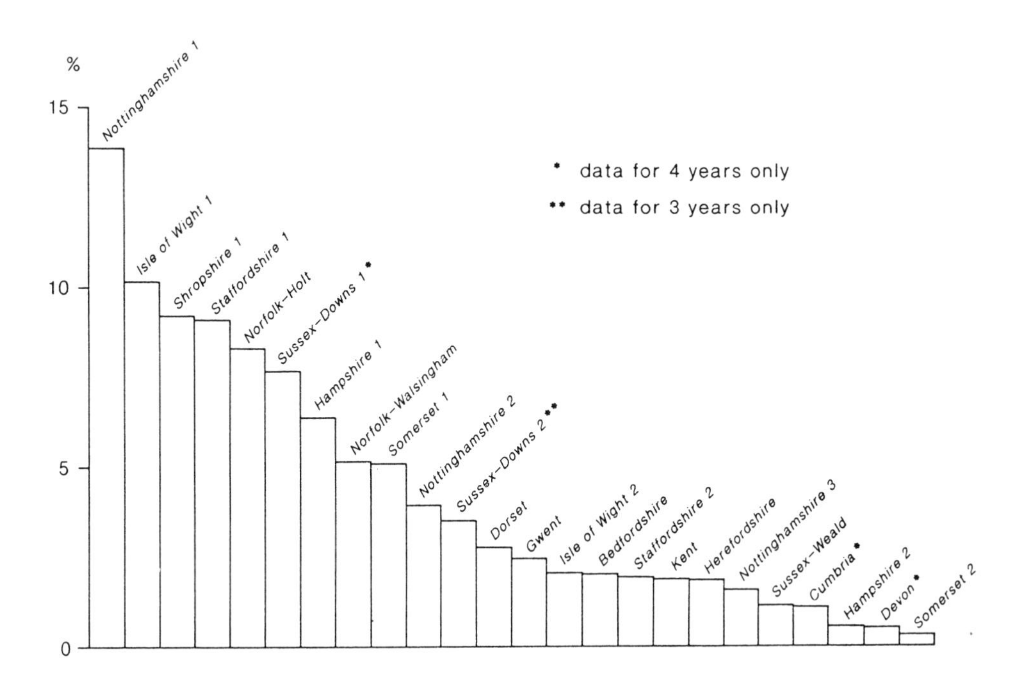

Figure 5. Extent of erosion on transects and landscapes (E1-E2) within transects.

Erosion was most widespread (mean >5 % of arable land per year) where the landscape was mostly under the plough and a wide range of crops was grown on easily worked soils in a 'rolling' landscape, as erosive rains falling at any time of year could cause rilling. Erosion was most extensive in those years when heavy rains fell in both autumn and late spring/early summer, as in 1983 and 1986. However, on the Sussex Downs erosion was dominantly in one crop type, autumn sown cereals. Most rain falls on the Downs in late autumn and early winter not long after the crop has been drilled (1). On the Downs autumn-sown cereals were the dominant crop. If the Sussex Downs and Kent, where very few winter cereal fields eroded, are excluded from the statistical analysis, there is a reasonable negative association between the extent of erosion and the percentage of the total number of rilled fields which were drilled to winter cereals (R^2=65.8 %). Finer textured soils were more resistant to erosion, and in the wetter west of the country were generally under grass, so rilling was not found extensively in these localities. Nor was erosion widespread on potentially erodible 'silty' soils in Kent because slopes were dominantly gentle (< 3 degrees) and cereal crops were not irrigated.

5. FREQUENCY OF RILLING

Over 50 % of the fields which eroded in Kent rilled twice or more, and in five other localities more than 35 % of the fields eroded more than twice (Fig. 6a). In most localities (8) between 25 and 30 % of fields eroded more than twice. Only in Cumbria was the figure lower than 20 %. A few fields eroded every year and many eroded three or four times in five years.

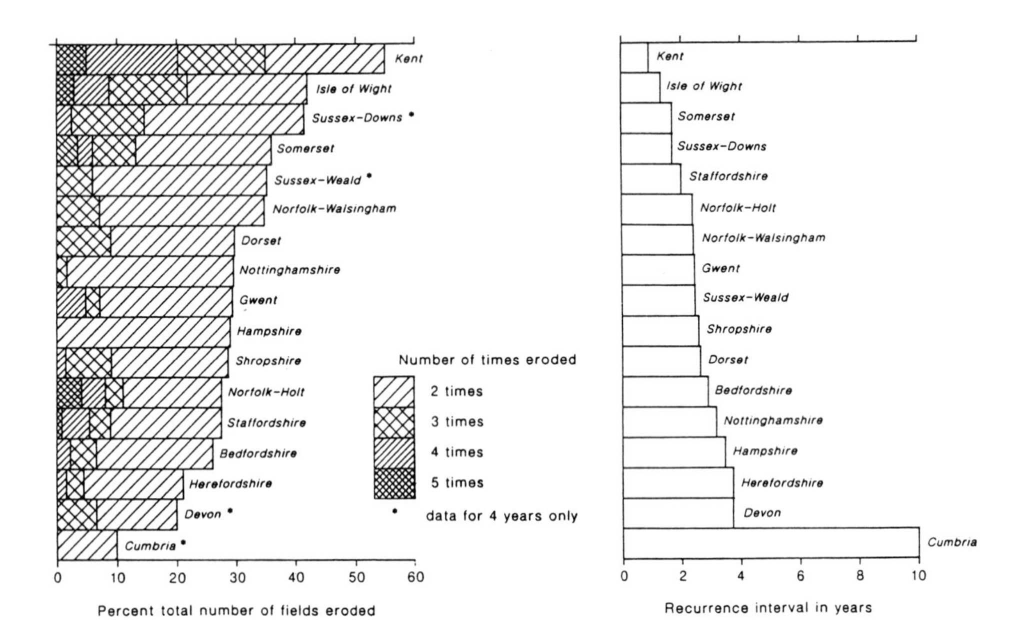

Figure 6. Frequency of erosion.
a) per cent fields eroded more than once.
b) recurrence interval.

In Kent fields eroded on average every 0.9 years (Fig. 6b) ; irrigation was commonly used there and more than one crop could be grown in a year. In 4 other localities the average recurrence interval of erosion, that is, how often on average a field eroded, was less than 2 years, everywhere else except Cumbria it was between 2 and 4 years. Even where erosion was not extensive, therefore, small parts of the landscape were vulnerable to frequent erosion.

6. RATES OF RILLING

Of the 17 localities four have mean erosion rates per field (the mean of the annual values of erosion per field) considerably higher than those found elsewhere, of c 4 m^3 ha-1 yr-1 or more (Fig. 7). Three of these localities have mostly 'silty' soils, the Isle of Wight mostly coarse loams. In Staffordshire and Shropshire rates were also higher than most, both transects covering mostly sandy soils. Irrigation was widespread in Kent, as already noted, as it was of field vegetables in the Isle of Wight ; in Staffordshire, Shropshire and Nottinghamshire occasional fields of sugar beet and potatoes were irrigated. Most localities had mean rates per field below 1.5 m^3 ha^{-1} yr^{-1}. Rates were lower where soils were of heavier texture and more resistant to erosion or where slopes were less steep (3).

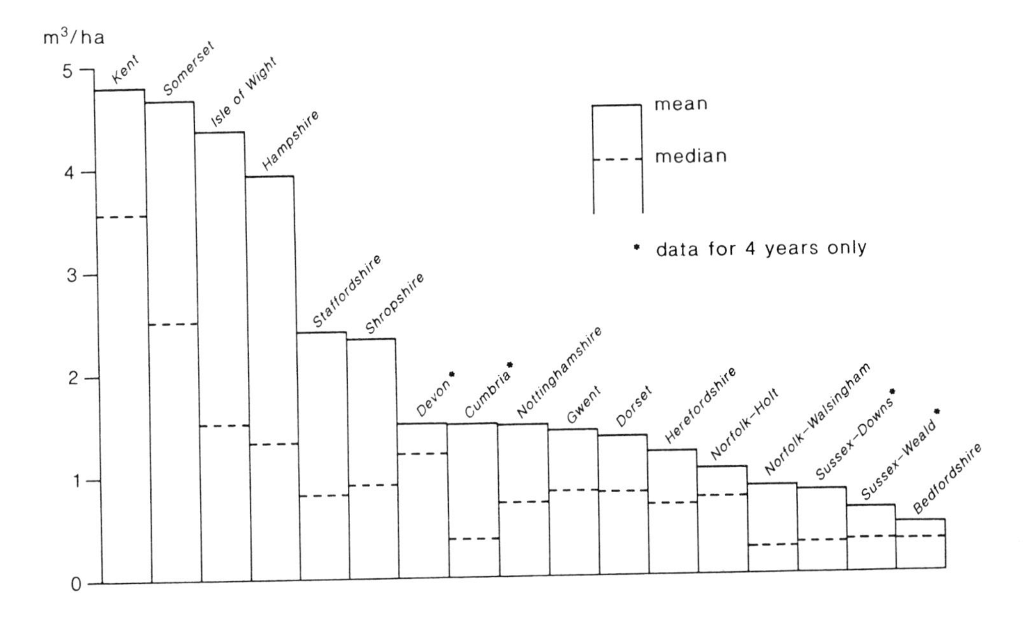

Figure 7. Mean and median rates of erosion per field in 17 transects.

Median rates were much lower than the mean rates, often about half or less, reflecting the marked left skewness of the data (2 - 3). Except in Kent, Somerset and the Isle of Wight they were below 1.5 m^3 ha^{-1} yr^{-1}.

Although median rates of erosion were in general lower when most eroded fields were drilled to winter and spring cereals, the variance explained was not great (R^2=26.1%), although significant at the 5 % level.

Assessing mean erosion rate (m^3 ha^{-1}) as the total volume of soil eroded divided by the total area of the rilled fields in the landscape, gave rates lower than the mean erosion rate per field and the order of ranking changed slightly (Fig. 8), although the data were well correlated (R^2=87.5 %). The most erodible landscapes were still those with 'silty' soils, followed by the coarse loams of the Isle of Wight and the sands of Shropshire, Staffordshire and Nottinghamshire. The fine silty soils of Somerset also had high erosion rates.

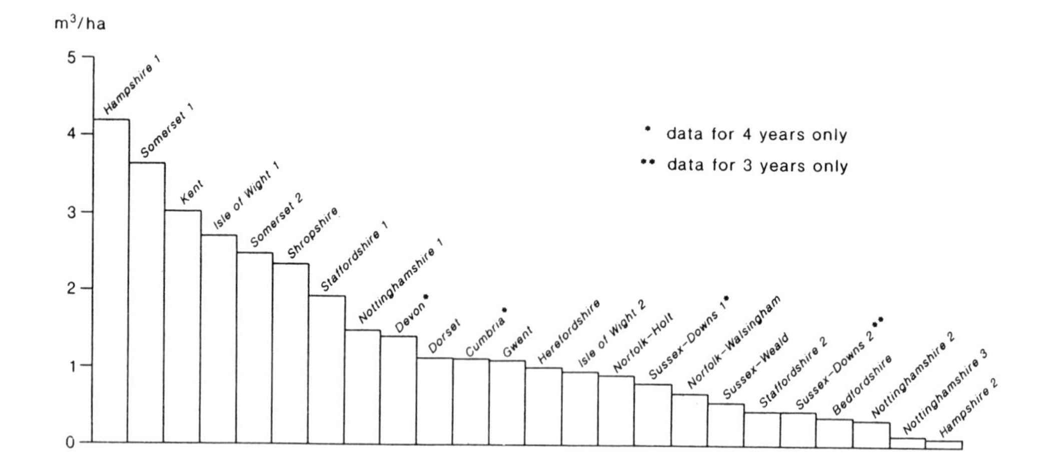

Figure 8. Mean rates of erosion for fields in transects or landscapes (E1-E3) within transects.

7. EROSION IN THE LANDSCAPE

The extent of erosion was generally greatest in sandy landscapes but volumes eroded were often highest where soils were 'silty'. The Isle of Wight with its mostly coarse loamy soils also ranked highly. Combining these measures of extent and rates best expresses the erodibility of landscapes (Fig. 9). The light loams of the Isle of Wight eroded most, but five other landscapes all having either sandy or 'silty' soils also eroded at rates higher than 0.1 m^3 ha^{-1} yr^{-1} for the whole of the arable land. Six landscapes eroded at rates lower than 0.01 m^3 ha^{-1} yr^{-1} and the range in rates between most and least erodible landscapes was two orders of magnitude.

8. DISCUSSION

The data on extent, frequencies and rates described above are conservative ones as there can be no guarantee that all eroded fields were identified, and measures of volumes transported in rills can only be approximate. However, it seems unlikely that many rilled fields were missed as many more fields other than the ones which eroded were checked and if erosion had occurred between the time the transect was photographed and the time of the field visit, these fields were generally found. Boardman's more intensive field monitoring on the South Downs (1990 and pers. comm.), which includes a large part of the monitored aerial photograph transect, gives similar answers to those obtained in this scheme (4).

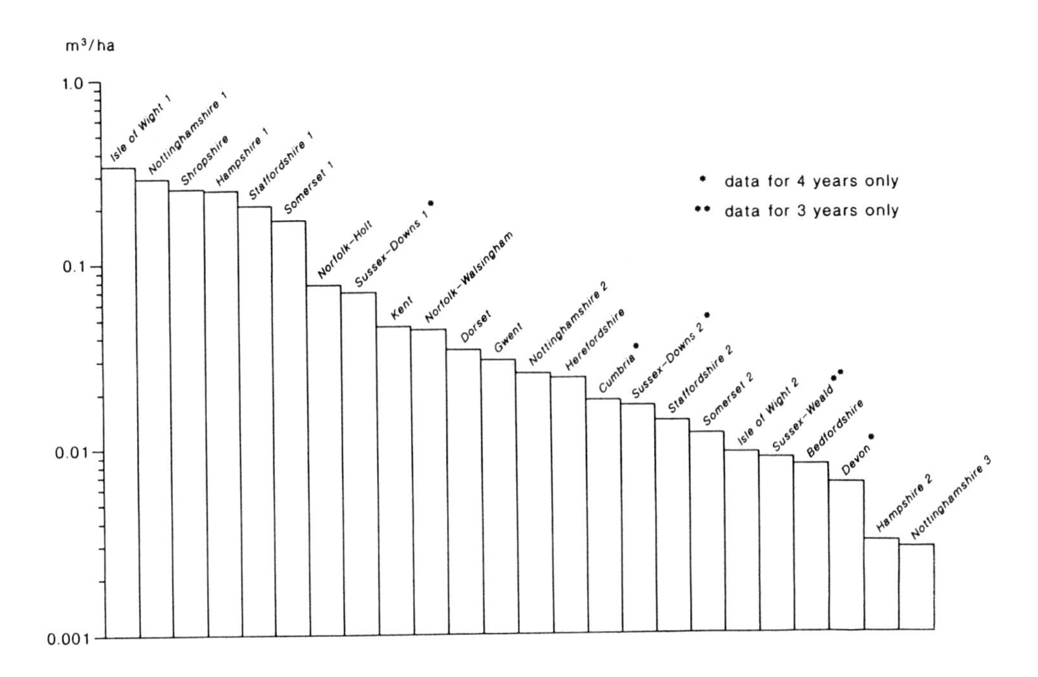

Figure 9. Rates of erosion for arable land on each transect and landscapes within transects.

It is noteworthy that results from the Staffordshire and Shropshire transects which have similar sandy soils and land use and were only 6km apart, were very similar. Rates of erosion measured in the field by different people are often within a range of one-half to twice the mean values and relate reasonably well to rates obtained on plots on similar soils (4). Although erring on the conservative side, therefore, it seems unlikely that the extent, frequencies and rates given here are much on the low side.

The monitored five year period did not have an unduly large rainfall event within it. Maximum rates of erosion for a field could be as high as 77.3 $m^3 ha^{-1}$ on the sandlands of Staffordshire, the result of a severe summer storm, and were more than 30 $m^3 ha^{-1}$ in fields in Nottinghamshire, Somerset, Shropshire, Kent and Hampshire. On the Sussex Downs where erosion was also widespread the maximum rate recorded for a field between 1983 and 1986 was only 7.1 $m^3 ha^{-1}$. This compares with 202 $m^3 ha^{-1}$ measured by Boardman in a field on the South Downs in late 1987 (6). The total amount eroded on the Downs was 7.4 times larger than measured in crop year 1983 (ie the year of harvest) (1), the year when most erosion was recorded in the national monitoring scheme, and the median rate in crop year 1988 was 2.9 times higher than in 1983. The mean median rate for catchments on the South Downs for the five year period 1983 to 1987 was 1.0 m^3 ha^{-1} yr^{-1} compared to that for 1983-1988 of 1.6 m^3 ha^{-1} yr^{-1}. Compared to 1983 the number of rilled fields increased by a factor of 1.4. A large scale erosion event, therefore, not only increased the area affected by erosion but also considerably increased median rates of erosion over the longer term.

Besides giving valuable data on the magnitude, frequency and extent of erosion such a field-based monitoring scheme can give other useful information indicating where in fields erosion is most likely to occur, which has implications for the assessment of erosion using plots, and why farmers in general do not consider rilling a problem.

Erosion often occurred only in valley floors or only on slopes (Fig. 10). Of localities with dominantly heavier textured soils (clays, fine silty and fine loamy) only in Dorset was erosion more prevalent on slopes than in valley floors. Where soils were mostly 'silty', coarse loamy or sandy, rills were dominantly on slopes. Mean median annual erosion rates were generally higher where most erosion was on slopes, but the variance explained, although significant at the 5 % level, was low (R^2=23.6 %).

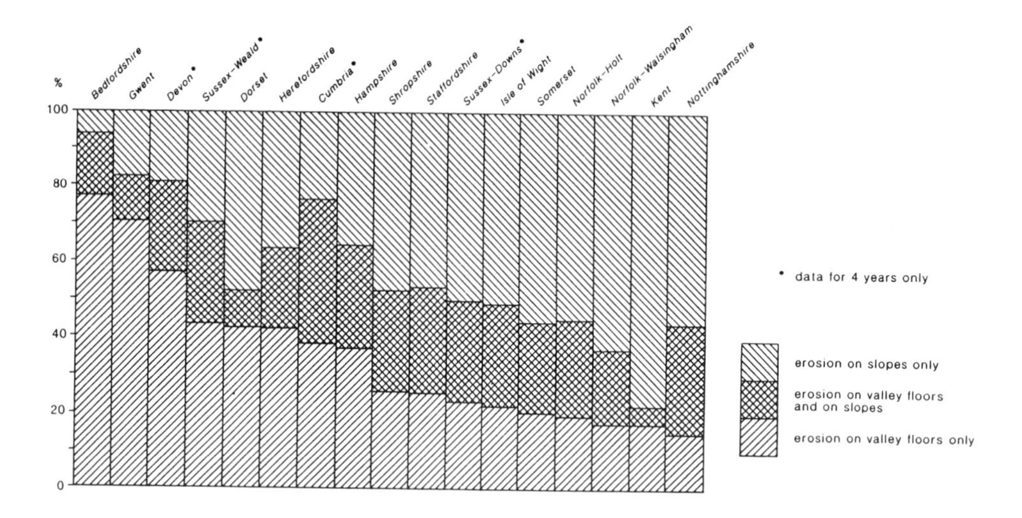

Figure 10. Rills and field morphology.

Erosional thresholds were different depending on the soil's resistance to erosion. Where soils were more resistant, runoff was needed from both slopes to give a sufficient (higher) head and velocity to initiate incision in the valley floor. On erodible soils the (lower) velocities attained on valley sides were sufficient to cause incision, but often insufficient for rills to form across footslopes or in valley floors. To be representative, plots to monitor rill erosion on heavier textured soils need to be much larger than those on lighter soils and should include a valley floor. In other words, field-based studies are more appropriate when investigating rilling, not plots.

The area of the field covered by rills and deposits was often minute. Only in Kent were mean and median values greater than 1.0 %. Highest values in each locality ranged from 0.3 % of the field in the Sussex Weald to 18.5 % in a field on the Isle of Wight. The next highest value was 9.7 % for a field in Nottinghamshire. The correlation between the mean median annual rate of rilling and the mean median annual area of the field affected by erosion was good (R^2=80.2 %). For 5 % of the field to be affected by erosion, rilling has to transport 22.1 $m^3 ha^{-1}$ of soil, a very high rate for England and Wales, and usually equated with the formation of a large gully or many large rills and large depositional fans which had buried much crop. The chances of a field eroding frequently at this rate are very small, and the farmer therefore does not think it worthwhile combatting erosion.

The average of median annual values of rates of erosion, rather than the average of the mean annual values, always correlated best with other parameters of interest, such as percentage of land under cereals or morphology.

9. CONCLUSIONS

This is probably the first field-based project monitoring water erosion over large areas of ground. It shows that in certain localities in England rilling was widespread and occurred often. Rates could be high, equivalent to a surface lowering over a whole field of c. 0.4 mm yr^{-1}. It is not possible to compare this data with that from other countries because elsewhere erosion data is largely derived from plot-based studies.

Where erosion and its impacts were known to be a problem before the project started in 1982, such as the sandlands of the west Midlands and Nottinghamshire, the coarse loams of the Isle of Wight and Norfolk and the 'silty' landscapes of Somerset and Hampshire, or became obvious during the period of the monitoring project, as on the chalk downs (7 - 8), the extent of arable land covered by eroded fields was greater than 5 % per annum. Elsewhere, erosion covered a smaller part of the landscape, often less than 2 % of the farmed land.

Erosion in fields recurred on average surprisingly freqently, in most localities more often than once every 3 years, regardless of the extent or rates of erosion, land use or soil type. Small parts of the landscape were very vulnerable to rilling therefore.

Erosion recurred most often in Kent where irrigation was an important factor promoting rilling. If irrigation was to become more widespread as a response to global warming, rilling would happen more often. Rates of erosion would also increase.

Median rates of erosion per rilled field per year are often half the mean rates, or less, and rates per field between localities vary by about an order of magnitude. However, the variation in amounts transported within landscapes, that is, the total amount eroded divided by the total area of the arable land of the landscape, can be as great as two orders of magnitude.

During the period of monitoring there were no exceptional rainfall events. Such events, as happened over the South Downs in late 1987 (6), increased markedly the median annual rate of erosion, not only for crop year 1988 but also that for the monitored six year period. The figures given here for the magnitude, frequency and extent of erosion in the 17 localities are conservative ones, therefore.

A field-based project such as this explains why farmers do not consider erosion much of a problem. And also shows that rilling in valley floors is an important phenomenon which needs to be taken account of more in erosion studies. Where rilling occurs is closely related to soil type, where soils are more resistant to erosion more erosion is confined to valley floors.

ACKNOWLEDGEMENTS

Dr. J. Boardman collected data on erosion for an area of the South Downs which included the aerial photograph transect. He also commented on an earlier draft of this paper. For carrying out both these tasks I am most grateful to him.

REFERENCES

1. Boardman, J. (1990) : Soil erosion on the South Downs : a review. In : J. Boardman, I. D. L. Foster and J. A. Dearing (Eds.). Soil Erosion on Agricultural Land. John Wiley and Sons, Chichester, 87-105.
2. Evans, R. (1988) : Water Erosion in England and Wales 1982-1984. Report to Soil Survey and Land Research Centre, Silsoe.
3. Evans, R. (1990) : Water erosion in British farmers' fields - some causes, impacts, predictions. Progress in Physical Geography 14, 199-219.
4. Evans, R. and Boardman, J. : Assessment of water erosion in farmers' fields. First International ESSC Congress, Silsoe, April 6-10, 1992. Symposiun Proceedings. In press.
5. Ssew. (1983) : Soil Map of England and Wales. Soil Survey of England and Wales, Harpenden.
6. Boardman, J. (1988) : Severe erosion on agricultural land in East Sussex, UK, October 1987. Soil Technology 1, 333-348.
7. Robinson, D. A. and Blackman, J. D. (1990) : Soil erosion and flooding. Land Use and Policy 7, 41-52.
8. Stammers, R. and Boardman, J. (1984) : Soil erosion and flooding on downland areas. The Surveyor 164, 8-11.

Farm Land Erosion: In Temperate Plains Environment and Hills
S. Wicherek (Editor)

Regional assessment of runoff and erosion risk. Example of the Nord/Pas-de-Calais region, France

D. King, Y. Le Bissonnais and R. Hardy

INRA-SESCPF - Ardon - 45160 Olivet, France.

SUMMARY

The aim of this study was the assessment of regional runoff and erosion risk. Such work is almost impossible when using small-scale data (e.g. 1:250,000) as one cannot use a mechanistic approach to the problem. It is thus necessary to use large-scale (e.g. 1:5000 to 1:25,000) data for selecting and classifying the factors that influence runoff and erosion, e.g. soil characteristics, land use, rainfall and morphology. Large-scale studies showed that erosion risk can be evaluated by studying spatial units like elementary catchment areas, which can be said to form spatial integration units (SIU). However, as such catchment areas generally are too small to be shown on a standard 1:250,000-scale map, the SIU is taken to be an association of several contiguous elementary catchment areas that have a similar morphology and soil variability.

The Nord/Pas-de-Calais region in France was studied because of the locally strong erosion of its soil. A geographical database was created that included runoff as the main factor. Most data came from other bases, e.g. on agricultural statistics, meteorology and geology. New data were obtained from Spot images and a soil survey. In all, 78 SIUs were defined on the basis of hydrological and morphological homogeneity. Each was characterized by the areal percentage of factors that large-scale work had indicated to favour runoff. The runoff risk for each SIU was then estimated, assuming that a high percentage of one or more of these factors creates a statistically high risk of erosion.

The statistical and cartographical representation of the various factors selected from the database, and their integration into SIU_s, indicate the main trends; a further advantage of this method is that quantitative data now are available for comparison between different large areas.

RESUME

L'objectif de l'étude est d'indiquer à l'échelle d'une région, les zones où les risques d'érosion sont les plus importants. A cette échelle, une approche mécaniste des phénomènes n'est pas envisageable. Par contre, en se basant sur des études menées à plus grande échelle (bassin versant élémentaire, secteur de référence), on peut sélectionner et hiérarchiser les facteurs à l'origine des phénomènes de ruissellement (nature des sols, pratiques agricoles, précipitations, type de modelé).

Les études à grande échelle (du 1/5 000 au 1/25 000) ont montré également qu'il faut juger des risques d'érosion non de façon ponctuelle mais sur des Unités Spatiales d'Intégration que sont les Bassins Versants Elémentaires. Ceux-ci constituent une unité trop fine à l'échelle du 1/250 000. L'Unité Spatiale d'Intégration (USI) est alors définie en regroupant les Bassins Versants Elémentaires contigus en Grands Bassins Versants.

La région Nord-Pas-de-Calais a été sélectionnée pour l'importance des phénomènes d'érosion observée. Une base de données géographiques a été constituée prenant en compte les principaux facteurs à l'origine du ruissellement. La plupart des informations sont extraites de bases de données existantes (SCEES, Météorologie Nationale, BRGM). De nouvelles données ont été acquises d'une part à l'aide du satellite SPOT, d'autre part au cours du levée de la carte des sols à l'échelle du 1/250 000. 78 Unités Spatiales d'Intégration ont été délimitées en tenant compte des limites hydrologiques et des limites de paysage. Chacune d'entre elles est caractérisée par le pourcentage de surface des différentes variables reconnues favorables pour le ruissellement lors des études à grande échelle. Un fort pourcentage de surface pour une variable ou plusieurs variables simultanées signifie une forte probabilité d'avoir un grand nombre de Bassins Versants Elémentaires sensibles à l'érosion.

Les présentations statistiques et cartographiques des différentes variables sélectionnées dans la base de données, puis intégrées par USI permettent de visualiser les grandes tendances régionales. L'intérêt est de disposer pour chacune d'entre elles de valeurs quantitatives facilitant des comparaisons rigoureuses.

INTRODUCTION

Despite moderate slopes and rainfall intensity, several areas with silty soil of northwestern Europe are subject to erosion, which in places can be quite strong. Several studies have been carried out on small, individual catchment areas (1 - 2 - 3 - 4) that have provided a first understanding of the mechanisms causing runoff, and of its concentration (5). They showed the complexity of the space/time interactions between the factors causing runoff, such as the structural instability of soils, agricultural practices, catchment morphologies and rainfall characteristics.

The main aim of this study was to provide a method for the regional assessment of erosion risk, based on the detailed analysis of processes at the scale of an entire region, in this case the Nord/Pas-de-Calais region that covers about 1,200,000 ha. A further objective was to provide decision makers with a tool for future land-use planning, infrastructure maintenance and water-development work. A third objective was to verify the accuracy of a local model when extrapolated to a regional scale, and to identify any possible discrepancies.

The first part of this paper presents a brief bibliographical review of the evaluation methods in the field of soil erosion, and a review of the knowledge on erosion processes in the Nord/pas-de-Calais region. The second part presents the data available for this region and the method for their spatial integration. The third part discusses one of the cartographic scenarios produced.

BIBLIOGRAPHICAL REVIEW

Methods for the evaluation of erosion phenomena

Various attempts have been made to describe erosion risk, which were based on the universal soil-loss equation (USLE) (6 - 7). This approach calls for a multiplying combination of the factors causing erosion phenomena, such as the erosivity of rain and the erodibility of soil, as well as slope angle, land use, and conservation techniques. The USLE equation was developed and calibrated in the

USA, for which reason it requires adaptation to the specific conditions found in each region (8 - 9). In addition, a quantitative and formalized method should be available for comparing various situations and for a critical evaluation of the results (10). Various authors have noted the interpretation difficulties when correlating between the factors that are used in the USLE equation (11 - 12).

This shows the limits of an approach that is statistically based on a restricted number of sites. Ideally, a mechanistic model should be proposed for the phenomena, which should then be integrated in a mapping process (13). Today, most studies simply superpose the data covers describing the spatial variability of factors, leading to a list of combined cartographic units to which an arithmetic equation is applied (14). Faced with the diversity of precision in such maps, Giordano (15) proposed to plot the data on a standard grid whereas Albaladejo et al. (10) suggested to use landscape units that are known to be homogeneous for a large number of factors. However, the relationship between neighbouring units, in particular the location of surface flow within functional units, is seldom taken into account when such generalizing work is done (5).

Review of the mechanisms seen on a large scale

Before any evaluation can be made at the scale of a region, the causal mechanisms and the complexity of their relationships must be explained. To this end, four areas of about 4000 ha, designated as reference areas, and 16 isolated catchment areas of about 300 ha each, were selected for monitoring. Thorough studies of a similar landscape to that of the Nord/Pas-de-Calais region, e.g. the Pays de Caux, the Picardy, have been ongoing for several years, and we briefly present the main results of this work that could serve as general guide lines for our risk-evaluation work.

Runoff phenomena and the simultaneous appearance of signs of linear erosion, e.g. rills and gullies, are the result of a highly complex interaction between four main factors that are (16) : the structural stability of the soil (mainly depending on soil texture) ; farming practice ; morphology of the elementary catchments; and rainfall. This complexity is the result of the variability in space and time of such interactions. For instance, for runoff to start it is necessary to have a conjunction between a certain type of farming practice on a certain type of soil, and a moment when rainfall is sufficiently strong to cause the degradation and closure of the soil surface (17 - 18). A further cause for concentrating runoff, is the spatial interaction between the geographical distribution of plots on which runoff is likely, and the morphology of the elementary catchment areas (19). When these conditions amplify each other, surface-flow rates can reach a sufficiently high level to produce, on certain types of soil, the appearance of first rills and then gullies.

It is difficult to quantify all possible interactions (20), but we can retain two essential points for a regional evaluation. The first is that the place and manifestation of erosion phenomena are distinct from the places where runoff starts, the latter being the underlying cause of the former. The second is that it is possible to rank the factors that cause erosion; in particular, it was found that the percentage of surfaces that might be subject to runoff within small elementary catchment areas, is statistically related to the volume of exported earth corresponding to the observed signs of linear erosion (3 - 4 - 5).

Based on these results, a model was proposed for evaluating the susceptibility of a plot of land to runoff. This combines the degree of instability of the surface structure of the soil as related to its texture (21), with the type of farming practice (22). A geographical superposition of these two types of data, is only valid when calculated as a percentage area of spatial units of hydrological functioning, i.e. elementary catchment areas.

In summary, the relatively detailed large-scale studies show that it is necessary to combine data from various thematic sources, particularly soil and farming practice, and that such combinations are invalid for isolated points that are taken out of the context of their surroundings. This understanding led to the realization that the work had to rely on data-integration methods, which should be based on Spatial Integration Units (SIU) i.e. the elementary catchments.

DATA AND METHODS

Analysis of available data ; creation of the geographical database

As the first elements for understanding the mechanisms of runoff and erosion were demonstrated in the field at large scales, it was possible to envisage their immediate evaluation on a regional level. However, the investigation of all agricultural land at different times of the year, or even over several years, is technically and financially impossible for a subject as complex and strongly targeted as this one. Geographic combination of the data is one of the key points in understanding large-scale phenomena, and it was thus necessary to design combination methods that would take account of the type of data easily available at this scale, but that in particular should integrate field data (23).

To examine the feasibility of regional evaluation, we drew up a list of all data available in public databases, as well as a list of further data to be collected. Agricultural statistics are available for the type of cultivation on the scale of municipalities ('communes'). The national meteorological service has a secondary network of 52 weather stations that covers the Nord/Pas-de-Calais region. A database is available of soil samples analysed for farmers (24), but is considered to give insufficient information on the spatial variability of the soil cover. A 1:250,000-scale mapping effort was thus launched, in particular for defining "surface texture" characteristics. Finally, a set of Spot satellite data was programmed for the entire region, in order to obtain an instantaneous view of the area of interest (25).

Each of the data-sets ("themes") collected formed the basis of a data cover in the Arc/Info geographic information system (GIS). Most data provided information on surfaces, except for the meteorological data that had to be interpolated. It is obvious that the geometrical nature of the elementary data will be different for each theme (Fig. 1) : meteorology provides point data, agricultural statistics are by 'commune', remote sensing data are in pixel format, soil data are by map unit, etc. Such data thus cannot be directly combined, as they correspond to different levels of resolution. Moreover, each theme has specifics that must be defined before any interpretation can start. As an example, we will discuss the agricultural statistics and the soil map.

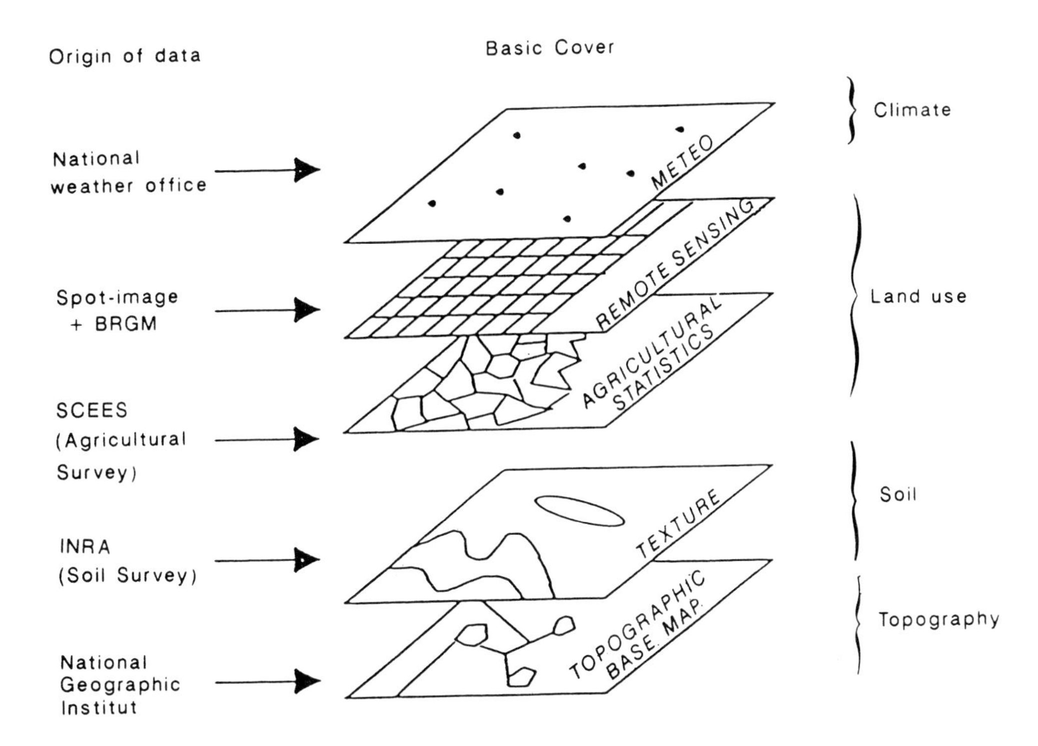

Figure 1. The types of data combined within a geographic database, and their origin.

The agricultural statistics is updated about once every decade. The data from the last census (1987-1988) were sufficiently recent to be used in our study. The statistics produced are calculated on the basis of farms in the same commune, which is determined by the location of the main farm buildings. Therefore, the number of hectares counted for a certain commune may not correspond to the precise surface area of farming in this commune, but rather to the surfaces worked by the farms that are registered for this commune. A particularly evident bias thus exists for small-sized communes, as the surface percentage of a certain crop is calculated in relation to the total surface area of the commune. The problem is even more acute in the case of communes having but one registered farm, in which case the data are kept secret.

The soil map was drawn up at a scale of 1:250,000 and contains so-called "complex" map units. The latter correspond to associations of various soil types within one geographic unit. For that reason, usually not enough data are available for precise location of specific types of surface texture, as the map only gives a surface percentage of each type of texture found within a map unit.

From these two examples, it can be seen that the geographic boundaries of each theme provide a level of spatial resolution that does not necessarily correspond to the required level of geographic resolution. In the case of the soil map, the spatial resolution is too coarse for showing the true spatial variability of a character, whereas the agricultural statistics provide a level of spatial resolution that is larger than that of elementary catchment areas. This raises the problem of partial statistical incompatibility between data with coarse and fine resolution, forcing a bias towards the larger mesh of the coarse data.

Method for geographic combination of data covers

As said above, the large-scale studies have shown that it is necessary to consider runoff and erosion mechanisms at the level of spatial integration units, ideally elementary catchment areas (ECA). However, the analysis of old and new data shows that most of these data, with the exception of remote sensing and agricultural statistics, have a much coarser level of geographic resolution than is acceptable. On a regional scale, the geographic-combination method of the various data covers is reproduced in a similar fashion to that of the large-scale studies, estimating the surface percentage (by integration unit) of the factors that might favour runoff. The main difference lies in the fact that, for practical reasons (the partial statistical incompatibility mentioned above and computer time), it is not possible to work at the level of the smallest functional spatial units, i.e. ECAs, but that much larger spatial-integration units (SIU) are needed. We thus defined 78 SIUs, taking account of the hydrological boundaries of the main catchment areas and of the morphological landscape boundaries as known by the regional experts (Fig. 2). Masks were used for excluding urban areas and grass-covered valley bottoms, both of which are known to be mostly unaffected by runoff and erosion. This subdivision made it possible to group all ECAs of a same type in large regional sub-units.

Based on the SIU cover, each thematic cover was then "integrated" (26). This operation involved the calculation of the surface percentage, within a given SIU, for each of the characteristics of a thematic cover (4). This method reduces all data to the same level of spatial resolution. Regardless of whether the data come from satellite-image pixels, or from statistical data on communes, they are all transformed into a surface percentage.

Moreover, the difficulties encountered when using agricultural statistics on the commune level are largely eliminated, as the SIUs encompass a large number of adjacent communes. The percentage of plots registered as belonging to an SIU, but in fact lying outside it, now has become very small when compared with the total surface area of the SIU. In addition, the data coming from soil associations and corresponding to surface percentages of the various textures, are directly summed within a particular SIU.

Using this method, it is thus possible to combine data of various origin and resolution, as well as to introduce results obtained at a large scale, in particular the relationship between observed erosion features and the surface percentage of the various factors in each ECA. However, it should be kept in mind that geographic combinations at a small scale have a mostly statistical value (27). They help to identify areas where a conjunction can exist of soil, farming and climate factors that are most likely to favour erosion. In that case, there is a high probability that most elementary catchment areas in the same SIU will combine the necessary conditions for the starting of concentrated runoff.

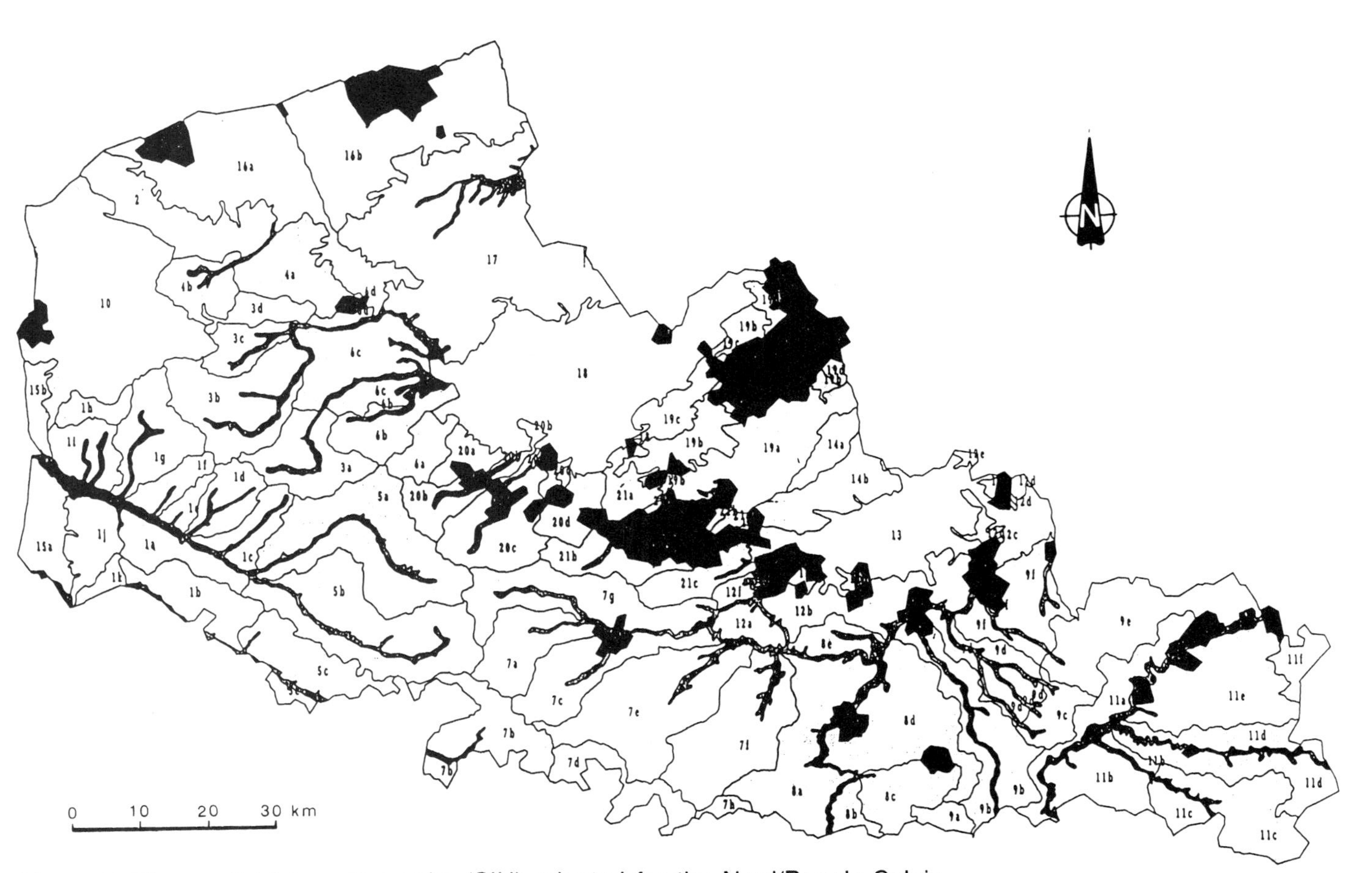

Figure 2. The spatial-integration units (SIU) adopted for the Nord/Pas-de-Calais region, northwestern France. In black, the areas masked for urban communities (pop. density of over 1000 inhabitants/km2) and for valley bottoms with forests and grass.

RESULTS

Several scenarios were tested, either for studying the triggering phenomena during a given season, or for obtaining a general overview (23). Here, we will only discuss the latter of these two possibilities. The results were first statistically analysed by projecting the variables two by two in a mathematical space. A single- or multi-factorial geographic projection then enabled zoning of the highest risks (26).

Statistical results

- *Soil vulnerable to surface crusting versus yearly tilled soil*

The objective was to see whether situations exist for which the joint presence of several factors can multiply the risk of erosion. The first analysis was made by combining the quantitative variables "percentage of soil vulnerable to surface crusting" and "percentage of yearly tilled soil". Each spatial integration unit (SIU) was plotted on an X-Y grid defined by these two variables (Fig. 3). The correlation coefficient turned out to be significant at 0.53, notwithstanding some dispersion of the points. We can see that only one SIU exists with more than 40 % of its surface vulnerable to rain, versus less than 45 % of yearly tilled soil. Only two SIUs have less then 35 % of vulnerable soil, against more than 55 % of yearly tilled soil. This reflects a coherence in the relationships that generally exist between farming practice and the variables of the physical environment.

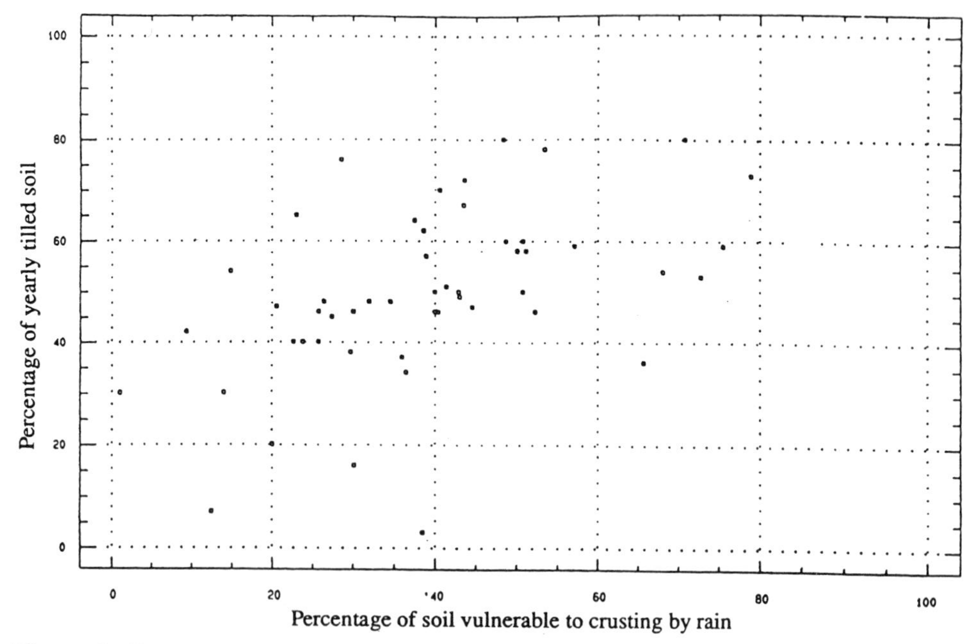

Figure 3. Percentage of "yearly tilled soil" surface (from agricultural statistics of 1987/88) against the percentage of "soil vulnerable to crusting by rain" (from the 1:250,000-scale soil map), by spatial integration unit (each point represents an SIU).

- *Yearly tilled soil versus annual rainfall*

The relationship between average annual rainfall and the percentage of yearly tilled soil is much lower than that of first example (Fig. 4). The cloud of SIU points has no discernible structure, and the only points of note are the absence of high percentages of yearly tilled soil above 900 mm of annual precipitation and the absence of low percentages of yearly tilled soil below 700 mm. This should not be interpreted in terms of direct cause-and-effect, but rather as a more general relationship between all characteristics of the landscape. For example, A fairly dense cluster, most of which fall in SIUs of the northeastern part of the Nord/Pas-de-Calais region, is seen in the field of Figure 4 that covers less than 25 % of yearly tilled soil and >800 mm annual precipitation. An increase in rainfall generally corresponds to an increase in elevation, which implies the presence of other soils that will have their own specific types of farming practice. Based on these results, we adopted thresholds of 700 and 900 mm of annual precipitation for the final combination work.

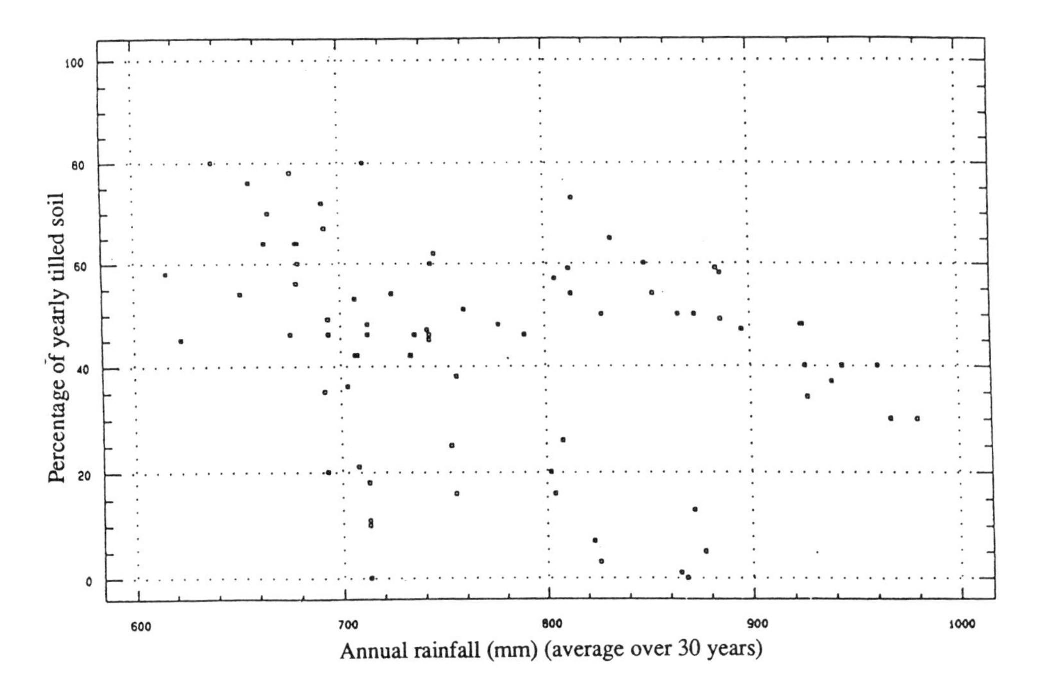

Figure 4. Percentage of "yearly tilled soil" surface (from agricultural statistics of 1987/88) against the "annual rainfall" (from the interpolation of data from 52 weather stations), by spatial integration unit (each point represents an SIU).

Cartographic results

The SIUs were first ranked according to the two surface-percentage variables of "yearly tilled soil" and "soil vulnerable to surface crusting", the latter including soils that are highly to moderately vulnerable. The relationships between these two variables, discussed in the preceding paragraphs, caused the empty spaces on Table 1, as well as the high values around the diagonal in this table. The shaded space at the top of the table shows areas where there is no erosion due to very flat landscape, forests or urban areas.

Table 1.
Comparison between the variables "percentage classes of yearly tilled soil surfaces" (from agricultural statistics of 1987/88) against "percentage of soil vulnerable to crusting by rain" (from the 1:250,000-scale soil map). The values are expressed as percentages of the total surface of the area studied.

		Percentage classes of yearly tilled soil surfaces				
		< 30 % ▽	30-45 % ▽	45-55 % ▽	55-65 % ▽	> 65 % ▽
Percentage of soil vulnerable to crusting by rain	Flat area, urban area, etc.	16,6	3,1	13,5	11,2	
	(1) low * >>	3,9	1,2	.	.	.
	(2) average * >>	1,1	4,5	5,1	0,9	
	(3) high >>	1,1	1,8	7,1	5,1	3,7
	(4) very high * >>	.	0,1	8,2	6,0	6,1

* The percentage classes of surfaces in each SIU that are vulnerable to crusting by rain, are defined as follows :

(1) 0-25% of soil highly vulnerable (SHV) and less than 50% of soil moderately vulnerable (SMV), per SIU.

(2) 0-25% of SHV and more than 50% of SMV, per SIU
25-35% of SHV and less than 50% of SMV, per SIU

(3) 25-35% of SHV and more than 50% of SMV, per SIU
35-45% of SHV and less than 50% of SMV, per SIU

(4) 35-45% of SHV and more than 50% of SMV, per SIU
more than 45% of SHV, per SIU

On the map, it would have been difficult to show all 19 fields in the table that have at least one SIU, and several fields thus had to be grouped. Such grouping was done on the basis of the thresholds discussed above, which correspond to surface percentages for each SIU, but have no direct mechanistic relationship with the intensity of erosion. Thus, the grouping of several fields primarily is for didactic reasons, helping to make the reading of the final map (Fig. 5) easier. As opposed to the large-scale work, such grouping is not possible when using just the values of observed erosion.

No ranking was made between the two variables "yearly tilled soil" and "soil vulnerable to surface crusting". The aim was to show the opposite poles as clearly as possible. These poles are formed by areas where SIUs have >35 % of very vulnerable soil and >55 % of yearly tilled soil, and by the areas where the SIUs have only <35 % of vulnerable soil and relatively little (<45 %) yearly tilled soil. The first areas, red on Figure 5, thus contain elementary catchment areas where the probability is high to find the two joint mechanisms that can trigger erosion. The other pole corresponds to areas with a low probability that erosion will occur, coloured green. Between the two, are orange-coloured areas that designate a moderate probability of erosion.

The factor of "annual rainfall" was the third variable to be superposed on the map. This factor gave cumulated annual precipitation data for the past 30 years on average. This makes it possible to make a relative judgement on which are the "dryer" and "wetter" SIU_s. The minimum and maximum thresholds of 700 and 900 mm define three classes of SIU on the map. For the moment, not enough data are available to choose more precise threshold values, which would require a more thorough analysis of the significance of such average values, when compared with a frequency analysis of rainy periods in different seasons, i.e. according to different surface conditions.

Each SIU was assigned a colour from the classification shown in Table 1. The "annual rainfall" classes then were superposed by varying the intensity of the colours earlier defined. For instance, an SIU with a high probability for potential runoff (red), will be deep red where the annual rainfall exceeds 900 mm, medium red where the rains are between 700 and 900 mm/year, and pale red over areas with a rainfall of less than 700 mm/year. This enables reading of the map two levels. The colour itself shows the probability of a simultaneous presence of the two factors "yearly tilled soil" and "vulnerability of soil to crusting by rainfall", the shade of the colour providing information on the amount of rainfall.

Discussion of the integration method

Integration of the main spatial units (SIUs) has the advantage that it brings all data back to the same level of resolution. In this way it becomes possible to study the variables that might influence runoff, based on the surface percentages that are represented by the variables within a given SIU. This type of reasoning agrees with the results of more detailed field, indicating that a sufficiently large surface must be present in each elementary catchment area for erosion processes to be triggered. However, the integration process also masks the internal spatial variability within $SIU_{S.}$ It would thus be advisable to add further criteria for the evaluation of spatial dispersion to the criterion of "surface percentage". A problem that still needs solving is to know whether a figure of 50 % of soils vulnerable to crusting by rain in an SIU means that 50 % of such soil types are found within each elementary catchment area, or whether two or more elementary catchment area sub-units exist, each of which has quite different percentages of such soil. By subdividing the region into

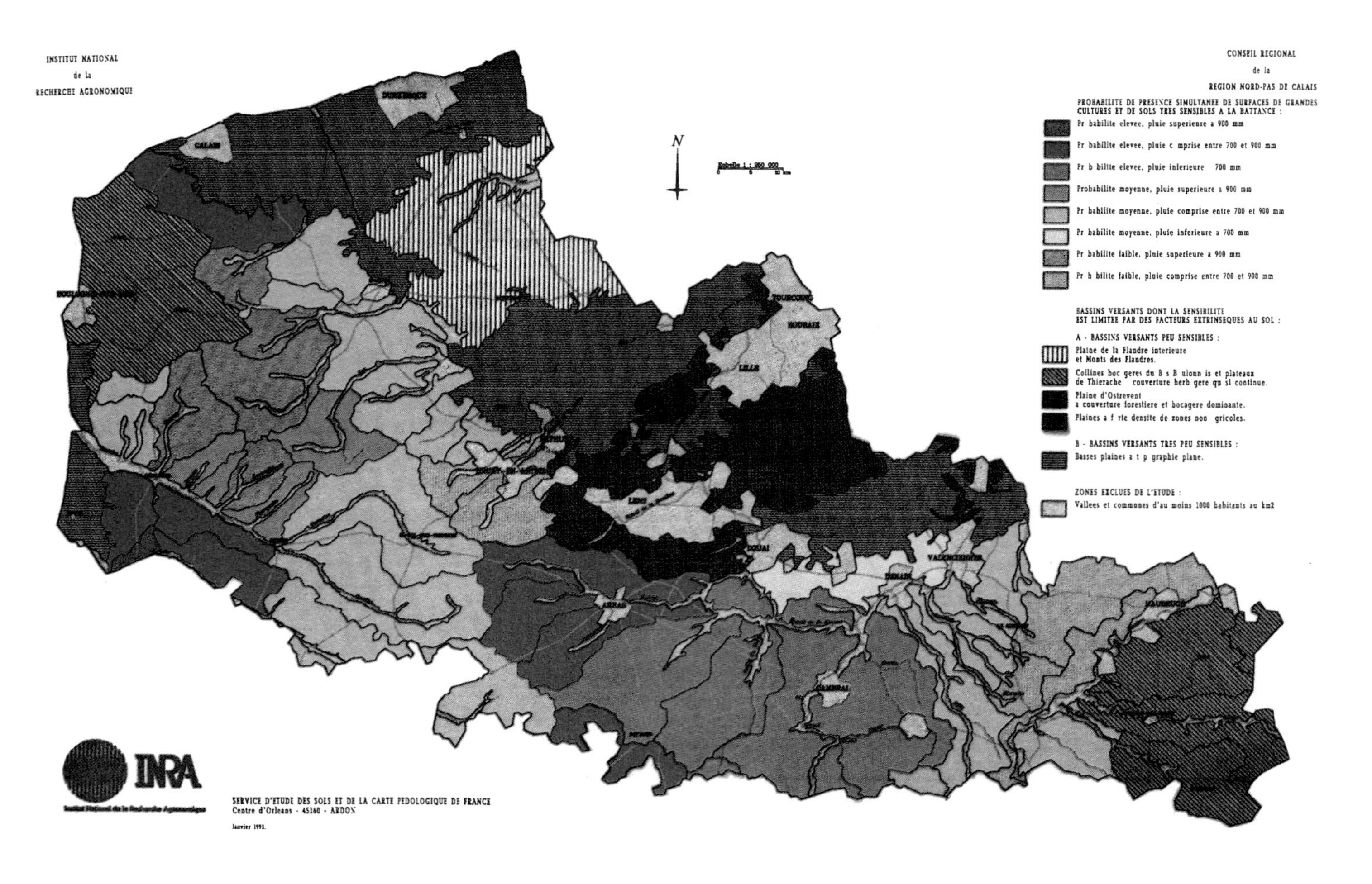

Figure 5. Sketch map of the soil-erosion risk in the Nord/Pas-de-Calais region, combining the factors "soil texture", "land use" and "annual rainfall" (see text for explanation of colours and shades).

major catchment areas with homogeneous landscapes, it was assumed that the main areas have the same characteristics of internal variability. Nonetheless, a quantitative evaluation of this internal variability should be carried out in order to be able to evaluate and validate (or not) this subdivision.

The method of spatial integration is a means for changing spatial scales with the help of numerical techniques. The same argument could be used for the time scale. The variable of "annual rainfall" corresponds to an integration of time elements such as hours and days. A finer analysis of such elements is feasible, e.g. monthly rather than yearly intervals, intensity rather than quantity, but it would inescapably raise the problem of assessing short rainfall events that generally are random in a geographical sense.

CONCLUSIONS

The regional assessment of runoff and erosion risk, is based on the understanding of large-scale mechanisms and on the analysis of data that are easily available for large areas. Erosion phenomena in northern France are related to the spatial dynamics of runoff within elementary catchment areas (ECA).

It is important that erosion risk is not evaluated in discrete points, but rather within spatial-integration units (SIU_S). An ECA is too small at the standard map scale of 1:250,000, and SIUs thus were defined by grouping several contiguous ECAs into major hydrological catchment areas that have a homogeneous landscape. Such major areas are not really functional units for the erosion phenomena as studied, but should be seen as statistical units, indicating those areas where ECAs combine several factors that might trigger erosion. The map of the different variables that were selected from the database and integrated by SIU, shows the regional trends. The main interest of this method, is that it provides quantitative values that facilitate comparison between SIU_S. A potential criticism of this approach, is that the selection is governed by the limited amount of data available at this scale. However, its main advantage is that it highlights the risk-triggering mechanisms. Another advantage is that the formalism of such work at a regional scale makes for easy updating, for instance in the case where the action of certain factors, e.g. rainfall, would be further defined. In addition, an integration into smaller SIUs may be possible when using data from successive satellite images. This would lead to a more detailed visualization of land use in time and space, than is now possible with the data from agricultural statistics, which have a subjective bias and are often out of date.

ACKNOWLEDGEMENTS

This work was made possible through a grant from the Regional Council of the Nord/Pas-de-Calais region. Thanks are due to R. Hardy, M. Eimberck, P. Chery, J. Maucorps, J. Daroussin, C. King, M. Sarrazin, A. Lindor and for their participation in this work. Thanks are also due to H.M. Kluyver for translating and editing this paper.

REFERENCES

1. Beuve, P., (1989) : Essai d'évaluation de la sensibilité à l'érosion de sites du Nord du Bassin de Paris. Prise en compte des caractéristiques de la couverture pédologique. DAA ENSA Rennes Chaire de Science du Sol. INRA Laon. 56 p.
2. Le Ber, F., (1989) : Essai d'évaluation de la sensibilité à l'érosion dans le Nord du Bassin Parisien. Prise en compte des caractéristiques des systèmes agraires. DEA INA-PG/Université Paris XI. INRA Laon. 34 p.
3. Ludwig, B., (1989) : Essai d'évaluation de la sensibilité à l'érosion de sites du Nord du Bassin Parisien. Prise en compte des caractéristiques morphologiques. DEA, Université Louis Pasteur (Strasbourg I) UFR de Géographie.INRA Laon. 69 p.
4. Chery, P., (1990) Modélisation spatiale de la sensibilité au ruissellement et à l'érosion. Recherche sur la combinaison de données cartographiques du milieu. Région Nord-Pas de Calais. DEA de Pédologie. Université Paris VI. INRA Orléans. 47 p.
5. Auzet, V., Boiffin, J., Papy, F., Maucorps, J. & Ouvry, J.F., (1991) : An approach for the assessment of erosion form and erosion risk on agricultural land in the northern Paris Bassin, France. In : "Soil erosion on agricultural land". Eds. : Boardman J., Foster I.D.L., Dearing J.A., Wiley & Sons, Chichester. 383-400.
6. Hudson, N., (1971) : Soil conservation. Eds. : BT Batsford Limited, London.
7. Wischmeier, W.H., Johnson, C.B. & Cross, B.V., (1971) : A soil erodibility monograph for farmland and construction sites. Journal of soil and Water Conservation, 20. 189-193.
8. Bergsma, E., (1986) : Report of soil erodibility studies in Java, South Sumatra and Thailande. In : "The erodibility studies were part of the research management and Land Use Zoning (ILWIS)". Doc-poly. ITC-RCR, Bakosurtanal. 13 p.
9.Rubio, J.L., (1988) : Erosion risk mapping in areas of the Valencia Province (Spain). In : "Erosion assessment and modelling". Eds : Morgan R.P.C. and Rickson R.J., CEC, Brussels, 3-24.
10. Albaladejo Montoro, J., Ortiz Silla, R. & Martinez-Mena Garcia, M., (1988): Evaluation and mapping of erosion risks: an example from S.E. Spain. Soil Technology, (1) 77-87.
11. Madsen, H.B., Hasholt, B. & Platou, S.W., : (1986) The development of a computerized erodibility map covering Denmark. In "Soil Erosion in the European Community, Impact of Changing Agriculture". Eds. : Chisci G. and Morgan R.P.C., CEC, Balkema, Rotterdam. 143-154.
12. Auerswald, K., (1988) : Erosion hasard maps for Bavaria. In "Erosion assessment and modelling". Eds. : Morgan R.P.C. and Rickson R.J., CEC, Silsoe, 41-54.
13. Morgan, R.P.C., (1980) : Mapping soil erosion risk in England and Wales. Assessment of Erosion. Eds. : De Boodt and Gabriels, Wiley & Sons. Chichester.
14. Biagi B., (1986) Development of a database of land characteristics and computerized analysis of actual and potential land degradation risks. In : "Soil Erosion in the European Community, Impact of Changing Agriculture". Eds. : Chisci G. and Morgan R.P.C., CEC, Balkema, Rotterdam. 125-141.
15. Giordano, A., (1988) : A first approximation of soil erosion risk assessment in the southern countries of the European Community. In : "Erosion assessment and modelling". Eds. : Morgan R.P.C. and Rickson R.J., CEC, Silsoe, 25-39.
16. Auzet, V., (1987) : L'érosion des sols par l'eau dans les régions de grandes cultures : aspects agronomiques. Eds. : Ministère de l'Environnement et Ministère de l'Agriculture. CEREG, CNRS. 60 p.

17. Monnier, G., Boiffin, J. & Papy, F., (1986) : Réflexions sur l'érosion hydrique en conditions climatiques et topographiques modérées. Cas des systèmes de grandes cultures de l'Europe de l'Ouest. Cahier ORSTOM, série Pédol. 12, 123-131.
18. Boiffin, J., Papy, F. & Eimberck, M., (1986) : Influence des systèmes de culture sur les risques d'érosion par ruissellement concentré. Agronomie 8, 663-673.
19. Daix, C., (1991) : Analyse du ruissellement en fonction de la dynamique spatiale des états de surface du sol. Approche à l'échelle du bassin versant élémentaire en Pays de Caux. Mémoire ENITA, Dijon. 35 p.
20. Eimberck, M., (1989) : Facteurs d'érodibilité des sols limoneux : réflexions à partir du cas du Pays de Caux. Cahier ORSTOM série Pédol., Vol 25. 81-94.
21. Le Bissonnais, Y., (1988) : Analyse des mécanismes de désagrégation et de mobilisation des particules de terre sous l'action des pluies. Thèse Univ. Orléans. 216 p.
22. Papy, F. & Boiffin, J., (1989) The use of farming systems for the control of runoff and erosion. In : Soil erosion protection measures in Europe. Soil Technology Series 1. 29-38.
23. King, D., Hardy, R. & Le Bissonnais, Y., : (1991) Evaluation spatiale de la sensibilité à l'érosion hydrique des terres agricoles de la région Nord-Pas-de-Calais. Eds. : INRA SESCPF - Conseil Régional de la Région Nord-Pas-de-Calais, 208 p.
24. Mathieu, R., (1991) : Elaboration d'une base de données géographique sur l'environnement en vue de l'estimation des risques d'érosion à l'échelle régionale. Réflexions méthodologiques et résultats pour la région Nord-Pas-de-Calais. Internal Report INRA-SESCPF, Orléans, 50 p.
25. King, Ch., Le Bissonnais, Y., Daroussin, J. & Malon, J.F., : (1991) Appui de la télédétection pour l'évaluation spatiale de la sensibilité à l'érosion hydrique des terres agricoles de la région Nord-Pas-de-Calais. Internal Report BRGM-INRA, 80p.
26. Daroussin, J., : (1991) Kaleïdos : une boite à outils complémentaire pour ARC/INFO. Proceedings INRA seminar : "gestion de l'espace rural et SIG". Paris, 309-320.
27. Legros, JP. & Bornand, M., (1989) : Systèmes d'Information Géographique et zonage agropédoclimatique. In: "Le zonage agropédoclimatique" - Commission d'Agrométéorologie de l'INRA. 23. 96-111.

Papy, F. & Douyer, J., (1991) : Influence des états de surface du territoire agricole sur le déclenchement des inondations catastrophiques. Agronomie, 11, 201-215.

Farm Land Erosion: In Temperate Plains Environment and Hills
S. Wicherek (Editor)

Assessment of Soil Losses in Brie (France) : Measuring suspended loads in rivers with a graduated monitoring network

M.J. Penven [a] and T. Muxart [b]
with collaboration of A. Andrieu and S. Chambolle, C.N.R.S.-URA 14

[a] Univ. Paris 8 et C.N.R.S.-URA 141, Laboratoire de Géographie Physique, 1 Pl. A. Briand, 92190 Meudon, France

[b] C.N.R.S.-URA 141, Laboratoire de Géographie Physique, 1 Pl. A. Briand, 92190 Meudon, France

SUMMARY

The main purpose of the "Land Use" group of the C.N.R.S. (Centre National de la Recherche Scientifique) PIREN-SEINE research programme is to establish the effects of land use changes in the Bassin Parisien on the hydrological regime and on some water physical and chemical characteristics. The Grand Morin catchment has been chosen for its social and economical modifications are representative of those recently noticed in the Bassin Parisien.

The effects of agricultural activities on the water quality form the first subject of the research. In this connection, *the concentrations of suspended load* - as a result of soil erosion in cultivated lands - carried out in the rivers were selected as an indicator of these modifications.

Actually, the aim is to understand the relationships between, on the one hand the variations of suspended load and flow and, on the other hand the physical and socio-economical factors which produce these changes.

According to the latters, the catchment has been divided into several *functional space units of different scales.* Their sizes vary from parcels of land drained by tubes, through first order catchment, to the whole Grand Morin watershed.

The hydrological system inputs (climatic data) and outputs (water discharge, suspended load, chemical polluants) are measured and the environmental and socio-economical factors are characterized for each space unit.

A network of nine stations has been fitted out in 1991 by the C.G.E. (Compagnie Générale des Eaux) and the C.N.R.S.- URA 141. It is completed by two stations already at work : the C.E.M.A.G.R.E.F. (Centre National du Machinisme Agricole du Génie Rural des Eaux et des Forêts) station at Mélarchez, located on a tributary of the Grand Morin, and the C.G.E. one at Neuilly sur Marne. In each of them, water heights are measured, water discharges are calculated and water samples are taken for chemical analysis.

Three of the C.N.R.S. stations, related to increasing drainage area (successively 90 ha, 5 and 20 km^2), are located on the river Vannetin, a tributary of the Grand Morin. The fourth one is located at the outlet of a drainage tube (drained surface of 5 ha) in the river Vannetin.

Some preliminary results obtained by the C.N.R.S. - URA 141, during the winter 1991-1992, are presented in this paper.

RESUME

Dans le cadre du programme PIREN-SEINE du C.N.R.S., l'un des quatre groupes de recherche a pour problématique de déterminer les effets des changements de l'utilisation de l'espace - et des pratiques et techniques qui y sont associées - sur le régime hydrologique et sur certaines caractéristiques physico-chimiques des eaux du bassin versant du Grand Morin. Ce dernier est en effet représentatif des types de changements socio-économiques ayant affectés récemment le Bassin Parisien .

Le travail a tout d'abord porté sur l'étude de l'impact des activités agricoles sur la qualité des eaux des rivières et les concentrations en *matières en suspension (MES) d'origine agricole* ont été choisies comme un indicateur de celle-ci.

L'objectif essentiel est de comprendre les modifications de la teneur en MES, couplée au débit, qui sont fonction des combinaisons de facteurs explicatifs appartenant au milieu physique et aux systèmes socio-économiques .

Afin de répondre à la complexité du problème étudié, un découpage de l'espace *en niveaux d'analyse hiérarchisés* a été effectué. Ces niveaux correspondent à un ensemble d'unités spatiales fonctionnelles emboîtées, combinant activités humaines et données du milieu physique. Leur taille s'échelonne depuis les unités élémentaires (drains agricoles, bassins versants élémentaires) jusqu'à l'échelle globale du Grand Morin. Les variables de sortie des hydrosystèmes (débits, MES mais aussi polluants chimiques) sont mesurées dans des stations installées à l'aval de chaque unité spatiale considérée. Pour chacune d'entre elles, les entrées (données climatiques), les facteurs physiques (texture du sol, évolutions de son état de surface et de sa structure) et socio-économiques sont caractérisés et leurs évolutions suivies.

En 1991, neuf stations de mesures formant un réseau hiérarchisé ont été équipées et sont gérées par la C.G.E. (Compagnie Générale des Eaux) et le C.N.R.S. (URA 141). Il s'y ajoute deux stations déjà existantes : celle du C.E.M.A.G.R.E.F. à Mélarchez et celle de la C.G.E. à Neuilly sur Marne. Dans toutes les nouvelles stations les hauteurs d'eau sont mesurées, les débits calculés et des prélèvements d'eau effectués pour analyses.

Trois des stations du C.N.R.S. sont situées sur le ruisseau du Vannetin, affluent du Grand Morin, et correspondent à des surfaces drainées croissantes (successi-vement 90 ha, 5 et 20 km^2). La quatrième est disposée au débouché d'un drain agricole (superficie drainée de 5 ha). Les premiers résultats obtenus par le C.N.R.S., pendant l'hiver 1991-1992, sont présentés.

INTRODUCTION

The catchment basin of the river Seine is an intensively cultivated region where the occurrence of soil erosion due to agricultural uses is evident, as shown by numerous other researches (1 - 2 - 3 - 4 - 5 - 6 - 7 - 8).

Our research concerns the Brie région (east of Paris) ; it is included within the framework of a research programme aiming at modeling the variations in the quality of the surficial water resources on various scales. As a matter of fact, the Seine river basin is heavily used and occupied by man and the problem of water resources and of their quality represents a great challenge. This preoccupation has lead to the creation of the "PIREN-SEINE" (Interdisciplinary Research

Programme on SEINE ENvironnement) which seeks to understand the functionning of the river Seine system, in its physical and chemical as well as in its biological aspects, taking into account the local socio-economic activities.

One of the four research groups of the programme is in charge of studying the variations in the hydrological regime and in the water quality (suspended particulate matter, nitrate, phosphate, pesticides) of the Grand Morin river - a subtributary of the Seine river - and of its tributary, the Vannetin stream. First, the aims and approach of this group will be exposed, followed by the report of the initial results obtained on the scale of two small units (agricultural drain and first order river) of the agrosystem of the Vannetin catchment area. In this paper, only the data concerning the water discharge and the suspended load will be presented ; the latter data are dependant, in rural areas, on the soil losses from cultivated fields.

1. AIMS AND APPROACH : THE MONITORING NETWORK

1.1. Aims

The "Land use" group of the PIREN-SEINE programme has oriented its researches towards the study of the impacts of the changes in the land use on the regime and on some physical and chemical characteristics of the water in the Grand Morin catchment basin.

The upstream part of the catchment basin of this river is wholly occupied by agriculture specialized in cereal production, while urbanization is expanding rapidly over the downstream part, nearer to Paris. In this respect, the Grand Morin river catchment is representative of the socio-economic changes which have occurred in the Paris basin in recent years. In a first stage, our researches have been devoted to the study of the impact of rural activities on the streams; the concentrations in suspended particulate matter produced by agriculture have been selected as indicators, amongst others, of the quality of the water.

1.2. Approach and monitoring network

The catchment basin of the Grand Morin (about 1200 km^2) corresponds to a relatively complex area which is not possible to consider as a whole. It was thus decided to divide this catchment area into several graduated levels. These levels represent functional spatial nested units where specific human activities and elements of physical environment are mixed. Their size rank from elementary units (drained agricultural field, first order drainage area) to the global scale of the Grand Morin river basin (Fig. 1).

In 1991, nine monitoring stations were equipped, five by the Compagnie Générale des Eaux (CGE) [On behalf of the Syndicat des Eaux de l'Ile de France and with the financial help of the Water Agency of Seine Normandy] : one on the Marne river - upstream the Grand Morin/Marne confluence -, three on the Grand Morin and one on its tributary the Aubetin) ; four by the CNRS - Laboratory of Physical Geography (on another tributary of the Grand Morin, the Vannetin stream) [With the financial help of PIREN-Seine programme]. Choice of sites and of material to be used was made in close concertation between these two bodies with the help of a third body, CEMAGREF, whose one station (Mélarchez) of the Orgeval basin has been included within the monitoring network. To this network was added the automatic analyzing station of the drinking-water production plant belonging to the Compagnie Générale des Eaux at Neuilly-sur-Marne, located downstream of the Grand Morin / Marne confluence.

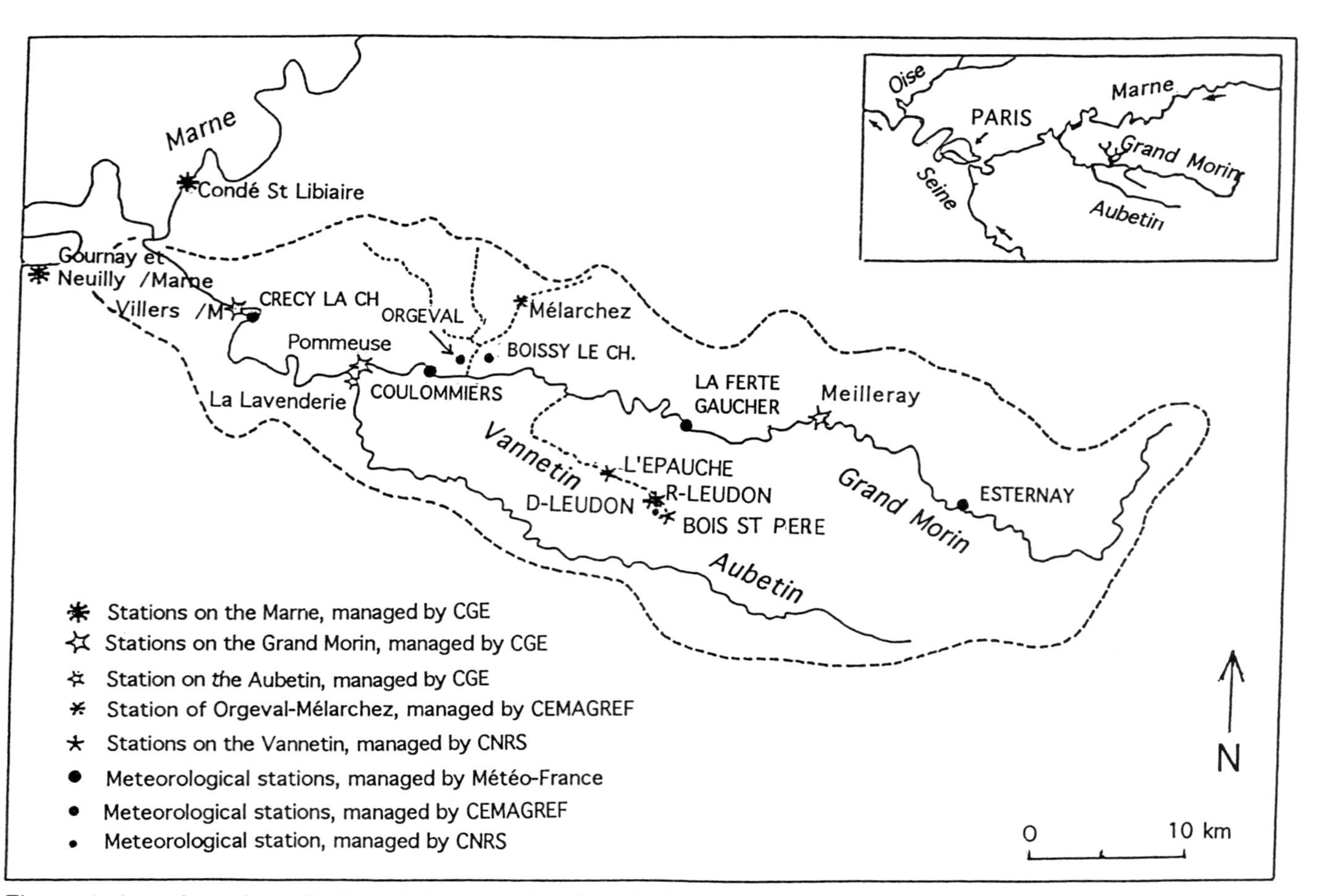

Figure 1. Location of monitoring stations on the Grand Morin river and on its tributaries, and on the Marne river near its confluence with the Grand Morin.

1.3. The upstream drainage area of the vannetin stream and its monitoring equipment

The monitoring stations are situated within the drainage basin of the Vannetin stream. The catchment is characterized by gentle slopes (3 to 4 % maximum at the most) and is organized in small basins, the floors of which are covered by silty colluvium. Silty to clayed superficial loams are spread over the other parts of the drainage basin whose basement is constituted by decalcified clays. The material used to be characterised by the presence of a temporary water table, near to the surface ; this is why a network of plastic drains was installed by farmers all over the catchment basin, at approximately 60 to 80 cm under the surface, which allowed the soils to dry out quickly. This area is very typical of the open field landscape of the Brie plateau and is mostly used for the production of winter cereals (over 50 % of the area each year) and peas ; in addition, sunflower, maïze and beets are less cultivated.

Within the catchment area of the Vannetin stream, four monitoring stations were installed and equipped in 1991 and 1992, following the pattern of a comprehensive system of graduated nested basins. Three of these stations are located upstream the first small village, called Leudon en Brie, on :

- a collector of both agricultural pipes and spring draining the Vannetin headstream basin, i.e. a 90 ha area (BOIS SAINT PERE) ;
- an agricultural drain (D-LEUDON) collecting water from a network of pipes under a 5 ha field ; its outlet is located immediately upstream of the following station on the Vannetin stream ;
- the Vannetin stream itself (R. LEUDON), upstream of the confluence of its first tributary ; the correspondance drainage area covers about 4 km^2 .

A fourth station (L'EPAUCHE) is installed in the Vannetin stream bed, downstream three villages (Leudon en Brie, Chartronges and Choisy en Brie). The catchment basin is approximately of 20 km^2.

The monitoring equipment of each of these stations consists of :

- a device measuring the water level ; its location differs in each station : for the main collector at BOIS SAINT PERE it is in a "contraflux"-type tube, for the drain (D-LEUDON) it is in a plastic container with a "V"-type outlet, and for the river beds at R-LEUDON and L'EPAUCHE it is located in a regular section covered with concrete;
- an automatic water sampling device which starts working, in each station, when the water level overflows a given threshold.

A rain gauge with a 0.2 mm accuracy has been installed at the R-LEUDON station.

2. RESULTS OF THE 1991-1992 WINTER SEASON

This paper deals with the only noticeable flood of the 1991-1992 winter season. The results given here will be only those from the stations in the drain of the 5 ha field (D-LEUDON) and in the Vannetin stream near Leudon (R-LEUDON), since all the other water samples had not yet been analyzed. The first water sampling occurred at both stations, on March 24, as soon as the flood began.

2.1. The conditions of removal of suspended particulate matter

They remained low at each station :

- for the agricultural drain (D-LEUDON, Fig. 2), curves of suspended particulate matter show five peaks ; maximum value is 79.3 $mg.l^{-1}$ registered at the beginning of the rising flood of the second discharge phase ;
- for the Vannetin stream (R-LEUDON, Fig. 3), several peaks are apparent. The highest maximum concentration value reaches 51 $mg.l^{-1}$; it occurred during the third phase of flow (the most important one) , i.e. later than in the drain.

During the 1992 March flood, the suspended loads appear to have been low; concentration values in the drain have always been higher than those in the water discharged by the stream

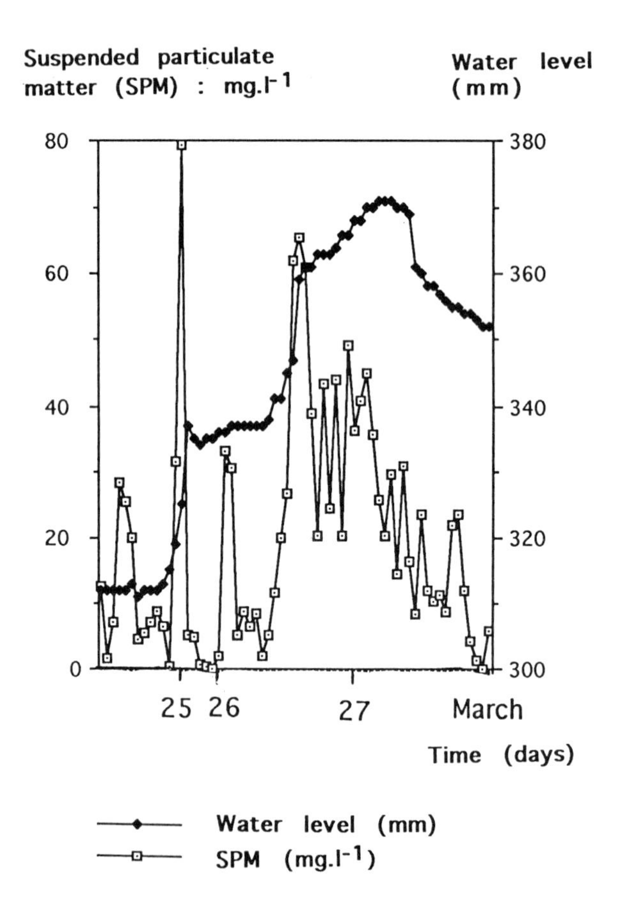

Figure 2. Variations in water levels and suspended particulate matter concentrations in the drain at D-LEUDON.

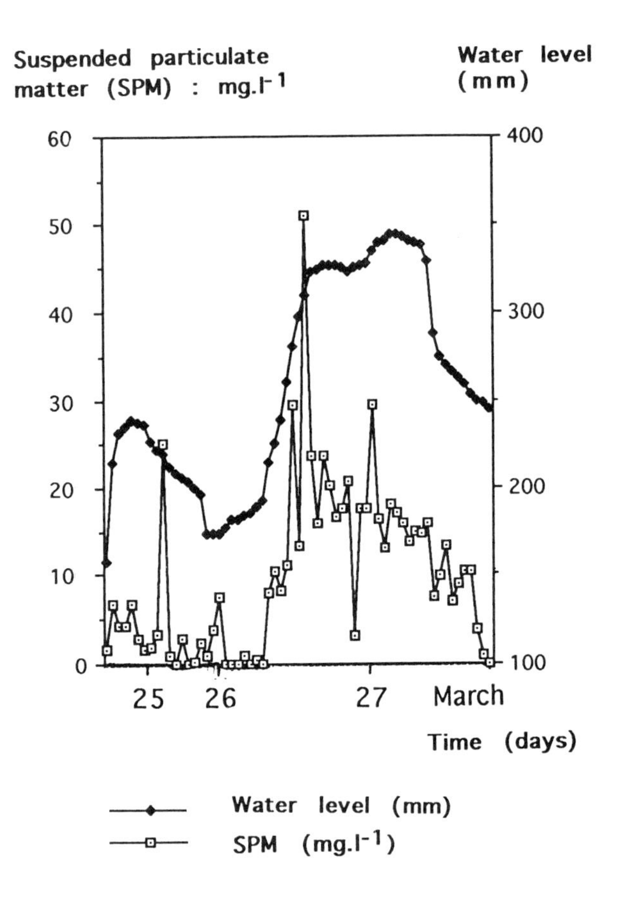

Figure 3. Variations in water levels and suspended particulate matter concentrations in the Vannetin stream at R-LEUDON.

2.2. The conditions of removal of suspended particulate matter

They depend on the development of both the rainfall event and the flood.

Characteristics of the rainfall event.

The 1991-1992 autumn-winter seasons were marked by a drought : the total amount of rain registered at Leudon station reached 322.8 mm over the time period running from 1991 September 1 to 1992 March 31, while the average value for this period is 407.8 mm in the catchment basin of the Grand Morin [According to an expert report (M. Tabeaud in (9) : the average value is calculated over an approximate 30 years period at 8 stations located in the catchment basin of the Grand Morin]. The previous summer season was, furthermore, especially dry. Winter rainfalls occurred in six periods alternating with long dry intervals. Two observations can be applied to all these events : a daily amount frequently inferior to 1 mm and a very low intensity.

The rainfall event which generated the first noticeable flood in March began on February 28 (Fig.4) after ten absolutely dry days. It had three phases :
- from February 28 to March 7 : 7.4 mm,
- from March 10 to March 15 : 12.4 mm,
- from March 20 to April 3 : 61 mm.

The two highest daily precipitated amounts occurred on March 23 and 26 with, respectively, 12.4 mm and 13.4 mm. The monthly amount of rain was above the average recorded on the catchment basin of the Grand Morin (55.9 mm) [According to an expert report (M. Tabeaud in (9) : the average value is calculated over an approximate 30 years period at 8 stations located in the catchment basin of the Grand Morin].

Intensities recorded during this rainfall event were low : the 13.4 mm of rain-water for the whole day of March 26 corresponded to a uninterrupted rain, the intensity of which was never greater than 0.4 mm per 12 min. The maximum intensity recorded reached 2.8 mm over 12 min. on March 23.

The low rainfall intensities may explain, together with the dryness of the winter season, the relatively weak response of the stream and of the drain to these rainfall events.

The water levels and water discharges.

The water levels remained moderate with a maximum value of 346 mm in the Vannetin stream, corresponding to a rise of 220 mm. Hydrograms of the drain (Fig. 4 A) and of the Vannetin stream (Fig. 4 B) are somewhat parallel, indicating a flood occurrence in 3 successive periods : the first one started during the afternoon on March 23, the second one during the evening on March 24, and the third and most important one by the end of the morning on the March 26. At the beginning of the flood, the water in the drain rose later (7 hours) than that of the stream, although afterwards they rose faster in the drain than in the stream-bed (2 hours earlier on March 24 and 26).

The flood peaks of the drain and of the stream did not coincide during the first two periods (Tab. 1) : intervals of 8 and 4 hours respectively. The third period was more complicated, presenting two phases : a first one was feeble in the drain and more obvious in the stream ; a second phase was well marked at both sites.

The flood of March 1992, characterized by moderate water levels, made clear the fundamental part played by agricultural drains in the variations in the water level of the Vannetin in its upstream catchment area. At the beginning of the flood the drains went into action later than the stream, allowing the beginning of a decrease of the water level in the stream-bed ; contrarily, during the next periods, the flood peaks were much steeper owing to a better correspondance in time between the drain and the stream.

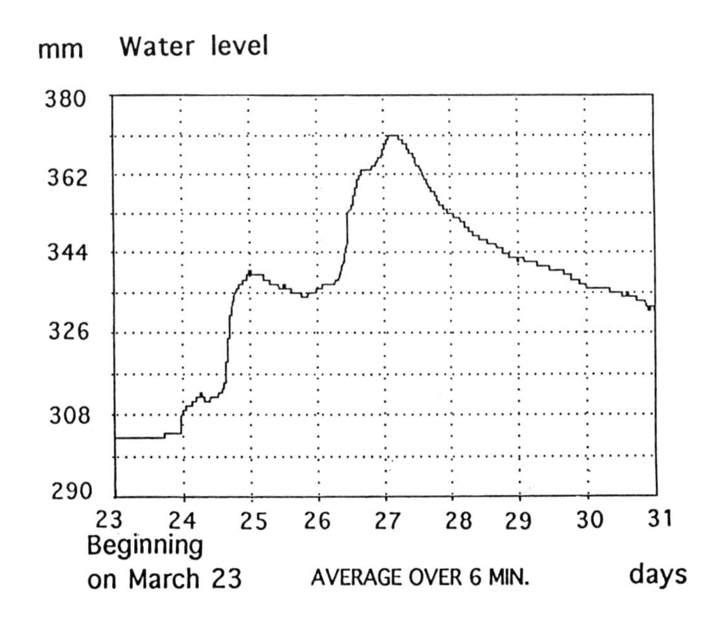

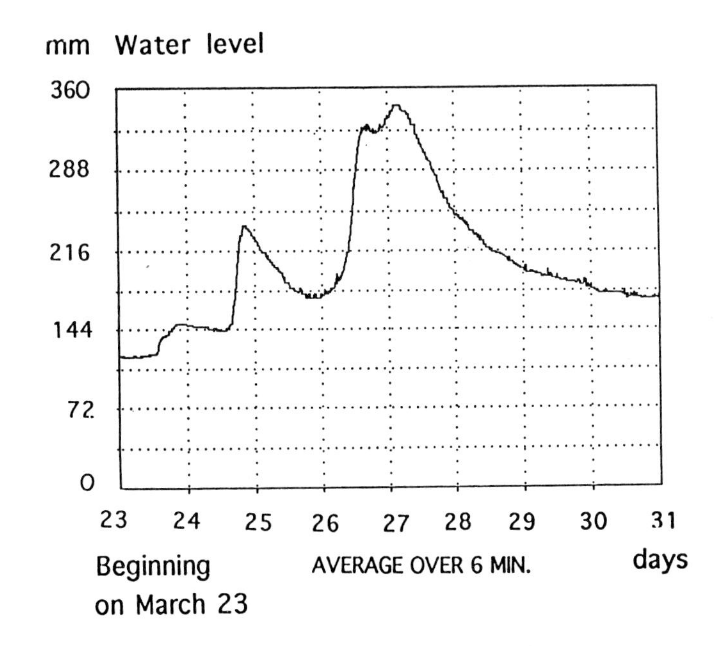

Figure 4 A. Hydrograms of the drain at D-LEUDON station.

Figure 4 B. Hydrograms of the Vannetin stream at R-LEUDON station

Table 1
Correspondance between suspended load peaks and water discharge peaks

STATION	Suspended peaks date ; hour	load conc. : mg/l	Water peaks date ; hour	discharge level : mm
drain : D - Leudon	24-3 ; 05h14	28.3	24-3 ; 06h48 to 06h48	313
drain : D - Leudon	24-3 ; 16h17	79.3	24/25-3 ; 23h48 to 0h24	340
drain : D - Leudon	26-3 ; 02h02	33.0	26-3 ; 02h06 to 06h30	317
drain : D - Leudon	26-3 ; 14h27	65.5	26-3 ; 16h18 to 20h	363
drain : D - Leudon	27-3 ; 23h02	49.3	27-3 ; 02h06 to 05h36	371
stream R - Leud.	N° sampling		23-3 ; 21h20	152
stream R - Leud.	24-3 ; 18h27 and 20h42	6.8	24-3 20h12 to 21h18	240
stream R - Leud.	25-3 ; 22h57	7.5	26-3 ; 01h30 to 04h42	183
stream R - Leud.	26-3 ; 13h57	51.0	26-3 ; 17h to 17h30	329
stream R - Leud.	27-3 ; 0h27	29.3	27-3 ; 03h30 to 04h	346

(*) Since sampling depends on a threshold value of 155 mm minimum in water level, no sampling occurred durint this first flood peak.

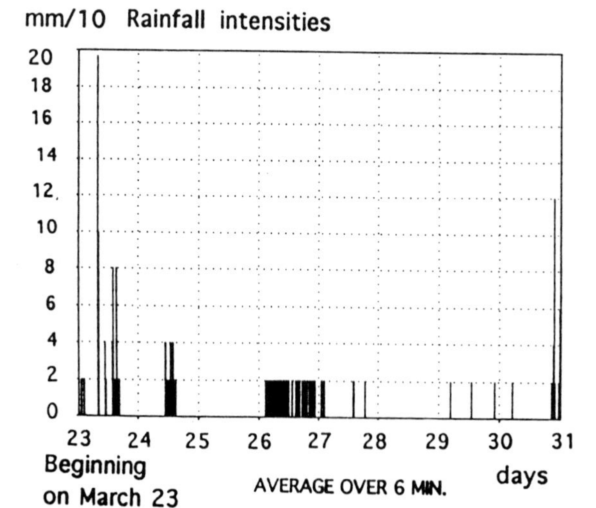

Figure 4 C. Rainfall measurements at R-LEUDON rain-gauge.

The flood started during the third phase of the rainfall event after a total amount of 34.8 mm of rain-water ; the efficient rain occurred on March 21,22 and 23 (Fig. 4 C), with a 15 mm total amount and a maximum intensity of 2.8 mm per 12 min. The second peak of the flood appeared after a 8.4 mm rainfall (maximum intensity : 0.8 mm per 12 min.) ; the third peak rose after a 7.4 mm rainfall only (maximum intensity : 0.4 mm per 12 min.). During this flood, water levels correspond to moderate flow rates (Table 2).

Table 2.
Water levels and flow rates in the drain and in the stream

STATION	Water level	Water discharge (l/sec) (*)
drain : D-LEUDON	24-03 : 312	0.12
drain : D-LEUDON	24-03 : 340	1.2
drain : D-LEUDON	26-03 : 365	2.0
drain : D-LEUDON	7-03 : 371	2.5
stream : R-LEUDON	23-03 : 152	22.0
stream : R-LEUDON	24-03 : 240	67.0
stream : R-LEUDON	26-03 : 329	152
stream : R-LEUDON	27-03 : 346	> 160.0

(*) The flow rates in the stream are estimated from the provisional gauge curve established by P. Rosique (CEMAGREF).

2.3. Comparative analysis of water level peaks and of suspended load peaks

Agricultural drain.

Curves in figure 2 show that the evolution in suspended loads over time displays five maxima which precede systematically the flood peaks in the drain (table 1). Incidentally, the 79.3 mg.l^{-1} value measured on March 24, at 16 h.15, most probably does not represent any maximum value since sampling in the drain stopped after this time.

Vannetin stream.

The same phenomenon can be noticed in the stream (table 1 and fig.3). However, there is one exception to this statement : a value of 25 mg.l^{-1} for suspended particulate matter was measured on March 25 at 01h 57 when the flood was decreasing. In this case, the suspended load peak was probably caused by the agricultural drains coming progressively into action in the area upstream from the station. As a matter of fact, 31 outlets of draining pipes (120 to 160 mm in diameter), plus 3 main collectors (250 to 300 mm in diameter) were located along the stream channel. At the beginning of the functioning of the Vannetin hydrosystem, inputs from these collectors arrive progressively along the stream and are diluted when they reach the R-LEUDON station several hours later than the suspended load peak from D-LEUDON station, located a few metres upstream from the R-LEUDON site.

During the whole of the March flood, the values of suspended loads remained low. They were higher in the water collected by the drain than that flowing in the stream. In both cases, the suspended load peaks occurred before the water level peaks.

3. DISCUSSION

The 1992 March flood was caused by moderate rainfalls with low intensities. In these conditions, the state of the soil surface was not very much damaged, either in the fields devoted to winter crops, rather uncovered in March, or in the bare fields ploughed in autumn or where spring crops (peas) had been sown, although in this later case, the special preparation of the soil (thin clods) usually leads to a greater tendancy to crusting. The results were that runoff was so low that no particulate matter reached the stream and that rain water infiltrated into the soil. Infiltration lasted throughout the rainfall event, feeding the underground network of agricultural draining pipes. As a result, an average flood occurred in both the monitored drain and the stream bed.

The flood in the stream largely depended upon the flood in the drains and the suspended loads measured in the stream water were mainly supplied by the drainage network. In fact, even if the first small peak in the stream may be explained by the removal of particulate matter already in the stream bed, the second peak shows definitely the dependance on the agricultural drainage, as stated above. Furthermore, the higher suspended loads in the drain tend to confirm this origin. Finally, the fact that the water discharges were low and that rather dense vegetation covered the bed of the stream, explains why the banks and the bed did not supply much material.

Observations made during the spring season of the previous year show that this is not always the case. In March 1991, monitoring stations had not yet been installed. A high flood occurred in the afternoon of March 21 after a light morning rain which ended by a downpour between 12h 30 and 15h. While turbid water was flowing through the agricultural drains, the stream was carrying a heavy brown water ; the concentrations of suspended particulate matter in the Vannetin must have been much higher than those measured at the same place, in 1992, when the water colour was only yellowish. In 1991, the suspended load in the stream was not only supplied by the drainage network, but also partly by the soil erosion on slopes. The consequence of the heavy noon rain was, in fact, that the soil crusted widely and that the fields were rainwashed. Rill wash also occurred in the tracks of agricultural machines, and may also have supplied suspended particulate matter to the stream. In a place which was especially fragile (overflow from a road-side trench), a field was even actively eroded by a broad rill carrying loams which were partly deposited downslope and partly discharged into the stream flow. On this occasion, the banks and the bed of the stream provided also a high quantity of particulate matter. Observations made after the flood show the obvious erosion of some parts of the bed of the stream (a several cm deep notch appeared in the bed and in its banks). As a matter of fact, the channel, straightened and cleaned out in 1990, was not yet covered by vegetation and therefore was subject to hydraulic erosion. Furthermore, sediment deposits were noticed in some places in the bed, which might have behaved as sediment supply during the next 1992 flood.

CONCLUSIONS

The informations provided by the results from the drain (D-LEUDON) and the Vannetin stream (R-LEUDON) stations are as yet partial since they need to be completed by the analysis of results from the intermediate-scale level (BOIS SAINT PERE) and the upper-scale one (L'EPAUCHE). In spite of the absence of these later data, some conclusions may be drawn.

The 1992 March flood is an illustration of the fundamental part played by the agricultural underground drainage network in the output of particulate matter (especially clayed material) during a rainfall event characterized by low intensities falling on soils which have suffered a rather heavy drought and which do not show extended sign of soil sealing. The only type of soil erosion was then constituted by the discharge of suspended particulate matter by the draining pipes. The quality of the water of the streams is thus largely determined by that of the drainage collectors (10). This type of flood illustrates the basic - i.e. minimum - operating system of soil losses in catchment basins in an open field land area, on the drained soils of the Brie region.

However, there are more complex situations, where other processes causing the removal of particulate matter in the system of a catchment area act simultaneously and interfere with each other. The 1991 March flood episode is one example of this.

The examination of the whole of the results from all the stations in the catchment basin should lead to a better understanding of the operating mechanisms on the various spatial nested scales and should allow us to complete the model of sediment delivery by the hydrosystem as it has been constructed by Gafrej and Leviandier (11 - 12).

AKNOWLEDGEMENTS

We are pleased to aknowledge P. Rosique (CEMAGREF) who made the provisional gauge curve of the Vannetin stream at R-Leudon station.

REFERENCES

1. Auzet A.V. (1987) : L'érosion des sols cultivés en France sous l'action du ruissellement. Ann. de Géographie, 537, pp. 529-556

2. Auzet A.V. (1988) : L'érosion des sols par l'eau dans les régions de grandes cultures : aspects agronomiques. Ministère de l'Environnement/ Ministère de l'Agriculture/ CEREG-URA 95-CNRS, 60 p.

3. Auzet A.V. , Boiffin J., Papy F., Maucorps J. and Ouvry J.F. (1990) : An approach to the assessment of erosion forms and erosion risk on agricultural land in the northern Paris Basin, France. In "Soil erosion on agricultural land". Ed. John Wiley and sons, London, pp. 383-400

4. Boiffin J. , Papy F. and Peyre Y. (1986) : Systèmes de production, systèmes de culture et risques d'érosion dans le pays de Caux. Ministère de l'Agriculture. 154 p. + Annexes.

5. Boiffin J. , Papy F., Eimberck M. (1988) : Influence des systèmes de culture sur les risques d'érosion par ruissellement concentré (II), Agronomie, vol 8 (9), pp. 745-756

6. Jarry F. (1988) : Genèse du ruissellement sur terrains agricoles. Une approche expérimentale par simulation de pluie et télédétection. Thèse, Université Paris 7, 97p.

7. Papy F. and Boiffin J. (1988) : Influence des systèmes de culture sur les risques d'érosion par ruissellement concentré (I), Agronomie, vol 8 (8), pp. 663-673

8. Wicherek S. (1990) : Paysages agraires, couverts végétaux et processus d'érosion en milieu tempéré de plaine de l'Europe de l'Ouest. Soil technology, Catena, V.3, n° 2, pp. 199-208.
9. Penven M.J., Muxart T, Mussot R., Roussel I., Tabaud M., Louahab Y. and Fauchon N. (1991) : Le bassin versant du Grand Morin. Premières données et installation du réseau de mesures. Rapport PIREN-SEINE III/91/05, Paris
10. Arlot M.P. (1989) : Caractérisation et limitation de l'impact du drainage agricole sur la qualité des eaux. CEMAGREF- Mission "Eau-Nitrates", 3 fasc., 174 p.
11. Gafrej R. and Leviandier T. (1991) : Modélisation statistique et conceptuelle des matières en suspension. Journée thématique de l'A.G.F., in : Influences des modifications des structures agraires sur l'érosion des terres. 1er juin 1991, St Cloud.
12. Gafrej R. and Leviandier T. (1992) : Statistical and conceptual modelling of sediments transport in a small catchment. European Geophysical Society, 17th General Assembly : Erosion and sediment transport processes and pathways. Edingburgh, 6-10 april 1992, 7 p., 11 fig.

Farm Land Erosion: In Temperate Plains Environment and Hills
S. Wicherek (Editor)

Gully typology and gully control measures in the European loess belt

J. Poesen

National Fund for Scientific Research, Laboratory for Experimental Geomorphology, K.U.Leuven, Redingenstraat 16 bis, B-3000 Leuven, Belgium

SUMMARY

In the intensively cultivated loess belt of Europe gully typology has been studied for erosion risk mapping and as an aid to the selection of gully control measures. Two main gully types occur in these loess landscapes : *ephemeral gullies* and *bank gullies* (i.e. gullies associated with steep steps (= banks) in the topography).*Ephemeral gullies* form where overland flow concentrates, i.e. in natural drainage-lines (thalwegs of zero order basins) or along (or in) linear landscape elements (e.g. drill lines, plough furrows, parcel borders, access roads, etc.). Ephemeral gullies differ in their cross sectional shape which can be described by their width-depth ratio (w/d). The w/d of ephemeral gullies is largely controlled by the thickness and resistance properties of the soil horizons. Ephemeral gullies in extremely erodible calcareous loess are usually deep with a w/d <<1. Results from field studies indicate that when soil moisture content exceeds 20 % shear strength at saturation (measured with a torvane after artificial saturation of the top soil) is a good indicator of the resistance of the soil horizon to concentrated flow erosion. Quantitative data on the *erosion resistance* of different horizons of a typical loess soil profile indicate the Bt horizon to be 3 to 4 times more resistant than the A2 and A3 horizon. The risk of ephemeral gully incision can be assessed by measuring the *thickness* of the soil horizons and their *shear strength at saturation. Bank gullies* form where a wash line, a rill or an ephemeral gully crosses an earth bank (e.g. a terrace, a lynchet, an exploitation talus, a sunken road bank, etc.). Bank gullies are not so much affected by overland flow intensity, but by other processes such as piping and mass movement.

The development of *ephemeral gullies* can be prevented by several means : i.e. by soil compaction, by preventing subsoiling, by conservation tillage and by establishing grassed waterways or of erosion resistant access roads in concentrated flow zones. The formation of *bank gullies* can be avoided by eliminating overland flow in the catchment, by eliminating banks in the landscape or by establishing grassed buffer zones just upslope of banks. Once these gullies have formed they can be controlled either by preventing runoff from flowing through the headcut or by applying biotechnical control measures. Testing structural measures to control a bank gully reveals the potential of *geomembranes*.

INTRODUCTION

Large areas of France, Belgium, the Netherlands and Germany are covered by loess which is part of a loess belt that covers the southern parts of the European lowlands and which extends further eastwards into Poland and the Ukraine (1). Soils in this loess belt are intensively cultivated and seriously affected by physical degradation (i.e. surface sealing, crusting and compaction) and water erosion (2 - 3). Soil loss by water results not only from splash erosion (4), interrill and rill erosion (5 - 6 - 7) but also from gully erosion (8 - 9). Referring to interrill and rill erosion, these silt loams with a geometric mean particle size (Dg : 10) ranging between 0.016 and 0.025 mm belong to the world's most erodible soils, particularly when their structural stability is low (e.g. due to a low organic matter content) (Fig. 1). As gully erosion frequently occurs in the loess belt it needs to be found out what makes these soils so susceptible to gully erosion. This paper starts with an examination of the different gully types that form in the intensively cultivated European loess belt. Factors influencing gully erosion are analyzed in order to map erosion risks and to select and apply measures of gully prevention and control. Particular attention is paid to the resistance properties of different loess soil horizons to concentrated flow erosion.

RESULTS

Typology

A gully has been defined as a channel resulting from erosion and caused by the intermittent flow of water usually during and immediately following heavy rains. These channels are deep enough to interfere with, and not to be obliterated by, normal tillage operations (12). In this study rills and gullies are distinguished by a critical cross sectional area of one foot^2 (= 929 cm^2) (13).

The typology of gullies formed on intensively cultivated land with rolling topography in the Belgian loess belt has been studied by Poesen (8) and by Poesen and Govers (9). Two main gully types have been recognized based on their location in the landscape, their morphology as well as on the dominant erosion process leading to their formation : i.e. ephemeral gullies and bank gullies. Figure 2 illustrates their characteristic location in the landscape.

Ephemeral gullies form where overland flow concentrates, i.e. either in natural drainage-lines (thalwegs of zero order basins or hollows) or along (or in) linear landscape elements (e.g. drill lines, plough furrows, parcel borders, access roads, etc.). They are continuous, temporary channels which are often erased by ploughing (Fig. 3).

Ephemeral gullies result from hydraulic erosion by concentrated flow. This implies that sediment detachment and removal is basically a function of flow intensity (14 - 15). Soil removed by these gullies represents approximately 35 % of total soil loss (due to interrill, rill and gully erosion) within catchments (calculation based on data collected by Van Daele (16)).

Practical implications suggest to subdivide ephemeral gullies according to their width-depth ratio (w/d). Wide ephemeral gullies with a w/d >> 1 cause important crop damage. In addition, a high percentage of total soil lost through gullying consists of fertile topsoil with a high organic matter and fertilizer content. These gullies are, however, easily erased by conventional tillage. On the other hand, deeper ephemeral gullies with a w/d = 1 or < 1 cause relatively little crop damage and the percentage of total loss of fertile topsoil is less than for the other gully type. These gullies, however, are not easily erased by conventional tillage and often heavy equipment is required to reshape the thalweg.

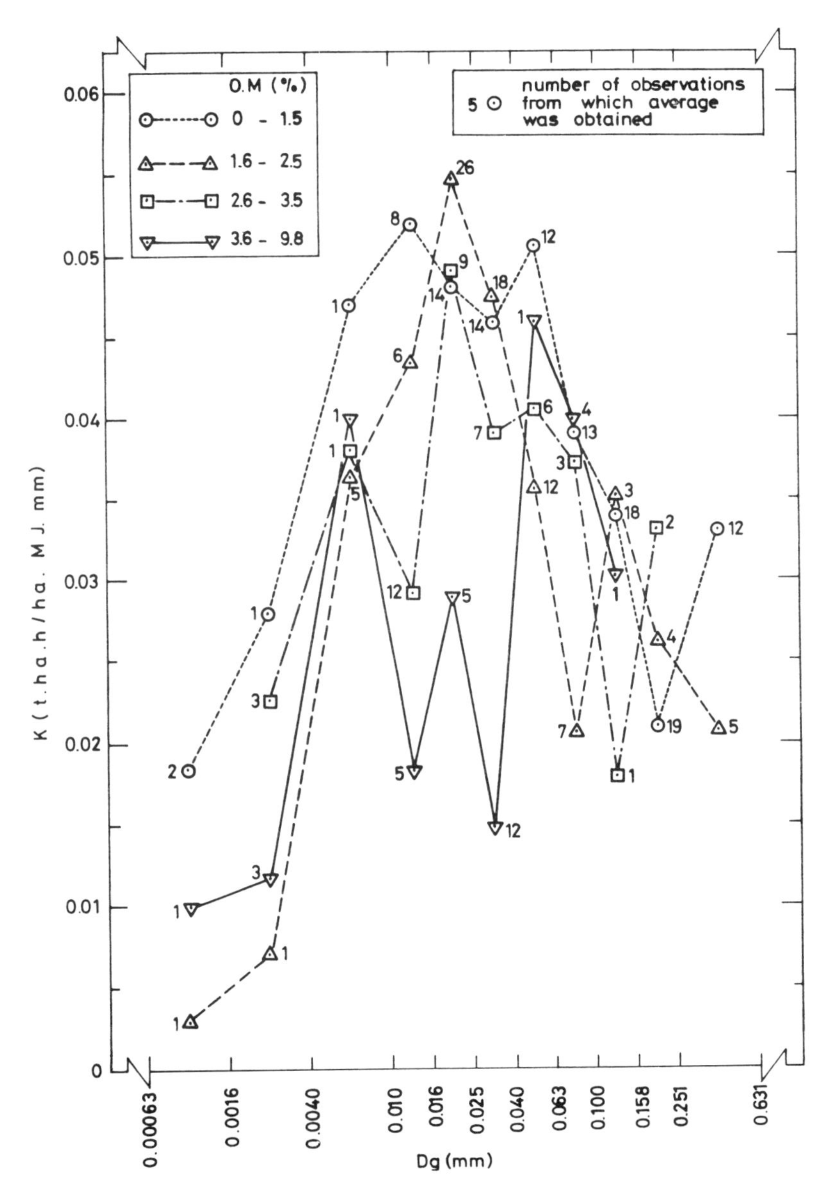

Figure 1.
Relation between geometric mean particle size (10) and organic matter content (OM) of 279 topsoils containing less than 10 % of rock fragments and their mean interrill and rill erodibility expressed by the K values of the Universal Soil Loss Equation. Source of K data which are calculated using field plot measurements is given in Poesen (11). Number of soils from which average K value was obtained for each Dg and OM class is indicated with each datapoint. Loess soils from the European loess belt have a Dg ranging between 0.016 and 0.025 mm.

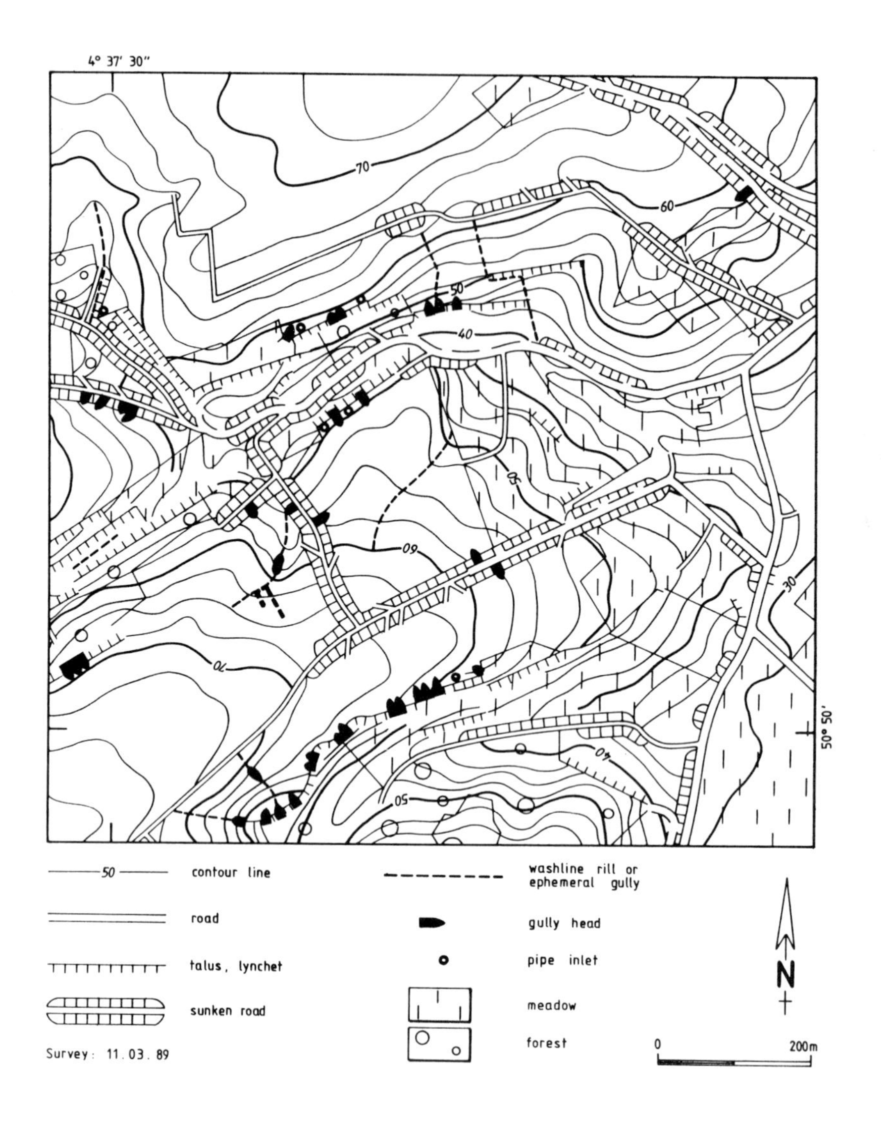

Figure 2. Excerpt of topographic map (Korbeek-Dijle and Leefdaal, Belgium) illustrating typical locations of ephemeral gullies and bank gullies in the European loess belt. All gully heads located on a bank are bank gullies.

Figure 3.
Ephemeral gully (width-depth ratio is approximately 2) erased by ploughing (Leefdaal, Belgium, March 1992).

A combination of various factors affecting flow width, flow intensity as well as resistance properties of the soil horizons to concentrated flow erosion determines the w/d of an ephemeral gully. The effects of local thalweg slope gradient and of a resistant soil horizon (i.e. a plough pan) on the width, the depth and the w/d ratio of an ephemeral gully is illustrated in Fig. 4.

If no erosion resistant soil horizon occurs at shallow depth in the profile the thalweg slope gradient does not seem to affect gully width (w) significantly, but has a significant positive effect on the depth (d) and therefore on the w/d ratio of the gully. The depth of an ephemeral gully increases rapidly beyond a critical slope of 0.03 - 0.04 which is in line with findings of Savat and De Ploey (17) and Govers (18) for rills. However, Schumm et al. (19) reported opposite trends for the effect of valley slope on w, d and w/d of simulated channels. If an erosion resistant soil horizon, such as a Bt horizon, a plough pan or a fragipan is present at shallow depth in the profile, gully depth remains small and therefore gully width and w/d are high even if the thalweg gradient is steep. If, on the other hand, calcareous loess (C horizon) outcrops, typical deep ephemeral gullies with a w/d << 1 develop (Fig. 5).

High intensity, low-frequency rainstorms precipitating on an initially dry and freshly cultivated top soil are likely to cause wide ephemeral gullies with a w/d >> 1 (Fig. 6). This is due to the fact that the high overland flow discharges generated during these storms will wet flat bottomed depressions in the landscape over a considerable width. Since the strength of the soil top layer is substantially reduced by wetting, the erodibility of this material is very high. Flow incision will be reduced at shallow depth if a less erodible layer is exposed by erosion, e.g. a soil layer with a high moisture content (20). On the other hand, low intensity rainstorms cause runoff to flow over a limited width in the thalwegs, often causing the formation of gullies with a w/d = or < 1 (e.g. Fig. 3).

Bank gullies form where a wash-line, a rill or an ephemeral gully crosses an earth bank (e.g. a terrace bank, a lynchet, an exploitation talus or a sunken lane bank ; Fig. 2, 7 and 8).

These discontinuous erosion features cause considerable soil losses : approximately 404 m^3 of soil were lost by this process in a 100 ha survey area (Fig. 2). If the volume of a particular bank gully is divided by its catchment area, a much higher average soil loss of 45 m^3 ha-1 was obtained in another but similar survey area (8). As bank gullies cannot be obliterated by conventional tillage operations these gullies are usually permanent landscape features. Field measurements revealed that, contrary to ephemeral gullies, erosion in these bank gullies is less controlled by overland flow intensity : see poor correlation ($r = 0.085$) in figure 9 between the eroded volume of 9 selected bank gullies and a measure of flow intensity (i.e. the product of the longitudinal gradient and catchment area of the gully) divided by a measure of soil resistance (i.e. penetration resistance measured with a pocket penetrometer).

Other processes such as piping and mass movement (i.e. slumping and soil fall) which are less controlled by the catchment size but depend more on the local site characteristics, such as biotic holes and cracks, control the development of bank gullies. This implies that prediction of the location and the volume of bank gullies is more difficult than it is for ephemeral gullies.

Resistance of different loess soil horizons to concentrated flow erosion

Quantification of the erosive power of overland flow is possible through the use of various flow intensity measures such as shear stress, shear velocity and unit stream power. Little or no quantitative information is available on the resistance of various soil horizons on loess to concentrated flow erosion. Such information, however, is crucial to predict A) the width-depth ratio of ephemeral gullies and the related type of environmental damage they cause and B) the retreat rate of the head of bank gullies (21). Therefore, field studies were undertaken to detect which soil parameters best reflect the resistance of loess loam soil horizons to erosion by concentrated flow.

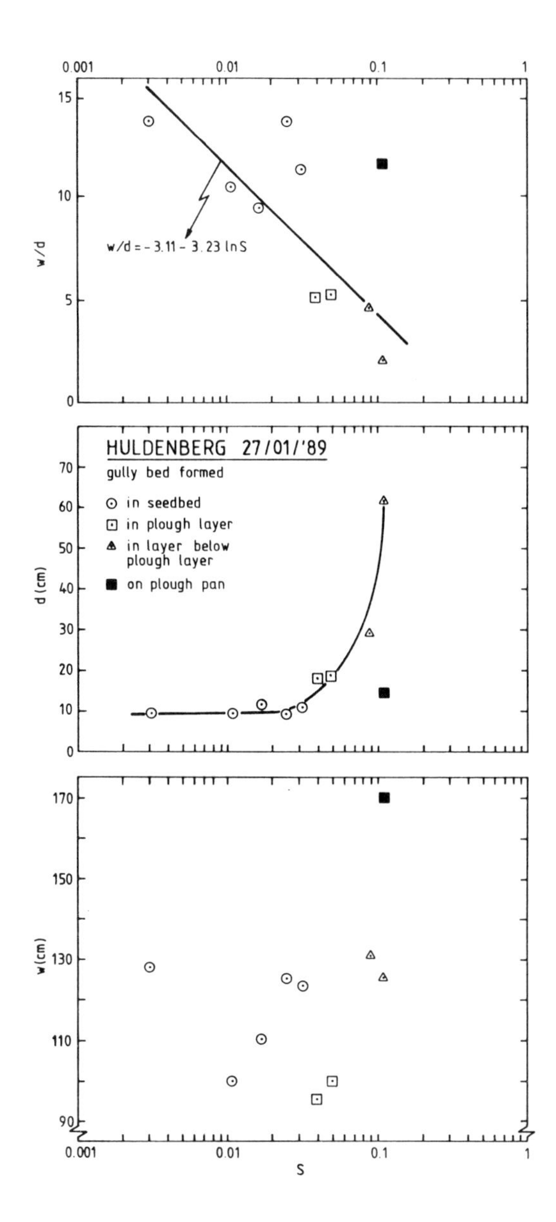

Figure 4.
Effect of thalweg slope gradient (S) on the width (w), the depth (d) and the width-depth ratio (w/d) of an ephemeral gully formed in soil horizons with different erosion resistance. All data refer to one and the same gully. Overland flow discharge during gullying was approximately the same in all gully cross-sections (after Poesen and Govers, 1990).

Figure 5. Ephemeral gully (Korbeek-Dijle, Belgium, March 1991). Note the change of its width-depth ratio (w/d) as the gully starts eroding calcareous loess : w/d is approximately 2 in the foreground due to the presence of part of an erosion resistant Bt horizon below the plough layer and drops to about 0.2 in the background due to the presence of calcareous loess at shallow depth in the profile.

Figure 6.
Ephemeral gully (width-depth ratio is approximately 16) formed during a high intensity low frequency summer storm. (Leefdaal, Belgium, August 1992). Note that this site is the same as shown in Fig. 3.

Figure 7.
Bank gully developed in a sunken lane bank (Leefdaal, Belgium, March 1991).

Figure 8. Bank gully developed in a sunken lane bank (Heverlee, Belgium, May 1986). Due to a thick deposit of calcareous loess at shallow depth in the soil profile, this bank gully has become more than 3 m deep.

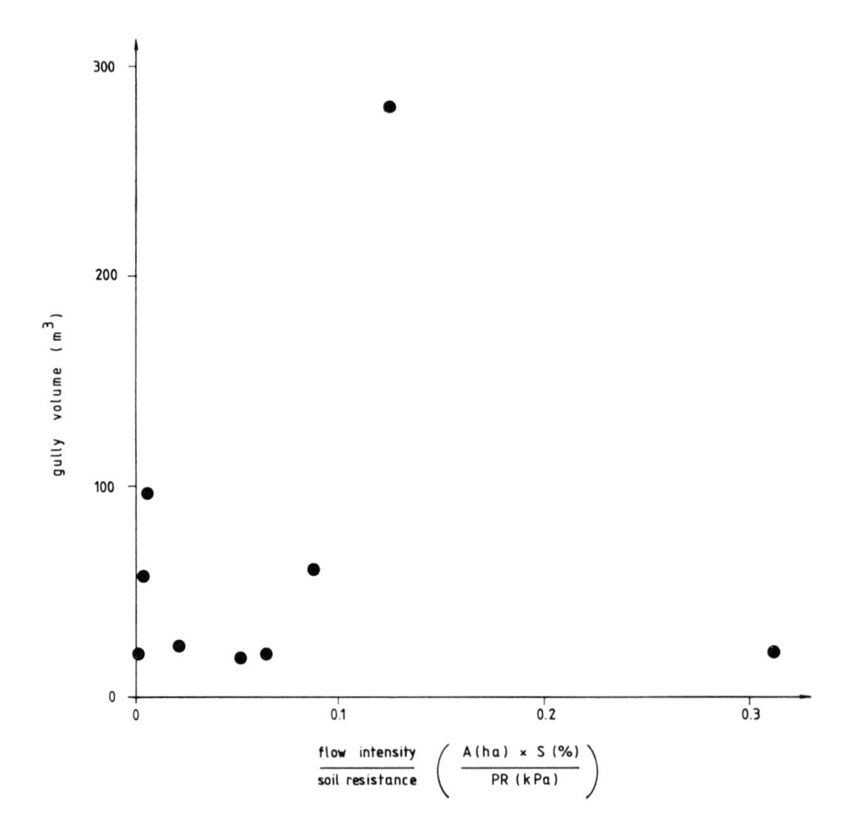

Figure 9. Eroded volume of 9 selected bank gullies (Korbeek-Dijle, central Belgium) and the corresponding ratio between flow intensity (assessed by the product of catchment area (A) and longitudinal gradient (S) of the gully) and soil resistance (assessed by the average penetration resistance of the soil horizons). $r = 0.085$. A is used here as a surrogate for overland flow discharge as the two are usually strongly positively correlated.

Poesen and Govers (9) investigated the relation between erosion resistance of horizons from loess soils and 14 soil properties. They found that only three properties were significant indicators. Out of these, soil shear strength measured under particular conditions was found to be a reliable and relevant indicator of erosion resistance. Shear strength was measured with a torvane in a moist period (autumn : moisture content > 20 %), after artificial saturation of the undisturbed soil horizon, the soil porosity consisting of essentially textural pores. Shear strength at saturation adequately reflects the condition of the soil horizon when being eroded by concentrated flow at that time of the year. Soil shear strength values obtained in this way are not only a function of intrinsic soil properties but also of the initial soil moisture content: shear strength at saturation is positively correlated with initial soil moisture content (9). In dryer periods (e.g. summer) much smaller differences in shear strength at saturation were found between the different soil horizons.

Therefore, shear strength measurements should be conducted during those periods of the year when concentrated flow events are likely to occur.

Figure 10 illustrates a typical shear strength profile of an undisturbed soil profile under deciduous forest which developed on a calcareous loess. Figure 11 depicts typical grain-size distribution curves for calcareous loess and decalcified loess from a textural B-horizon. As expected, shear strength (c) measured before artificial saturation of the soil horizons is higher than at saturation. Lowest c-values are found in the A2 and A3 horizons while highest c-values are recorded in the transition zone between the Bt1 and Bt2 horizon. At saturation a maximum c-value of 14.5 kPa was measured which is similar than the value for a Bt horizon observed elsewhere (9). The relative high c-values in the A1 horizon are caused by the soil binding effects of roots.

Figure 10 also features the corresponding critical shear velocities (U*cr) for flow detachment. U*cr was calculated using the equation (Rauws and Govers, 1988) :
U*cr = 0.86 + 0.56 c (1) with U*cr = critical effective shear velocity (cm s-1) for rill generation : U*cr = (g Rg S)0.5 with Rg = hydraulic radius due to grain resistance ; c = shear strength of the top soil (5 mm thick) measured at saturation (kPa).

These threshold conditions match relatively well with those described by Torri et al. (23 - 9). Such a U*cr profile allows one to predict which part of the soil profile is erodible for a given flow event. In addition, such U*cr profile helps not only to explain the spatial distribution of loess soil profiles which developed after deforestation but also to predict the type of gullies formed by concentrated flow erosion.

On slopes steeper than 2° to 3° concentrated overland flow often produces U*values larger than 3 to 3.5 cm s-1 (18). Since these values exceed the U*cr value for a typical undisturbed A horizon (Fig. 10), rills (and gullies) will rapidly develop and remove the A horizon. This is in accordance with the soil map of the Belgian loess belt in which the absence of an A horizon is indicated on most steep slopes. The detachment of the undisturbed Bt horizon however only occurs if U* exceeds 5-7 cm s^{-1}, which is less probable in rill flow but more probable in gully flow during high intensity rains. This too is in accordance with the soil map of the loess belt indicating that approximately 90 % of the area outside the valley bottoms in the central Belgian loess belt still has a complete (or partial) Bt horizon. Tillage of soil horizons, however, will reduce their average shear strength (24) and make these horizons more vulnerable to concentrated flow erosion. In the remaining 10 % of the central Belgian loess area the Bt horizon has been removed by quarrying for brick earth or by gully erosion.

If, during moderate rainfall events, concentrated flow with sufficient erosive power occurs on a freshly tilled soil surface or on an undisturbed A horizon, a narrow rill or gully with a width-depth ratio smaller than or about equal to 1 will erode vertically until a more resistant horizon (e.g. a plough pan, a Bt horizon or a fragipan) is reached. The gully will then widen to form an (ephemeral) gully with a w/d >> 1. If concentrated flow erodes calcareous loess (C horizon) a rapid formation of deep gullies with a w/d << 1 is to be expected because of the high erodibility of the calcareous loess (Fig. 5 and 8).

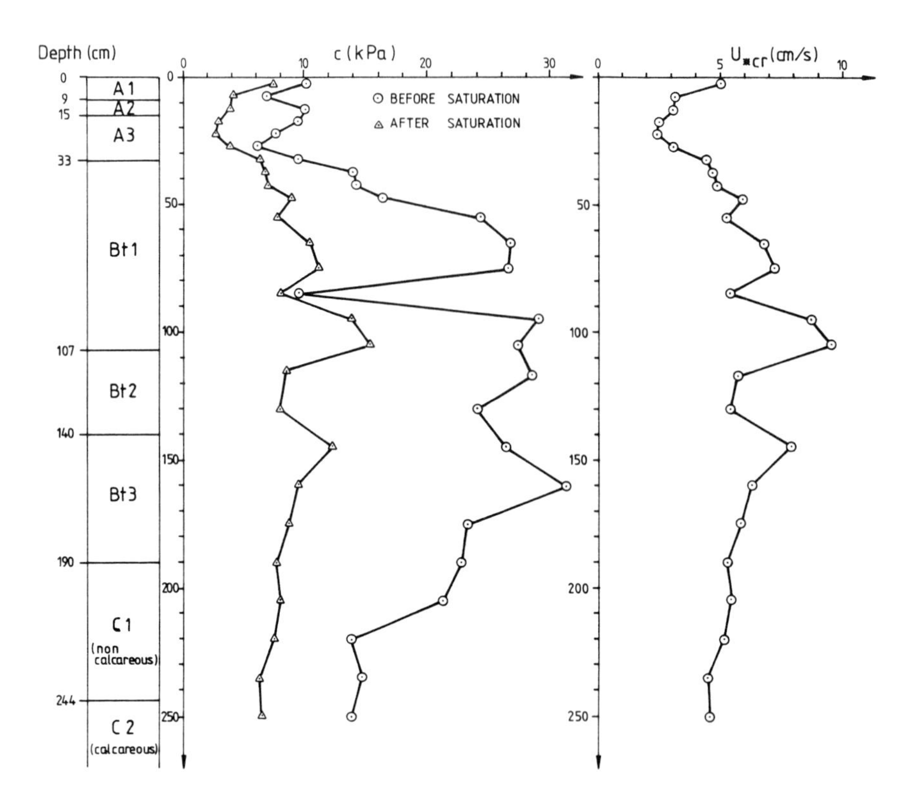

Figure 10. Soil shear strength (c) measured with a torvane, before and after saturation, versus depth for an undisturbed soil profile with a textural B horizon (well drained and slightly degraded) on loess under forest (Bierbeek, Belgium, March 1991) (Haplic Luvisol (F.A.O. classification), Glossic Hapludalf (USDA classification)). Antecedent gravimetric moisture content equalled 23 %. Each datapoint represents an average of 3 to 4 measurements. Also shown are critical flow shear velocity values (U*cr) calculated using soil shear strength at saturation and equation (1).

Calcareous loess (Fig. 11) in Belgium contains up to 16 % calcium carbonate which is present primarily as detrital grains (25). Field observations reveal that the presence of $CaCO_3$ in the silt and clay fractions of the loess increases its susceptibility to gully erosion which is in accordance with observations made by several investigators : Peele et al. (26) observed that the presence of finely dispersed calcium carbonate increased the instability of aggregates while Barahona et al. (27) found that the content of clay and silt sized calcium carbonates in Spanish soils had a strong positive effect on its interrill erodibility.

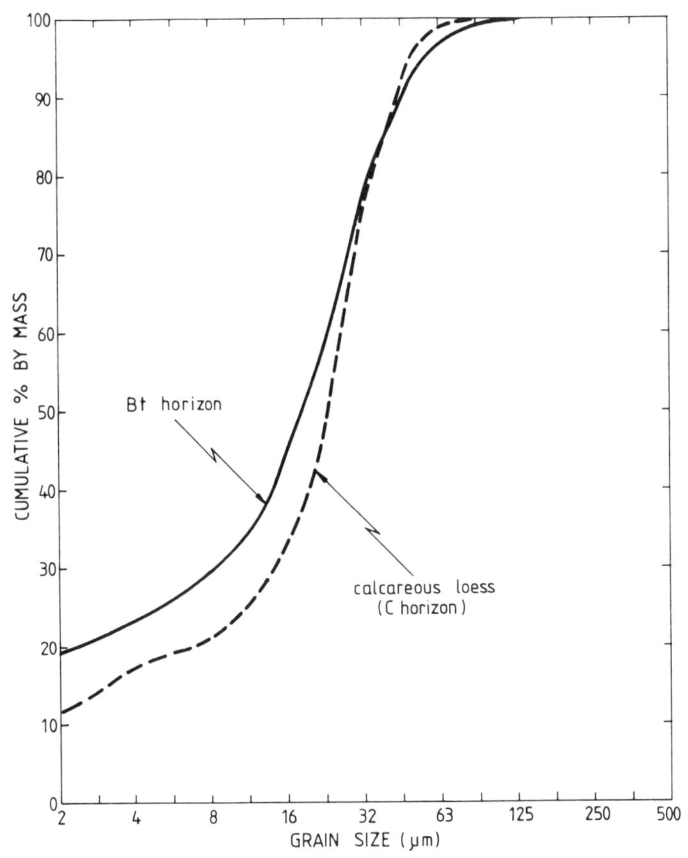

Figure 11.
Grain-size distributions of a typical calcareous loess (C-horizon) and a decalcified loess from a textural B-horizon (Bt) (Kesselt, Belgium) after physical and chemical dispersion.

Prevention and control measures

A basic principle for erosion control in general, but for gully control in particular is that 'prevention is better than cure'. Since both gully types described differ widely in the processes causing them, prevention and control measures also have to be different (8 - 9).

Ephemeral gullies

Ephemeral gullies can, in principle, be prevented by either preventing overland flow from gathering in the thalwegs of hollows or along (or in) linear landscape elements (e.g. by increasing the infiltrability of the soils) or by increasing the resistance of the soil top layer to concentrated flow erosion.

The first solution can be achieved in several ways : e.g. by improving the top soil structure, by mulching, etc. However, deep subsoiling, as has been proposed by Fullen (28) on loamy sands in order to break the plough pan, is not recommended on all slopes or depressions where overland flow might concentrate.

Field observations in central Belgium clearly reveal that the destruction of compact soil horizons by subsoiling leads to an increased risk of the formation of ephemeral gullies with a width-depth ratio = 1 or < 1.

An approved resistance of the top soil in the thalwegs to concentrated flow erosion can be achieved in different ways : i.e. by compacting the soil top layer, by applying no-tillage, by establishing grassed waterways or by constructing erosion-resistant access roads in concentrated flow zones. Compaction of loamy topsoils in the thalweg has been successfully tested by Ouvry (29) on gentle sloping valley bottoms in northern France. The efficiency of compaction in reducing ephemeral gully development, however, will largely depend on the moisture content at both the moment of compaction as well as at the moment of concentrated flow occurrence. No-tillage has also been successfully tested in central Belgium (9). Although no-tilled field plots produced more runoff as the conventionally tilled plots ephemeral gully development was negligible on no-tilled winter barley field plots while it was quite important (up to 20 ton ha-1 yr-1) on conventionally tilled winter barley plots during the same year. This difference is attributed to the mulch effect as well to the higher bulk density and shear strength values of the topsoils on the no-tilled plots. A grassed waterway in combination with fascines running across the thalweg was found to reduce ephemeral gully erosion by approximately 50 % in the loess belt of SW-Germany (Baade et al. (30)). In some situations, the construction of erosion-resistant paved access roads in thalwegs conducting large runoff volumes during and after heavy rainfall has been a suitable solution to decrease the risk of ephemeral gully formation in central Belgium.

Bank gullies

Several measures can be applied to prevent the formation of bank gullies : 1) preventing runoff from flowing across the bank, 2) eliminating banks in the landscape (e.g. in the framework of a land-consolidation programme), or 3) applying biotechnical measures, i.e. putting the land upslope of a bank under permanent grassland (which leads to an increase of infiltration rates, surface roughness and shear strength of the top soil) and by reinforcing the banks with deep-rooting species in order to reduce the risk of mass movements. Often, only the third type of measure will be feasible.

Once a bank gully has developed, two possible control measures can be taken : 1) preventing overland flow from flowing through the headcut, and 2) stabilizing the gully by structural measures and accompanying revegetation.

The first option can be achieved by improving the gully catchment in such a way that no runoff is produced or by diverting the surface runoff above the gully area. Runoff elimination in an intensively cultivated catchment, however, is difficult to achieve since the response of the entire catchment has to modified, which is seldom the case, while runoff diversion would create similar problems elsewhere. Nevertheless, the second option often remains the only feasible one.

A bank gully in Central Belgium was selected in order to test structural gully control measures (8 - 9). A first attempt to stop erosion was made by a plug of loose rocks installed in the gully head to prevent the headcut from further migration. The advantages of the loose rock plug are the low costs and its high porosity, which avoids excessive water pressure. After installation, however, erosion continued in the contact zone between the loess loam of the gully head and the rock plug. In order to prevent further erosion and the formation of a hollow which would undermine the headcut a geomembrane was used. A geomembrane is a thin, flexible sheet of synthetic material with a very low permeability (31). A reinforced polyethylene tarpaulin of 190 g m-2 was installed on the bottom of the concentrated flow channel upstream of the gully head and extended over the rock plug in a downslope direction.The installation of this geomembrane proved to be a very effective way to conduct the concentrated overland flow safely over the gully head without eroding and undermining the gully head. When applying this simple technique to control a gully head, care should be taken to extend the geomembrane sufficiently far downslope where the erosive forces can be safely dissipated on any kind of structure without creating new erosion problems.

CONCLUSION

Based on their location in the landscape, their morphology and their dominant erosion process, two main gully types can be distinguished in the cultivated loess belt that covers extensive parts of Europe : i.e. ephemeral gullies and bank gullies. Contrary to interrill erosion and partly also to rill erosion, the knowledge of the soil profile (type and thickness of horizons) and of shear strength values of the (un)disturbed loess soil horizons (consisting essentially of textural pores) at saturation in zones of the landscape where overland flow concentrates is essential when predicting gully risk and gully type. This study provides quantitative information on the erosion resistance of a typical loess soil profile. Testing of structural measures to control a bank gully reveals the potential of geomembranes.

ACKNOWLEDGEMENTS

I would like to thank Dr. J. Deckers for providing pedological information on the soil profile in Bierbeek, Dr. G. Govers for help during field work, Dr. D. Goossens for providing information on the spatial distribution of loess soil profiles in the study area and Dr. K. Bunte for critical reading of a previous draft of this paper.

REFERENCES

1. Pye, K., (1984) : Loess. Progress in Physical Geography, 8:176-217.
2. De Ploey, J., (1989a) : Soil erosion map of western Europe. Catena Verlag, Cremlingen-Destedt, Germany.
3. Oldeman, L.R., Hakkeling, R.T. and Sombroek, W.G., (1990) : World map of the status of human-induced soil degradation. Wageningen. International Soil Reference and Information Centre, Nairobi. United Nations Environment Programme I11.
4. Poesen, J., (1986) : Field measurements to validate a splash transport model. Zeitschrift für Geomorphologie Supplement Band, 58:81-91.
5. Poesen, J. and Govers, G., (1986) : A field-scale study on surface sealing and compaction on loam and sandy loam soils. Part II. Impact of soil surface sealing and compaction on water erosion processes. In Callebaut, F., D. Gabriels and M. De Boodt (eds.) Assessment of Soil Surface Sealing and Crusting. State University of Gent, Belgium, 183-193.
6. Govers, G. and Poesen, J., (1988) : Assessment of the interrill and rill contributions to total soil loss from an upland field plot. Geomorphology, 1:343-354.

7. Govers, G., (1990) : Rill erosion on arable land in central Belgium : rates, controls and predictability. Catena, 18:133-155.
8. Poesen, J., (1989) : Conditions for gully formation in the Belgian loam belt and some ways to control them. Soil Technology Series, 1:39-52.
9. Poesen, J. and Govers, G., (1990) : Gully erosion in the loam belt of Belgium: Typology and Control measures. In Boardman, J., I. Foster & J. Dearing (eds.). Soil Erosion on Agricultural Land. J. Wiley, Chichester, U.K., 513-530.
10. Shirazi, M. and Boersma, L. (1984) : A unifying quantitativeanalysis of soil texture. Soil Science Society of America Journal, 48:142-147.
11. Poesen, J. (1988) : A review of the studies on the mechanisms of incipient rilling and gullying in the Belgian Loam Region. In Ijioma, C.I., Anaba, S. and Boers, T.M. (eds.). Proceedings of the International Symposium on Erosion in S.E. Nigeria. Federal University of Technology, Owerri, Nigeria, 13-21.
12. Soil Science Society of America, (1984) : Glossary of soil science terms. Madison, Wisconsin, U.S.A.
13. Hauge, C.J., (1977) : Soil erosion definitions. California Geology, 30:202-203.
14. Foster, G.R. and Lane, L.J., (1983) : Erosion by concentrated flow in farm fields. In Ruh-Ming Li, P.F. Lagasse and Simons, Li & Associates (eds.) Proceedings of the D.B. Simons Symposium on Erosion and Sedimentation, Colorado State University, Fort Collins, 9.65 - 9.82.
15. Thorne, C.R., L.W. Zevenbergen, E.H. Grissinger and J.B. Murphey, (1986): Ephemeral gullies as sources of sediment. Proceedings of the Fourth Federal Interagency Sedimentation Conference, 1:3-152 - 3-161.
16. Van Daele, K., this volume : Assessment of factors affecting ephemeral gully erosion in cultivated catchments of the Belgian Loam Belt.
17. Savat, J. and De Ploey, J., (1982) : Sheetwash and rill development by surface flow. In Bryan, R. and A. Yair (eds.), Badland Geomorphology and piping. Geo Books, Norwich, 113-126.
18. Govers, G., (1985) : Selectivity and transport capacity of thin flows in relation to rill erosion. Catena 12, 35-49.
19. Schumm, S.A., Mosley, M.P. and Weaver, W.E., (1987) : Experimental Fluvial Geomorphology. Wiley Interscience.
20. Govers, G., Everaert, W., Poesen, J., Rauws, G, De Ploey, J. and Lautridou, J.P., (1990) : A long-flume study of the dynamic factors affecting the resistance of a loamy soil to concentrated flow erosion. Earth Surface Processes and Landforms, 15:313-328.
21. De Ploey, J., (1989b) : A model for headcut retreat in rills and gullies. Catena Supplement, 14:81-86.
22. Rauws, G. and G. Govers, (1988) : Hydraulic and soil mechanical aspects of rill generation on agricultural soils. Journal of Soil Science, 39:111-124.
23. Torri, D., M. Sfalanga, G. Chisci, (1987) : Threshold conditions for incipient rilling. Catena Supplement, 8:97-105.
24. Luk, S.H., H. Chen, G.Q. Cai and Z.J. Jia, (1989) : Spatial and temporal variations in the strength of loess soils, Lishi, China. Geoderma, 45:303-317.
25. Pye, K., (1983) : Grain surface textures and carbonate content of late Pleistocene loess from West Germany and Poland. Journal of Sedimentary Petrology, 53:973-980.
26. Peele, T.C., O.W. Beale and E.E. Latham, (1938) : The effect of lime and organic matter on the erodibility of Cecil clay. Soil Science Society of America Proceedings, 3:289-295.
27. Barahona, E., Quirantes, J., Guardiola, J.L. and Iriarte, A., (1990) : Factors

affecting the susceptibility of soils to interrill erosion in South-eastern Spain. In Rubio, J.L. and Rickson, R.J. (eds.) Strategies to combat desertification in Mediterranean Europe. Commission of the European Communities, EUR 11175 : 216-227.
28. Fullen, M.A., (1985) : Compaction, hydrological processes and soil erosion on loamy sands in east Shropshire, England. Soil and Tillage Research, 6:17-29.
29. Ouvry, J.P., (1987) : Ruissellement et erosion des terres. Bilan des travaux campagne 86-87. Association régionale pour l'étude et l'amélioration des sols, Boisguillaume, France.
30. Baade, J., Barsch, D., Mäusbacher, R. and Schukraft, G., this volume. Field experiments on the reduction of sediment yield from arable land to receiving watercourses (N-Kraichgau, SW-Germany).
31. Veldhuijzen van Zanten, R., (1986) : Geotextiles and Geomembranes in civil engineering. Balkema, Rotterdam 592 pp.

Farm Land Erosion: In Temperate Plains Environment and Hills
S. Wicherek (Editor)
1993 Elsevier Science Publishers B.V.

The extinction of some perennial grass vegetation and the degradation of chernozem due to anthropo-zoogenial factors, in some steppes of Ukrainian SSR

J. G. Ray, A. P. Travleev and N. A. Belova,

Department of Ecology and Soil Science, Dniepropertrovsk State University, Ukraine, CIS

SUMMARY

Chernozem is the stable soil type of a climax steppe. Natural steppes are characterized by dominant perennial grasses.*Stipa lessingiana* Trin. et Rupr., *Koelaria cristata* (L) Pers., Festuca valesiaca L., etc. are the most dominant perennial grasses in the Ukrainian steppes. They are the final conquerors of these steppes and have a high resistance towards degradational factors. They not only protect the soil from degradation forces of erosion but contribute to the "life" of the biogeocenoce. Never before has a study been aimed at revealing the influence of these perennial grasses in maintaining the stable properties of chernozem and their reaction towards various grades of degradational factors in the Ukrainian steppes. Four different steppes of the same climatic region on the watersheds of the same climatic region on the watersheds of the Samara tributary of the Dniepr were studied for this purpose. The first, second and third sites are protected from grazing but have different soil profiles. The fourth site is a totally steppe degraded by many years of overgrazing. Overground and underground phytomass, physico-chemical properties of soil, etc. of the different biogeohorizons and the species structure were determined in all the four sites. The morphology of the root systems and the scanning electron microscopy of the rhizospheres of all the perennial grasses in different soil systems were also determined. A comparative analysis of the results shows that physical qualities of soils are altered quickly due to continuous overgrazing but a corresponding change in chemical properties are not immediately noticeable. Species structure, root/shoot ratio, etc. correlate with the extent of damage. The extinction of *Stipa, Koelaria,* etc. was noted in the highly destroyed system. It is clear that the gradual extinction of these perennial grasses was due to a change in the physico-chemical qualities of soils and the difficulty of the root systems of these plants to adjust to the changed environment. Comparative morphological studies of root systems and SEM observations of rhizospheres also support this reasoning. The final conclusion is that overgrazing is a vicious circle - destruction of phytocenoce affects the geocenoce and vice versa, which in the long run leads to the extinction of perennial grasses and to the total collapse of the system.

RESUME

Le chernozem est un sol stable typique du climat de la prairie. La prairie naturelle se caractérise par la présence de *Stipa Lessingiana Trin et Rupr., Koelaria cristata (L) Pers., Festuca vallesiaca L.,* etc... Les végétaux qui offrent une grande résistance aux facteurs de dégradation, ne protègent pas le sol de l'érosion mais contribuent à l'existence de biogéocoenoce. Dans le même contexte climatique, trois prairies différentes ont été étudiées, elles se situent dans le bassin de la rivière Samara, affluent du Dnieper. Les deux premières sont protégées des pâturages, la troisième a subi un surpâturage durant de nombreuses années. Des études concernant les parties aériennes et souterraines de la phytocoenoce, les propriétés physico-chimiques des différents biogéohorizons, la morphologie du système racinaire et l'étude des rhizosphères ont été réalisés.

Il apparaît que les caractères physiques du sol sont rapidement altérés en raison du surpâturage tandis que les changements des propriétés chimiques demeurent difficiles à mettre en évidence. La disparition de *Stipa, Koelaria,* etc. caractérise les secteurs très dégradés.

Le surpâturage est donc destructeur de la phytocoenoce ; il affecte la géocoenoce et réciproquement.

INTRODUCTION

Steppe biogeocenoce is a unique and very delicate natural system in the biosphere. Unfortunately the steppes have now become one of the most degraded ecosystems in the world. Towards the end of last century Dokuchaev (1) himself had remarked upon the pathetic ecocatastrophe of natural steppes in the East European Region. The pioneer steppe ecologists Pachosky (2), Keller (3), Lavrenko (4), etc. have found that virgin steppes are gradually disappearing from the face of the earth. Steppe according to Dokuchaev is a vegetation on stable chernozem and chernozem a stable soil under a climax steppe vegetation. Steppes are dominated by perennial thick-sod grasses, the root systems of which play a major role in the formation and stability of chernozems.

In Eastern Europe, Ukraine has the largest area of steppes and chernozems. Steppes occupy 40 percent - the southern half - of the territory of Ukraine (5). But natural steppes are now limited only to certain reserves. The major reasons for degradation of steppes are excessive agricultural usage of chernozems and overgrazing by livestock. Many years of anthropo-zoogenial manipulations of steppes have resulted in complete degradation of the natural qualities of chernozem and vegetation. In the Dniepropetrovsk Region alone, which is situated in the central steppic zone of Ukraine, about 600 hectares of chernozems need immediate reclamation procedures (6). From the overgrazed Ukrainian steppes the typical dominant perennial grasses like Stipa lessingiana Trin. et Rupr., Koelaria cristata (L) Pers., Festuca valesiaca L., etc. were found to have disappeared.

In order to understand the exact influence of all the degradative factors on the stability and existence of steppe biogeocenoces we have conducted an investigation - structural and functional characteristics of vegetation as well as physico-chemical, morphological and micromorphological peculiarities of soil systems - of certain steppe biogeocenoces on the Mid-Dniepr watersheds on loess and red-brown clay parent materials, under direct and indirect anthropo-zoogenial factors of degradation. A special attempt was made to learn the specific role of perennial thick-sod wild grasses as stabilizers of the system.

SITE DESCRIPTION

The investigation was carried out in the steppe sites in the experimental territory of the Prisamara Biogeocenological Monitoring Research Station (PBMRS) of Dniepropetrovsk State University (DSU) at Andreevka (Prisamara). The following were the four sites examined ; Site N° 201-R (Ethalon) - Reserve virgin steppe with typical ordinary chernozem on loess, Site No. 201-R1 (KBG) - Reserve steppe with ordinary chernozem (eroded) on red-brown clay, Site No. 201-R2. (Popas) - Reserve steppe with ordinary chernozem (eroded) on loess and Site N° 201-R3 (Tirlo) - Totally degrated steppe with highly degraded chernozem due to many years of uncontrolled overgrazing. The Prisamara steppes belong to the Mid-Dniepr watershed plain vegetation in the South-East of Ukraine - in Novomoskovsk district Dniepropetrovsk region.

The region has a climate and edapho-topographical factors typical for a dry steppe zone. Annual atmospheric precipitation (200-500 mm/year) and atmospheric humidity (30-60 percent) are very low. The evaporation rate is very high. The physico-geographical, geomorphological, geological and bio-climatic peculiarities of the area show that all natural components and factors of the area are well suited for the existence of stable steppe biogeocenoces in the region. But at present almost all of the natural steppes in the area except those in the experimental territory of PBMRS have been degraded due to intense agricultural manipulations, overgrazing, industrial activities, etc.

METHODS OF INVESTIGATION

Ecomorphological examination of vegetations were conducted according to the methods of Albitskaya M.A. (7) and Belgard A.L. (8), and the principles and concepts of Clements (9), Whittaker (10), Ricklefs (11), etc. Quantitative and qualitative investigations of roots in different soil horizons were determined in accordance with the instructions of Uzbeck (12). Ultramorphology of root-soil interphase was observed using a scanning electron microscope (SEM). Samples for observation were prepared without any chemical treatment of fixation and staining. Hydrophysical and physico-mechanical studies of soils were mainly done according to the methods of Arunushkina E.V. (13) and Jackson M.L. (14). Thin soil sections were prepared and described using the methods and concepts of Yarilova and Parfinova (15) as well as according to the methods of Fitzpatric E.A. (16).

RESULTS AND DISCUSSION

a. Anthropo-zoogenial impact on ecomorphological characteristics of vegetation

In order to uncover the impact of disturbances on the stability of steppe vegetation of the area, we have examined the species richness, population density, ecobiomorphic characteristics and relationships of species, etc. of the four different units of vegetation under different forms/grades of disturbances in the area.

Species structure is one of the most important indicators of pertubations within a system. It is a well established fact that high disturbances will reduce the total number of species in a system, whereas moderate disturbances an alter the relative abundance of species without changing their number (17). Any disturbance of species structure relationships in a stable system will induce new successive events and thus make the system unstable (18).

Of the four steppe units which we have examined, Ethalon is the least disturbed of the virgin sites. The reserve steppes KBG and Popas are also virgin but under exodynamic and endodynamic, direct/indirect, degradative influences respectively. The last site, Tirlo, represents a totally degraded steppe due to direct zoogenial degradations over many years.

Comparison of species structure of the four sites shows that a change in species number occurred only in the site where degradation was extreme. All three virgin sites, irrespective of the indirect/direct exogenic/endogenic degradative influence, have an average of 40-42 spp/100 m^2. But in the totally overgrazed site Tirlo, the species number has been reduced sharply to about half of that of the virgin sites, only 24 spp/100 m^2.

We have given special attention to the content of grass species especially that of the thick-sod perennial grasses in the area. Since these grasses are the basic species and they play a major role in the formation of chernozems - by providing more organic mass to soils and by physically protecting the aggregate soil structure. At the KBG site an increase in total percentage of grass species is visible but the site is far behind the other reserve steppes in thick-sod perennial grass species content. The KGB site differs from the other two virgin sites, mainly in that the soil forming material at the site is red-brown clay, not loess. Therefore, this change in soil condition is acting as an endodynamic disturbance in the system - acting against the luxuriant growth of thick-sod perennial species. This is definitely a sign of dangerous instability and appears to be a threat to the development and existence of stable steppe biogeocenoces, perennial thick-sod grasses, the major builders and protectors of chernozems.

A comparison of the percentage of different eco-biomorphic forms in the various sites also uncovers certain interesting facts. Though the total species number has been reduced to about 50 percent in the degrated Tirlo site, in the total percentage of perennial species there is not much difference between the different sites. This suggests that the eco-biomorphic relationships of species in the community are not easily changed by natural degradations, but on the other hand if the degradations were direct anthropo-technogenial types, there should have been a totally different eco-biomorphic relationship.

In a stable vegetation the realized niche of individual species are very close to their fundamental niches. Therefore the stability will be visible in the vertical as well as horizontal groupings (stratification and mosaicity) of species. Any serious disturbance -endodynamic or exodynamic - will therefore be visible in its stratification and mosaics. Therefore an examination of these characteristics provides a useful preliminary guide to plant environment relationships which in turn helps to give a clear summary of the structural characteristics of plant communities. The stratification pattern in the Ethalon site is typical of a normal natural steppe. But at the KBG site and Popas deviation of stratification pattern is visible. In KBG usual third stratum species have moved to the first or second strata. In Popas an abnormal separation of grasses and dicot herbs into different layers is visible. At Tirlo, there is no stratification at all. Most of the plant mass -overgrazed, disabled plant parts - remains at 0-5 cm layer. This clearly explains the instability of the system due to anthropo-zoogenial influence.

b. Root Characteristics of the three perennial grasses in different soil systems :

Stipa, Koelaria and Festuca are the three most common perennial dominant grasses of the East European steppes. Different species of these plants differ in their microclimatic requirements and accordingly dominate in different localities in various ways. In the different sites which we have examined, Stipa lessingiana Trin.

et Rupr. is the most dominant species whereas Koelaria cristata (L) Pers. and Festuca valesiaca L. are sub-dominants. The ability of the root systems to adapt to varied local conditions and to periodic or accidental changes determines to a great extent the success of a species in the local and consequently in the general community. The specific interaction of root system with soils enables this range of tolerence. So we have examined in detail the morphological peculiarities of the root system and the specific root-soil interactions of each of the plants in the four different soil situations to understand the exact reasons for their success or failure in diverse systems.

General morphological examination of root systems of the three different perennial grasses shows that they differ in their root morphology. Stipa lessingiana has the deepest and widest growing root system. Festuca valesiaca has a more superficial root system ; its roots are hair-like and are dark-coloured. Koelaria cristata has a root system similar in morphology to that of Stipa lessingiana but lesser in depth and breadth of growth and quantity. At the Ethalon site all three plants have more or less their original pattern of root systems. At Popas the root systems of all plants show a slight increase in growth and quantity, but at KBG, on the contrary, root systems of all plants show a reduction in their normal growth and quantity. Thus it is evident that the roots of all these three grasses show their maximum development when the sub-soil is loess, and when it is red-brown clay normal development is hindered.

At Tirlo, where Festuca valesiaca is the only perennial thick-sod species, root deformation is visible. Most of the thin roots of the plant at this site were found to be concentrated in 0-2 cm depth as a thick mat, as if they were prevented from piercing into deeper layers. It was also found that where the soil is cracked the roots of these grasses pierce deep into the soil through the cracks. A general increase in the quantity of roots was found for Festuca at this site. These observations suggest that the root system of Festuca valesiaca is very flexible in thriving in disturbed soil systems. The increase in quantity of roots suggests an increased compensatory rate for destroyed roots. The shift of roots to the surface and their downward growth through cracks point to flexibility in absorption strategies. The Stipa and Koelaria grasses do not show such adaptation and that might be one major reason for their disappearence from disturbed fields.

SEM observations of root-soil interphase is a recent focus of research and progress thus far has been limited by a lack of adequate techniques for direct observation. Using a new method which we have developed, we could see the actual undisturbed root-soil environments. We found that in Stipa and Koelaria root hairs are perennial and it seems that they are produced continuously (Photo 1). Not much difference was observed between the ultramorphology of young and old roots in these plants, but in Festuca it was found that the old roots shed their hairs and only the young roots remain absorptive with enough hairs. In Stipa the gap between the root and soil was found to be occupied by mycorrhiza and the perennial root hairs provide them with a perch.

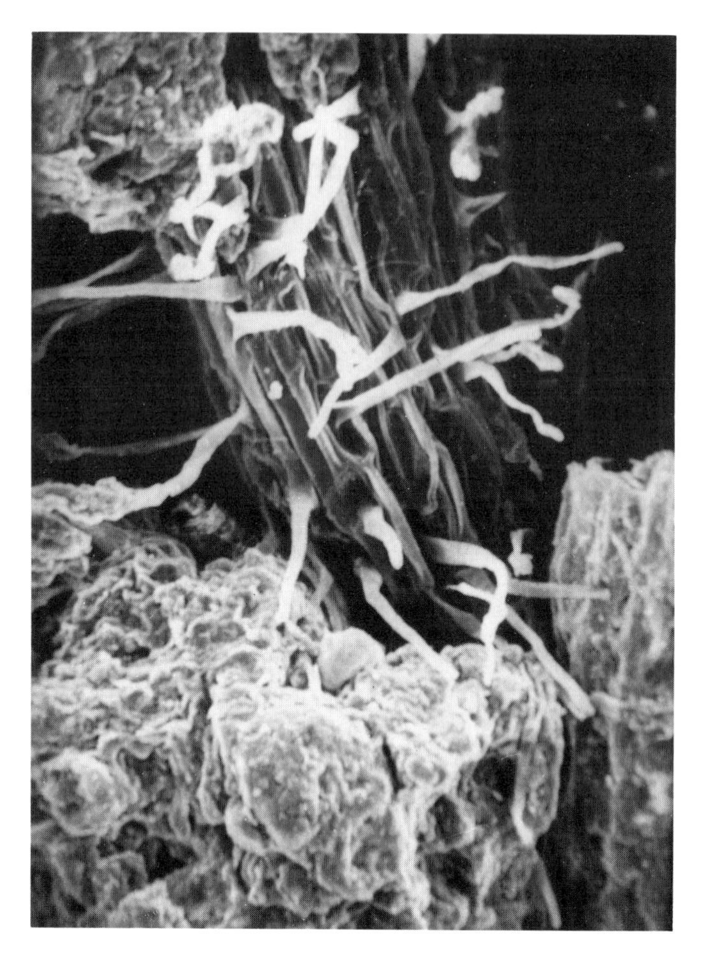

Photo 1.
Undisturbed surface of an ancient perennial root of Stips lessingiana Trin. et Rupr. showing root hairs of different ages (continuous root hair formation).

From the observations, it is possible to conclude that Stipa and Koelaria have permanent root-soil arrangements - with permanent root hair-mycorrhizal set-up. So in natural undisturbed soil systems they can compete efficiently with all others. But when the soil is degraded physically, perennial grasses such as these cannot cope with the disturbance due to a complete failure of their permanent absorptive set-up. Festuca valesiaca, however, due to a lack of such a permanent set up, is not an efficient competitor in natural systems. But in degraded systems this "inefficiency" coupled with an increased turnover rate of roots enables them to survive degradation and thrive to a certain extent.

PRODUCTIVITY CHARACTERISTICS

The total phytomass content and the contribution to it from the overground living part, the dead mass and the roots, and the distribution pattern of the phytomass in different biogeohorizonts are very important in understanding the ecological and evolutionary process in natural systems.

Among the four different sites examined, the site on red-brown clay has the least total phytomass. The ratio of overground living mass, litter and root mass is 3 : 2 : 5 at the site. At the Ethalon site the ratio is 3 : 1 : 6, at Popas it is 2 : 1 : 7 and at Tirlo it is 1 : 4 : 5. These ratios prove that there is a shift (translocation) in the position of the phytomass in the Popas site and KBG site according to differences in local conditions and degradations. At Popas the shift of phytomass is from the aboveground to underground region whereas on the contrary the phytomass has shifted from underground to overground at the KBG site which is antagonistic to the development and stability of normal chernozems.

In the pattern of distribution of aboveground living mass and underground root mass in different biogeohorizons, the three different reserve steppes do not show much difference ; a more or less pyramidal fashion of distribution is visible at all the sites. But at Tirlo the pattern has very much changed. 98 percent of the total overground mass and 60 percent of the total root mass are concentrated in 0-10 cm height/depth in the site. This is the climax of a degradation as a slight erosion can remove the whole of the biomass from the system ; especially when the soil is trampled by animals, the chance of such an accelerated erosion is very high and hence a collapse is inevitable (Photo 2 and 3).

SOIL CHARACTERISTICS

A comparison of morphological peculiarities and mechanical content of the four different soils helped us to find the basic cause of the occurrence of vertisol in the region. All of the four different soils which we have studied are more or less similar in total content of clay fraction. But cracking and reversal of humus-rich surface horizon (vertisol morphology) was found only in degraded plots where natural vegetation has degraded and disappeared. In virgin steppes where the vegetation is dominated by perennial thick-sod grasses like Stipa and Koelaria, their highly developed root systems and the soil biota provide a perfect granular structure to surface horizonts and physically protect the soil from cracking and inversion of horizons. Humus accumulates more and more at the surface horizonts and in the absence of illuviation and lessivage a humus-rich "A" horizon of chernozem develops on a carbonated parent material. On the other hand, when the vegetation is destroyed and/or the soil unphysically/exogenously degraded due to trampling by animals, deep cracks appear and thus reversal of horizon - vertisol development - occurs. Especially when the sub-soil is red-brown clay this will result from even a slight disturbance.

Photo 2.
Accelerated erosion of chernozem in the graded steppe (Tirlo).

HYDROPHYSICAL CHARACTERISTICS

The stability of a soil system depends on a stable tilth. The tilth is the sum total of physical organization of the soil which makes it suitable for optimum plant growth. The tilth develops and evolves with the vegetation and in a stable biogeocenoce the soil remains tilthy. Any exodynamic factor that negatively affects the tilth parameters leads to destructive successions in the plant world and consequently animal world aswell. Bulk density and porosity, maximum hygroscopic capacity, available water content, field moisture, etc. were determined at all the sites. A comparison of the four sites shows that all the virgin sites have more or less the same good tilth at the surface layers. At the Tirlo site it was found that the bulk density sharply increases towards lower horizons and maximum is seen at the mid-horizon. This definitely acts as a barrier for the growth of deep-rooted steppic perennial plants.

Photo 3.
Luxuriant growth of perennial grasses (Stipa and Koelaria) and their root mass in normal steppe with non-degraded chernozen (Ethalon).

In the content of water-stable aggregates, physical degradation is also visible. At Tirlo the water-stable aggregates are less than 30 percent of the total aggregates (1-2 mm size) whereas the percentage is above 60 percent in the surface soils of all the other sites. In the available water content the soils do not differ much. From the observations it is possible to conclude that the physical organization of the soils of the steppe zone does not change quickly if there is no exogenous direct degradative influence on them. At Tirlo the increase in bulk density and a consequent increase in resistance to penetration of roots cannot be the only reason for the disappearence of perennial grasses. Particularly when the soil cracks the bulk density does not correspond to lack of aeration and mechanical barrier for roots. But animals trample the soil which losses its granular structure. Roots are damaged as well. So this may also be a major reason for their disappearence as was proved by other observations.

CHEMICAL CHARACTERISTICS

The soil of a zone usually do not differ in its chemical properties especially when the vegetation, climate and parent materials are uniform. But properties like pH, percentage of humus, available nitrate and phosphorus, water soluble ions, etc. vary according to local differences in degradative influence. An increase in pH of surface soils is visible at Popas and at KBG and this due to erosion. In humus content both these sites show degradations. At the KBG site an abnormal increase of soluble salts in the lower horizonts was found. This is an inherent characteristic of the red-brown clay. At Tirlo this type of an abnormal increase in salt concentration is visible, due to a secondary degradative process that the soil has underground there.

MICROSTRUCTURE

A comparison of microstructure of the surface horizons of the first three virgin sites shows a more or less similar microorganization. But at the Tirlo site the soil has lost all its microorganization and the plasma and skeleton are seen together as a pressed and cemented mass. The soil material has a layered elementary microorganization. Thus it is certain that exogenous degradative factors like overgrazing can totally destroy the microorganization of chernozems.

CONCLUSIONS

The natural steppe biogeocenoces on the Mid-Dniepr watersheds of the Ukraine are under the influence of direct/indirect anthropozoogenial and technogenial degradative factors and the units on red-brown clay and those under overgrazing pressure are under-going local catastrophic succession. In steppes on loess material, slight soil erosion induced by indirect anthropo-zoogenial factors is overcome by a positive reaction, by shifting more phytomass further underground as roots. This takes place as an inevitable regulatory process and is possible due to specific physico-chemical characteristics of loess. But in steppes on red-brown clay the vegetation fails to react thus and on the contrary an opposite shifting of more mass to the aboveground region takes place due to inherent physico-chemical characteristics of the material and this induces degradative successions. Perennial dominant grasses like Stipa and Koelaria act as the major structural builders and protectors of chernozems. When the soil system is degraded, in the absence of a compensatory increase in root growth and due to an inability to change the absorption strategy, the permanent root-soil set-up of these plants fails and they disappear from the degraded systems. Therefore these types of anthropo-zoogenial degradations - uncontrolled overgrazing - create a vicious circle. Destruction of phytocenoce affects the geocenoce and vice versa which in the long run leads to the extinction of perennial grasses and to the total collapse of the system.

REFERENCES

1. Dokuchaev V. V. (1936) : Our Steppes - Now and Then. Govt. Agr. Publ. Moscow-Leningrad. (in Russian).

2. Pachosky I. K. (1922) : Basis of Phytocenology. Kherson (in Russian).
3. Keller B. A. (1931) : (ed) Stepp - Central Chernozemic Region Govt. Agr. Publ., Moscow-Leningrad. (in Russian).
4. Lavrenko E. M. (1940) : Steppes of the USSR. In Vegetations of the USSR. Vol. II. Moscow AS-USSR (in Russian).
5. Krupskov N. K and Palupan N. E (1979) : Atlas of the Soils of Ukraine. "Urashai" Publ. Kiev (in Russian).
6. Belova N. A. (1987) : Ph. D. Thesis, Dniepropetrovsk State University, 1987 (in Russian).
7. Albitskaya M. A. (1969) : Artificial forests in Steppe Zone, Kharkov. pp. 155-208. (in Russian).
8. Belgard A. L. (1971) : Steppe Forest Management, Forest. Industr. p. 336, Moscow (in Russian).
9. Clements F. E. 1916, Plant Succession - An analysis of Dept. of Vegetation. Carneg. Inst. Wash. Publ. 242, pp. 1-512.
10. Whittaker R. H. (1967) : Biol. Rev. 49. pp. 207-264.
11. Rickleffs R. E. (1990) : Ecology, W. H. Freeman & company, New York, 896 p.
12. Uzbeck I. H. (1987) : "Pochvovegenie", pp. 101-107 (in Russian).
13. Arunushkina E. V. (1970) : Guide of Soil Chemical Analysis MSU Publ., Moscow. (in Russian).
14. Jackson M. L. (1958) : Soil Chemical Analysis, Const. London.
15. Yarilova E. A. and Parfinova (1977) : Guide to Micromorphological Investigation of Soils, "Nauka" Publ. Moscow. 195 p. (in Russian).
16. Fitzpatric E. A. (1984) : Micromorphology of Soils, Chapman and Hall London.
17. Mooney H. A. and Gordon M. (eds) (1983) : Disturbance and Ecosystems Springer - Verlag, Berlin.
18. Chapin I. V. (1983) : Disturbance and Ecosystem, Springer-Verlag, Berlin.
19. Travleev A. P. and Travleev L. P. (1979) : Sputnik - Geobotany, Soil Science and Hydrology, DSU Publ. Dniepropetrovsk (in Russsian).

Farm Land Erosion: In Temperate Plains Environment and Hills
S. Wicherek (Editor)

Assessment of Soil Erosion in Quebec (Canada) with Cs-137

C. Bernard [a], and M.R. Laverdière [b]

[a] Ministère de l'Agriculture, des Pêcheries et de l'Alimentation du Québec, Service des Sols, 2700 rue Einstein, Sainte-Foy, Québec, G1P 3W8, Canada.

[b] Université Laval, Département des Sols, Sainte-Foy, Québec, G1K 7P4, Canada.

SUMMARY

The universality of its fallouts and their strong retention by soil particles makes cesium-137 an interesting and reliable tracer of the soil movements going on since the mid 1950's. It is used in a research program on soil erosion in Québec (Canada). The influence of time and of dilution of fallouts by soil tillage on the variability of the relationship between soil and cesium losses has been studied. In a second time, the Cs-137 spatial redistribution was used to estimate soil losses have in 63 fields ranging from 0,2 to 1,5 ha and presenting various conditions of soil texture, slope and land use. An average soil loss of 5,7 t ha^{-1} yr^{-1} was estimated. The redistribution pattern of Cs-137 down the slopes revealed an alternation of soil loss and deposition.

RESUME

Le césium-137 constitue un excellent indicateur des mouvements de sol ayant cours depuis le milieu des années 1950, en raison de l'universalité des retombées et de la forte rétention de cet élément par les fractions fines du sol. Cet isotope est utilisé dans le cadre d'un programme de recherche sur l'érosion hydrique au Québec (Canada). La variabilité de la relation entre la perte de sol et celle de césium dans le temps et en fonction du niveau de dilution des retombées par le travail du sol a d'abord été étudiée. Un deuxième volet portait sur l'étude de la redistribution spatiale du Cs-137 pour estimer les pertes de sol dans 63 champs d'une superficie de 0,2 à 1,5 ha et présentant une large gamme de textures, de pentes et d'utilisations du sol. La perte nette de sol a été estimée à 5,7 t ha^{-1} an^{-1} enmoyenne. Le patron de redistribution du Cs-137 le long des pentes indique une alternance d'arrachement et de redéposition du sol.

1. INTRODUCTION

Intensification and specialisation of agriculture over the last decades has resulted, in Québec as elsewhere in the world, in various forms of soil degradation. Since the area without severe limitation for agriculture in this province is limited to 2,4 Mha, i.e. less than 2 % of the total territory, soil degradation must be seen as a serious threat to agriculture.

Recent measurements have shown that soil erosion may exceed agronomically tolerable levels under Québec agro-climatic conditions (1 - 2). However, despite the severity of its agronomic and environmental impacts, soil erosion has not been studied very extensively. In addition, most of the past studies were carried over rather short periods, not exceeding 3 years and often limited to the vegetative season, and thus neglected the snowmelt period where more than 50 % of the annual losses may occur (3 - 4). In all cases, slope length was limited to 15 m. Consequently, the long-term trend of erosion, at the full-field scale and over the complete hydrological cycle still has to be assessed under Québec conditions.

For this purpose, a research program using Cs-137 as a soil movement indicator was initiated. This isotope originated from atomic device testing in high atmosphere in the late 1950's and early 1960's. Once fallen back with precipitations, Cs-137 is very strongly retained by the fine fractions of soil, moving thereafter almost exclusively with soil particles. This characteristic makes Cs-137 an interesting and reliable indicator of soil movements having occurred in the last 35 years or so (5).

2. MATERIALS AND METHODS

In the first stage of our research program, the variability of the relationship between cesium and soil losses by erosion was investigated. Three applications of 3700 Bq m^{-2} of Cs-134 were done on plots and incorporated at two different depths (5 or 20 cm), resulting in well contrasted surface concentrations of Cs-134. Each cesium application was followed by two 15-minutes simulated rains, with 63 and 42 mm h^{-1} intensities. Losses of soil and Cs-134 were recorded for each rainfall (6).

In the second stage, the severity of soil erosion was estimated in 63 fields presenting various conditions of soil texture, slope gradient and land use. Sampling was done every 30 m along transects laid down the dominant slope. The areal activity in Cs-137 (Bq m^{-2}) of the sampled fields was compared to that of non-eroded forested sites. The magnitude of soil movements was considered to be proportional to the variation of the Cs-137 areal activity, i.e. net loss in case of reduced Cs-137 activity and net gain in case of increased activity (7).

3 RESULTS

3.1 Relationship between cesium and soil losses

As reported by some authors, the eroded soil is commonly enriched in fine particles and nutrients, as compared to the original soil (8 - 9). The same trend was noted for Cs-134 concentrations, the eroded sediments showing concentrations 1,8 to 10,8 times those of the plot soils (6).

Figure 1 illustrates how the Cs-134 concentration of sediments varied in time and was influenced by the depth of incorporation. The addition of cesium resulted in concentrations increasing with time (treatments 1 and 3). The tillage depth, by controlling the level of dilution of the added Cs-134, influenced strongly the resulting cesium concentration at the soil surface and on the eroded material. Sediments originating from shallow-tilled plots (treatment 3) consistently exhibited higher cesium concentrations that those from deep-plowed plots (treatment 1). The results of treatment 2 confirm this influence. For example, the Cs-134 concentration of sediments was significantly reduced after application 2, despite the addition of cesium, as a result of deep plowing.

These results suggest that a kilogram of eroded soil may carry variable amounts of cesium.

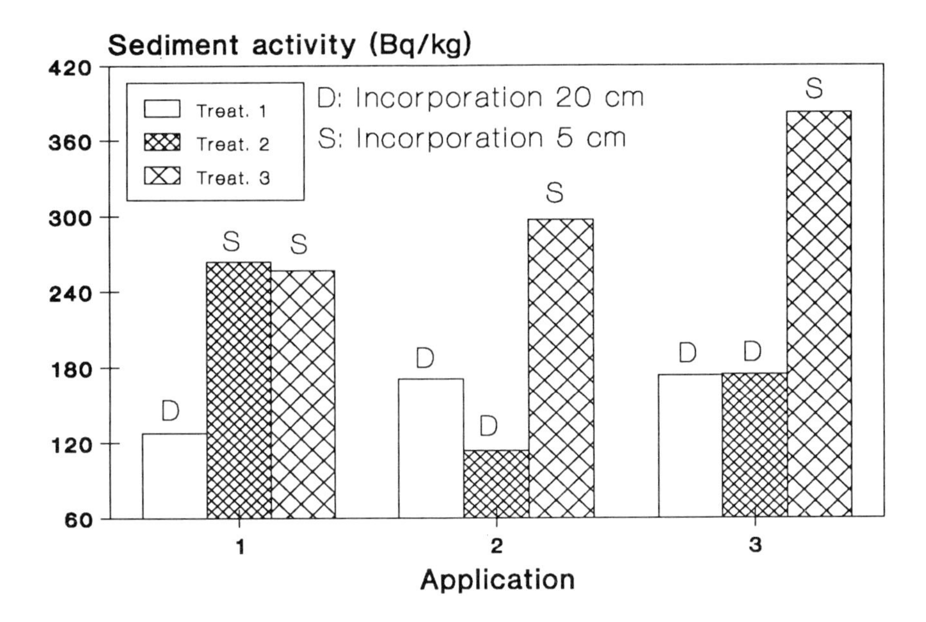

Figure 1. Variability of the activity of eroded soil in Cs-134 (Adapted from Bernard et al., 1992).

Figure 2 presents different relationships between soil and cesium losses as they were developped in this project. Curve 1 illustrates the overall relationship :
$SL = 0,039\ Cs^{0,535}$ $R^2 = 0,61$
where SL : Soil loss (kg m^{-2})
Cs : Cs-134 loss (Bq m^{-2})

Curves 2 and 3 show the temporal evolution of the relationship between the first and the third application. The lower slope of curve 3 is the consequence of the higher cesium concentration of the soil and of the eroded material, after 3 applications. Under this condition, a given soil loss results in a larger Cs-134 loss or, at the opposite, a given Cs- 134 loss is associated with a smaller soil loss.

Curves 4 and 5 reflect the influence of tillage depth and of the resulting level of dilution of the applied Cs-134. Curve 4 (shallow incorporation) shows a lower slope than curve 5 (deep incorporation), as a result of higher Cs-134 concentrations at the soil surface and on the eroded material. Here too, a given Cs-134 loss may be associated to variable soil losses.

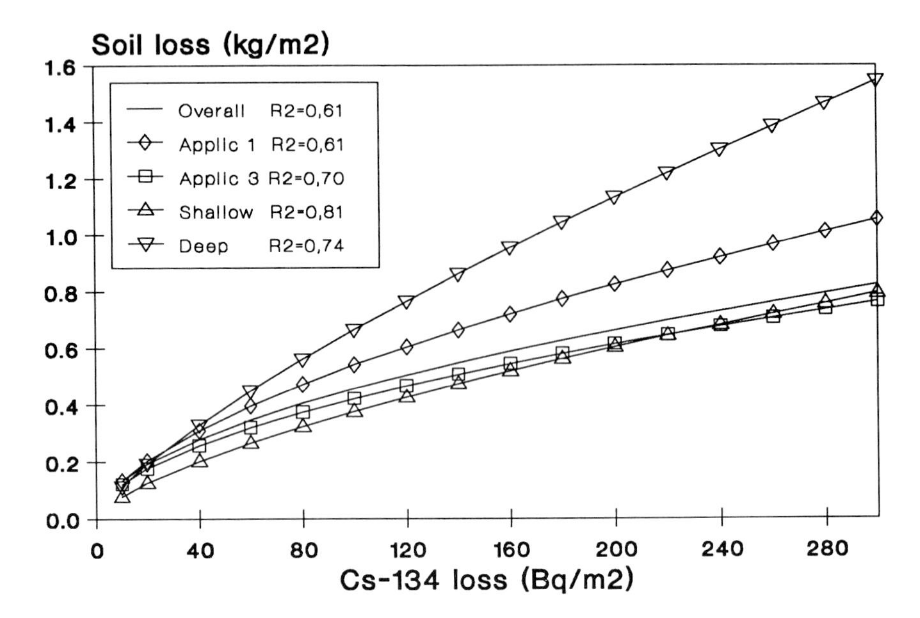

Figure 2. Relationships between soil and Cs-134 losses (Adapted from Bernard et al., 1992).

These results are in agreement with a conceptual model proposed by Kachanoski and DeJong (10), describing the temporal evolution of the soil activity in Cs-137 as a function of additions of cesium, dilution of the fallouts by tillage and removal by erosion. However, as stated by these two authors, the hypothesis of a soil loss directly proportionnal to Cs-137 depletion in the plow layer is still acceptable, for erosion rates currently encountered.

3.2 Estimates of soil losses at the field scale

The 63 fields sampled in the second stage of our research program showed an areal activity ranging from 1580 to 4470 Bq m^{-2}, with an overall average of 3280 $\pm$ 680 (mean $\pm$ standard deviation). This last figure translates into an average soil loss of 5,7 t ha^{-1} yr^{-1}. Table 1 reports on the influence of the studied factors on erosion rates.

The mean soil loss did not differ statistically for 3 of the 4 studied soil textural groups. For group 2 however, it was significantly lower, with an average of 11,7 t ha^{-1} yr^{-1}. This low value was the consequence of the very low Cs-137 levels measured in the fields of this textural group located on horticultural farms.

The effect of slope gradient was more apparent, although not statistically significant. Average soil losses of 4,1, 6,1 and 7,2 t ha^{-1} yr^{-1} were estimated for <2%, 2-5 % and >5 % slopes respectively. Sutherland and DeJong (11) and Kiss et al. (12) also reported Cs-137 activities decreasing with an increased slope gradient. Fields from dairy farms lost an average 3 t ha^{-1} yr^{-1} while the loss was 10,9 t ha^{-1} yr^{-1} on horticultural farms. Ritchie and McHenry (13) and Rogowski and Tamura (14) have also reported a relationship between the density of the vegetative cover and the level of the Cs-137 losses.

Table 1.
Mean Cs-137 activities and net soil losses on Orléans Island, Québec
(From Bernard and Laverdière, 1992)

Textural group	Slope	Use[z]	Cs-137 ($Bq\ m^{-2}$)	Net soil loss ($t\ ha^{-1}\ an^{-1}$)
---------- Overall ----------			3280 ± 680[y]	5,7
1	---	---	3460a ± 470	3,4a[x]
2	---	---	2860b ± 810	11,7b
3	---	---	3480a ± 630	2,6a
4	---	---	3550a ± 380	2,1a
---	<2%	---	3390a ± 630	4,1a
---	2-5%	---	3270a ± 730	6,1a
---	>5%	---	3160a ± 710	7,2a
---	---	D	3460a ± 520	3,0a
---	---	H	2930b ± 820	10,9b

[z] D: dairy farm H: horticoltural farm.
[y] Mean ± standard deviation
[x] Figures followed by the same letter are not statisticaly different at the 0,05 level.

The analysis of the Cs-137 level of the 307 composite samples collected revealed the spatial redistribution pattern of this isotope down the slopes (Figure 3). Three distinct sections can be distinguished. First, there is a fast decrease of the Cs-137 level in the first 30 m. Between 30 and 150 m approximately, the Cs-137 level still decreases, but at a smaller rate and following an irregular and oscillating pattern, suggesting an alternation of deposition and transport. Finally, beyond 150 m, the Cs-137 areal activities increase, indicating that an important fraction of the material eroded from upslope positions was redeposited.

These samples were also classified into soil movement classes. Two thirds of the samples had experienced a net soil loss, with a maximum near 40 t ha^{-1} yr^{-1} and an average of 12,6 t ha-1 yr^{-1}. The other ones exhibited a Cs-137 aerial activity in excess of that measured under non-eroded forested sites. They were therefore considered as having experienced net soil deposition, with an average of 7,9 t ha^{-1} yr^{-1}.

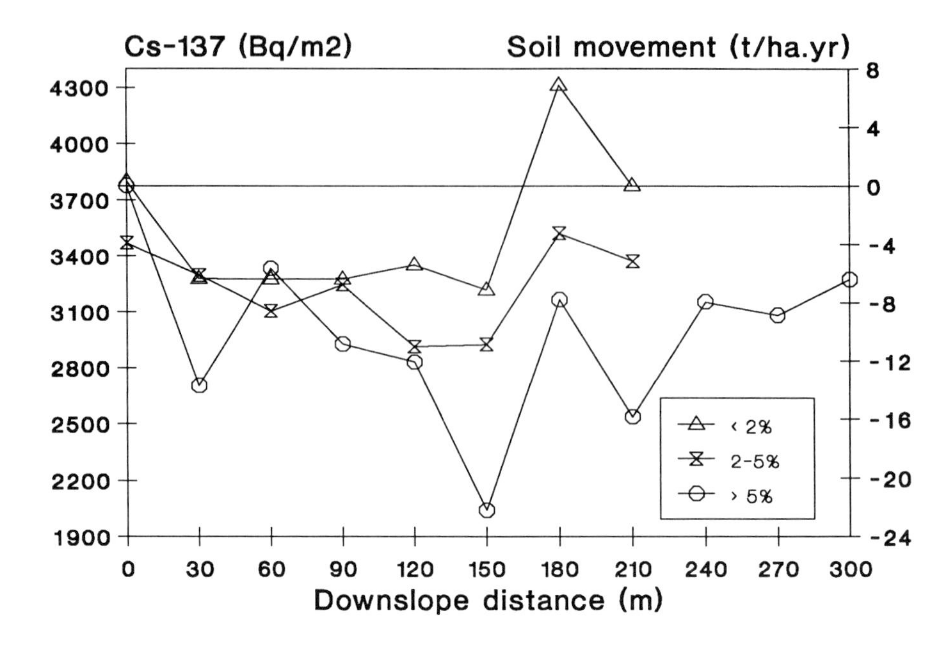

Figure 3. Downslope redistribution of Cs-137 (From Bernard and Laverdière, 1992)

The slope gradient and the land use had a significant influence on the distribution of the samples into soil movement classes. On <5% slopes and on dairy farms, a larger proportion of the samples showed a net gain of material while a larger proportion of those from >5% slopes or horticultural land use showed a net soil loss (7). These results suggest that the low net soil loss rates measured under so-called 'non-erosive' conditions may result not only from low erosion rates but also from the redeposition of an important fraction of the material eroded from upslope positions.

4. CONCLUSIONS

The measure of the spatial redistribution of Cs-137 made it possible to assess the long term trends of soil erosion under agro-climatic conditions typical of southern Québec. This technique highlighted the dynamic nature of the erosive process.

In the coming years, the same technique will be used to study erosion on complex topographies and on long and almost level slopes where erosion is considered unimportant. Given the possibility offered by this technique to distinguish areas of net loss from those of net accumulation, soil movements will also be investigated at the watershed scale.

5. REFERENCES

1. Salehi, F., Pesant, A.R., Lagacé, R. (1991) : Validation of the Universal soil loss equation for three cropping systems under natural rainfall in southeastern Quebec. Can. Agric. Eng. 33:11-16.
2. Pesant, A.R., Dionne, J.L., Genest, J. (1987) : Soil and nutrient losses in surface runoff from conventional and no-till corn systems. Can. J. Soil Sci. 67:835-843.
3. Kirby, P.C., Mehuys, G.R. (1987) : The seasonal variation of soil erosion by water in southwestern Quebec. Can. J. Soil Sci. 67:55-63.
4. Dubé, A. (1975) : L'eau et l'érosion des sols. Ressources 6:8-15.
5. Ritchie, J.C., McHenry, J.R. (1990) : Application of radioactive fallout cesium-137 for measuring soil erosion and sediment accumulation rates and patterns : A review. J. Environ. Qual. 19:215-233.
6. Bernard, C., Laverdière, M.R., Pesant, A.R. (1992) : Variabilité de la relation entre les pertes de césium et de sol par érosion hydrique. Geoderma 52:265-277.
Bernard, C., Laverdière, M.R. (1991) : Variation spatiale de l'activité en césium-137 et ses répercussions sur l'estimation de l'érosion hydrique. Pedologie 40:299-310.
7. Bernard, C. (1992) : Etude de l'érosion des sols de l'Ile d'Orléans à l'aide du césium-137. Université Laval, Dép. des Sols. Thèse de doctorat. 135 p.
8. Young, R.A., Olness, A.E., Mutchler, C.K., Moldenhauer, W.C. (1986) : Chemical and physical enrichments of sediment from cropland. Trans. ASAE 29:165-169.
9. Sharpley, A.N. (1985) : The selective erosion of plant nutrients in runoff. Soil Sci. Soc. Am. J. 49:1527-1534.
10. Kachanoski, R.G., De Jong, E. (1984) : Predicting the temporal relationship between soil cesium-137 and erosion rate. J. Environ. Qual. 13:301-304.
11. Sutherland, R.A., De Jong, E. (1990) : Estimation of sediment redistribution within agricultural fields, using caesium-137, Crystal Springs, Saskatchewan, Canada. Applied Geography 10:205-221.
12. Kiss, J.J., De Jong, E., Rostad, H.P.W. (1986) : An assessment of soil erosion in west-central Saskatchewan using cesium-137. Can. J. Soil Sci. 66:591-600.
13. Ritchie, J.C., McHenry, J.R. (1978) : Fallout Cesium-137 in cultivated and noncultivated North Central United States watersheds. J. Environ. Qual. 7:40-44.
Rogowski, A.S., Tamura, T. (1970) : Erosional behavior of Cesium-137. Health Physics 18:467-477.

BERNARD, C., LAVERDIERE, M.R. (1992): Spatial redistribution of Cs-137 and soil erosion on Orléans Island, Québec. Can. J. Soil Sci. 72: (In press).

Farm Land Erosion: In Temperate Plains Environment and Hills
S. Wicherek (Editor)

Combination of single storm erosion and hydrological models into a geographic information system

H.Chakroun[a], F. Bonn[a] and J.P. Fortin[b]

[a] Centre d'applications et de recherches en télédétection (CARTEL), Université de Sherbrooke, Sherbrooke, Québec, J1K 2R1, Canada.

[b] INRS-Eau, C.P 7500, Sainte Foy, Québec, G1V 4C7, Canada.

SUMMARY

This paper proposes a methodology for mapping and monitoring soil loss at regional scale for single storm events. The concept is based on interfacing a hydrological model to a modified version of the widely known universal soil loss equation. The hydrological model called Hydrotel is characterized by a modular structure and the ability to simulate spatially distributed processes allowing the discretization of the watershed into regular grids. The privileged tools of remotely sensed data, digital terrain models and other ancillary data are used in the estimation of physiographic factors and in the different modules of the hydrologic model. A geographic information system allows integration of the different factors responsible for erosion, which are derived from both the hydrological simulation and the processsing of the various data. A soil loss map can then be generated for the storm event.

RESUME

Dans cette étude, on se propose de développer une méthodologie de calcul des pertes de sol basée conjointement sur un modèle hydrologique et un modèle d'érosion. Le modèle hydrologique proposé s'appelle HYDROTEL : c'est un modèle discrétisé et conçu selon une approche modulaire spéciale pour effectuer des simulations hydrologiques en utilisant les outils privilégiés de la télédétection et des modèles numériques de terain. Pour le modèle d'érosion on utilise l'équation de perte de sol modifiée (MUSLE : Modified Universal Soil Loss Equation) qui tient compte de l'érosivité de la pluie lors d'un événement en associant l'effet des précipitations et du ruissellement ; les débits de pointe et les écoulements estimés par HYDROTEL sont utilisés à cet effet. Un système d'information géographique (SIG) permet l'intégration des différentes données issues de la simulation hydrologique et d'autres sources en vue de calculer les pertes de sol.

1. INTRODUCTION

Many experimental studies have delt with the water erosion mechanisms and non-point source pollution at the plot scale ; the use of rain simulators and sets of managements related to the most important factors that affect soil loss follow certain laws which stay available for experimental conditions. However, there have been few studies that attempt to spatialize these laws at a regional scale, at least at the watershed level. This should include the spatial variability of major factors that affect soil loss.

Erosion modelling methods treating watersheds as distributed systems require, like every large scale environmental modelisation, a huge quantity of data processed by powerful calculation means. Since the coming of remote sensing satellites,there has been a revolution in the storage and transmission of data. More over, the rapid progress of computer techniques in both memory capacity and speed of calculation gives more opportunities to create and manage environmental geo-referenced databases.

2. PROSPECTION OF SOIL LOSS MODELISATION AT REGIONAL SCALE: THE CONTRIBUTION OF SATELLITE DATA AND GIS

The integration of multi-source data, representing major factors that affect erosion, into a Geographic Information System (GIS) leads to a wide production of soil loss maps in many regions of the world. The most common model used was the Universal Soil Loss Equation (USLE) because of its ability to use remotely sensed data in georeferenced databases (1).

In the USLE and other erosion models, the process is described by the same factors which can be devided into :

1- first order factors : these are physical factors describing the terrain, the land-use and the management practices.

2- second order factors : these are issued from the combination of climatic conditions and physical characteristics providing a quantification of the hydrologic behaviour of the basin.

Determining the first-order factors at a regional scale is becoming more and more accurate by means of remotely sensed data available at various resolutions (30m for the Landsat Thematic Mapper and 10 m for the Spot High Resolution Visible imaging instruments) and other digitized data like Digital Elevation Models (DEM). On the other hand, second order factors have been scarcely taken into account when modelling soil loss at regional scale ; this is essentially due to the erosion models considering the runoff effect on detachement and transportation developped up to now. Those models treat watersheds as lumped systems, that is, their properties are spatially averaged and no attempt is made to describe the topology of the watershed and its stream network (2).

Therefore, some recent attempts have been made to consider the spatial variability of the basin by affecting explicitely its surface elements to their reference in the geographic space. The most promizing model is the ANSWERS model (Areal Non-point Source Watershed Environemental Response Simulation, (3) that estimates sediment yield distributed in time and space for single events using observations on a regular grid to determine parameter values. Also, the AGNPS model (Agricultural Non Point Source pollution, (4) is being integrated to the large Water Erosion Prediction Project (WEPP) which represents a new departure for factors based erosion prediction technology (5).

For a good estimation of erosion-sediment yield production, the use of a hydrological model provides the required hydrologic-hydraulic inputs at the required points in time and space (6). In fact, factors affecting rainfall or runoff have a direct influence on erosion and sedimentation; more over the complex detachement, transportation and sedimentation of soil particles is related to the whole watershed in order to identify major sediment sources and the basin erosion-sedimentation history.

3. Hydrotel : a new hydrological model adapted to an erosion study

Hydrotel (7) belongs to the new generation of deterministic distributed hydrological models that predict and simulate the evolution of the hydrologic cycle with a good use of remotely sensed data and digtal elevation models. Various types of data are integrated spatially to Physitel, a complementary program to Hydrotel that prepares the set of physiographic, meteorologic and landuse data. The spatial structure of Hydrotel is shown in Fig. 1.

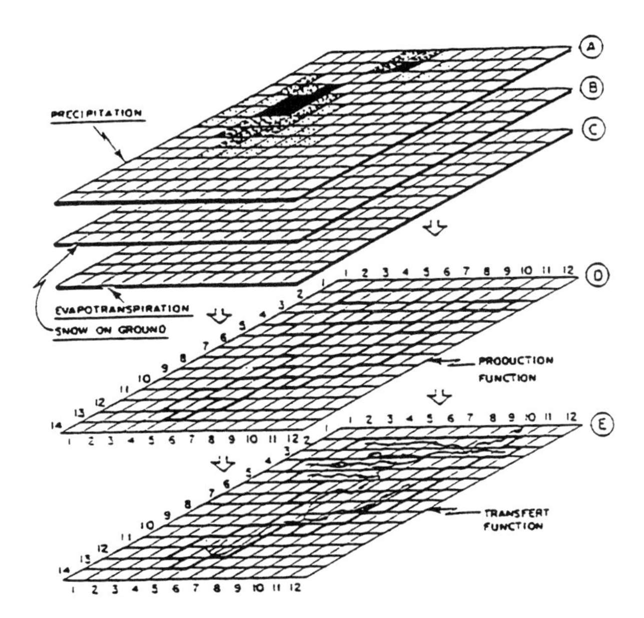

Figure 1. Spatial structure of the Hydrotel model

Many specialized tasks allow the extraction of useful information to accomplish the hydrologic simulation by processing the DEM : slopes, aspects and the surface drainage structure are derived from the integrated topographic data, leading to the automatic definition of the whole watershed and the drainage network as well as its subdivision into subbasins corresponding to the Homogeneous Hydrologic Units (HHU). The determination of the HHU's is based mainly on the permanent physical features of the watershed (topographic patterns, soil types and drainage efficiency) (7). Fig. 2 shows an example of a DEM processed by Physitel to determine the Homogeneous Hydrologic Units.

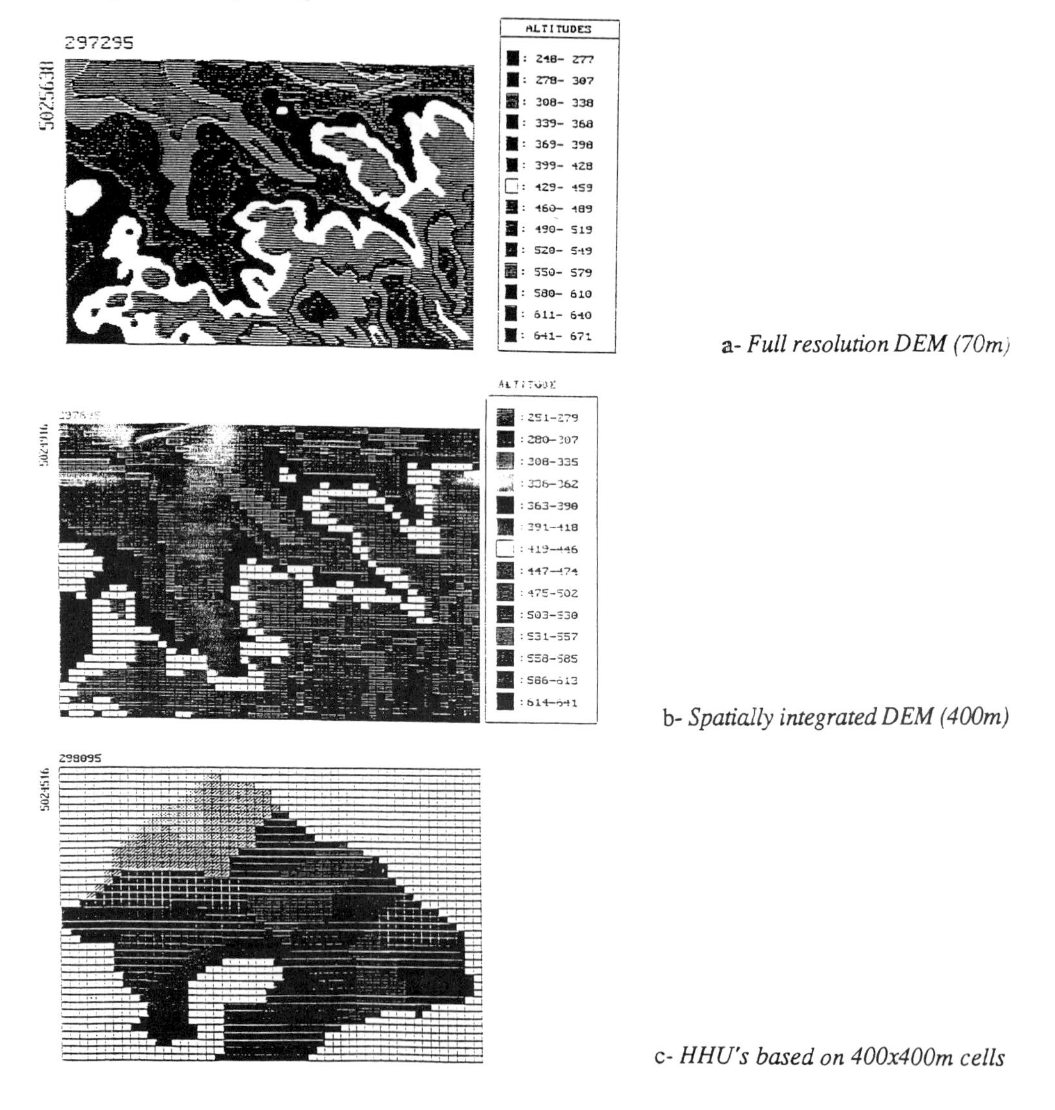

a- *Full resolution DEM (70m)*

b- *Spatially integrated DEM (400m)*

c- *HHU's based on 400x400m cells*

Fig.2 : *Homogeneous hydrological units (HHU) defined from the integration of cells.*

The input, intermediate and final sets of variables like rainfall, snowcover, snowmelt, evapotranspiration, soil moisture and ground water are spatially discretized, as are also surface and subsurface runoff and channel routing. It is then possible to keep track of what happens anywhere in a given basin at any time step. This is especially interesting in the case of the runoff amounts relative to each cell and which will be used in the erosion distributed model.

The hydrologic simulation acquires a set of data relative to land-use classes characterized by quantitative attributes like the albedo and the vegetation height used in the evapotranspiration equations. The different types of soils characterized by their hydraulic properties control the generation of runoff. Both land-use and soil types are integrated spatially to Physitel whether they are available through remotely sensed data or from another source.

4. INTERFACING HYDROTEL WITH SOIL LOSS EQUATIONS : CHOICE OF RESOLUTION

When choosing the grid size to make the hydrologic simulation both first order and second order factors must be taken into account: it is usually considered that the smallest cell size produces the most accurate representation of the first order factors, but this depends of the terrain : for example, rolling terrain with small to moderate relief can better be represented by a larger cell-sized DEM than can a dissected catchement with sharp ridges, ravines and abrupt changes in slope or elevation (8).

When considering the second order factors in choosing the resolution to discretize the watershed, the main criterium is that the number of cells should be high enough to represent the spatial and temporal variations of the various processes on the watershed as well as of its physical characteristics. On the other hand the computation time is importatnt in the choice of the number of cells, thus a compromise has to be reached. It is possible to calibrate the model using a given resolution and then prepare more sets of different resolutions, proceed to simulations and notice the changes, if any. The comparison of simulated to measured runoff with each resolution leads to specify an optimum resolution for the watershed and the process under study.

5. SPATIALIZATION OF THE SOIL LOSS EQUATION FACTORS

The principle of the Universal Soil Loss Equation (USLE) is maintained in the present study because this model provides a good example of the use of remotely sensed data and other types of geo-referenced libraries ; subsequentely, all the factors can be spatially overlaid pixel by pixel using a Geographic Information System .

The Modified Universal Soil Loss Equation (MUSLE, (9)) is one of the erosion models that integrate the hydrologic behaviour of a watershed allowing the spatialization of the second order factors. It is a modification of the USLE where the rainfall erosivity factor has been replaced with one having separate terms for rainfall and runoff (6)

The general expression of the MUSLE is :

$$A_{st} = R_m \, K \, LS \, C \, P$$

where A_{st} is the amount of soil loss for a single storm event,
R_m is the rainfall-runoff erosivity factor,
K is the soil erodibility factor,
LS is the slope-length or topographic factor,
C is the cover and management factor
and P is the supporting conservation practice factor (10).

5.1. The erosivity factor

The rain erosivity factor R_m is expressed by

$$R_m = 0.5 \, EI_{30} + 3.5 \, V_u \, s_{pu}^{1/3}$$

where I_{30} is the storm's maximum 30-minute intensity in mm/hr
E is the storm energy in MJ/ha ;
V_u is the runoff volume in mm
and s_{pu} is the peak rate of runoff in mm/hr.

The expression suggested by Foster *et al.*, (1981) for E is :

$$E = S \, (e_k \, D(V_k)), \; k=1,p$$

where e_k is the rainfall energy by unit rainfall
and $D(V_k)$ is the depth of rainfall for the kth increment of the storm hyetograph divided into p parts and expressed in mm, with k = 1,...p.

The expression of e_k for a rain having an I_k intensity during the kth increment is given by :

$$e_k = 0.119 + 0.0873 \log_{10}(I_k) \quad \text{for } I_k < 76\text{mm/hr}$$

$$e_k = 0.283 \quad \text{for } I_k > 76\text{mm/hr}$$

Rainfall intensities and depths can be spatialized if there are enough meteorological stations in the basin; when the stations are well distributed along the basin it is possible to assign precipitation and depth values to each of the watershed cells using an interpolation method that takes into account the vertical gradient of precipitation and the difference in altitude between a particular grid and the altitudes of the stations used in the interpolation process. In the case of a lack of stations the rainfall erosivity factor is considered uniform on the whole basin, but this assumption generates more errors especially during summer storms. Meteorological radars can help to locate the strongest storms if available, or topographic gradients can be applied to the rainfall data, but the latter are more accurate on a statistical basis for a long time than on a single storm event.

Runoff volume and peak rate runoff are given by the hydrologic module of Hydrotel which computes surface and subsurface runoff available for transfert to the outlet of the basin. At each time step of the simulation, it is possible to save the runoff amounts relative to every grid, their sum being equal to the runoff volume V_u, and the maximum amount observed at a given step during the storm event and reported to the time unit corresponding to the peak rate of runoff s_{pu}.

5.2. The soil erodibility factor

A digitized soil map leads to the production of a geo-referenced file affecting the soil type to each grid of the basin, inherent soil erodibility (K) factors are then assigned to the soils from the nomographs established by Wischmeier and Smith (11) or from field tests using rain simulator or experimental plot data.

5.3. The topographic factor

The application of a finite differences interpolation technique on a digitized topographic map creates a digital elevation model, the processing of the DEM by the methods of slope and aspect calculation offered in the geographic information system leads to the determination of the slope factor S. The slope length / is derived from the aspect values of each cell and used in the calculation of the length factor L. S and L are given by :

$$S = 65.41 \sin^2 O + 4.56 \sin O + 0.065$$

$$L = (/22.5)^m$$

where m is the slope length exponent variying as a function of.the slope angle O.

5.4. The cover factor C and the supporting practices factor P

The C factor is determined from the classification of satellite images such as Landsat or Spot after a radiometric correction to eliminate the atmospheric influence and a geometric correction to overlay the image to other types of geo-referenced data ; the classification process assigns each pixel to a landuse class, the C factor is then affected to the classes according to the ground cover rate and the physiological state of the vegetation. Some vegetation indices such as the TSAVI (Transformed Soil Adjusted Vegetation Index, (12)), derived from red and near-infrared reflectance of satellite imagery, can give a realistic evaluation of the cover factor (13). However, for a single storm, it is the cover at the outbreak of the storm that counts, and the imagery requested must then be taken as close as possible to this event. If the crop calendar and land-use is relatively stable, an image from a previous year at the same phenological stage can be adequate.

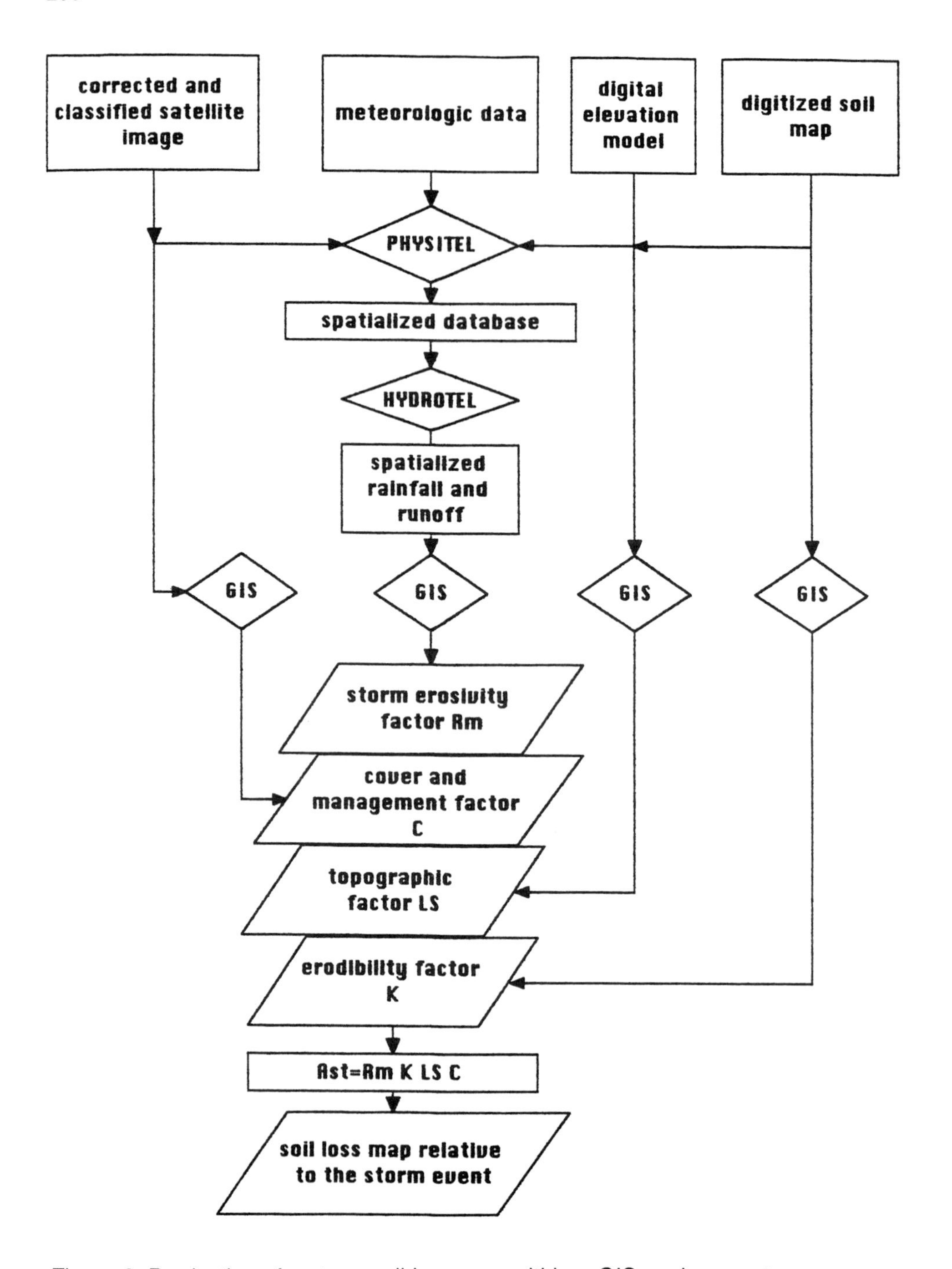

Figure 3. Production of a strom soil loss map within a GIS environment.

The supporting practice factor P is not easily derived from remotely sensed data, it can be determined from a terrain study or using an updated map of conservation practices.

6. PRODUCTION OF A STORM SOIL LOSS MAP

Once integrated to the GIS, the various factors represented by layers are multiplied according to the MUSLE expression, thus an event-based soil loss map is generated. Erosion factors can be modified by simple software manipulations leading to the production of a set of scenarios relative to various types of land cover and storm events. The methodology described above is illustrated in the overall flowchart in Fig. 3. Addition of a delivery ratio based on the proximity and relative altitude to the talweg for each cell can also help to predict the sediment load in streams (14).

7. CONCLUSION AND RECOMMANDATIONS

Integrating multisource data into GIS for targeting non-point source pollution is being more and more used ; the first attempts have been made with the USLE but it is evident that more accurate models describing the erosion process are needed. The methodology proposed above is one of the new orientations to monitor the soil loss at the regional scale and for a time scale adapted to the need of the modeller. However, more research to determine the best way of interfacing a distributed hydrologic model to an erosion model is needed, the most important problems to solve being the optimal resolution of the basin discretization and the relative accuracy in determining the different factors responsible of the soil degradation.

AKNOWLEDGEMENTS

Special thanks for Marie-Hélène de Sève and Patrick Marceau for the generation of the digital elevation model at INRS-Eau, to the Canadian International Development Agency who provided a fellowship to H. Chakroun and to NSERC grant OGP 0006043 to F. Bonn.

REFERENCES

1. Pelletier, R. E. (1985) : Evaluating non-point pollution using remotely sensed data in soil erosion models. *Journal of soil and water conservation*, vol 40, n° 4, pp. 332-335.
2. Maidment, D.R. (1991) : GIS and hydrologic modelling. *First Int. Symp./workshop on GIS and Environemental Modelling*, Boulder, Colorado, Sept.
3. Beasley, D.B. and HugginS, L.F. (1982) : ANSWERS (areal non-point source watershed environemental response simulation), user's manuel. EPA-905/9-82-001. *U.S. Environmental. Protection Agency*, Chicago, Ill. 54 pp.

4. Young, R.A., Onstad, C.A., Bosch, D.D. and Anderson, W.P. (1987) : AGNPS, agricultural non-point source pollution model : a large watershed analysis tool. *Conservation Resources Report* 35. U.S.Dept. Agr., Washington, D.C. 77 pp.
5. Laflen, J.M., Leonard, J.L., and Foster, G.R. (1991) : WEPP : a new generation of erosion prediction technology. *Journal of soil and water conservation,* January-February, 34-38.
6. Johnson, H.P. and Brakensiek, D.L. (1982) : Hydrologic modelling of small watersheds. *ASAE technical editor.*
7. Fortin, J. P., Villeneuve, J.P., Benmouffouk, D., Montminy, M., Blanchette, C. (1991) : Physitel 2.0 user's guide. *Scientific report,* INRS-Eau n° 287, 77pp.
8. Panuska, J.C., Moore, I.D. and Kramer, L.A. (1991) : Terrain analysis: integration into the agricultural nonpoint source (AGNPS) pollution model. *Journal of soil and water conservation*, Jannuary-February, 59-64.
9. Foster, G.R., Meyer, L.D. and Onstad, C.A. (1977) : An erosion equation derived from basic erosion principles. *Transactions of the ASAE*, 20 (4), 678-682.
10. Hession, W.C. and Shanholtz, V.O. (1988) : A geographic information system for targeting nonpoint-source agriculture pollution. *Journal of soil and water conservation,* May-June, 264-266.
11. Wischmeier, W.H. and Smith, D.D. (1978) : Predicting rainfall erosion losses : a guide to conservation planning. *U.S.D.A, agricultural handbook 537*, 58 pp.
12. Baret, F., Guyot, G., Major, D. J., (1989) : TSAVI : A vegetation index which minimizes soil brightness effects on LAI and APAR estimation." *Proceedings, Canadian symposium on remote sensing and IGARSS 89*, vol. 3, pp. 1355 - 1358.
13. Cyr, L., Bonn, F., Pesant, A. (1992) : "Remote sensing and vegetation indices for evaluating protective land cover " *Presented at the International. Seminar on Erosion in Plains and Hills*, ENS St Cloud, France, and submitted to CATENA.
14. Anys, H. (1991) : "Utilisation des données de télédétection dans un système d'information géographique pour l'étude de l'érosion hydrique du bassin versant de l'Oued Aricha, Settat, Maroc"*Mémoire de M.Sc., Département de géographie et télédétection, Université de Sherbrooke*, 83 p.
15. Fortin, J.P. and Bernier, M. (1991) : La transformation des données acquises par télédétection en données utiles pour le modèle hydrologique Hydrotel. *Télédétection et gestion des ressources.* Vol. VII, Paul Gagnon (ed), l'AQT, 181-188.
16. Fortin, J.P., Villeneuve, J.P., Benoit J., Blanchette, C., Montminy, M., Proulx, H., Moussa, R., and Bocquillon, C. (1991) : Hydrotel 2.1 user's guide. *Scientific report,* INRS-Eau n° 286, 171 pp.

Farm Land Erosion: In Temperate Plains Environment and Hills
S. Wicherek (Editor)

The use of caesium-137 to investigate soil erosion and sediment delivery from cultivated slopes in the Polish Carpathians

W. Froehlich [a], D. L. Higgitt [b], D. E. Walling [c]

[a] Institute of Geography and Spatial Organisation, Polish Academy of Sciences, Dept. of Geomorphology and Hydrology, Research Station, Frycowa 113, 33-335 Nawajowa, Poland.

[b] Department of Geography, Lancaster University,Lancaster, LA1 4YB, United Kingdom.

[c] Department of Geography, University of Exeter, Exeter, EX4 4RJ, United Kingdom.

SUMMARY

Information concerning the spatial patterns of erosion and deposition on cultivated hillslopes and the associated sediment delivery ratios is difficult to obtain using conventional soil loss monitoring techniques. The use of the fallout radionuclide caesium-137 as a sediment tracer offers considerable potential for elucidating patterns of soil redistribution and sediment delivery. This paper presents the results of an investigation of soil erosion and sediment delivery on a cultivated hillslope within the Homerka catchment in the Polish Carpathians, based on ^{137}Cs measurements. The ^{137}Cs inventories of soils in this region reflect inputs from both 'bomb'- and Chernobyl-derived fallout, and interpretation of the redistribution of radiocaesium must take account of the different timescales associated with the two sources and the complications introduced by Chernobyl fallout. The results indicate that the individual terraced fields are associated with substantial rates of soil redistribution, but that little of the eroded soil leaves the fields. The results provided by the ^{137}Cs measurements are consistent with other process-based measurements undertaken on the slopes and with available evidence concerning the dominant sources of suspended sediment transported by local streams. A general model of soil loss and sediment delivery from cultivated slopes in the Polish Carpathians is proposed.

RESUME

Sur les versants cultivés des Carpathes polonaises (bassin de Homerka), il est difficile d'obtenir par les méthodes informatiques classiques des données concernant la répartition spatiale de l'érosion et des dépôts. L'utilisation du Cesium 137 comme marqueur, permet de pallier ces carences. L'approche présentée doit tenir compte des retombées de l'accident de Chernobyl.

Les résultats obtenus indiquent que sur les replats limités par des rideaux, une importante mobilisation des sols est réalisée mais que peu de matériel quitte les champs.

Les résultats fournis par l'utilisation du Cesium 137 concordent avec d'autres mesure réalisées sur les pentes et avec celles qui concernent les transports en suspension dans les cours d'eau locaux. Les auteurs proposent un modèle de perte en sol et de transport de sédiment sur les pentes cultivées des Carpathes polonaises.

INTRODUCTION

The cultivated slopes of the Flysch Carpathians are characterized by a mosaic of field plots of various sizes, separated by terraces and intersected by networks of unmetalled roads. The agricultural practices on such slopes have been shown to promote soil erosion (1), but predicting the pattern of erosion and sediment delivery is complicated by the fragmentation of the slope into field plots (2). Identifying the spatial variability of erosion intensity across these slopes must be seen as a key research problem, since such information is required both to assess the potential impact of soil loss on crop productivity and to investigate the contemporary evolution of cultivated slopes and the delivery of sediment to the fluvial system.

Previous measurements undertaken on field plots in the Polish Carpathians have demonstrated that dispersed sheet wash is the dominant erosion process. Local rill erosion can occur during heavy rainfall, and earth flows have been observed on some potato fields (3 - 4). Erosion rates as high as 22 t ha^{-1} $year^{-1}$ have been recorded in potato fields, whilst typical rates under winter crops, pasture and forest are 2.4, 0.1 and 0.03 t ha^{-1} $year^{-1}$, respectively (5). The network of unmetalled roads increases the drainage density and represents an important source of suspended sediment delivered to the stream channels (6). It appears that the delivery of sediment from slopes to channels in cultivated Carpathian catchments is dominated by concentrated runoff from furrows, unmetalled roads and gullies and only to a limited extent by dispersed overland flow from field plots.

Most of the existing data on soil erosion have been generated by investigations involving slopes of uniform land use or experimental plots (1 - 4 - 7 - 5) and cannot readily be extrapolated to assess the spatial variability of erosion intensity on actual slopes. The use of Gerlach troughs, for example, is hampered by problems of defining the contributing area and of potential disturbance at the trough edge (2). Furthermore, troughs cannot be installed permanently within cultivated fields. There is, therefore, a problem in extrapolating point-specific measurements to larger landscape units and if the patterns of erosion and sediment delivery within small catchments are to be meaningfully assessed, a research method which can be applied to provide spatially consistent data is required. The use of the fallout radionuclide caesium-137 (^{137}Cs) as a sediment tracer offers potential for meeting this requirement.

The ^{137}Cs technique for assessing soil erosion has been described elsewhere (8 - 9 - 10). Its main advantage is that it permits the investigation of rates and patterns of erosion integrated over the time period since global fallout of ^{137}Cs commenced (1954), on the basis of a single site visit. Data aquisition is less expensive and time-consuming than in the case of most conventional monitoring techniques, and, since most existing estimates of erosion rates represent only short observation periods, data obtained from ^{137}Cs measurements could provide a means of validating these results against longer timescales. It also provides the basis for identifying the main suspended sediment sources within a catchment (cf. Walling and Bradley, 1990). Such information is fundamental to effective soil protection and reduction of sedimentation in downstream reservoirs.

In order to investigate patterns of soil erosion and sediment delivery on cultivated hillslopes in the Polish Carpathians, a research project has been undertaken in the Homerka experimental catchment, where traditional soil erosion and suspended sediment transport monitoring have been undertaken since 1971. Investigations exploiting the use of 137Cs as a sediment tracer commenced in 1984.

THE STUDY AREA

The 19.6 km^2 Homerka drainage basin (Fig. 1) is developed on flysch rocks of varying resistance to weathering and it can be subdivided into two parts, representing the montane headwater and the lower foothill zones. The montane zone, rises to 1060 m a.s.l. and is underlain by resistant sandstones and shales. This area is characterized by steep slopes (15-30°) and shallow skeletal soils. Most of the hillsides are forested and the forests are accessed by a dense network of unmetalled roads and lumber tracks. The foothill zone reaches a maximum altitude of 350 m a.s.l. and is underlain by shale-sandstone flysch series which weather to produce silty clay soils which support agricultural land use. The mosaic of arable fields on the slopes is served by a dense network of unmetalled roads which in many places are sunken below the level of the surrounding land.

The lower valley floors are occupied by meadows and permanent pasture. The mean annual precipitation in the montane headwaters exceeds 1000 mm, whilst in the lower zone it is about 900 mm. The equivalent values of mean annual air temperature are 5° and 7.5°C respectively.

In order to permit detailed investigation of erosion and sediment delivery from a representative cultivated zone in the lower part of the basin, an area of 26.5 ha located on the boundary of the forest and the agricultural areas has been designated an 'experimental slope' (Fig. 1). The slope is 500-700 m long and convexo-concave in form. The silty clay soils increase in depth towards the base of the slope. The slope is subdivided into numerous field plots (Fig. 2) which are cultivated across the slope. The plots are separated by terraces and furrows and by the unmetalled roads which traverse the area from the watershed to the stream channel. During times of heavy rainfall, these unmetalled roads act as channels for surface runoff and in many places they are deeply incised into the slope. When progressive incision of these tracks makes access to the fields difficult, farmers are forced to make a new track parallel to the old course. The zone of concentrated flow and accelerated erosion is thereby gradually enlarged at the expense of the cultivated land. The length of unmetalled road traversing the experimental slope is 3.3 km, equivalent to a density of 11.9 km km^{-2}. The density for the overall basin is 5.3 km km^{-2}.

METHODS OF INVESTIGATION

The application of radiocaesium measurements to studying soil erosion and sediment delivery within the Homerka basin began in 1984. The Chernobyl disaster in 1986 produced a substantial increase in ^{137}Cs inventories in the area and introduced problems in making comparisons between samples collected before and after the incident. Measurements of the caesium-134 activity of soils and sediments can be used to apportion the total ^{137}Cs activity between bomb- and

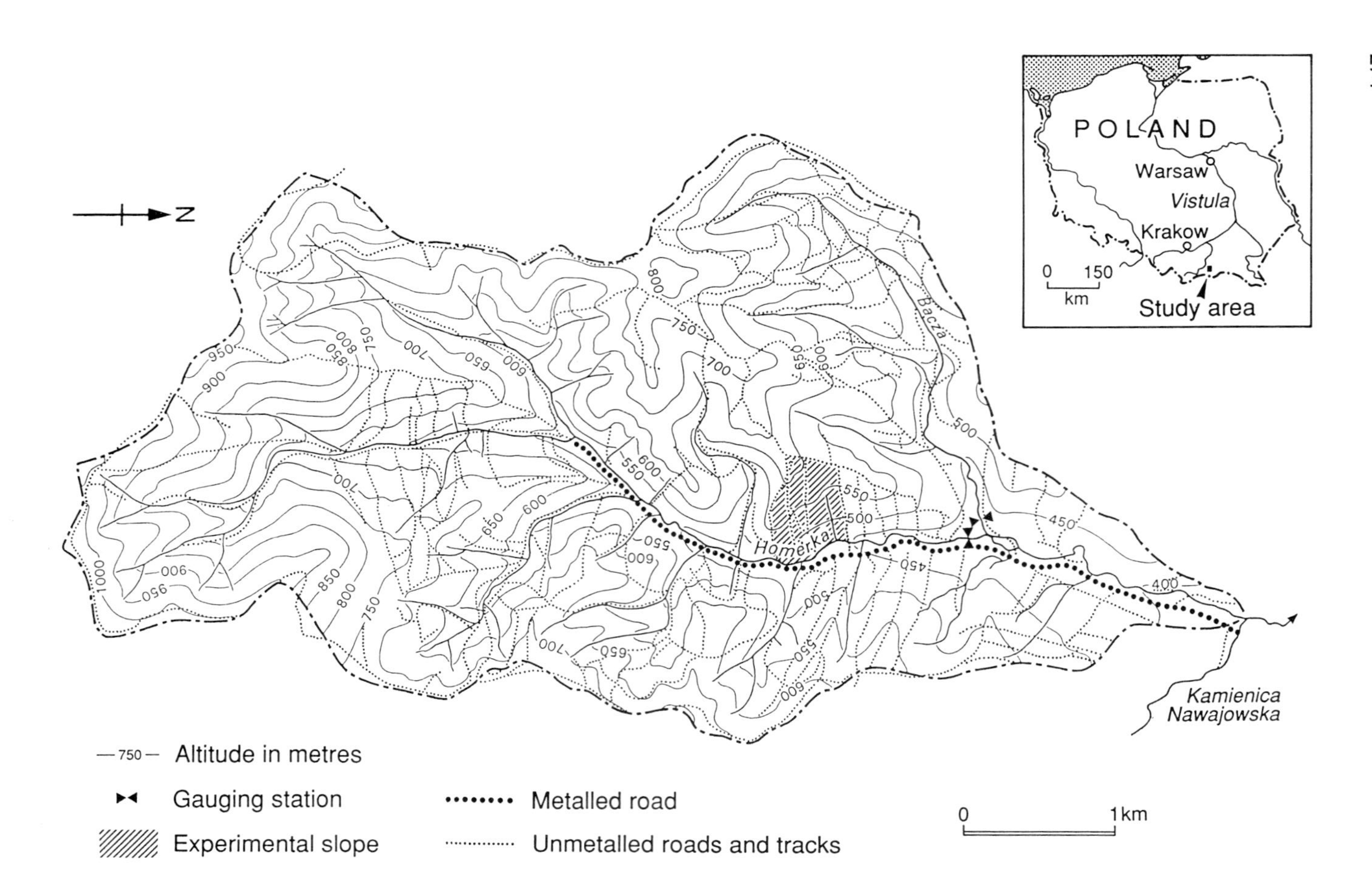

Figure 1. The Homerka basin and the location of the study area in Poland.

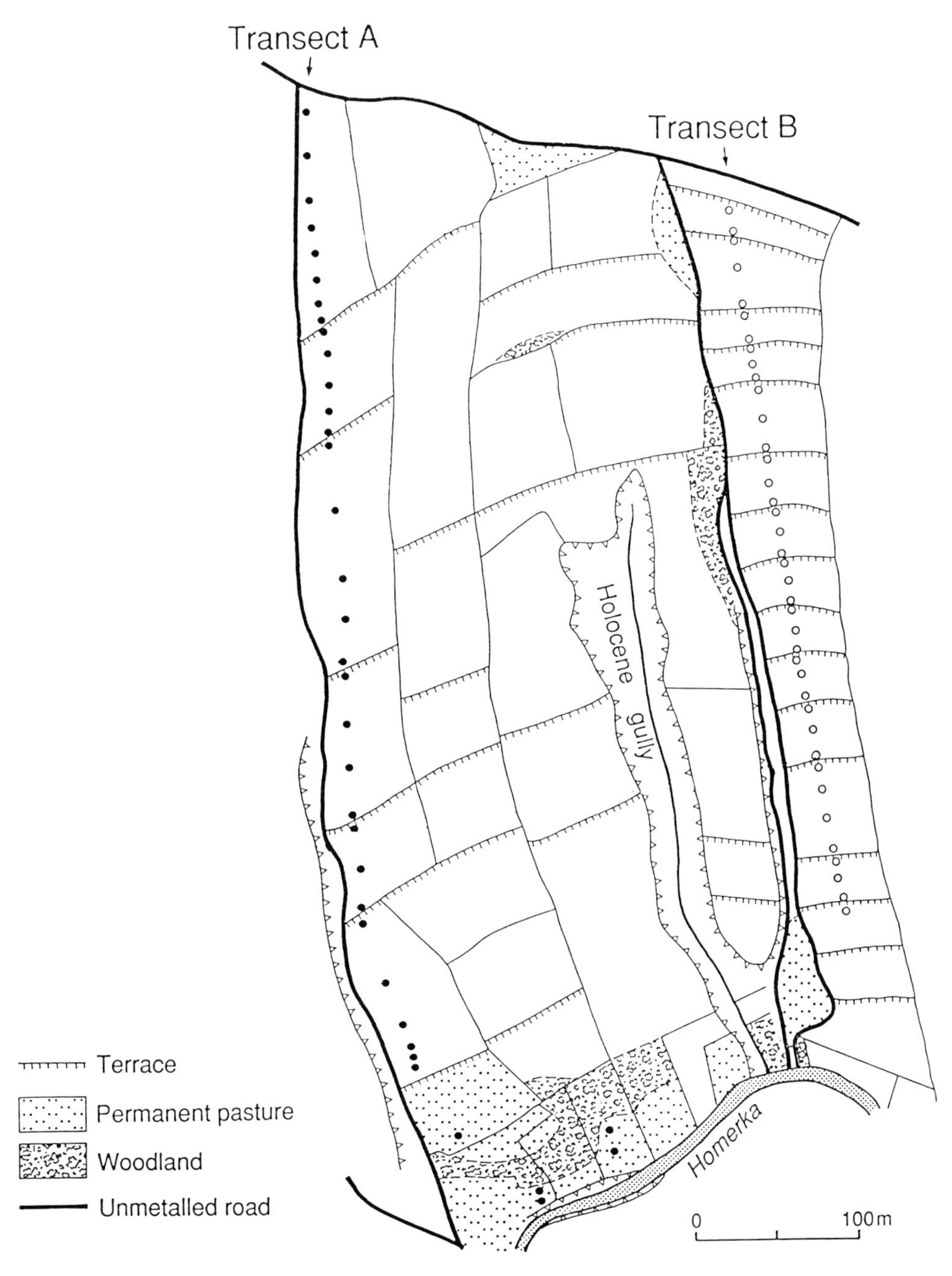

Figure 2. The two sampling transects on the experimental slope used for the collection of soil cores for radiocaesium measurements to investigate soil erosion and redistribution.

Chernobyl-derived fallout, but some of the results presented in this contribution relate only to samples collected betwen 1984 and 1986, prior to the Chernobyl incident, in order to simplify the interpretation.

The use of ^{137}Cs measurements to evaluate rates and patterns of soil erosion is commonly based on measurements of either the total radiocaesium inventory (Bq m^{-2}) of the soil profile or the vertical distribution of the inventory within the soil profile. A 75mm diameter cylindrical steel corer was used to collect soil cores to depths of up to 50 cm and in most cases these were sectioned at 5 cm intervals prior to analysis. All samples were dried, desaggregated and sieved to pass a 2 mm mesh prior to analysis by gamma spectrometry in the Department of Geography at the University of Exeter Information on suspended sediment sources within the Homerka basin was assembled using the 'fingerprinting' approach advocated by Peart and Walling (11). Caesium-137 measurements were used to fingerprint the various potential sources. Samples of surface material were collected from a range of potential sources, including areas of forest, pasture and arable cultivation, unmetalled roads, gully walls and channel banks. In order to take account of grain size effects and to permit direct comparison with suspended sediment samples, the <0.063mm fraction of the source materials was separated by sieving prior to gamma spectrometry analysis. Bulk samples of suspended sediment were collected from the main gauging station on the Homerka stream during flood events. The suspended sediment was recovered from the samples by sedimentation and centrifugation and the >0.063 and <0.063 mm fractions were separated by wet sieving.

SOIL REDISTRIBUTION INDICATED BY CAESIUM-137 MEASUREMENTS

The soil erosion study focused on the area of the experimental slope. Two downslope transects crossing the field plots and their associated terraces were sampled (Fig. 2). Transect A comprises several relatively long field plots up to 100 m in length and with gradients between 12° and 18°, separated by terraces. Transect B is steeper and comprises much shorter field plots, again separated by terraces. Bulk soil cores were collected from within the field plots to a depth of 35 cm. On the upslope edges of the terraces, cores were taken to a depth of 50 cm and the upper 25 or 30 cm portions were sectioned into 5 cm increments. Information on baseline radiocaesium inventories necessary to interpret the subsequent pattern of redistribution was obtained from a series of three sectioned soil profiles representing areas of undisturbed grassland with minimal slope on the watershed of the experimental slope. All samples were collected during 1988 and therefore contain both 'bomb'- and Chernobyl-derived radiocaesium fallout.

The total ^{137}Cs inventories of the three 'input' sites ranged from 5302 $\pm$ 114 to 7226 $\pm$ 134 Bq m^{-2}. The ^{134}Cs contents of the same profiles ranged from 1597 $\pm$ 76 to 1916 $\pm$ 96 Bq m^{-2}, providing a mean ^{134}Cs input (corrected to May 1986, the period of Chernobyl deposition) of 1782 Bq m^{-2}. In this area, the pre-existing 'bomb-test' inventories of ^{137}Cs were approximately doubled by Chernobyl fallout. The variability of the estimates of the reference inventory for the experimental slope noted above, which largely reflects local variability in the receipt of Chernobyl fallout, and the uncertainties involved in separating 'bomb' and Chernobyl contributions to the overall ^{137}Cs inventory necessarily impose limitations on the interpretation of the radiocaesium measurements, as compared to areas which received little or no Chernobyl fallout (12). As a result, these measurements are most useful for providing information about the general pattern of erosion and soil redistribution operating over the slope as a whole, rather than the detailed rates and patterns of erosion and sediment transfer within individual fields.

Taking account of these limitations, the most striking feature of the data collected from the transects is the elevated levels of radiocaesium in the cores collected from the terrace edges, when compared to average levels within the plots (Fig. 3, Table 1).

Table 1.
Radiocaesium inventories associated with cores collected from the field plots and the terraces.

	Radiocaesium inventory (Bq m-2)			
	Mean	Min.	Max.	CV.
FIELD PLOTS				
Caesium-134	1804	306	2919	38.6
Total caesium-137	7937	2160	10128	29.2
Bomb caesium-137	3789	-	6082	37.6
TERRACES				
Caesium-134	2092	547	6359	60.4
Total caesium-137	12737	6173	19438	30.9
Bomb caesium-137	8553	3875	15877	35.5

The total ^{137}Cs inventories of the individual cores collected on the terrace edges were in nearly all cases substantially greater than those associated with cores collected from within the plots, reflecting soil loss within the fields and deposition on the terraces at the lower boundary of the fields. Cores collected from immediately below the terraces have similar total ^{137}Cs inventories to mid-plot samples, although they sometimes contain slightly higher levels of ^{134}Cs. Thus, although it is likely that some of the eroded sediment may cascade over the terrace onto the adjacent downslope plot during major runoff events, this would not seem to be of major significance. The occurrence of deposition on the terraces is further substantiated by the increased depth to which radiocaesium is found at these locations (cf. Fig. 3). The average depth of cultivation in these fields is of the order of 20 cm and, assuming that the associated mixing would distribute radiocaesium to this depth, the terrace edge profiles evidence an average deposition of ca. 10-15 cm. If it is assumed that this deposition has occurred during a period of about 35 years, since the first occurrence of significant bomb fallout, average rates of deposition may be estimated at ca. 4 mm $year^{-1}$. The radiocaesium measurements therefore indicate that appreciable rates of soil redistribution are occurring within the fields, in response to both erosion processes and the movement of soil associated with tillage practices. However, a major proportion of the eroded soil is redeposited on the terraces and is not transported beyond the field system. These conclusions are consistent with the results of long-term monitoring of sediment movement on the experimental slope reported by Froehlich (2), which suggested that, although significant rates of erosion occur on the cultivated fields, these do not

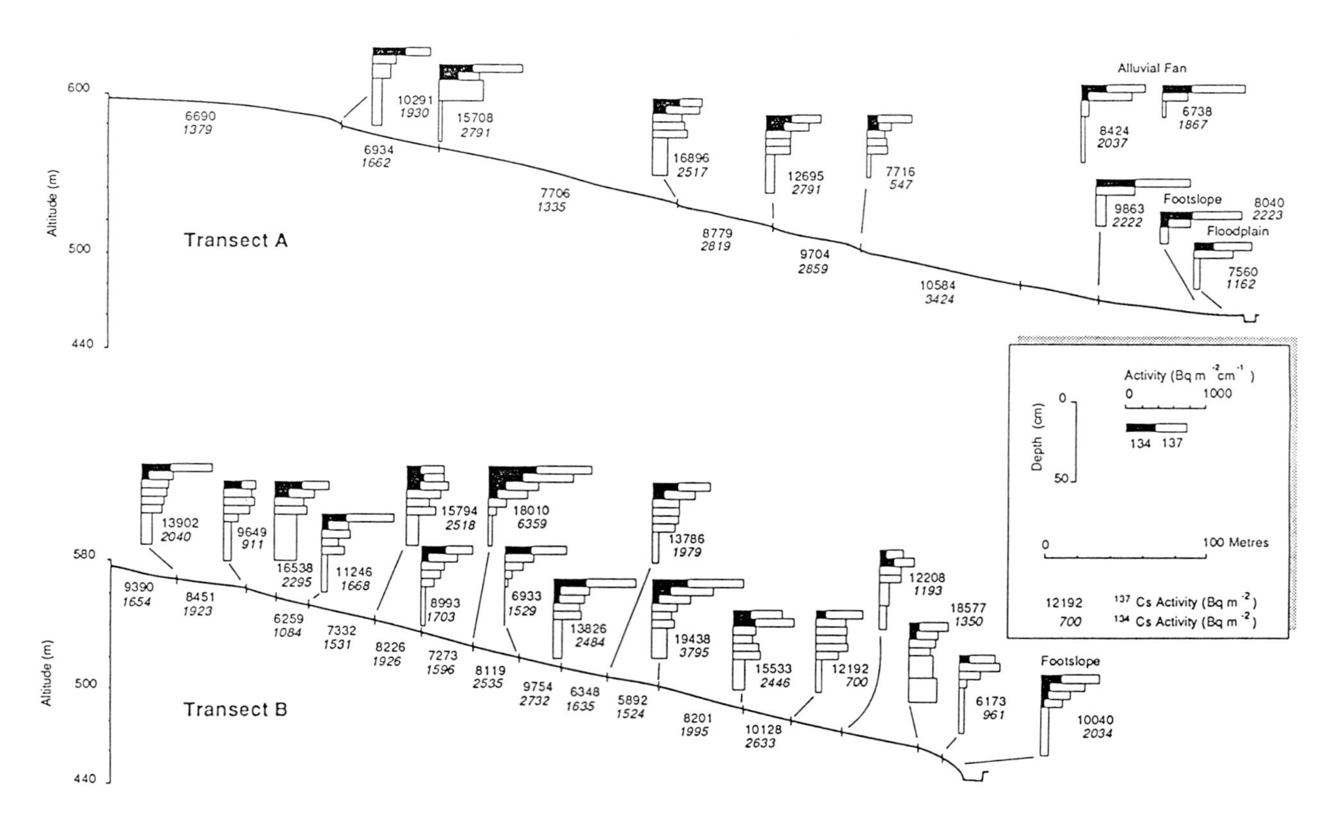

Figure 3 Topographic profiles along the two sampling transects transects shown in Figure 2. The mean radiocaesium inventories measured in the individual field plots, the inventories measured on the terrace edges and the vertical distribution of radiocaesium in the terrace edge cores are also shown.

represent a significant source of sediment transfer to the streams. The major sources of suspended sediment appeared to be the unmetalled roads and the active gullies.

FINGERPRINTING SUSPENDED SEDIMENT SOURCES

The inferences concerning the major sources of suspended sediment within the Homerka basin noted above and the limited importance of the cultivated fields as sediment sources, despite the evidence for significant erosion rates in these areas, were further investigated using ^{137}Cs as a 'fingerprint' tracer. Measurements of the ^{137}Cs content of the < 0.063 mm fraction of suspended sediment collected from the Homerka stream at the main gauging station in the pre-Chernobyl period, indicated a range between 6.3 and 22.6 mBq g^{-1}, with a mean of 11.9 mBq g^{-1} and a standard deviation of 4.4. Comparison of these values with typical values for potential source materials (Table 2, Fig. 4) suggest that they closely match those associated with material collected from the surface of unmetalled roads. Sediment eroded from the surface of forest and pasture areas is very unlikely to represent an important sediment source, since its ^{137}Cs content is substantially higher. Material eroded from cultivated areas and from gully and channel banks is similarly unlikely to constitute a major source, because the range of ^{137}Cs levels associated with these materials extends well above that representative of suspended sediment. The evidence provided by the radiocaesium fingerprints therefore suggests that the major source of the suspended sediment transported by the Homerka stream is the unmetalled roads, which occur throughout both the forested and the agricultural zones of the basin. The importance of the unmetalled roads as the main source of suspended sediment is further substantiated by analysis of the available information on the runoff processes operating within the basin which is discussed below.

Table 2.
Caesium-137 concentrations associated with suspended sediment, silt deposited in the stilling basin above a drop structure, and potential suspended sediment sources within the Homerka basin.

Samples	Mean concentration (mBq g-1)	Standard deviation (mBq g-1)
Suspended sediment <0.063mm	11.9	4.4
Suspended sediment >0.063mm	0.8	0.8
Silt from drop structure	3.7	5.3
Channel bank material	13.5	15.3
Gully bank material	16.5	17.9
Surface material from unmetalled roads	3.8	6.3
Surface material from cultivated fields	21.7	12.0
Surface material from pasture	49.0	27.6
Surface material from forest	57.5	38.0

Existing evidence relating to the generation of storm runoff within the Homerka basin indicates that the frequency of occurrence of surface runoff on unmetalled roads and in gullies and the furrows between fields is considerably greater than on the cultivated areas. This is further emphasized on Figure 4 where

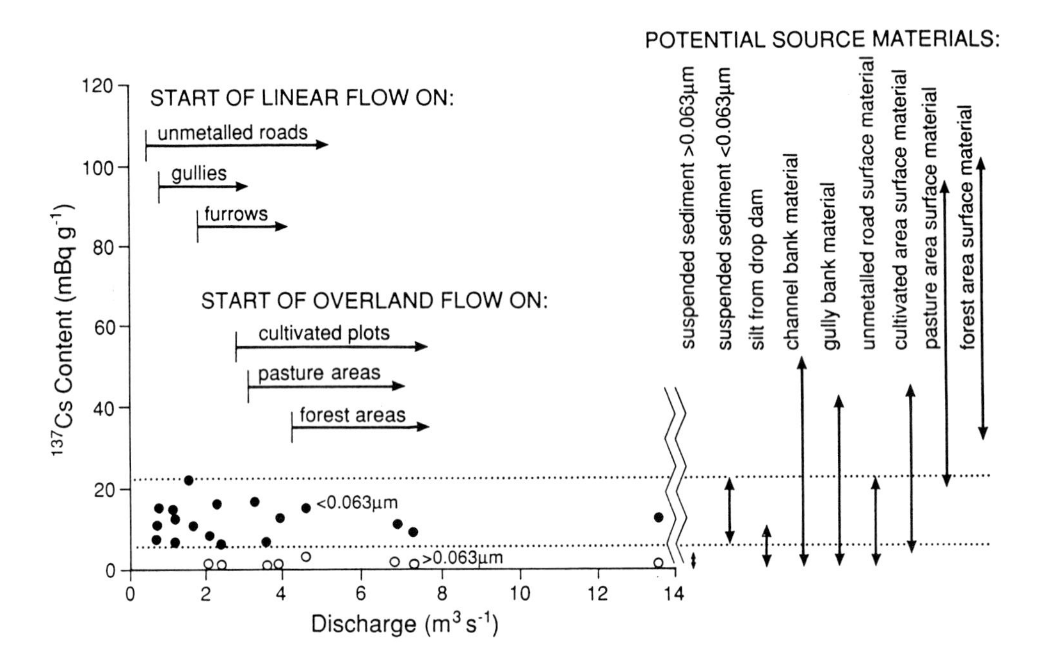

Figure 4 The use of ^{137}Cs measurements to fingerprint suspended sediment sources. The relationship between the ^{137}Cs content of suspended sediment and discharge, the discharge thresholds associated with the occurrence of storm runoff from various sources within the basin, and the range of ^{137}Cs concentrations associated with suspended sediment and potential source materials are illustrated.

the typical discharge levels at the main gauging station on the Homerka stream associated with initiation of linear flow on the unmetalled roads and in the gullies and furrows, and of overland flow on the cultivated plots and within the areas of pasture and forest are shown. The relationship between the ^{137}Cs content of suspended sediment and the discharge of the Homerka stream presented in Figure 4 shows no evidence of a shifts associated with the incidence of overland flow contributions from the cultivated plots and pasture and forest areas of the basin and with the initiation of runoff from furrows between the fields. This again strongly suggests that surface runoff from the unmetalled roads, and perhaps also gullies, represents the major source of suspended sediment transported by the stream.

This evidence for the lack of importance of cultivated fields as sediment sources is supported by other field observations which suggest that direct delivery of sediment to the channel from cultivated fieldsd is likely to be restricted, since the pasture areas which exist at the foot of the slopes and bordering the stream channel would act as efficient sinks for sediment transported by overland flow (2). Furthermore, any sediment delivered from the cultivated fields to the channels is likely to move via the furrows, which are in turn connected to the unmetalled roads. Concentrations observed in the furrows are, however, much smaller than those measured on the unmetalled roads and in the stream channels, again emphasizing the limited connection between the cultivated fields and the river channels. Some sediment from the cultivated fields is, however, transported into the unmetalled roads and gullies by deflation processes during the winter and this will be delivered to the streams during the spring floods (2 - 6).

The >0.063 mm fraction of the suspended sediment is characterized by very low ^{137}Cs activity, ranging from 0.0 to 2.5 mBq g^{-1}, with a mean value of 0.8 mBq g^{-1}. These low values reflect both the preferential association of radionuclides with the finer fractions ([13]) and the dominance of unmetalled roads and gullies as sediment sources. Some samples of the finer sediment were recovered from sediment basins located above the drop structures constructed along the Homerka stream (Fig. 1, Table 2). These were characterized by a relatively low ^{137}Cs content ranging from 0.1 to 10.7 mBq g^{-1}, which conforms with the range associated with both the >0.063 mm fraction and the < 0.063 fractions of suspended sediment. This in turn confirms the longer-term representativeness of the ^{137}Cs fingerprints of the suspended sediment samples.

CONCLUSIONS

Although the presence of both Chernobyl and 'bomb'fallout in the soils of the Homerka catchment complicates the interpretation of radiocaesium measurements, there is clear evidence that the cultivated areas are characterized by significant rates of erosion and that the intensity of erosion and deposition processes varies markedly across the individual field plots and agricultural terraces. This variability reflects both natural factors and changes in the pattern of land use and land partition through time. The agricultural terraces effectively restrict transport of sediment downslope and the cultivated areas are not a significant contributor to the overall suspended sediment output from the basin. The main sediment contributing areas are the footslopes and the unmetalled roads and gullies which may extend back to the divide. There is some connection between the field plots and the stream, through the network of furrows, gullies and unmetalled roads, but this linkage is limited and operates over only restricted zones during extreme runoff events. In view of this variability in erosional and depositional activity within the landscape of the Flysch Carpathians and the discontinuities within the sediment delivery system, traditional approaches to erosion monitoring involving point-specific measurements may provide unrepresentative results which should not be extrapolated to larger areas. The use of fallout radionuclides as sediment tracers offers opportunities to investigate the processes and sediment delivery linkages operating within the larger system.

ACKNOWLEDGEMENTS

The support of the Polish Academy of Sciences and the University of Exeter for the work reported in this paper is gratefully acknowledged. Thanks are also extended to the UK Natural Environment Research Council who supported DLH as a postgraduate student in the Department of Geography at the University of Exeter and to Mr G. Grapes for assistance with gamma spectrometry measurements.

REFERENCES

1. Gerlach, T. (1966) : Wspolczesny rozwoj stokow w dorzeczu gornego Grajcarka (Beskid Wysoki-Karpaty Zachodnie). Prace Geogr. IG i PZ PAN, 52, 111 pp.
2. Froehlich, W. (1982) : Mechanism transportu fluvialnego i dostawy zwietrzelin do koryta w gorskiej zlewni fliszowej. Prace Geogr. IG i PZ PAN, 143, 144 pp.
3. Gil, E. and Slupik, J. (1972a) : Hydroclimatic conditions of slope wash during snow melt in the Flysch Carpathians. Symp. International de Geomorphologie, Univ. Liege, 67, 75-90.
4. Gerlach, T. (1976) : Wspolczesny rozwoj stokow w Polskich Karpatach Fliszowych. Prace Geogr. IG i PZ PAN, 122, 116 pp.
5. Gil, E. (1976) : Splukiwanie gleby na stokach fliszowych w rejonie Szymbarku. Dokum. Geogr. IG i PZ PAN, 2, 63 pp.
6. Froehlich, W. (1991) : Sediment production from unmetalled road surfaces. In : Sediment and Stream Water Quality in a Changing Environment : Trends and Explanation, N. E. Peters and D. E. Walling (eds.), IAHS Publication n° 203, 21-29.
7. Gil, E. and Slupik, J. (1972b) : The influence of plant cover and use on the surface run-off and wash down during heavy rain. Studia Geomorph. Carpatho-Balcanica 6, 181-190.
8. Ritchie, J.C. and McHenry, J.R. (1990) : Application of radionuclide fallout caesium-137 for measuring soil erosion and sediment accumulation rates and patterns. J. Environ. Quality, 19, 215-233.
9. Walling, D.E. and Quine, T.A. (1990) : Use of caesium-137 to investigate patterns and rates of soil erosion on arable fields. In : Boardman, J., Foster, I.D.L. and Dearing, J.A. (eds.) Soil Erosion on Agricultural Land, Wiley, Chichester, 33-53.
10. Walling, D.E. and Quine, T.A. (1991) : Use of caesium-137 measurements to investigate soil erosion on arable fields in the UK ; potential applications and limitations. J. Soil Science, 42, 147-165.
11. Peart, M.R. and Walling, D.E. : (1988) Techniques for establishing suspended sediment sources in two drainage basins in Devon, UK : a comparative assessment. In : Sediment Budgets, Proc. Porto Alegre Symp., M.P. Bordas & D.E. Walling (eds.) IAHS Publ. n° 159, 41-55.

12. Higgitt, D.L., Froehlich, W. and Walling, D.E. (1992) : Applications and limitations of Chernobyl radiocaesium measurements in a Carpathian erosion investigation, Poland. Land Degradation and Rehabilitation, 3, 15-26.
13. Frissel, M. and Pennders, R. (1983) : Models for the accumulation and migration of 90Sr, 137Cs, 239,240Pu and 241Am in the upper layer of soils. In : Ecological Aspects of Radionuclide Release, P.J. Coughtrey (ed.), 63-72. Spec. Publ. Brit. Ecol. Soc. n° 3.

Farm Land Erosion: In Temperate Plains Environment and Hills
S. Wicherek (Editor)

The effect of water erosion and tillage movement on hillslope profile development : a comparison of field observations and model results

G. Govers [a], T.A. Quine [b] and D.E. Walling [b].

[a] National Fund for Scientific Research, Laboratory for Experimental Geomorphology, Catholic University of Leuven, Redingenstraat 16B, 3000 Leuven, Belgium

[b] Department of Geography, University of Exeter, Amory Building, Exeter, EX4 4RJ, Devon, United Kingdom

SUMMARY

The measured long-term and short-term erosion rates on two hillslope profiles, one in Huldenberg (Belgium) and one in Dalicott (United Kingdom) are compared with the results of simulations carried out using a simple water erosion model. The water erosion model predicts very well the observed variation of water erosion rates over the Huldenberg slope. However, when long-term profile development is considered, the agreement is much weaker. For both slopes, the agreement between simulation results and field observations could be significantly improved by adding a component of diffusive soil movement. However, transport rates associated with diffusion processes like rainsplash transport and soil creep are far too low to account for the rate of diffusion soil movement suggested by the model results. Limited literature data suggest that soil tillage may be the primary diffusion process on hillslopes. According to these data and the model results, soil redistribution by tillage may be as important as soil redistribution by water erosion processes on the field scale in many landscapes in Western Europe.

1. INTRODUCTION

Water erosion processes have been the subject of intensive research both in the field and in the laboratory over the last decades. As a result of such studies a wide range of information on the factors influencing water erosion processes is now available. This information has been readily incorporated into process-based erosion models. However, validation of the predictions generated by so-called spatially distributed erosion models has frequently been limited to consideration of total sediment output at the bottom of the slope or at the catchment outlet and until present there have been few attempts to test models against observed spatial variation in erosion rates. Furthermore such tests were generally limited to the qualitative comparison of spatial patterns (eg. 1). One of the reasons for the latter is undoubtedly the limited amount of data available on spatial patterns of erosion and deposition rates. In most cases, such data can only be obtained from tracer studies (eg. Caesium-137) or detailed plot and hillslope measurements. However, in some cases, evidence concerning the long-term development of a slope may be obtained by comparing its actual profile with the profile of adjacent woodland or pasture which may be considered to have remained stable over an extended time period. In this paper, documented erosion rates for two slopes, one in Belgium and one in the UK where data on both recent and long-term soil redistribution patterns are available, are compared with the output of a simple model of sediment movement over a hillslope. The model estimates soil redistribution by both water erosion and diffusion processes such as tillage and, therefore, allows an assessment of their relative importance to be assessed through comparison of actual and modelled patterns.

2. MATERIALS AND METHODS

a. The slopes

The experimental slope at Huldenberg is located in the loess belt of Central Belgium. It occupies a south-facing convex-concave slope with a maximum slope angle of up to 14 degrees (Fig. 1).

Soil texture is variable, but silt loams are dominant. Further details can be found in Govers (2) and Govers and Poesen (3). On the western side, the field is bordered by woodland which, according to historical-cartographic evidence, has remained undisturbed since at least 1850. The same data indicate that the experimental field was taken into cultivation between 1850 and 1865 (E. Paulissen, pers. comm.).

Soil redistribution rate data are available for this field at two timescales: long-term and short-term (1 year). Long-term redistribution rates within the field were inferred from measurement of the height difference between the field and the woodland along the western border. Data on short-term rill and interrill erosion within the field were collected between September 1983 and December 1984 and were discussed by Govers (2) and Govers and Poesen (3). For this study, rill erosion rates were calculated using data for only the westernmost 11m strip of the slope. Total water erosion rates for 15 transects were calculated by summing the measured rill and interrill erosion rates as reported in the study of Govers and Poesen. The data used in this study correspond to the period from September 1983 to the 3rd October 1984 and, therefore, represent the pattern of soil redistribution by water erosion between two tillage events.

The slope at Dalicott Farm in Shropshire, UK, is described by Quine and Walling (4). Again, soil redistribution rate data are available for this field at two timescales : long-term (c.200 years) and recent (35 years). Long-term soil redistribution rates have been determined from the height difference between the cultivated field and an adjacent are of undisturbed pasture. In addition, information on rates of soil redistribution over the last 30 years has been provided by caesium-137 (^{137}Cs) measurements taken along the profile.

b. The model

The model used does not intend to provide a detailed, process- based simulation of the erosion processes occurring within the field, but merely tries to take into account the effect of topography on the various processes responsible for soil movement. Functionally, it resembles the two-dimensional 'store' model proposed by Kirkby et al. (5 p. 85-91). However, different equations were employed to represent the processes. The model incorporates rill erosion, interrill erosion and soil movement by diffusion processes which include direct movement by splash erosion, soil creep and tillage. When the flow is below transport capacity, rill erosion rates are modelled as a function of slope angle and length :

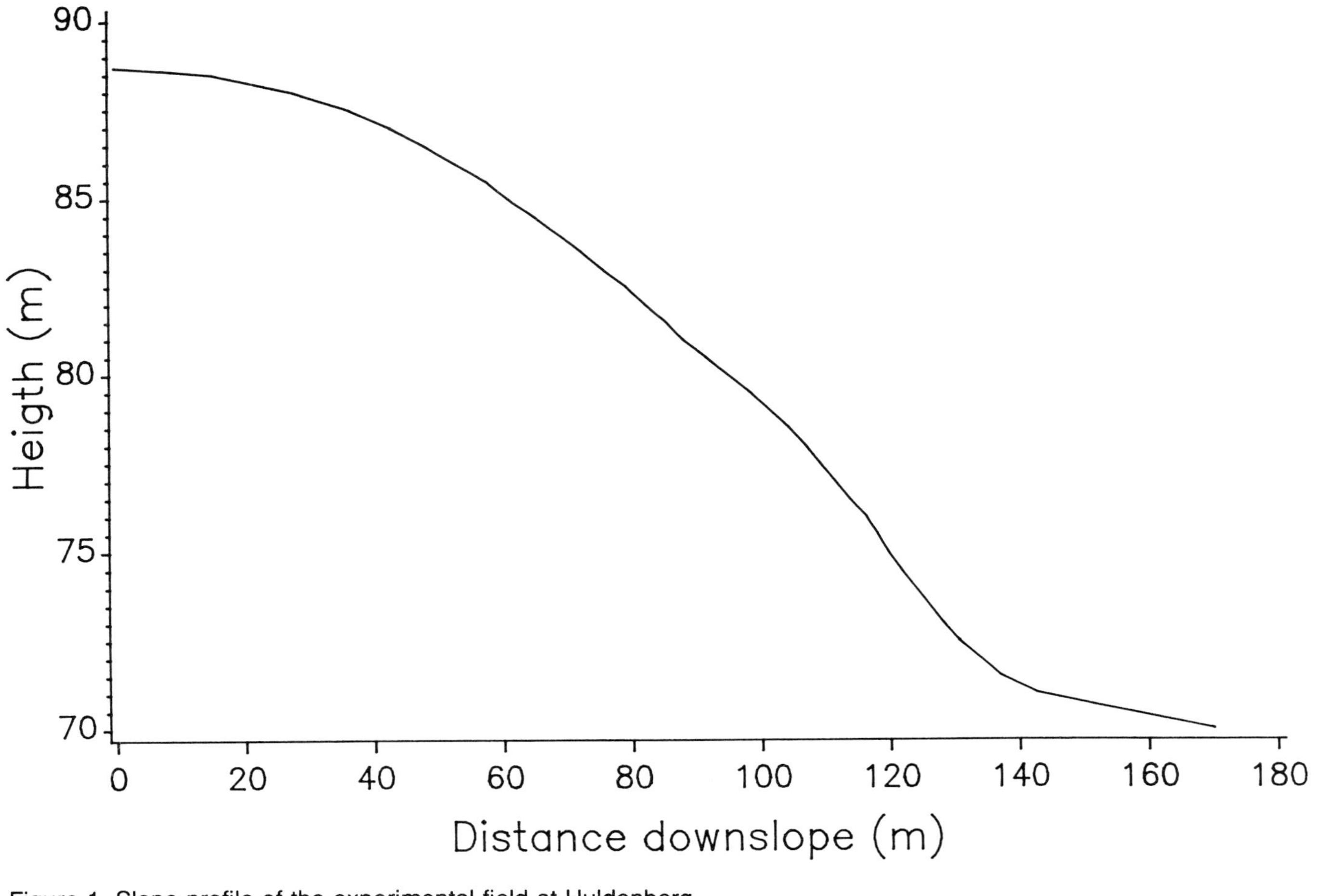

Figure 1. Slope profile of the experimental field at Huldenberg.

$Er = fb\ a\ S^b\ L^c$

where Er = the rill erosion rate per unit area (kg m^{-2})
fb = the dry bulk density of the soil (kg m^{-3})
S = the slope (sine)
L = the horizontal distance from the interfluve or field boundary
a, b and c are coefficients.

From a field study of rill erosion in the loam belt of Central Belgium, Govers (6) derived mean values of 1.45 and 0.75 for (b) and (c) respectively. As pointed out by Govers (6), these values correspond closely with values reported in the literature and with those derived from process-studies and they were therefore used in this study. The value of (a) was set to $3*10^{-4}$, and a dry bulk density of 1350 kg m^{-3} was assumed.

Detachment on interrill areas is thought to be dominated by raindrop detachment, while transport of sediment occurs by the flowing water. Interrill erosion is assumed to depend only on the local slope :

$Eir = fb\ d\ S^e$

where Eir = the interrill erosion rate per unit area (kg m^{-2})
d and e are coefficients.

Interrill erosion is thought to be essentially independent of slope length, because the runoff and sediment generated from interrill areas reaches the rill network relatively rapidly. A value of 0.8 was used for the interrill slope exponent (e). The same value was derived by Foster (7), but his equation also includes also a positive constant, suggesting significant interrill soil loss on a zero slope. However, the necessity of the latter is not always supported by experimental data (8). The slope exponent for interrill erosion is assumed to be lower than 1 because on steeper slopes interrill erosion becomes raindrop detachment-limited. The intensity of interrill erosion was supposed to vary proportionally with that of rill erosion. Data from the Huldenberg experimental field indicate that interrill erosion rates equal rill erosion rates on a 0.06 slope at a distance of c. 65m from the divide. The coefficient (d) was therefore assumed to be equal to 3.68*a.

The transport capacity on a given slope segment (Tc) is considered to be directly proportional to the potential for rill erosion :

$Tc = fb\ f\ Er$

where Tc = the transport capacity (kg/m)
f is a coefficient

This implies that the flow over the slope is not always transport limited. This modification of the approach of Kirkby et al. (5) was thought to be necessary as observations undertaken in several experiments have demonstrated that sediment concentrations in rills are often well below the transport capacity, especially when soils are wet and compacted (eg. 9).

Finally the model accounts for diffusion processes including transport by splash, soil creep and tillage by assuming that the resultant soil movement will be proportional to the sine of the slope angle :

$Qsd = g\ S$

where Qsd = the net soil flux due to diffusion processes (kg m^{-1})
g is a coefficient

Erosion rates are, therefore, calculated as follows :

$Esd = Qsd/\ x$

where Esd = the erosion rate per unit are attributable to diffusion processes (kg ^{m2})
x = the distance from the divide

The model can be run for the desired number of iterations, each of an arbitrary length of time. During each iteration, soil movement by both water erosion and diffusion processes is calculated and the slope profile is adjusted according to the calculated soil redistribution rates.

Due to the different formulation of the governing equations, rates of soil movement due to rill and interrill erosion are not directly comparable to those due to diffusion processes. However, some idea may be obtained by comparing fluxes over a slope profile of constant convexity, having a slope of 0.0 at the divide and a slope of 0.2 at 100 m from the divide. For each iteration, total rill and interrill erosion will then be equal to 61.6 kg/m. Total erosion due to diffusion processes will be only dependent on the value of (g). Thus, erosion rate due to diffusion processes will equal the water erosion rate when g=308. Of course, this value only holds for the assumed slope profile.

3. RESULTS

The model was first used to simulate the spatial pattern of soil redistribution observed for the Huldenberg slope on the 3rd October 1984. This observed pattern represents the net effect of water erosion over a period of one year. In this case the model output corresponded closely with the observed data when no diffusion soil movement was included (Figure 2, Table 1).

Table 1.
Results of regression of measured soil movement on predicted soil movement (hmeas = A*hpred + B)

Simulation n°	Site	Value of g	Number of Iterations	Regression Gradient	Equation Intercept	r^2
1	Huldenberg	0	22	1.03	-0.002	0.95
2	Huldenberg	0	130	0.96	0.11	0.39
3	Huldenberg	120	155	0.96	0.09	0.92
4	Dalicott	0	450	0.97	-0.02	0.68
5	Dalicott	325	325	0.97	-0.04	0.88
6	Dalicott	0	27	0.99	0.0	0.82
7	Dalicott	100	24	0.99	0.0	0.86

The value used for (e) ($3*10^2$) causes deposition to commence at a slope of c. 0.04 and at a horizontal distance of c. 145 m from the divide, which is in close agreement with field observations. Although this first test of the model provides only crude validation, it does indicate that the predicted spatial pattern is meaningful and consistent with observed erosion rates. *For all of the subsequent simulations the same values of the coefficients a,b,c,d,e and f were retained.* Only the coefficient (g) and the number of iterations were varied in order to obtain an optimum agreement between observed and predicted values.

The second simulation attempted to use the same water erosion model to simulate the long-term pattern of soil redistribution on the Huldenberg slope. As Table 1 and Figure 3 illustrate, the correspondence between observed and predicted rates is quite weak.

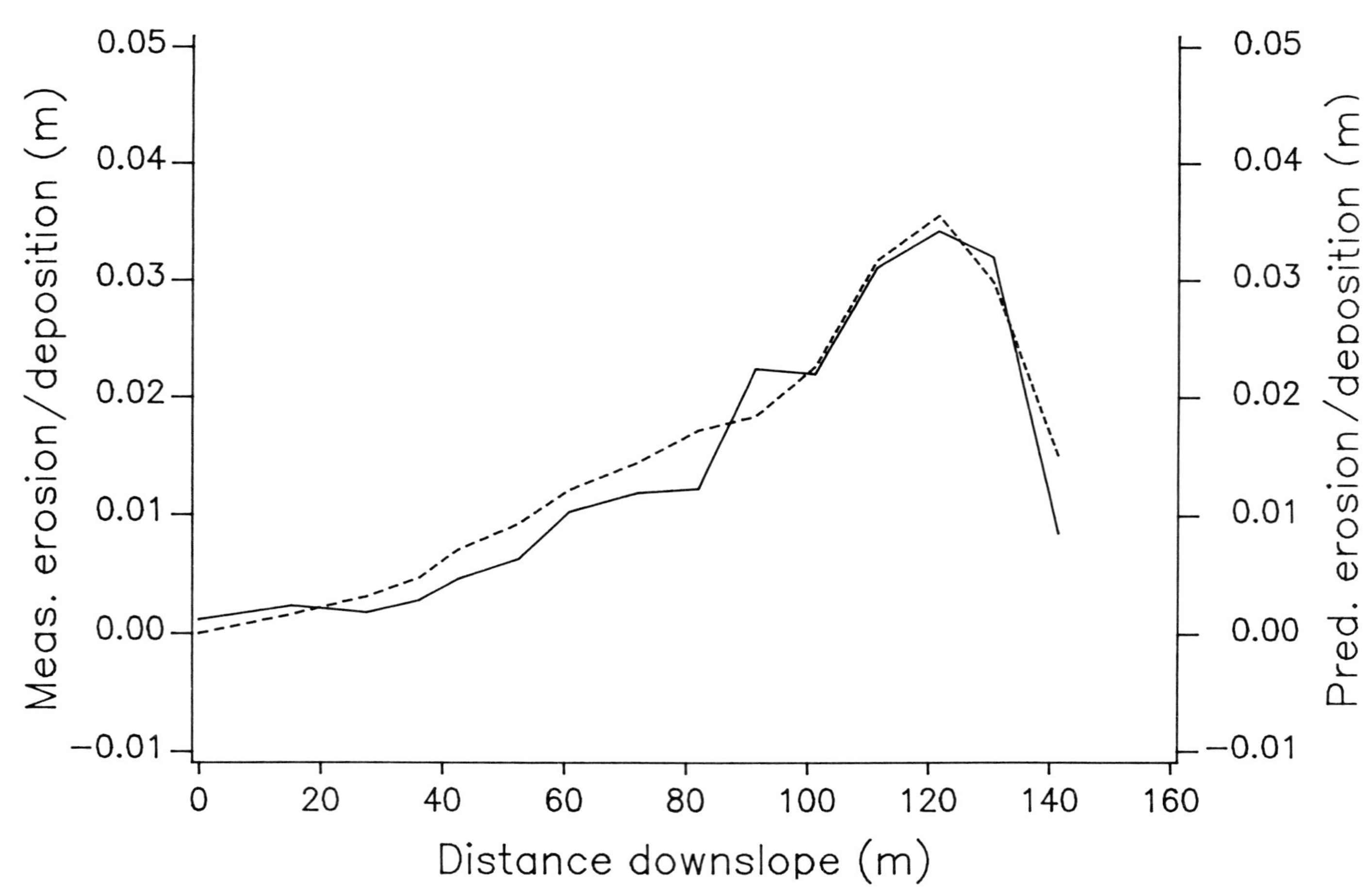

Figure 2. Comparison of measured (full line) and predicted (broken line) short-term (1 year) soil redistribution on the Huldenberg field slope. (Simulation 1 - g=0, 22 iterations).

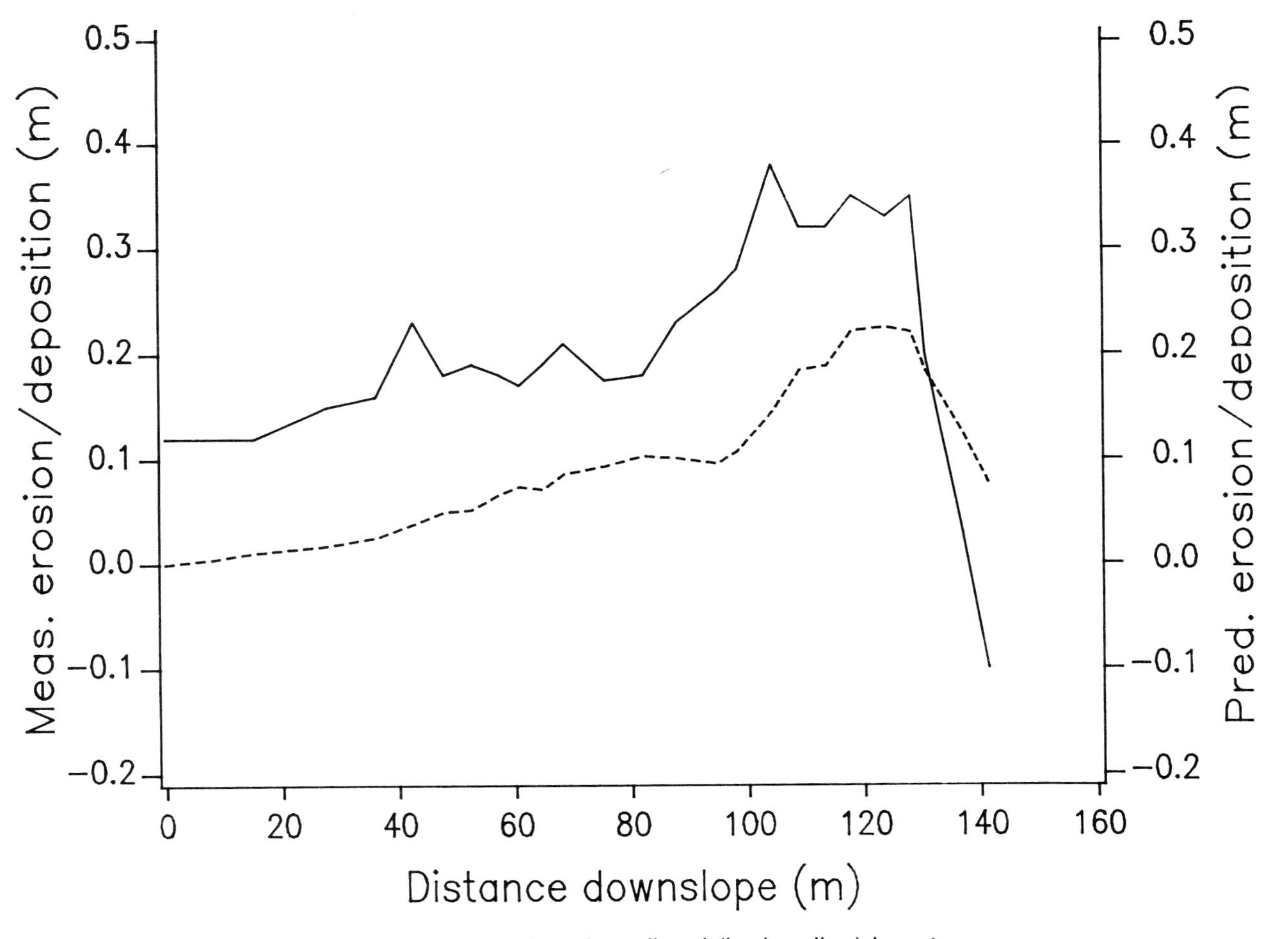

Figure 3. Comparison of measured (full line) and predicted (broken line) long-term soil redistribution on the Huldenberg field slope. (Simulation 2 - g = 0, 130 iterations).

Although the discrepancy seen at the upper boundary may be due to lowering of the field surface by compaction rather than soil loss, other differences between the observed and predicted distributions are not so readily explained. For example, the observed distribution shows a rapid change from high erosion rates to deposition over the last 10m of the slope, while the predicted distribution merely indicates a decrease in erosion rate. Furthermore, comparison of the woodland profile and the contemporary field profile suggests that, in the past, deposition started at a slope of 0.07-0.09 which is higher than would be expected if rill and interrill erosion were the sole cause of soil redistribution. It is therefore necessary to consider the possible influence of diffusion processes. This was done in the third simulation which incorporated some diffusive movement and achieved a dramatically increased correspondence between observed and predicted redistribution patterns (Figure 4 and Table 1).

Although improvement is most significant at the basal concavity, the pattern of soil redistribution over the whole hillslope is predicted better.

The fourth simulation used only the water erosion model to simulate the pattern of long-term (c. 200 year) soil redistribution over the slope at Dalicott, UK, and again a relatively poor correspondence was found (Figure 5 and Table 1).

The predicted soil erosion maximum is located too far downslope and deposition is observed to start further upslope than is predicted. The fifth simulation which achieved the best correlation with long-term soil redistribution at Dalicott assumed an important contribution from diffusive soil movement (Figure 6 and Table 1).

In contrast to the long-term rates at Dalicott, soil redistribution over the last 35 years (derived from caesium-137 measurements) is quite closely replicated by the model incorporating only water erosion and deposition (Simulation 6 - Figure 7 and Table 1).

However, the highest erosion rates occur much higher on the slope than is predicted. A closer correspondence was achieved in the seventh simulation which assumes that diffusion processes have played a significant role in soil redistribution (Figure 8 and Table 1).

The improvement in correlation is not as spectacular in this case.

4. DISCUSSION

The close correspondence between observed patterns of water erosion on the field slope at Huldenberg with those predicted by the water erosion model (Figure 2) suggests that, despite its simple structure, the model is capable of providing a reasonable simulation of the spatial variation of water erosion rates over a hillslope. It is therefore necessary to explain the relatively poor fit between model predictions and the long-term patterns of soil redistribution at both Huldenberg (Figure 3) and Dalicott (Figure 5). Although it is possible that discrepancies between the process descriptions used and their actual behaviour may account for some of the variation, it is unlikely that this is a major factor as the process descriptions used are supported by a wide body of literature. It is therefore apparent that processes other than rill and interrill erosion must be considered in simulating long-term soil redistribution.

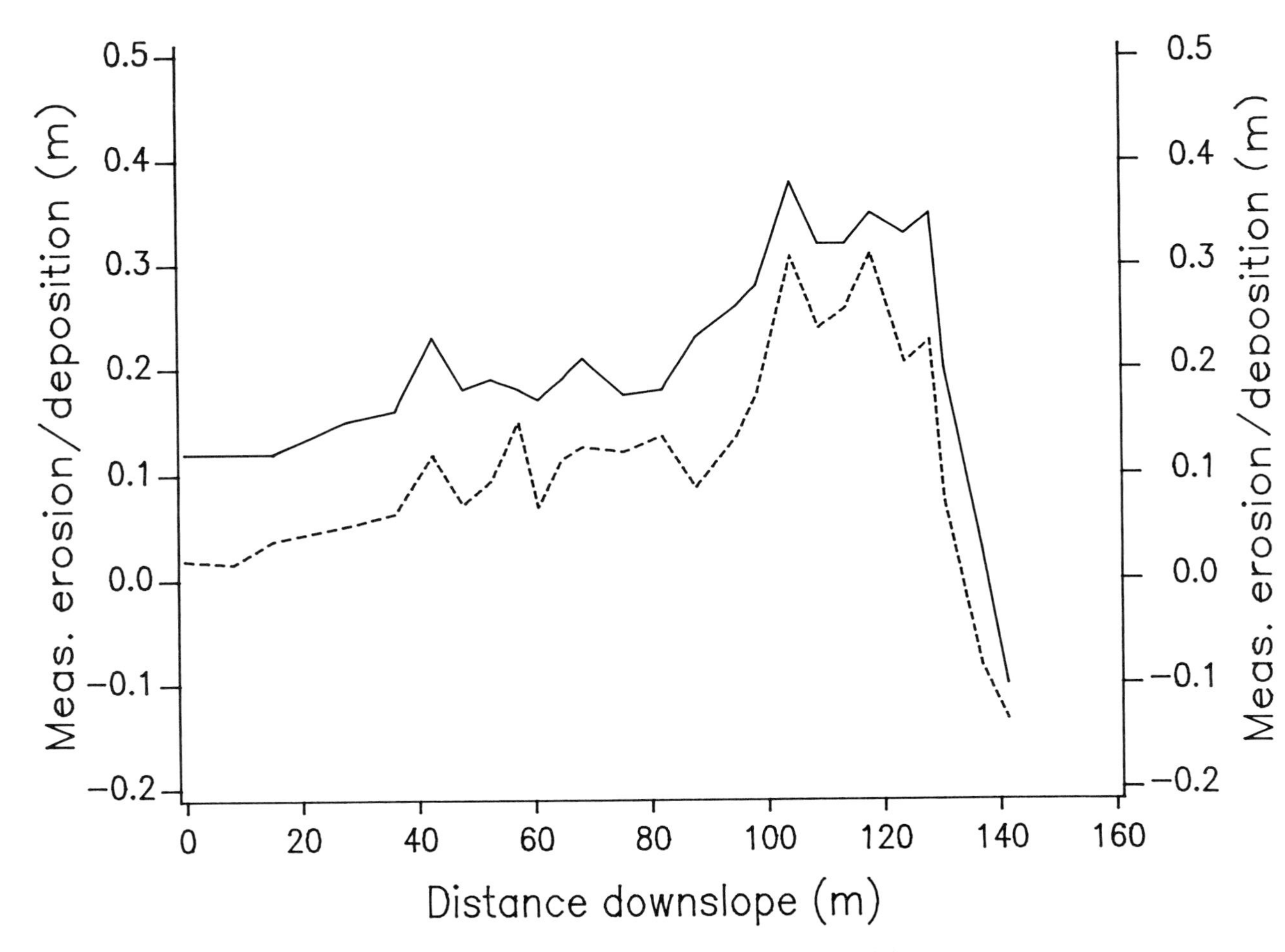

Figure 4. Comparison of measured (full line) and predicted (broken line) long-term soil redistribution on the Huldenberg field slope. (Simulation 3 - g = 120, 155 iterations).

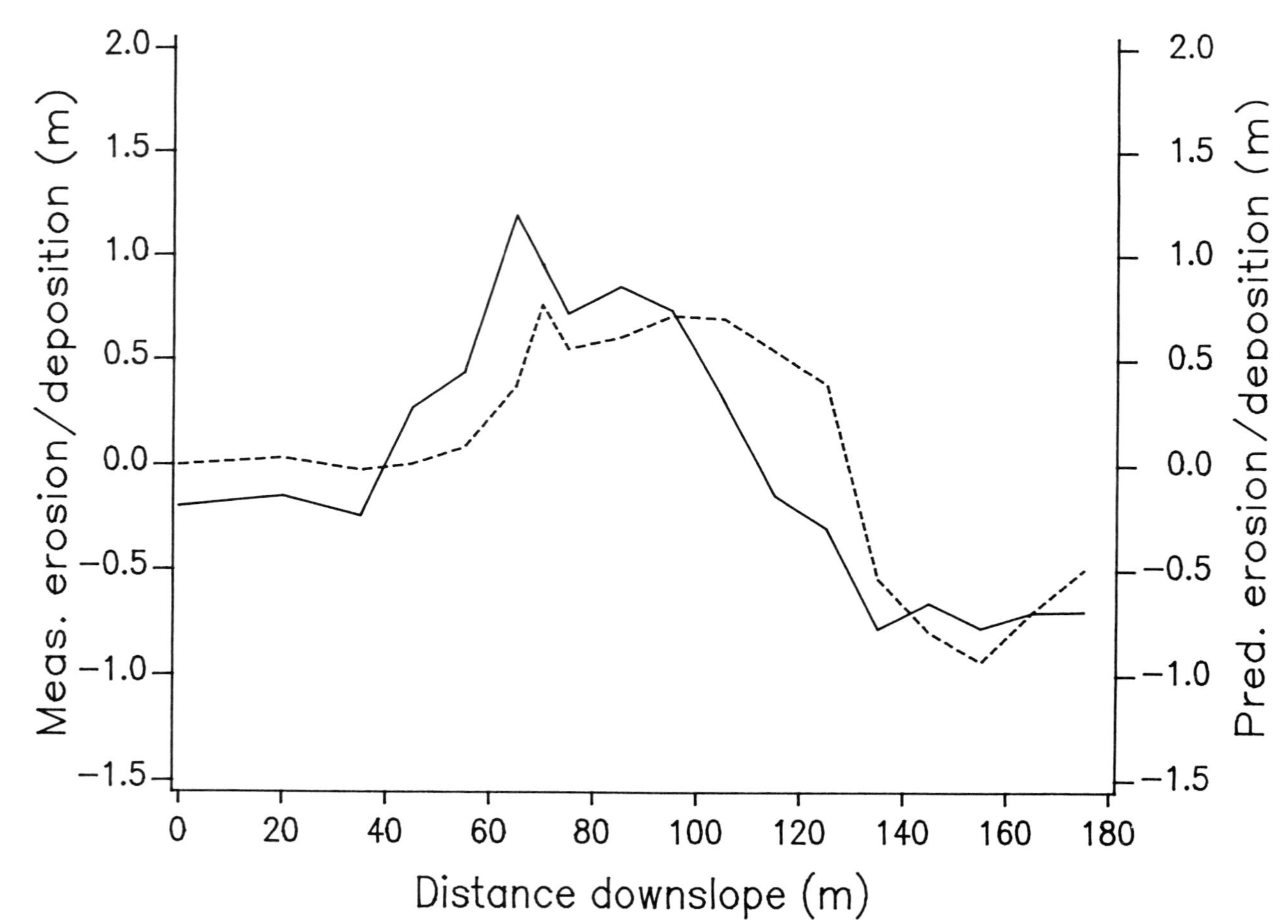

Figure 5. Comparison of measured (full line) and predicted (broken line) long-term (c.200 years) soil redistribution on the Dalicott field slope. (Simulation 4 - g = 0, 450 iterations).

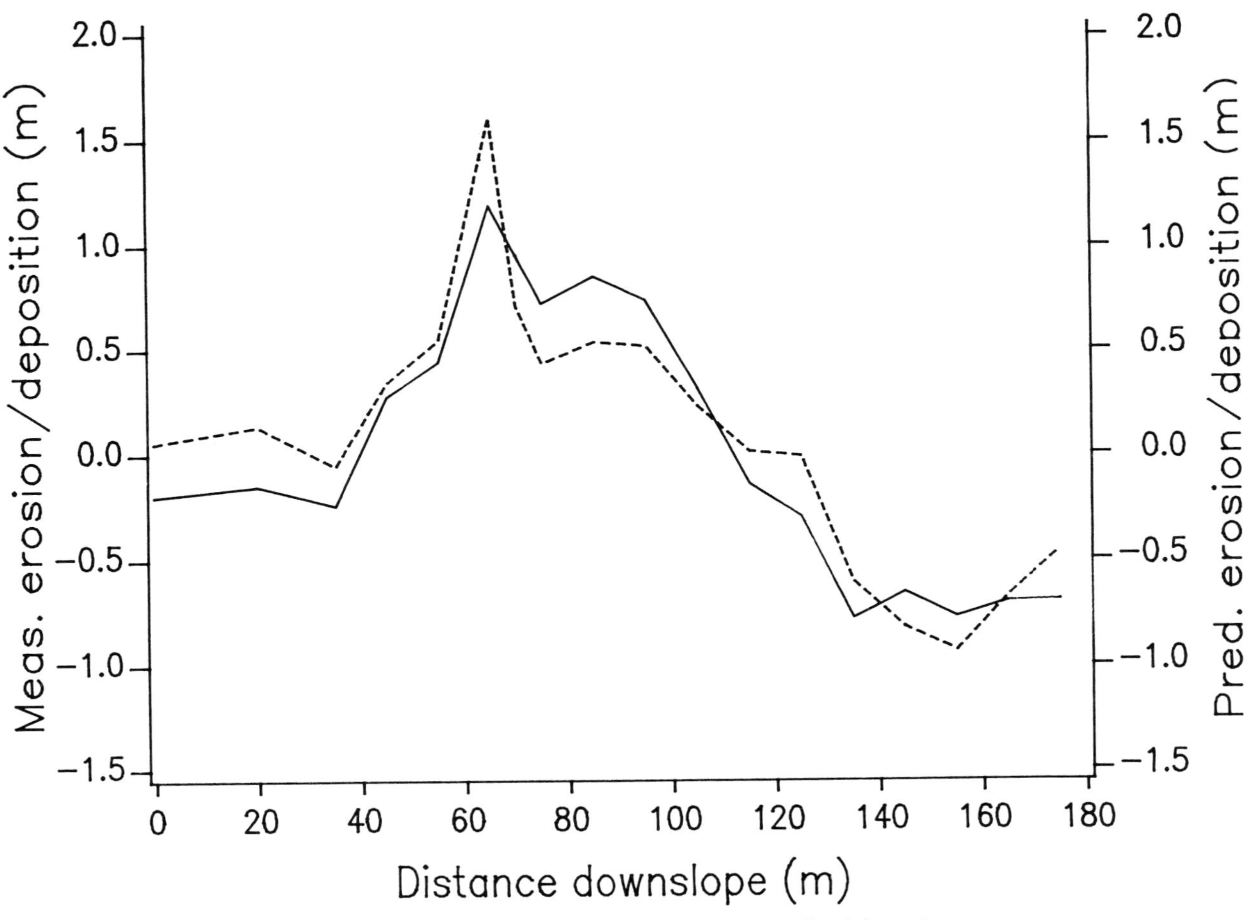

Figure 6. Comparison of measured (full line) and predicted (broken line) long-term (c.200 years) soil redistribution on the Dalicott field slope. (Simulation 5 - g = 325, 325 iterations).

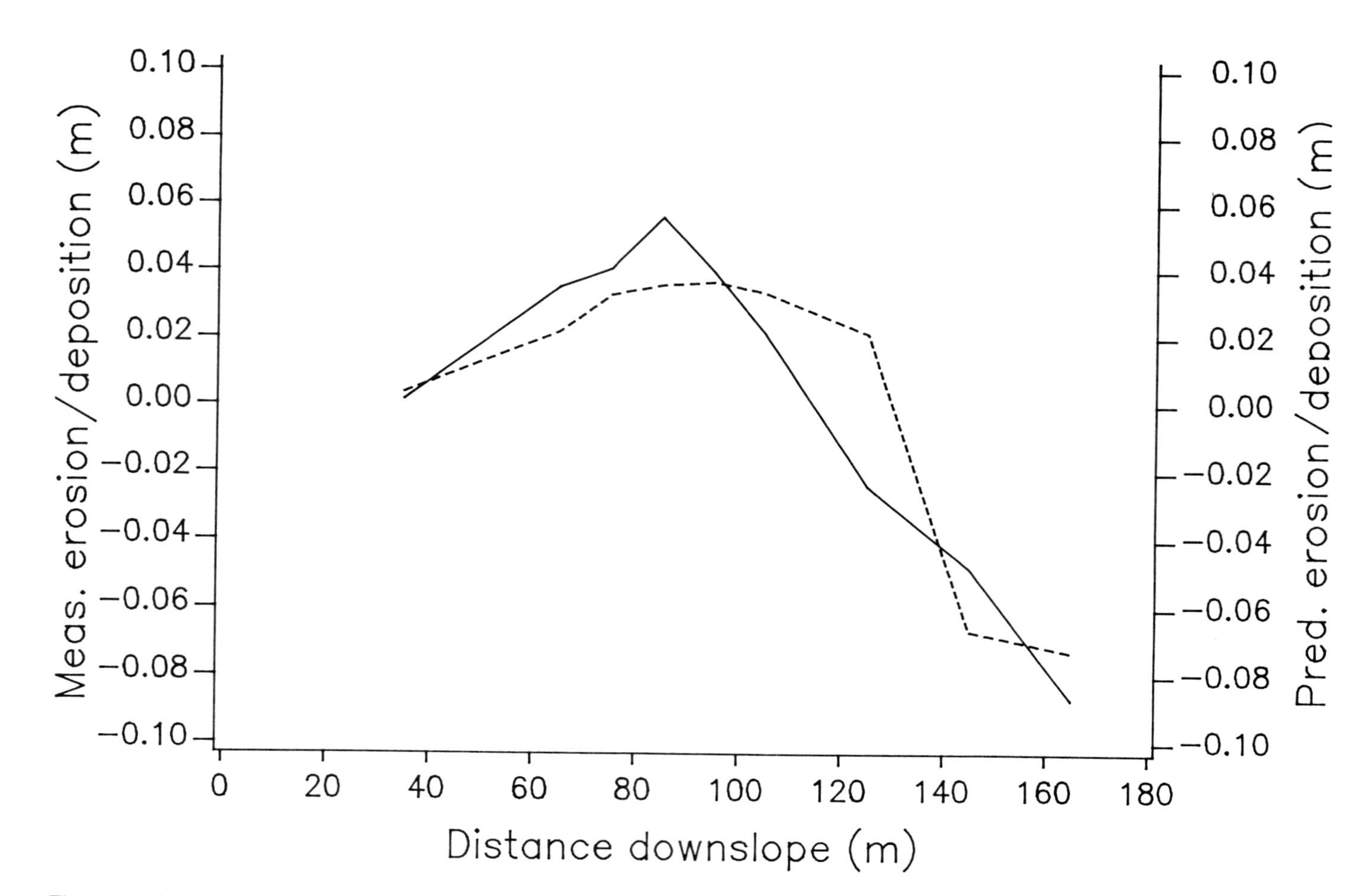

Figure 7. Comparison of measured (full line) and predicted (broken line) short-term (c.35 years) soil redistribution on the Dalicott field slope. (Simulation 6 - g = 0, 27 iterations).

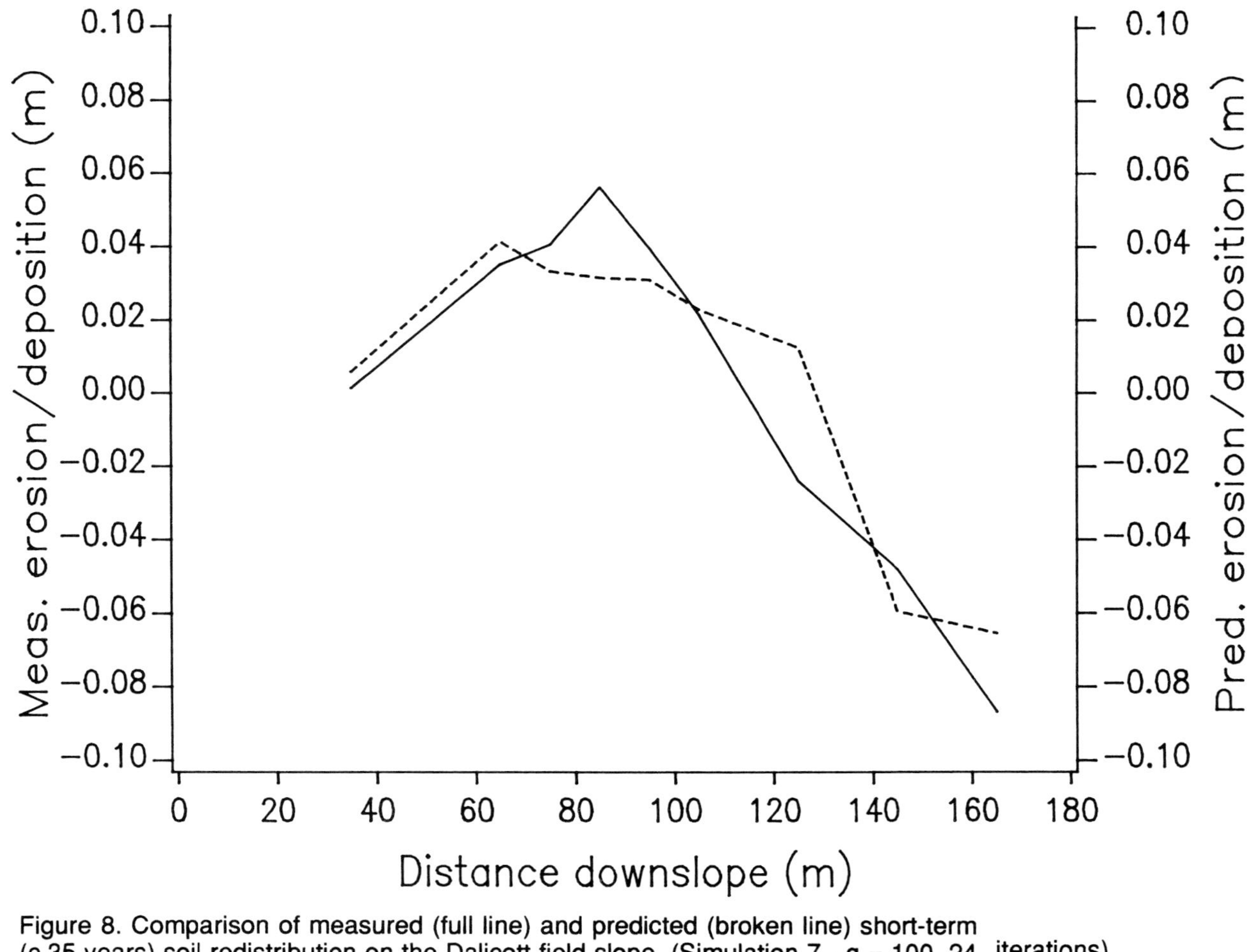

Figure 8. Comparison of measured (full line) and predicted (broken line) short-term (c.35 years) soil redistribution on the Dalicott field slope. (Simulation 7 - g = 100, 24 iterations).

This suggestion is supported by the improvement in model predictions achieved when diffusion processes are taken into account. Such processes will typically lead to high ablation rates on slope convexities and deposition in concavities. These are features which have been noted in studies of soil redistribution using caesium-137 (10 - 11). It is therefore necessary to consider the processes of soil redistribution which may be characterized by diffusion models.
A classical example of a diffusion process is direct downslope movement by rainsplash. However, although considerable amounts of soil are detached by rain, downslope movement by rainsplash is only of limited significance and is very unlikely to be responsible for the observed soil redistribution rates. On the Huldenberg slope a mean splash detachment rate of 15.2 kg m^{-2} was measured from September 1983 to October 1984. Applying the model of Poesen (12), this would correspond to a flux of c. 1.85 kg m^{-1} on a 0.2 slope. Corresponding ablation and aggradation rates are very low.

Soil creep on cultivated soils may also be described by a diffusion model. After each tillage operation, the bulk density of the arable layer is markedly decreased. During the growing season, increases in bulk density of c. 20 % may occur. Assuming that settling of the soil occurs in the vertical direction, while decompaction occurs normal to the soil surface, this may result in a net downward flux of c. 1.3 kg m^{-1} per growing season on a 0.2 slope, assuming an increase in bulk density from 1100 to 1350 kg m^{-3} and a plough layer of 0.25 m depth. Although some additional creep may be caused by the action of heavy tractor loads on a wet soil, the cumulative ablation and aggradation rates are still very much lower than those observed at Dalicott and Huldenberg.

Another process which may be simulated by a diffusion model is soil tillage. Experimental data on soil movement due to tillage are very scarce. The only study known to the authors is that by Revel et al (13), although others have indicated that the process may be important. On a slope of 18 % Revel et al measured an average net downslope flux of 55 kg m^{-1}. Assuming a 100m long slope with a constant convexity, this would result in a soil loss of 0.55 kg m^{-2} or 5.5 t ha^{-1} per tillage operation ! If this figure is also realistic for other soil types, this would imply that tillage movement is by far the most important diffusion process active on arable land. In many cases soil redistribution rates due to tillage could be of the same order of magnitude as water erosion rates. This type of soil movement should therefore be taken into account when a process-based explanation is sought for observed patterns of within-field soil redistribution.

5. CONCLUSION

Comparison of observed patterns of soil movement over a hillslope with those predicted by a simple model indicate that diffusion processes may contribute significantly to soil movement on arable land in Western Europe. The optimum simulation of long-term soil redistribution (c.200 years) on the slope at Dalicott (Figure 6) suggests that soil movement due to diffusion processes has been of the same order of magnitude as soil movement due to water erosion. This is illustrated in Figure 9 which also demonstrates that the maximum mean rate of soil loss due to diffusion processes is approximately 3 times the maximum due to water erosion.

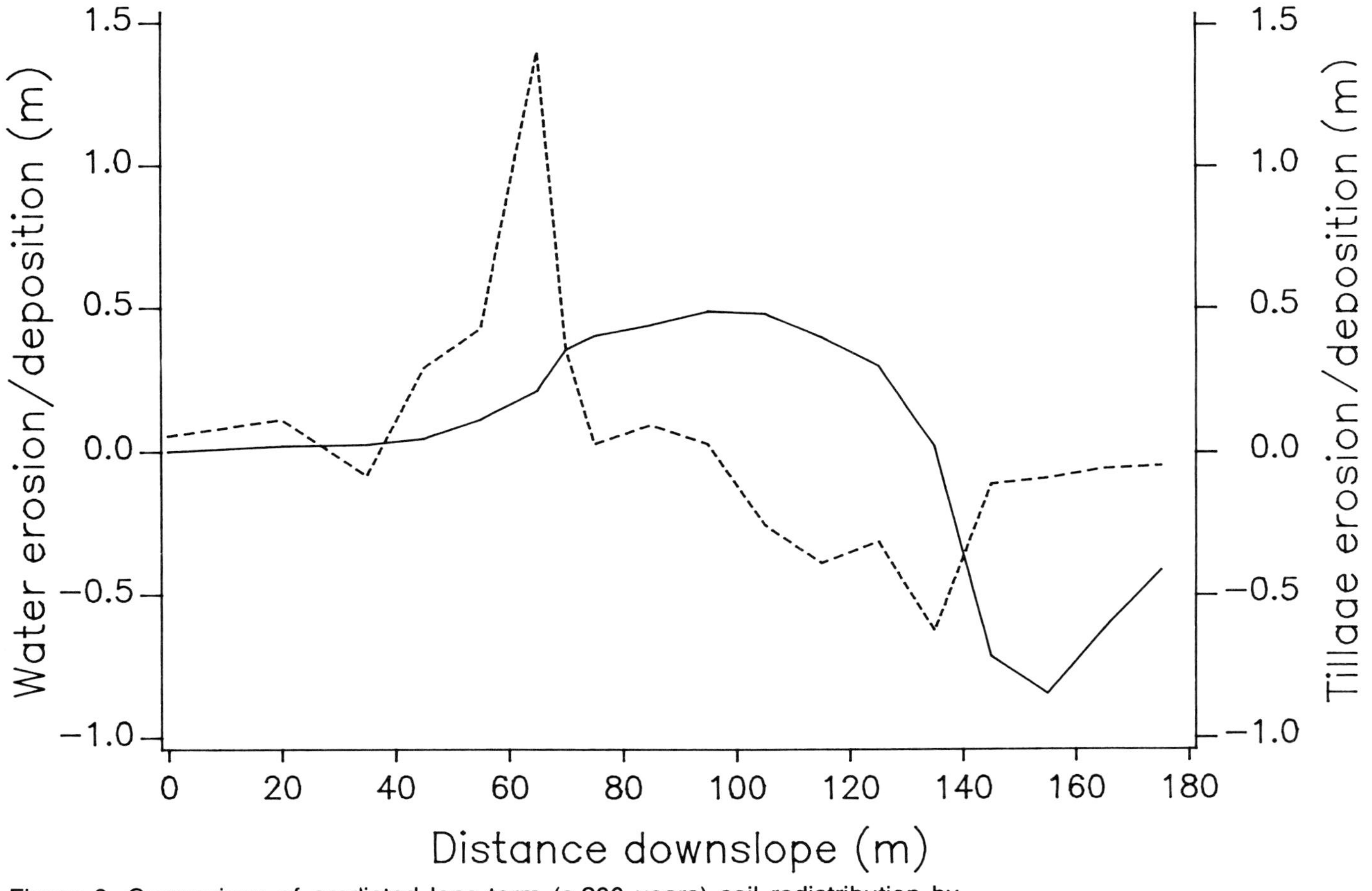

Figure 9. Comparison of predicted long-term (c.200 years) soil redistribution by water erosion (full line) and tillage (broken line) on the Dalicott field slope. (Simulation 5 - g = 325, 325 iterations).

Furthermore, although the pattern of soil redistribution over the last 35 years at Dalicott indicates a relative increase in the importance of water erosion, which is consistent with the widespread shift to high erosion hazard practices (eg. autumn sown cereals), soil redistribution by diffusion processes still represents a significant proportion of the total.

The available data indicate that the most significant diffusion process is probably soil tillage. The data of Revel et al. (13) as well as the results of the model simulations suggest that, within a field, soil redistribution rates due to tillage will be often of the same order as those due to water erosion. Although diffusion processes in general will not contribute directly to the export of soil from the field, it is clear that if one wishes to explain spatial patterns of soil movement at the field scale, tillage should be taken into account.

REFERENCES

1. Moore, I.D., Burch, G.J. and Mackenzie, D.H. (1988) : Topographic effects on the distribution of surface soil water and the location of ephemeral gullies. Transactions of the ASAE, 31 : 1098-1107.
2. Govers, G., (1987) : Spatial and temporal variability in rill development processes at the Huldenberg experimental site. In: R.B. Bryan (ed.), Rill erosion : processes and significance, Catena Supp., 8 : 17-34.
3. Govers, G. and Poesen, J. (1988) : Assessment of the interrill and rill contributions to total soil loss from an upland field plot. Geomorphology, 1: 343-354.
4. Quine, T.A. and Walling, D.E. (this volume) : Assessing recent rates of soil loss from areas of arable cultivation in the UK.
5. Kirkby, M.J., Naden, P.S., Burt, T.P. and Butcher, D.P. (1987) : Computer simulation in physical geography, J. Wiley, Chichester, 227 pp.
6. Govers, G. (1991) : Rill erosion on arable land in central Belgium : rates, controls and predictability. Catena, 18 : 133-155.
7. Foster, G.R. (1982) : Modelling the erosion process. In : C.T. Haan, H.P. Johnson and D.L. Brakensiek (eds.). Hydrologic modelling of small watersheds, ASAE Monograph n° 5, ASAE, St. Joseph, Michigan, pp. 297-380.
8. Watson, D.A. and Laflen, J.M. (1985) : Soil strength, slope and rainfall intensity effects on interrill erosion. Transactions of the ASAE, 28 : 98-102.
9. Govers, G., Everaert, W., Poesen, J., Rauws, G., De Ploey, J. and Lautridou, J.P. (1990) : A long flume study of the dynamic factors affecting the resistance of a loamy soil to concentrated flow erosion. Earth Surface Processes and Landforms, 15 : 313-328.
10. Walling, D.E. and Bradley, S.B. (1990) : Some applications of caesium-137 measurements in the study of erosion, transport and deposition. In : D.E. Walling, A. Yair and S. Berkowicz (Eds.), Erosion, Transport and Deposition Processes, IAHS Publication n° 189 : 179-203.
11. Walling, D.E. and Quine, T.A. (1991) : Use of 137Cs measurements to investigate soil erosion on arable fields in the UK : potential applications and limitations. Journal of Soil Science, 42 : 147-165.
12. Poesen, J. (1986) : Field measurements of splash erosion to validate a splash transport model. Zeitschrift für Geomorphologie N.F., Supp. Bd. 58 : 81-91.
13. Revel, J.C., Coste, N., Cavalie, J. and Coster, J.L. (1990) : Premiers résultats expérimentaux sur l'entraînement mécanique des terres par le travail du sol dans le Terrefort Toulousain (France). Cahiers ORSTOM, série Pédologie, 25 : 111-118.

Farm Land Erosion: In Temperate Plains Environment and Hills
S. Wicherek (Editor)

The value of caesium-137 measurements for estimating soil erosion and sediment delivery in an agricultural catchment, Avon, UK.

D. L. Higgitt [a] & D.E. Walling [b]

[a] Department of Geography, Lancaster University, Lancashire, LA1 4YB, United Kingdom.

[b] Armory building, Rennes Drive, Exeter Devon, EX4, 4RJ, United Kingdom.

SUMMARY

In recent years, the fallout radionuclide caesium-137 has found increasing application in investigations of soil erosion on agricultural land. By comparing caesium-137 inventories from different locations in fields with the baseline inventory for the area it is possible to assemble information on the rates and patterns of intra-field soil loss. When such work is extended to the catchment scale the technique offers the potential for identifying the principal sediment sources, pathways of transfer and storage areas and for estimating sediment delivery ratios. However, such an approach would require either substantial commitment to sample collection and analysis or compromise on the level of detail achieved. When sample coverage is reduced, questions concerning the representativeness of sample sites, the method of converting caesium-137 data into estimates of actual soil loss/gain and the confidence intervals associated with these estimates become pertinent.

An investigation of soil erosion and sediment delivery was undertaken in the 4 km^2 lake-catchment of Corston Brook, Avon. Traditionally, soil erosion has not been considered a major problem on the rolling Jurassic topography in this part of southern England but there has been a trend towards field enlargement, increased cultivation and use of autumn-sown cereals during the last decade. A network of caesium-137 measurements in the catchment suggest that erosion rates are highest on recently cultivated fields near the escarpment base. Caesium-137 profiles of lake sediments suggest that the catchment sediment yield has increased markedly since a previous survey in 1979, coincident with land use changes. The combination of measurements from soil cores and lake sediments provides a means of estimating catchment sediment budgets, but interpretation of the data is subject to a number of uncertainties and estimates remain tentative.

RESUME

Au cours des années récentes, les retombées nucléaires Caesium 137 ont été de plus en plus utilisées dans les recherches sur l'érosion des terres agricoles.

Si l'on compare les quantités de Caesium 137 présentes dans différents sites et la quantité type de la région, on parvient à obtenir des informations sur les taux et les types de pertes en terres. Etendue à l'échelle d'un bassin, cette technique permet potentiellement de répertorier les sources de sédiments, les flux de transport et les avis de stockage. Toutefois, cette approche impose une importante collecte d'échantillons et d'analyses sans laquelle le degré de précision atteint, sera très discutable. La révolution du nombre d'échantillons force à s'interroger sur la représentativité des sites d'échantillonnage, sur la méthode de conversion des données du Caesium 137 en estimations de pertes / gains de sols réels et sur le degré de confiance à conférer à ces estimations.

Une étude sur l'érosion du sol et la fourniture de sédiments a été menée dans un bassin lacustre de 4 km^2 situé à Corston Brook dans l'Avon. Jusqu'à une date récente, l'érosion des terres n'a pas constitué un problème majeur dans les secteurs doucement ondulés du Sud de l'Angleterre, cependant au cours des dix dernières années, la tendance a été à l'agrandissement des champs, au passage à une agriculture de plus en plus intensive et à l'utilisation accrue de céréales d'automne. Les mesures de Caesium 137 montrent que les taux d'érosion sont les plus élevés au bas des pentes Des traces de Caesium 137 dans les sédiments du lac indiquent une augmentation significative des apports par comparaison aux informations fournies en 1989 ; cela coïndice avec un changmeent des pratiques agricoles. Si des mesures concernant les sols et les sédiments du lac sont effectuées, leur interprétation est soumise à de nombreuses incertitudes ce qui rend les estimations provisoires.

INTRODUCTION

Caesium-137 is an artificial radionuclide originating in the global fallout of debris from the testing of nuclear weapons in the 1950s and 1960s. It possesses a number of distinctive tracer properties which can be exploited to investigate the erosion, transport and deposition of sediment within the landscape (1). Subsequent to its deposition as fallout, ^{137}Cs is strongly adsorbed to fine-grained particulate matter at the soil surface and its redistribution is associated with the erosional transport of sediment particles. These properties combined with a half-life of 30.1 years afford a valuble medium term tracer which can be used in erosion and sedimentation investigations. By comparing the ^{137}Cs contents of soil cores collected from different locations with a baseline value representing the total fallout to the region, it is possible to assemble information on the patterns and rates of soil loss.

Such an approach has proved most popular for investigating soil loss on arable fields (2 - 3 - 4), but has also been employed on rangeland (5 - 6 - 7) and towards deriving catchment sediment budgets (8 - 9). The ^{137}Cs is perhaps unique amongst other soil erosion research methods for its ability to be applied over several different spatial scales. Areas of temporary or permanent sediment storage, such as lakes and reservoirs, provide an opportunity to determine catchment sediment yields. Linked to ^{137}Cs measurements within the catchment this enables the source areas of eroded sediments to be identified and the nature of sediment delivery to be examined. Clearly, as the scale of the catchment increases the investigator requires more samples or accepts a loss of detail and/or confidence associated with erosion estimates.

The strategic aims of this study are threefold. First, it attempts to examine the pattern and rates of soil erosion within a small lake-catchment in Avon, UK. Second, it examines the potential for using the distribution of ^{137}Cs within lake sediment cores to estimate both the rate of sedimentation and the source of incoming sediment. Third, by relating these estimates of on-site erosion and sediment yield it is possible to test the hypothesis that recent land use changes within the catchment have contributed to an increase in erosion rates and to elucidate upon the nature of sediment delivery within the catchment. Given a finite limit on the detector time available for sample analysis, the representativeness of the sampling network and the confidence associated with the erosion estimates should also be examined.

STUDY AREA

The Corston Brook is a small tributary of the Bristol Avon, occupying a topographic low between the Cotswold Plateau and the Mendip Hills. The course of the brook is interrupted by two small ornamental lakes which provide the opportunity to examine the potential for using ^{137}Cs measurements to identify patterns and rates of soil erosion within the catchment and the pathways of sediment delivery to the upper lake. A previous study undertook continuous monitoring of the sediment yield between 1977 and 1979 (10).

The morphology of the 4.1. km^2 catchment is shown in Fig. 1. The southern side of the catchment comprises an escarpment rising to 168 m with two outliers forming the northwestern watershed. A nearly flat structural shelf extends up the catchment as far as the foot of Stantonbury Hill. The present channel lies within an incised valley reach in the lower part of the catchment. The topography of the catchment is strongly related to the underlying geological structure of the basin, which is composed of near-horizontal Jurassic strata. The Lower and Middle Lias clays form the lower escarpment slopes above the shelf and are overlain by Midford Sands and Inferior Oolite, a shelley limestone, both of which are porous. A line of springs occurs on the main escarpment at the junction between the Midford Sands and the impermeable Lias clays, forming a perennial source of water at the lower part of the escarpment. Seepage lines occur at the equivalent junction of the outliers during the winter and may contribute significantly to sheet erosion. Four relict landslips are located on the clay slopes and are most likely relict features related to the existence of former solifluction mantles, rather than the present slope form.

The soils of the catchment essentially reflect the underlying geology. The escarpment slope is dominated by stagnogleyic argillic brown earths (Curtisden Association), the structural shelf and the escarpment ridge by brown rendzinas (Sherborne Association) and the outlier slopes by calcareous pelosols (Evesham Association). Surface water gleys (Martock Association) are found on part of the Lias clay outcrop. All of these soils, with the exception of the Martock soils, are suitable for autumn cultivation (11), though, until comparatively recently, the catchment has been dominated by permanent pasture.

Figure 1 shows comparative maps of the landuse in 1977 (10) and 1988. In the 1977 survey 51.1 % of the catchment was under permanent pasture and 43.2 % under arable cultivation. Cultivated fields were largely restricted to the flanks of the escarpment and its outliers and to limited parts of the shelf, whilst permanent pasture was maintained around the water courses. By 1988 it can be seen that the land area under cultivation had increased significantly to 68.3 % of the catchment as the area of pasture was halved to 26.0 %. Cultivation has extended across most of the shelf area, the flat ground at the western watershed and on the lower flanks of the escarpment adjacent to the stream network. This extension has been accompanied by an increase in field sizes resulting from the widespread removal of hedgerows. Thus, the average effective slope length for runoff generation has been increased coincidental with an enlargement of the area of cultivated land adjacent

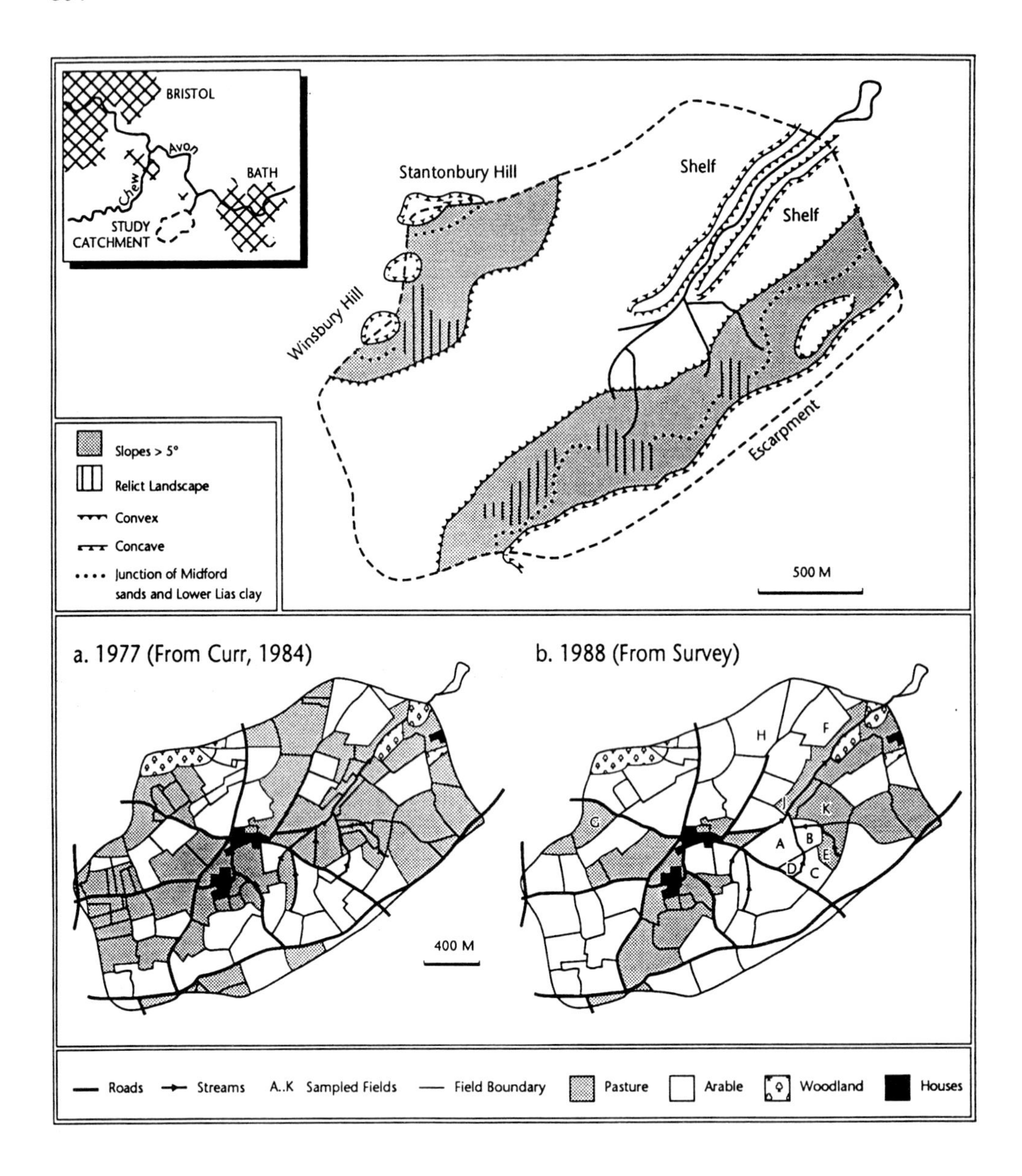

Figure 1. Morphology and land use in the catchment.

to the stream network. It may be expected that these landuses changes have resulted in increased soil erosion during the intermediate period.

SAMPLING STRATEGY AND METHODS

A number of recent papers have examined intra-field variation in rates of soil erosion obtained through a high density ^{137}Cs sampling network (12 - 4). On a larger spatial scale the density of sampling must be determined in the light of the practicality of sample collection and the restraint of time available for analysis. In this study about one half of detector time was dedicated to the analysis of lake sediment samples leaving a provision of about one hundred samples for the estimation of baseline input and the examination of the patterns of soil redistribution within the catchment.

Rather than covering the whole catchment with a sparse network, sampling was concentrated in areas considered to be the most critical to sediment transfer processes. It is clear from the morphological map (Fig. 1) that certain parts of the catchment are more likely to be sediment sources than others. Soil erosion is likely to be greatest on the steeper slopes and suspended sediment most likely derived from eroding areas relatively close to the stream network. Sampling was therefore concentrated on the cultivated fields of the main escarpment around the first order streams. The location of the fields sampled is annotated in Fig. 1b. It should also be noted that metalled roads and culverts provide an additional and potentially important linkage for runoff from fields relatively distant from the drainage network. Soil cores were also collected from fields on the shelf and from pastures on escarpment and outlier slopes. Baseline inventory was estimated from cores collected on flat undisturbed pasture from the summit of one outlier.

Samples were collected between September 1988 and July 1989 using a 75 mm diameter corer which was driven into the ground to a depth of 50 cm. The soil was transported to the laboratory where it was dried, disaggregated and passed through a 2 mm sieve prior to being packed into 1 l Marinelli beakers. Caesium-137 content was determined by gamma spectrometry using a HPGe detector linked to a multi-channel analyser. Measurement precision is typically around ± 5%. All samples have been corrected for radioactive decay to the end of 1988.

PATTERNS AND RATES OF SOIL LOSS

Baseline fallout to the catchment was determined from a flat pasture site located on the summit of Winsbury Hill. Eleven cores (to a depth of 30 cm) were collected at 10-20 m intervals. The ^{137}Cs inventories range from 2487 to 3890 Bq m^{-2}. The mean inventory (and its standard error) is 3099 ± 37 Bq m^{-2} and the coefficient of variation is 13.0 %. Such variability in ^{137}Cs inventories at this site could be due to a combination of uneven fallout deposition, variable uptake by soil and analytical errors. There is also evidence, on another part of the hilltop, of burrowing activity by moles which may promote heteorogenic redistribution of ^{137}Cs fallout.

A small amount of Chernobyl-derived ^{134}Cs was detected in the uppermost layers of an incremented profile from a pasture on the shelf (Field K). The estimated Chernobyl input of 62 ± 27 Bq m^{-2} represents a 2.0 % increase to the pre-existing ^{137}Cs inventory. However, no ^{134}Cs was detectable in full length cores. Recent studies have shown that a significant addition of Chernobyl radiocaesium to the local inventory inhibits the application of ^{137}Cs-derived erosion estimates (13 - 14), but this is not considered to be a problem in this study area.

The ^{137}Cs inventories of soil cores collected from locations within the catchment can be used to examine the relative pattern of soil loss. The densest area of sampling was undertaken in the five fields A-E on the flank of the main escarpment adjacent to the stream network. The distribution of ^{137}Cs inventories within these fields is shown in Fig. 2a with cross sections of some transects plotted in Fig. 2b. The ^{137}Cs inventories at each point are compared with the 95 % confidence interval for the baseline fallout. Individual cores with inventories outside this range have significantly different ^{137}Cs inventories from the baseline input.

The highest losses of ^{137}Cs are observed on the eastern transect of Field A. This is the steepest part of the field, though only 5°, and ^{137}Cs inventories decline to nearly 60 % depletion on the steepest part of the slope. High levels of ^{137}Cs depletion continue along this transect until ^{137}Cs enrichment is encountered at the base of the slope. Surface water ponding was observed in this part of the field in September 1988. On the transect in Field B ^{137}Cs depletion continues to the slope base, adjacent to a stream, suggesting that there is relatively little deposition within the field and that much of the eroded sediment is exported from the field via the stream. Parallel transects along a dry valley feature (swale) and adjacent spur were sampled in Field C, on the main slope of the escarpment. The highest rates of ^{137}Cs depletion occur at the head of the swale, but sites further down the swale have stable or accumulating ^{137}Cs inventories. Depletion on the spur is slighter, but there is no evidence of mid-slope deposition. This pattern of soil loss provides further evidence of the apparent importance of swales in focussing the transport of water and sediment, but of spurs experiencing the greatest net soil loss. Similar patterns have been observed in Shropshire (15) and Devon (16).

The final transect depicted in Fig 2b is from Field F located on the shelf close to the catchment outlet. The upper part of the field has a gentle gradient but the lower part steepens towards the incised valley reach. Curr (10) has documented two occasions between 1977 and 1979 when runoff occurred from the shelf into the valley bottom. The second of these storms on 27 December 1979 resulted in the erosion of a gulley (c. 0.5 m deep) along the western edge of Field F. Soil cores, collected some 20 m from the edge of the field, show high ^{137}Cs depletion on the upper part of the slope with accumulation in the lower part of the field. A colluvial fan extends beyond the gateway at the base of the field into a 20 m wide strip of pasture separating the cultivated field from the main stream channel. Samples on this fan reveal considerable ^{137}Cs enrichment, with the value of 5566 Bq m^{-2} being the highest recorded inventory in the catchment. The final core in the transect, adjacent to the stream channel, is close to the baseline inventory, suggesting that most of the sediment derived from the field was deposited within a relatively narrow zone at, or beyond, the base of the field rather than reaching the channel. Other sample sites located on the structural shelf indicate stable ^{137}Cs inventories.

Two transects were located on pasture slopes of the main escarpment (Field K) and the outlier of Winsbury Hill beneath the input sites (Field G). Both transects began at the seepage line at the junction of the Midford Sands and the underlying Lias Clay. Only the uppermost sample on the main escarpment shows significant ^{137}Cs depletion (2306 Bq m^{-2}), though there is considerable evidence of poaching by cattle, both at the head of the stream channel and around the base of the field. On Winsbury Hill none of the sites have a significant depletion of ^{137}Cs,

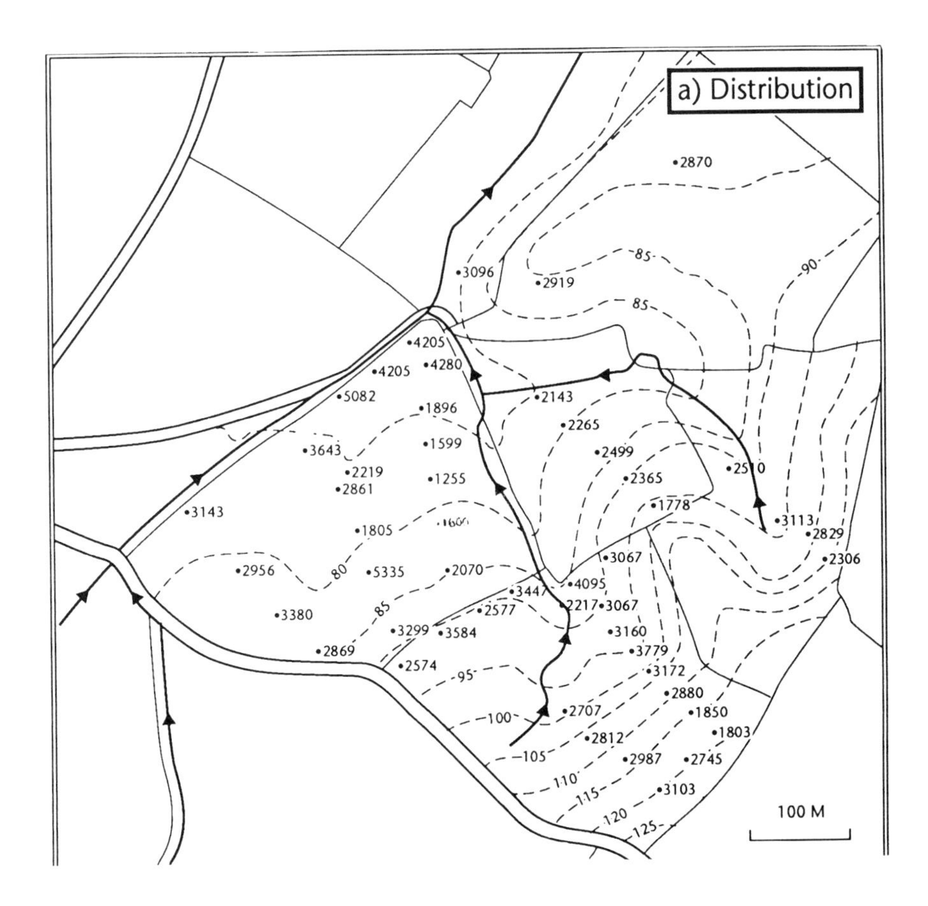

Figure 2a. Caesium-137 inventories on the lower escarpment : a) distribution.

but the core collected at the hedgerow boundary of the field has significant accumulation of 4176 Bq m^{-2}, suggesting that some sheet erosion is proceeding on the slope.

The description of the ^{137}Cs inventories across the catchment provides an indication of the relative rate of soil loss. A number of papers have proposed methods for equating ^{137}Cs depletion with actual soil loss using either empirical relationships or theoretical accounting procedures and recent developments have been reviewed by Frederick and Perrens (17) and by Walling and Quine (18). Partly based on Quine (19), Higgitt (15) has developed a mass balance procedure to enable quantitative estimates of the rates of soil loss, which takes into account the temporal variability of fallout deposition, radioactive decay, annual loss to cropping, mixing by ploughing and the exponential distribution of incoming fallout

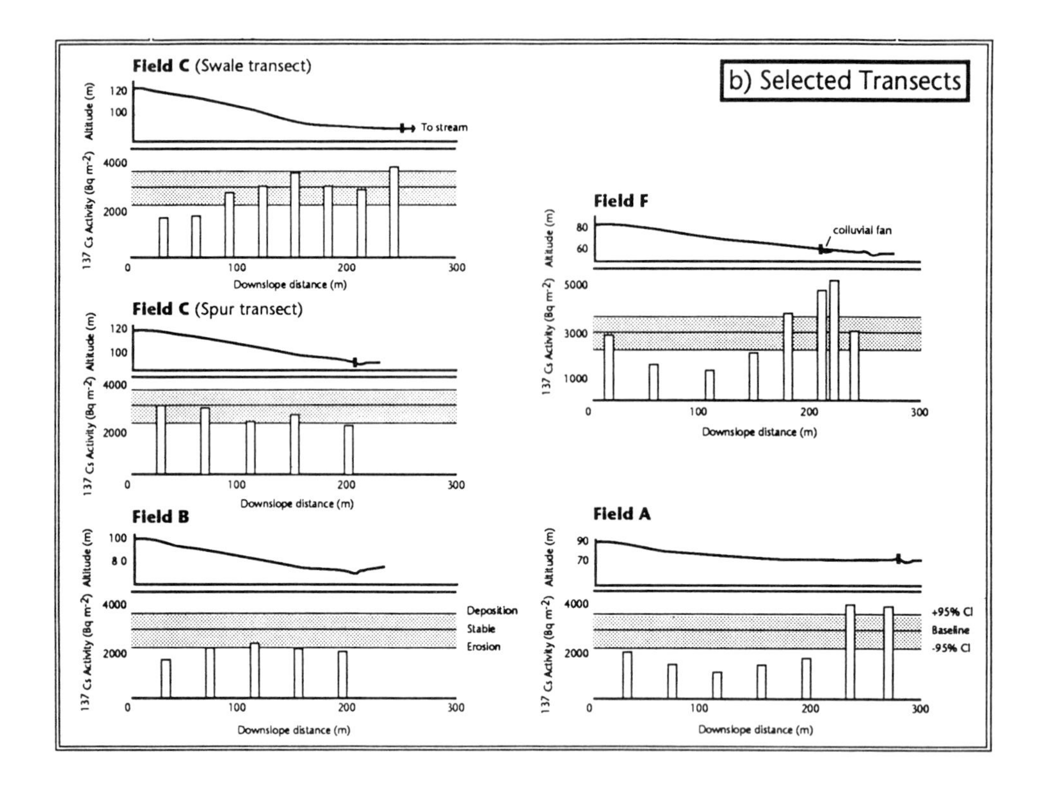

Figure 2 b. Caesium-137 inventories on the lower escarpment : (b) selected.

in the uppermost layers of the soil, prior to any mixing. The mass balance procedure produces estimates based on the assumption of constant erosion rate and annual cultivation over the fallout period. Estimates of average rates of soil loss from eroding sites have been compiled in Table 1, but in the absence of any independent measurements of soil loss it is difficult to assess the validity of the model. The most active rates of soil loss, in the order 8-10 t ha^{-1} $year^{-1}$ are found in fields at the escarpment base and the shelf edge, though it should be noted that these high rates only occur in limited parts of these fields. Furthermore, Fields A and B were converted from pasture to arable land use in the early 1980s suggesting that post-conversion soil loss rates have been extreme. More moderate but widespread soil loss is encountered on the main escarpment. The variability observed in the input cores and the limited coverage across the catchment means that all estimates of on-site erosion are subject to a high degree of uncertainty.

Table 1.
Estimated soil erosion rates in selected fields.

Location	Field	Number of samples	number of samples different from reference activity (A_R) N < A_R (sig.)	N > A_R (sig.)	Erosion Rate ($t\ ha^{-1}\ year^{-1}$) Mean	S.E
Base of Escarpment	A	20	12 (7)	8 (4)	8.1	4.7
	B	5	5 (3)	0		
Main Escarpment	C	14	9 (3)	5 (1)	3.3	3.0
Shelf Edge	F	8	4 (3)	4 (3)	10.7	9.5
Pastures	E G J K	11	8 (0)	3 (1)	0.9	0.4

CAESIUM-137 ACCUMULATION IN LAKE SEDIMENTS

The lake-catchment ecosystem provides an opportunity to examine the response of sediment yield and delivery processes to land-use changes, since change will be reflected in an influx of greater quantities of sediment into the lake or through a change in the properties of the sediment (20). The ^{137}Cs content of lake sediments is both a function of the ambient atmospheric fallout concentration at the time of their incorporation and of the sediment source area. Combined measurements of the ^{137}Cs contents of lake cores and catchment soil cores provide a potential for interpreting the nature of sediment delivery within the catchment and of relating this to medium term ecological change.

Principally, ^{137}Cs measurements have been used to quantify the rate of sedimentation in the lake based on the identification of the depth of peak ^{137}Cs concentration which is taken to represent the maximum fallout year of 1963. This approach is most reliable in lakes where autochthonous sedimentation is the dominant process (21) and the ^{137}Cs content of accumulating sediment closely reflects the prevailing atmospheric concentration of fallout. In lakes where a substantial part of the sediment is derived from the surrounding catchment, there will be both a lag in the time ^{137}Cs, initially deposited in the catchment, takes to reach the lake and a variation in the ^{137}Cs content of incoming sediment dependent upon its source area. If recent changes in land use have resulted in an increase to the sediment load delivered to the catchment outlet, this may be reflected in the shape of the ^{137}Cs distribution in lake cores.

The course of the Corston Brook is interrupted by two small ornamental lakes designed by Capability Brown and completed between 1760 and 1765. The lakes lie behind earth dams constructed across the incised valley and their size is such that a very large fraction of the sediment produced by individual flood events during their 225-230 year history has been trapped by the upper lake. The lake sediments therefore comprise a record of the accumulated sediment from individual events and of the relative importance of different sediment source areas throughout that time. Curr (10) undertook a survey of the depth of sedimentation in the lake basin by sounding with aluminium rods. Assuming a trap efficiency of 90 %, Curr (10) then calculated that the mean sediment yield to the lake was 111 t year^{-1}, which compared well with the measured sediment load of 100 t year^{-1} over the two year monitoring period.

If the land-use change documented within the last decade has resulted in an increase in the proportion of sediment derived from cultivated fields relative to that from channel banks, one would expect the ^{137}Cs distribution to initially follow the pattern of atmospheric fallout but for higher levels of ^{137}Cs to extend after the 1963 peak. Using the same principles by which the mass balance procedure calculates the annual input and loss of ^{137}Cs it is possible to estimate the relative levels of ^{137}Cs in suspended sediment derived from soils eroding at a given rate and from this simulate the expected ^{137}Cs distribution resulting from various combinations of catchment sources (15). Since bank material within the catchment is relatively coarse whilst recent cultivation has extended to clay loam soils at the base of the escarpment, a change in sediment delivery in favour of sediment derived from cultivated fields would be accompanied by an upward increase in clay content and decrease in sand content in lake sediments. Superimposed upon this trend, large magnitude flood events would tend to project coarse material further into the lake.

Six core sampling sites are indicated in Fig. 3. Four of these, closest to the stream inlet, were obtained as bulk samples after the surface water level of the lake had been drawn down and their inventories indicate net ^{137}Cs accumulation at these sites. Two sites were sampled in the deeper part of the basin using a specially designed coring device for shallow water. At each site, two cores (63 mm diameter) were obtained from an inflatable boat. The sediment was extruded from the core tubes on site and sectioned into 1 cm slices. Inventories are considerably higher further into the lake.

The distribution of ^{137}Cs within the middle sectioned lake core (Core A) is also illustrated in Fig. 3 and displays a peak in ^{137}Cs content in the lower to middle part with a subsequent decline towards the sediment-water interface. During the period of continuous monitoring, Curr (10) estimated, on the basis of channel bank erosion pin measurements, that the sediment load of the Corston Brook was derived in roughly equal proportions from bank and field sources. Samples of bank material collected within the catchment had little or no ^{137}Cs content, suggesting that any ^{137}Cs adsorbed on material in bank surfaces is rapidly removed. Comparison between the actual and a simulated distribution combining bank and field sources in equal proportions suggests an increasing contribution from surface soils moving up the core. Assuming that the peak ^{137}Cs horizon represents 1963 and the base of the profile 1954, the mean annual sedimentation rate at this point before and after 1963 is estimated as 1.3 and 1.7 cm year^{-1}, respectively. The sedimentation survey undertaken in 1979 (10) suggests that 2.75 m of sediment had accumulated at this site during the history of the lake, equivalent to an annual sedimentation rate of 1.25 cm year^{-1}. This rate is similar to the estimate of pre-1963 sedimentation derived from the ^{137}Cs profile.

The clay content of the core shows a marked increase in the upper 30 cm, which supports the hypothesis that the increase in the sedimentation rate has been due to an increase in the amount of sediment supplied from cultivated fields. The sand content in the cores decreases upwards with marked variations in the lower half of the core. These pulses of coarser material appear to represent individual large storms of which the largest may represent an extreme event in 1968.

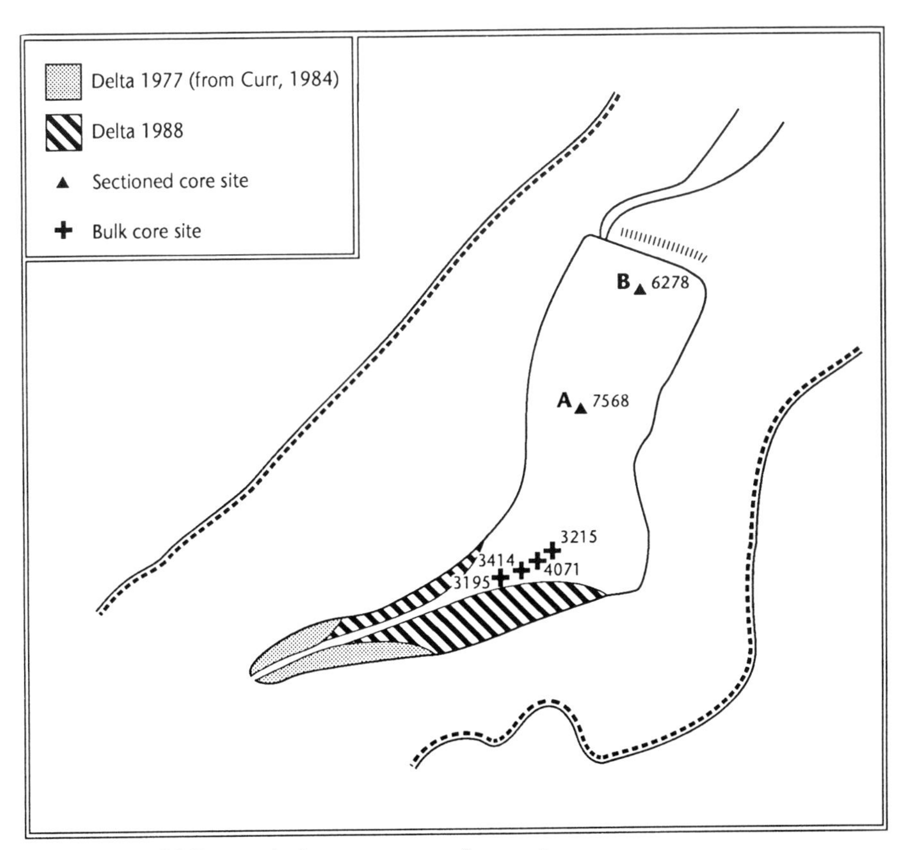

Caesium - 137, sand clay content of core A.

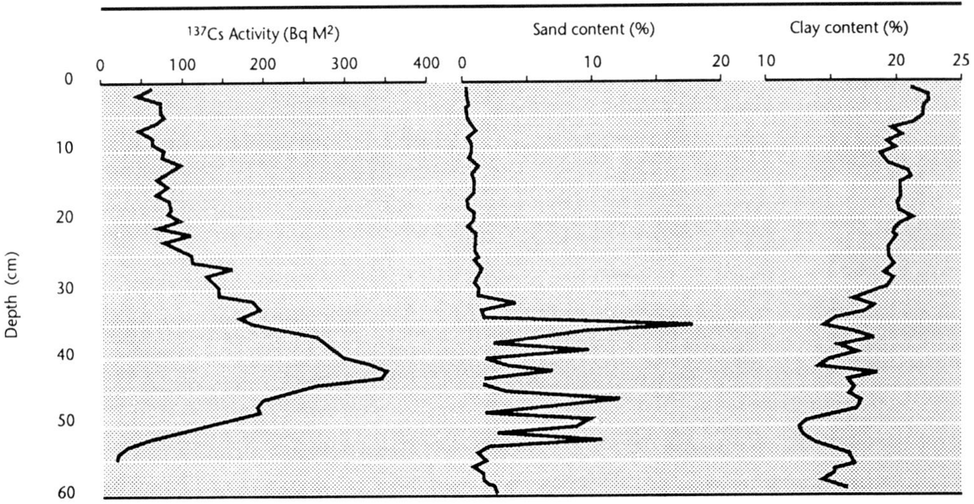

Figure 3. Location and inventories of lake sediment cores.

The evidence obtained from the ^{137}Cs and particle size distributions of the cores provides some support for the argument that there has been an increase in the rate of sedimentation in the lake in recent years caused by an increase in the rate of soil erosion on cultivated fields. However, the conclusions must remain tentative because of the problem of equifinality whereby a different combination of processes could account for a similar distribution of ^{137}Cs in the lake cores. The results suggest that there is considerable scope for using ^{137}Cs in lake-catchment based interpretations of recent ecological change, but that the method would be best used in conjunction with other independent techniques.

ESTIMATION OF A CATCHMENT SEDIMENT BUDGET

Figure 4. A tentative catchment sediment budget.

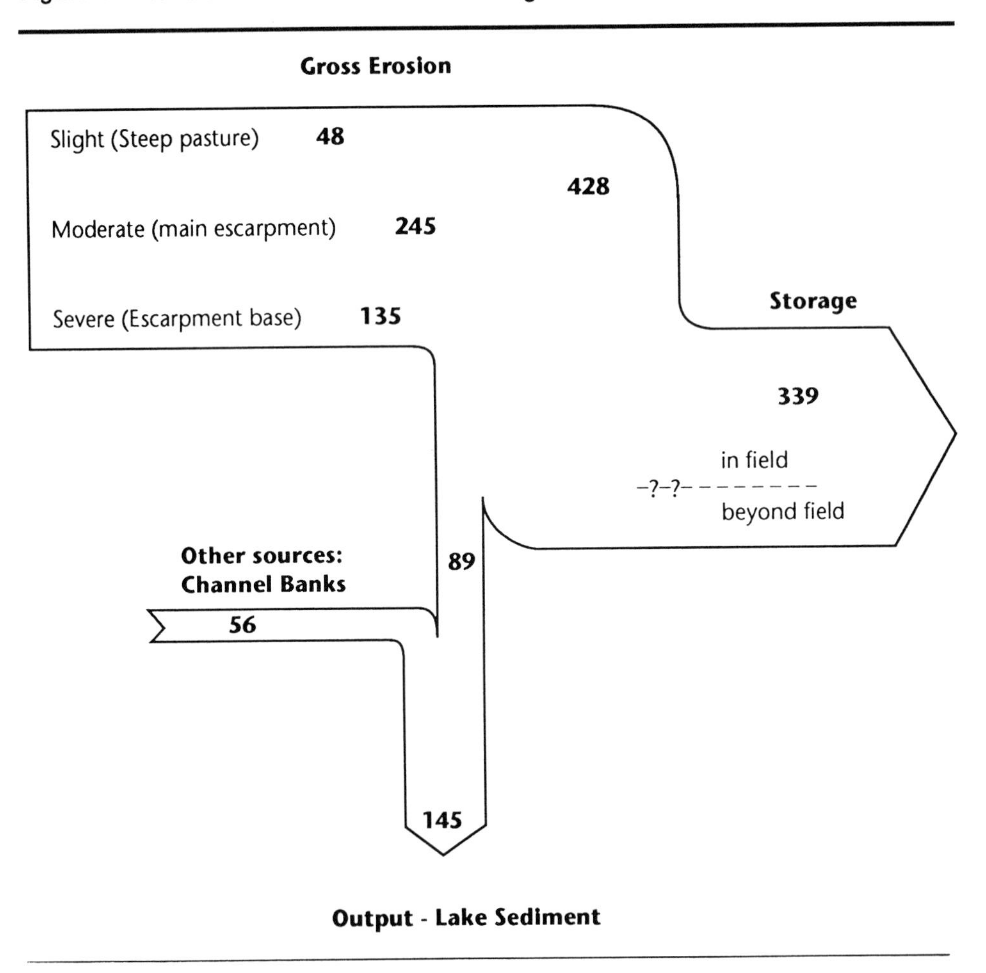

The absence of a complete coverage of ^{137}Cs inventories across the catchment limits any attempt to estimate the sediment delivery ratio of the catchment. In particular, it is difficult to assess the extent and importance of sediment accumulation at field bases without further sampling. Nevertheless, it is possible to produce a tentative sediment budget based on extrapolation of rates to similar morphological units.

About half of the catchment area has a gentle gradient defined by the structural shelf and the samples collected from Field H suggest that erosion in this area is negligible. Slight erosion (< 1 t ha^{-1} year^{-1}) occurs on steeply sloping pastures. The cultivated main escarpment slope has erosion rates of up to 5 t ha^{-1} year^{-1} and severe rates (>5 t ha^{-1} year^{-1}) are confined to arable land at the base of the escarpment or breaching the slope below the structural shelf. These areas are associated with high sediment accumulation downslope, but the exact areas of deposition are difficult to define from the limited samples. A tentative sediment budget for the catchment is produced in Fig. 4, partly using data from Curr (10). Allowing for the apparent increase in the sedimentation rate during the last decade, suggested by the ^{137}Cs profile, the contemporary output is estimated as145 t year^{-1}. The catchment sediment delivery ratio is therefore 30.0 %, whilst some 20.8 % of soil eroded from fields is delivered to the catchment outlet. The degree of within-field deposition relative to storage beyond field boundaries cannot be deduced from the data, but there are limited opportunities for channel storage. Of ten samples of channel bed material sampled within the catchment, only one contained detectable levels of ^{137}Cs. This occurred where there were visible signs of poaching by cattle and would most likely represent material derived from the edge of the pasture through bank collapse rather than storage of eroded material derived from upstream. Investigation of colluvial material is underway.

CONCLUSION

The land use changes which have occurred during the last decade at the study site are typical of agricultural land in S. England. Evidence from ^{137}Cs depletion in arable soils, the shape of the ^{137}Cs distribution and increased clay fraction in lake cores supports the hypothesis that these land use changes have resulted in increased soil erosion and sediment yield.

The use of combined catchment and lake sediment ^{137}Cs inventories appears to offer a convenient method for establishing the nature of sediment delivery in small catchments. However, three main areas of uncertainty remain.

First, the conversion of ^{137}Cs depletion into actual soil loss is problematic. There are no measurements of soil loss for this area but the mass balance procedure produces estimates that tend to be greater than expected rates. This may reflect inappropriate simulation of ^{137}Cs mobility in soil surface environments. Second, though the mass balance procedures can be used to predict the temporal variation in the ^{137}Cs content of eroded soil and the subsequent shape of the ^{137}Cs profile in a sedimentary sequence, equifinality is a problem, whereby different combinations of process activity rates could account for similar results. Soil loss estimation and profile interpretation would be boosted if ^{137}Cs measurements were supplemented by other independent techniques.

A third problem concerns the application of ^{137}Cs measurements over increasing spatial scales. Since most previous studies using ^{137}Cs measurements have been undertaken to investigate the potential of the technique, there has been little attention towards sampling strategies and the statistical errors associated with the results. If the technique is to be developed as a consultancy tool, further research is required to examine the representativeness of, and errors associated with, erosion estimates.

REFERENCES

1. Ritchie, J.C. and McHenry, J.R. (1990) : Application of radionuclide caesium-137 for measuring soil erosion and sediment accumulation rates and patterns. Journal of Environmental Quality, 19, 215-233.
2. McHenry, J.R. and Ritchie, J.C. (1977) : Estimating field erosion losses from fallout cesium-137 measurements. In : Erosion and Solid Matter Transport in Inland Waters, Proc. Paris Symp., IAHS Publ. n° 122, 26-33.
3. Sutherland, R. A. and de Jong, E. (1990) : Estimation of sediment redistribution within agricultural fields using caesium-137, Crystal Springs, Saskatchewan, Canada. Applied Geography, 10, 205-221.
4. Walling, D.E. and Quine, T.A. (1991) : The use of caesium-137 to investigate soil erosion on arable fields in the UK -potential applications and limitations. Journal of Soil Science, 42, 146-165.
5. Lance, J.C., McIntyre, S.C., Naney, J.W. and Rousseva, S.S. (1986) : Measuring sediment movement at low erosion rates using cesium-137. Journal of the Soil Science Society of America, 50, 1303-1309.
6. Loughran, R.J., Campbell, B.L., Elliot, G.L. and Shelley, D.J. (1990) : Determination of rates of sheet erosion on grazing land using caesium-137. Applied Geography, 10, 125-133.
7. Zhang, X., Higgitt, D.L. and Walling, D.E. (1990) : A preliminary assessment of the potential for using caesium-137 to estimate rates of erosion in the Loess Plateau of China. Hydrological Sciences Journal, 35, 243-252.
8. Brown, R.B., Cutshall, N.H. and Kling, G.F.)1981) : Agricultural erosion indicated by 137Cs redistribution : I. Levels and distribution of ^{137}Cs activity in soils. Journal of the Soil Science Society of America, 45, 1191-1197.
9. Loughran, R.J., Campbell, B.L. and Walling, D.E. (1987) : Soil erosion and sedimentation indicated by caesium-137 : Jackmoor Brook catchment, Devon, England. Catena, 14, 201-212.
10. Curr, R.H.F. (1984) : The sediment dynamics of the Corston Brook. Unpublished PhD thesis, University of Exeter.
11. Findlay, D.C., Colborne, G.J.N., Cope, D.W., Harrod, T.R., Hogan, D.V. and Staines, S.J. (1984) : Soils and their use in South West England, Soil Survey of England and Wales, Harpenden.
12. Quine, T.A. and Walling, D.E. (1991) : Rates of soil erosion in arable fields in Britain : quantitative data from caesium-137 measurements. Soil Use and Management, 7, 169-176.
13. de Roo, A.P.J. (1991) : The use of ^{137}Cs in an erosion study in South Limburg (The Netherlands) and the influence of Chernobyl fallout. Hydrological Processes, 5, 215-227.
14. Higgitt, D.L., Froehlich, W. and Walling, D.E. (1992) : Applications and limitations of Chernobyl radiocaesium measurements in a Carpathian erosion investigation, Poland. Land Degradation and Rehabilitation, 3, 15-26.
15. Higgitt, D.L. (1990) : The use of caesium-137 measurements in erosion investigations. Unpublished PhD thesis, University of Exeter.

16. Walling, D.E. and Bradley, S.B. (1988) : The use of caesium-137 measurements to investigate sediment delivery from cultivated areas in Devon, UK. In : Sediment Budgets, Proc. Porto Alegre Symp., IAHS Publ. n° 174, 325-335.
17. Fredericks, D.J. and Perrens, S.J (1988) : Estimating erosion using caesium-137 : II. Estimating rates of soil loss. In : Sediment Budgets, Proc. Porto Alegre Symp., IAHS Publ. n° 174, 233-240.
18. Walling, D.E. and Quine, T.A. (1990) : Calibration of caesium-137 measurements to provide quantitative erosion rate data. Land Degradation and Rehabilitation, 2, 161-175.
19. Quine, T.A. (1989) : Use of a simple model to estimate rates of soil erosion from caesium-137 data. Iraqi Journal of Water Resources, 8, 54-81.
20. Foster, I.D.L., Dearing, J.A. and Grew, R. (1988) : Lake-catchments : an evaluation of their contribution to studies of sediment yields and delivery processes. In : Sediment Budgets, Proc. Porto Alegre Symp., IAHS Publ. n° 174, 413-424.
21. Livingstone, D. and Cambray, R.S. (1978) : Confirmation of ^{137}Cs dating by algal stratigraphy in Rostherne Mere. Nature, 276, 259-261.

ACKNOWLEDGEMENTS

This study was undertaken as part of a Natural Environment Research Council Studentship at the University of Exeter, supervised by Professor D.E. Walling. I am especially grateful to Rick Curr (Bath College of Higher Education) for assisting the planning and operation of fieldwork. Field assistance was also provided by John Rowan and Emma Bradley. Matthew Ball (Lancaster University) prepared the diagrams.

Farm Land Erosion: In Temperate Plains Environment and Hills
S. Wicherek (Editor)

Le modèle E_S de Jan DE PLOEY. Discussion et nouvelles applications

B. Kaiser

Laboratoire de géographie physique, Université Paris 7 et UA 141 du CNRS
115 avenue de Paris, 78000 Versailles, France

RESUME

Jan de Ploey a proposé un indice E_S de susceptibilité à l'érosion qui, pour une unité fonctionnelle donnée au cours d'un temps t., traduit le quotient entre la masse de matière érodée et la quantité d'énergie dépensée. E_S s'exprime en S^2/m^2 ou Kg/joules. Selon les domaines morpho-climatiques et les types de processus érosifs les valeurs de E_S s'établissent entre 10^{-7} et 10^2 $5^2/m^2$.

Les applications au domaine alpin sont présentées ici pour deux types d'érosion contrastée : l'érosion torrentielle et la gélifluxion laminaire. On obtient E_S de l'ordre de 10^{-2} S^2/m^2 dans un cas et de 10^{-5} à 10^{-6} dans le second.

Le choix des hypothèses et des échelles de temps et d'espace doit toujours être explicite.

SUMMARY

Jan De Ploey proposed the E_S indicator for erosion susceptibility which, for a functional unit given during a lapse of time "t", translates the quotient between the mass of eroded material and the amount of energy spent. E_S is expressed in S^2/m^2 or in Kg/joules. According to morpho-climatic domains and types of erosive processes, the values of E_S are established between 10^{-7} and 10^{-2} $5^2/m^2$.

Applications in the alpine environment are presented here for two kinds of contrasting erosion : torrential erosion and laminar solifluction E_S is obtained, for 10^{-2} S^2/m^2 in one case and 10^{-5} to 10^{-6} in the second.

The choice of hypotheses and time and space scales must always be explicit.

INTRODUCTION

Lorsqu'il nous a brutalement quittés Jan De Ploey préparait un nouvel article sur le modèle Es de susceptibilité à l'érosion qui aurait pris en compte les multiples échanges suscités par ses premières publications sur ce sujet (1-3). Les quelques pages qui suivent se veulent une brève présentation du modèle Es et un appel à poursuivre la réflexion dans l'esprit d'ouverture qui caractérisait notre collègue disparu auquel nous espérons, ainsi, rendre hommage.

RAPPEL. QUELQUES FORMULES

L'idée de base est que l'érosion dépend de la mise en jeu d'une énergie mécanique sous les deux formes de l'énergie potentielle et de l'énergie cinétique des eaux qui s'écoulent sur un territoire donné. La susceptibilité à l'érosion, E_S, exprime la quantité d'énergie nécessaire pour éroder une masse donnée de matériaux. Elle traduit l'efficacité du travail géomorphologique.

E_S est exprimée par quatre formules qui se déduisent les unes des autres (1) :

$$Es = \frac{V}{M\,(gh + 1/2\ Uo2)} \quad (1)$$

$$Es = \frac{V}{M\,g\,(h + R.S/2)} \quad (2)$$

on pose $Uo = (g.\ R.\ S)^{1/2}$

avec
- . V volume érodé au sein d'une unité donnée A, en m^3
- . M volume des précipitations sur A pendant un temps t donné, en m^3
- . g accélération de la pesanteur, en m/s^2
- . h perte de charge, en m
- .U_0 vitesse de cisaillement des eaux courantes au fond, en m/s
- . R rayon hydraulique moyen de l'écoulement des eaux, en m
- . S pente moyenne des versants de la surface érodée AE incluse dans A, en %

$$Es = \frac{V}{M.g\left(h + \dfrac{AE0.6.\ M0.6.\ n0.6.\ S07}{2\ \ A0.6.\ t\ 0.6.\ W\ 0.6}\right)} \quad (3)$$

Cette formule fait appel aux lois empiriques de l'hydraulique pour un écoulement turbulent. Outre les symboles précédents on lit :
- . n indice de rugosité appliqué à la surface érodée AE, en $s/m^{1/3}$
- . w largeur du chenal ou somme des largeurs des chenaux élémentaires , en m
- . A superficie de l'unité fonctionnelle, en m^2
- . P hauteur des précipitations sur l'unité A considérée, en m (A.P = M)
- . AE superficie érodée incluse dans A, en m^2
- . t durée, en s

$$Es = \frac{AE}{A.P.g} \quad (4)$$

Cette formule simplifiée souligne l'aspect aréolaire de l'érosion. Dérivée de la formule (2) elle s'applique lorsque R.S/2 est négligeable devant h. Dans le cas d'une érosion en rigoles ou en ravines on prend V/h = AE (ou encore selon la géométrie des ravines V/h =1/2 AE ou V/h =3/4 AE).

Toutes les formules s'appliquent pour une unité fonctionnelle A donnée et pour une durée t spécifiée.

DISCUSSION

Le recours à l'analyse dimensionnelle

La susceptibilité à l'érosion, Es, a la dimension d'une masse divisée par une énergie. Dans le système M.K.S. A. les valeurs de Es s'expriment en *kg/joules* ou en s^2/m^2.

$$Es = \frac{V}{M\,(gh + 1/2\, U_0^2)} \qquad (1)$$

soit en analyse dimensionnelle

$$Es = \frac{L^3}{L^3\,(L.T^{-2}.L + L^2.T^{-2})}$$

L= une longueur
T= un temps

$$Es = \frac{1}{L^2.T^{-2}} = \frac{T^2}{L^2}$$

soit l'inverse du carré d'une vitesse {en unités M.K.S.A. s^2/m^2}

En multipliant numérateur et dénominateur par *Ma*, la dimension d'une masse, on obtient au dénominateur la dimension d'une énergie :

$$Es = \frac{Ma}{Ma.\,L^2.\,T^{-2}}$$

{en unités M.K.S.A. *kg/joules*}

Exemple : la valeur 4. 10^{-4} représente 4 kg/10 000 joules. Il faut 10 000 joules pour éroder 4 kilogrammes de matériau. En ramenant le numérateur à la valeur 1 on obtient 1kg/2 500 joules .

Une formule simple, une formule unificatrice ?

Modèle simple, le modèle Es doit permettre des comparaisons rapides entre des domaines lithologiques, orographiques ou morphoclimatiques différents, même si chacun d'eux est soumis à des *combinaisons de processus érosifs.* Les modèles plus élaborés testés en laboratoire et sur le terrain s'attachent usuellement à *une* modalité érosive donnée contrôlée par un nombre limité de variables. Le modèle Es correspond à un fonctionnement global. Là résident sa richesse et sa limite.

Les formules (1) et (2) de Es ont dans un premier temps été présentées comme valables pour un grand nombre de processus et de combinaisons de processus:
- l'érosion pelliculaire en nappe (sheetwash)
- l'incision linéaire en rigoles et ravines (rills and gullies)
- l'érosion par glissement de terrain (landslide)

- l'érosion par coulée de débris (debris-flow)
- l'érosion par reptation et gélifluxion (creep and gelifluction) etc..

Cependant Jan De Ploey a émis un doute et proposé une nouvelle écriture pour la formule (2) : une virgule remplace un plus.(4)

$$Es = \frac{V}{M\,g\,(h + R.S/2)} \qquad (2)$$

devient

$$Es = \frac{V}{M\,g\,(h\,,\,R.S/2)} \qquad (2\ bis)$$

avec

$$Es = \frac{V}{M\,g\,h} \qquad (5)$$

appliqué à l'érosion en rigoles et en ravines, aux glissements de terrain, aux "debris-flow", à la reptation et la gélifluxion

et avec

$$Es = \frac{V}{M\,g\,(R.S/2)} \qquad (6)$$

pour l'érosion pelliculaire en nappe.

Une gamme large de valeurs centrées à 1.10^{-5} kg/joules

Les valeurs de Es calculées par Jan De Ploey à partir de mesures personnelles comme à partir des travaux effectués par le Laboratoire de Géomorphologie Expérimentale de Leuven et des données de la littérature internationale vont de 10^{-2} à 10^{-7} s^2/m^2 avec une valeur moyenne proche de 1.10^{-5} s^2/m^2.

Une hiérarchie de valeurs se dégage selon les processus (1- 4) : à masse égale enlevée l'érosion en nappe est celle qui consomme le moins d'énergie (nombreuses valeurs de Es entre 10^{-2} et 10^{-3} s^2/m^2) et la reptation est celle qui demande le plus d'énergie (valeurs de Es entre 10^{-6} et 10^{-7} s^2/m^2 ; information orale au colloque du Réseau Erosion. Montpellier, France, 1990). Les calculs que nous avons personnellement effectués pour des versants et bassins-versants des Alpes françaises du nord se situent dans cette gamme de valeurs.

1° exemple : Es est compris entre 4.10^{-5} et $5{,}5.10^{-6}$ s^2/m^2 pour un versant affecté par de la gélifluxion laminaire entre 2000 et 3000 mètres d'altitude pour une superficie A d'environ 0,5 km^2. La première valeur s'applique à la tranche altitudinale supérieure, la seconde à la tranche inférieure. Il s'agit dans les deux cas de données moyennes correspondant à 10 années de mesure sur le déplacement des matériaux (5).

2° exemple : Es vaut $0{,}8.10^{-2}$ s^2/m^2 pour le bassin-versant de la Ravoire d'une superficie d'environ 10 km^2 soumis à une érosion torrentielle exceptionnelle les 31 mars et 1er avril 1981.

Cette valeur forte de Es pour un processus de ravinement - accompagné de glissements de terrain sur les berges- rejoint celles proposées par Jan De Ploey lorsque le calcul de Es est réalisé à l'échelle d'un événement bref et intense, et non à l'échelle annuelle ou pluri-annuelle. (4).

Choix de l'unité fonctionnelle. Choix des échelles de temps et d'espace

L'unité fonctionnelle A au sein de laquelle s'exerce l'érosion n'est pas sans ambiguïté dans le modèle Es. Est-elle une unité hydrologique, ce qui justifierait de prendre en compte la *totalité des précipitations* tombées sur sa surface (le volume M), ou bien est-elle une unité morphodynamique, et alors il conviendrait de ne prendre en compte que les "précipitations utiles" pour le travail géomorphologique? Celles-ci ne tombent que sur une partie de la surface. Quelle partie?

Cette question rejoint celle des échelles de temps et d'espace.
Les durées retenues dans les exemples cités par Jan De Ploey (1-4) vont de quelques heures (durée d'un événement météorologique et érosif particulier) à quelques milliers d'années, en passant par une seule année.

Autant sur la très longue durée il est possible d'admettre que toutes les précipitations tombées sur la surface A agissent sur l'érosion, directement, ou indirectement par altération des matériaux, autant pour une seule année, le choix du volume total M des précipitations pose problème.

La taille de l'unité fonctionnelle permet aussi de mieux interpréter les valeurs de Es. Il me parait discutable d'introduire des valeurs de Es calculées pour des unités très petites, de l'ordre de 1 m^2 (2).

S'il est difficile d'annoncer, d'ores et déja, quelles sont les échelles d'espace et de temps les plus pertinentes pour utiliser le modèle Es il convient, pour établir des comparaisons raisonnables de toujours préciser la durée et la dimension de l'unité fonctionnelle.

CONCLUSION

Le modèle Es de susceptibilité à l'érosion tire une grande partie de son intérêt de la confrontation des résultats à laquelle il conduit. Nous souhaitons que les nombreux chercheurs qui ont collecté de multiples données dont certaines sont utilisables dans les formules de Es (1) à (6) - volumes ou tonnages érodés, superficies concernées, épaisseur du décapage ou profondeur des incisions, volume des précipitations- acceptent de mettre en commun leurs mesures afin de rechercher les régularités inhérentes au travail géomorphologique et de hiérarchiser sur ces nouvelles données les principales différences dues aux contextes bioclimatiques et orographiques. Le travail de Jan De Ploey nous y incite.

REFERENCES

1 De Ploey J., (1990) : Modelling the erosional susceptibility of catchments in terms of energy, Catena, 17 : 175-183.

2 De Ploey J., (1990) : The Es model : a definition of slope and catchment erosional susceptibility. European Society for Soil Conservation, 2 + 3 : 22-30.
3 De Ploey J., (1991) : L'érosion de bassins-versants : analyse et prévision selon le modèle Es., Physio-géo, 22-23 : 7-12.
4 De Ploey J., (1992) : L'érosion de bassins-versants. Analyse et prévision selon les modèles. Bulletin de L'Association des Géographes français, 4 : 339-355.
Kaiser B., (1987) : Thèse de Doctorat d'Etat. Université Paris 7 : 201-310.
La solifluxion laminaire à Lanserlia ; mesure morphodynamique et contexte climatique.

Farm Land Erosion: In Temperate Plains Environment and Hills
S. Wicherek (Editor)

Towards improved hypothesis testing in erosion-process research

D.M. Lawler

School of Geography, The University of Birmingham, Edgbaston, Birmingham, B15 2TT, United Kingdom.

SUMMARY

Many advances in the understanding of physical systems flow from critical observation of the responses of system attributes to variations in the stresses applied by the controlling agents. Monitoring should ideally be of sufficiently-high temporal resolution to detect all important episodic and transient responses to facilitate strong process-inference. However, the traditional, manual, monitoring techniques available to erosion researchers simply reveal *net change* to an eroding surface since the previous measurement, and not the temporal distribution of that change within any given measurement interval which is so important to unravel the mechanisms responsible. To help solve this problem, I describe here the recent development of the Photo-Electronic Erosion Pin (PEEP) system. This inexpensive system allows, for the first time, quasi-continuous time series of erosion and deposition data to be collected automatically - a capability which is especially valuable in highly episodic systems and for remote sites which cannot be visited frequently. Measurement principles, simple calibration techniques, and advantages of the method are discussed. Brief results, from a river bank site eroding into lowland pasture farmland in the English Midlands, show how the PEEP system can quantify the erosional impact of *individual,* rather than aggregated, forcing events, reveal the full complexity of geomorphological change, and improve understanding of the driving processes.

RESUME

Une exigence fondamentale dans n'importe quelle science d'observation est la capacité de quantifier les qualités de variabilité temporelle du système. Toutefois, l'absence d'un système automatique destiné à enregistrer l'érosion a empêché un suivi précis (débits liquides et solides). Les techniques manuelles traditionnelles ne révèlent que l'évolution d'une surface érodée depuis la dernière mesure. L'auteur étudie ici le développement récent du système de Photo-Electronic Erosion Pin (PEEP). Ce système peu coûteux permet, pour la première fois, d'obtenir une série quasi continue enregistrée de renseignements sur les flux. Les principes de mesure, les techniques simples d'étalonnage, l'installation sur le terrain, ainsi que les avantages de cette méthode sont discutés. Les premiers résultats obtenus sur le pâturage dans les Midlands, montrent que le système PEEP permet de quantifier l'impact d'événements forcément individuels plutôt que collectifs ; il peut révéler la complexité entière du changement géomorphologique et également améliorer la compréhension des processus principaux.

1. INTRODUCTION

An increased understanding of a physical system often results from careful observation of its precise response to a series of induced or natural external stimuli. In many sciences techniques exist which allow the temporal complexity of both stimulus and response to be detected with appropriate resolution for strong inference of the processes at play. Workers on contemporary erosion problems, however, have only been able to draw upon automatic instrumentation for the continuous monitoring of changes in the controlling variables (stimuli), and not for the geomorphological changes (responses) themselves. The *manual* techniques available to monitor land surface advance or retreat (e.g. erosion pins (1) or repeated profiling (2 - 3 - 4) - see Loughran (5) for a review) - though useful - merely reveal the *net change* to an eroding surface *since the previous survey.* Hence very few data have been collected on the precise timing, frequency, magnitude and duration of individual erosion events - an especially unfortunate drawback in characteristically episodic erosion systems.

This mismatch, between data of high temporal resolution for the driving forces but low-resolution information for the eroding surface, poses a number of problems. First, there is a danger of underestimating the dynamism of landscape change, because conventional methods may not detect complex activity (e.g. scour-and-fill sequences) unless manual resurveys are exceptionally frequent. Secondly, our abilities to infer the driving processes and establish causality in the system - often at present based on comparing coarse, *temporally-aggregated,* erosional responses to groups of hydrological or meteorological events - are severely limited by a lack of data on the timing of erosion. For example, as Morgan (6) recognised, 'potentially useful information on whether most of the soil is eroded early or late in a storm is lost' with conventional methods. Thirdly, our abilities to test erosion hypotheses, build models, and ultimately generate theory is constrained, not least because model predictions of high temporal resolution can only be evaluated against low-resolution field data. These difficulties are clearly exacerbated when dealing with remote sites which cannot be visited frequently for remeasurement.

The aim of this short paper is to discuss a new technique - the Photo-Electronic Erosion Pin (PEEP) system - which provides a way of obtaining automatic, quasi-continuous, data on geomorphological change and thus strengthen process-inference and improve hypothesis testing. The solution offered is by no means a panacea for all types of erosion monitoring problems, but it does serve to illustrate the potential of one attempt. It is hoped that by highlighting this research need, further scientific attention will be attracted to this important area. The examples chosen stem from river bank erosion investigations, but the approach is applicable to many accelerated soil erosion studies concerned with the measurement of sheetwash, rill and gully processes.

2. THE PHOTO-ELECTRONIC EROSION PIN (PEEP) SYSTEM

2.1. Sensor design and measurement principle

The Photo-Electronic Erosion Pin (PEEP) sensor, recently developed in the School of Geography at The University of Birmingham, UK (7 - 8), is in effect a self-reading electronic erosion pin. It is a relatively simple and inexpensive instrument, consisting of a row of photovoltaic cells connected in series on a printed circuit board (or simple stripboard), with an additional cell at the front end to act as a reference. The board is housed within a transparent acrylic tube of 16 mm external diameter which is sealed and waterproofed at both ends (Photo 1) (8). A cable takes voltage outputs from the cells to the datalogger.

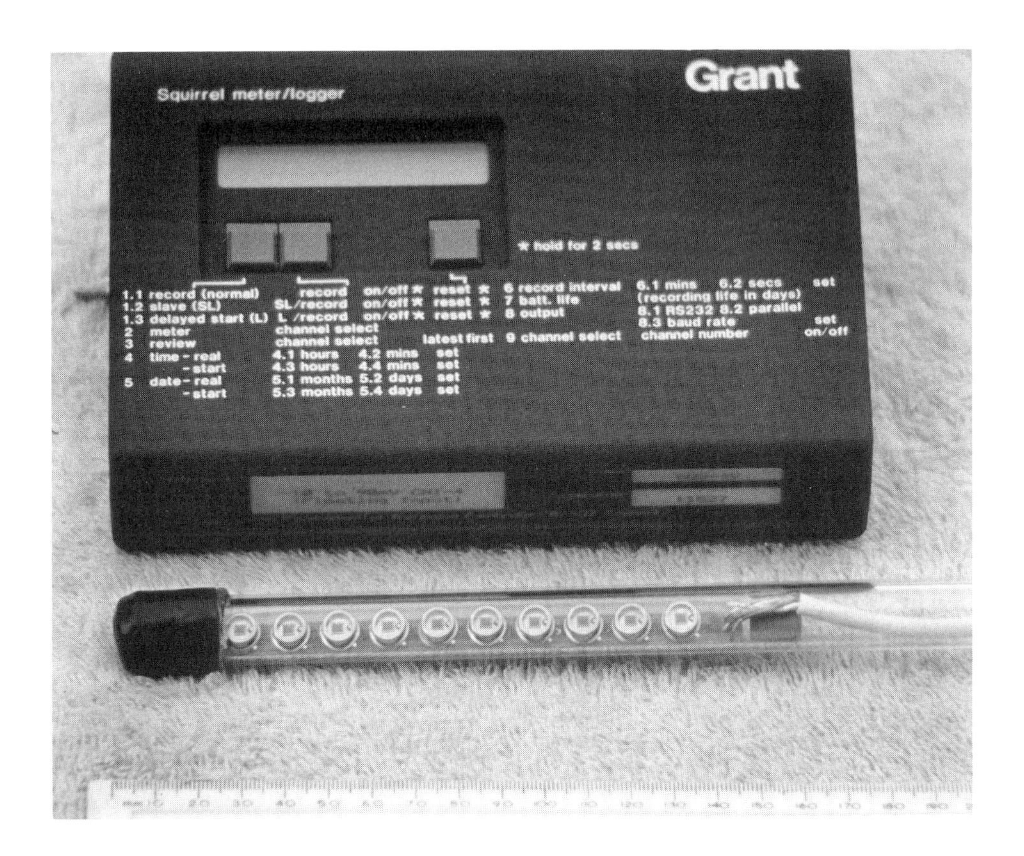

Photo 1.
A Photo-Electronic Erosion Pin (PEEP) sensor (prototype version) with datalogger. Only the front 23 cm of tube can be seen. The connecting cable leading away to the back end of the sensor is also visible.

The PEEP sensor is inserted into an eroding (or accreting) surface so that only the reference cell and the first of the cell series is exposed to light (Photo 2). Its cable is then connected to a nearby datalogger (or chart recorder or PC) equipped to accept an analogue voltage signal (as most are). As the monitored surface retreats, more photovoltaic cells are exposed to light; this increases the voltage output from the PEEP. Accretion of the surface causes photovoltaic outputs to drop. Inspection of the ramps, steps and troughs in the recorded data, therefore, reveals the magnitude, frequency and timing of individual erosion and deposition events with much greater precision than has hitherto been possible and, seemingly for the first time, it becomes feasible to collect a near-continuous record of contemporary geomorphological change.

Each PEEP sensor requires two logger channels to record voltage outputs although, to save logger channels/memory, one reference cell might feasibly serve a whole network of sensors in the locality. The total length of the sensor used here is 0.40 m, with an 'active length' of 0.10 m composed of ten photovoltaic cells (Photo 1), although these lengths can be varied to suit the application. No power supply is needed in the field, because the cells convert incident light energy directly into a measurable voltage. Outputs are in the form of diurnal cycles which zero at night : this means, of course, that nocturnal erosion cannot be detected till the following morning (a problem which can be solved by logger-activated bursts of artificial light at the instant of sensor-scanning). The PEEP system is fully described in Lawler (8), along with problems and prospects, and details of instrument availability can be provided by the author.

2.2. Instrument calibration and sensitivity

Laboratory calibration of PEEP sensors is first achieved under natural illumination by simulating known amounts of ground surface change around the tube. This is done simply by placing the sensor inside a light-tight box equipped with an orifice of diameter 16 mm at one end (Photo 3). The PEEP tube is progressively pulled through the orifice, exposing more solar cells to light, while recording the voltage output levels at each increment.

The instrument delivers a reasonably continuous response as more photosensitive material is disclosed to light, and encouragingly strong, linear, relationships are obtained between the length of PEEP tube exposed and sensor outputs (8). Regression analysis reveals typical coefficients of explanation of around 99.7 % for a sample of 250 scans, with standard errors for estimated tube length as low as 0.85 mm. Standard errors could be further reduced by averaging a number of instantaneous scans over a given sampling period, but appear to be sufficiently low for many soil erosion investigations (see surface-wash rates cited in Saunders and Young (9) and gully erosion rates given in Bocco (10). (As a general guideline, if the site is appropriate for the use of traditional erosion pins and profilometers then the PEEP system is also likely to be suitable.) In-situ calibrations in the field ('ground-truth measurements) are also possible by comparing occasional manual measurements of tube exposure with recorded outputs.

Photo 2 .
A prototype Photo-Electronic Erosion Pin emerging from the river bank face on the River Arrow at Studley, Warwickshire, UK. The reference cell and one other cell can be seen, and about 50 mm of tube is visible.

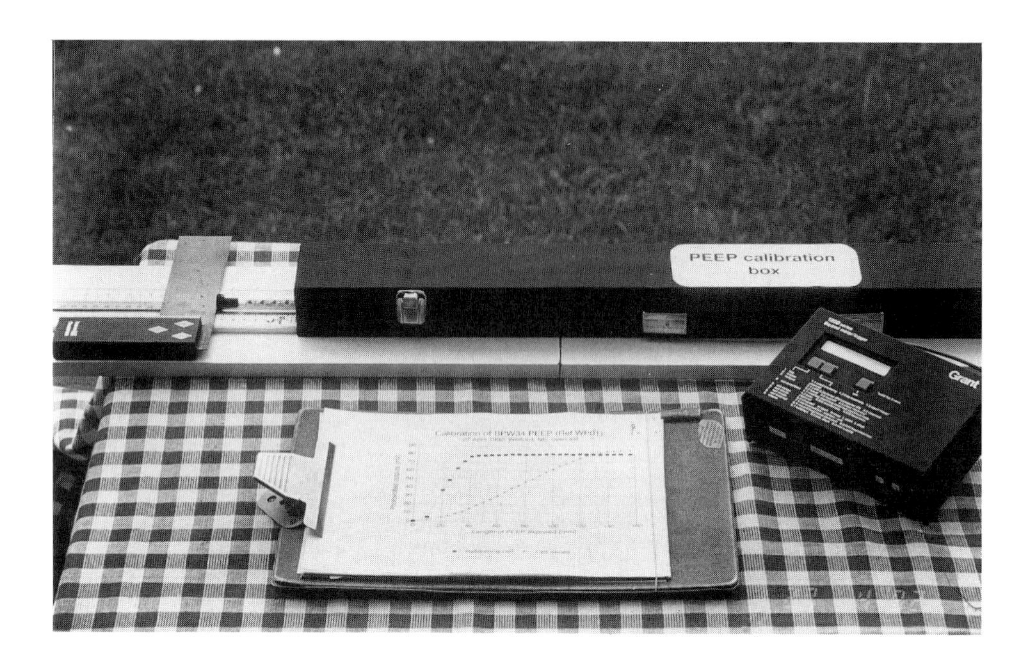

Photo 3.
The PEEP calibration box with sensor emerging from an orifice on the left side over a graduated scale for measuring the length of exposed tube. A datalogger can be seen to the right, with sample calibration curves in the foreground.

3. EXAMPLE RESULTS

3.1 Study area

PEEP systems have been tested in bank erosion projects on the River Severn in mid-Wales (11) and the River Arrow in Warwickshire (7 - 8), both in the U.K. Brief results are presented from the latter study to illustrate the type of benefit that automatic monitoring systems can achieve. Trials in other erosional/ depositional contexts (e.g. soil erosion plots, gully heads and sidewalls, desert dunes, beaches and coastal cliffs) would be very interesting.

Following initial field testing in Spring 1988, two prototype PEEP sensors were installed in the right bank of the meandering River Arrow near Studley, Warwickshire (UK grid reference SP 082635) in January 1989. This site has been actively eroding into floodplain pasture land (Photo 4), but has exhibited stabilizing tendencies recently. Drainage area here is 98 km^2, bank heights are around 2.2 m, and bank materials are relatively cohesive silt-clays.

The system has helped to define more closely the true temporal distribution of erosional and depositional activity - including on occasions the virtual moment of erosion (11), reveal the precise geomorphological response to individual (and not merely aggregated) hydrological and meteorological events, and improve the basis for inferring the driving processes (7 - 12). Results in this paper, however, are used simply to indicate the potential of the PEEP system for the testing of erosion hypotheses. Although the example is drawn from a river bank site, the approach is equally applicable to the more general soil erosion case. Data are derived from PEEP sensor 2, installed 0.2 m above typical winter flow level. The installation technique, which ensures minimal contact with, and disturbance to, the bank face, is fully described in Lawler (8).

3.2 Hypothesis testing : The example of erodibility changes

Bank erosion workers (and some soil erosion workers) have recently turned their attention to changing material erodibility over seasonal and subseasonal timescales. It has often been observed on river banks - especially cohesive ones - that similar flow events may effect vastly different erosional consequences (e.g. Fig. 1) (13 - 14). The scatter of Fig. 1 suggests that other processes are also at work. It appears that the resistance to fluid erosion changes over time, because of a varying intensity of antecedent preparation processes which alter the erodibility of a bank surface and prepare a supply of readily-entrainable material. Hitherto, however, such effects have been difficult to detect.

The PEEP system can be used to disentangle competing hypotheses of the controls of such transient erodibility peaks. The impact of a schematic double-peaked flow event will be considered as an example of discriminating between three simple hypotheses to account for variable bank response (Fig. 2). If it is assumed that bank erodibility is invariant through time, then, notwithstanding random scatter and measurement deficiencies, each hydrograph event, being similar in magnitude, should effect broadly similar amounts of bank retreat (Hypothesis A), as the hypothetical cumulative erosion trend in Fig. 2A shows. If preparation processes are active at the monitored site, however, then, assuming minimal time-lags in the system, the first of the two peaks should be associated with maximum erosion (Hypothesis B) because a surface layer of erodible bank material is available for entrainment. Minimal retreat will be effected by the second event because of the limited time for the fresh bank surface to be weathered and sediment supplies renewed (Fig. 2B). Finally, if the site is more sensitive to fluvial pre-wetting than subaerial weathering (15) then the second event of the

Photo 4.
The bank site on the River Arrow, Warwickshire in which two prototype PEEP sensors were installed. The datalogger is protected in the housing to the left, and cables to the PEEPs are trained underground to eliminate fouling of bank surfaces (8). Flow is towards camera.

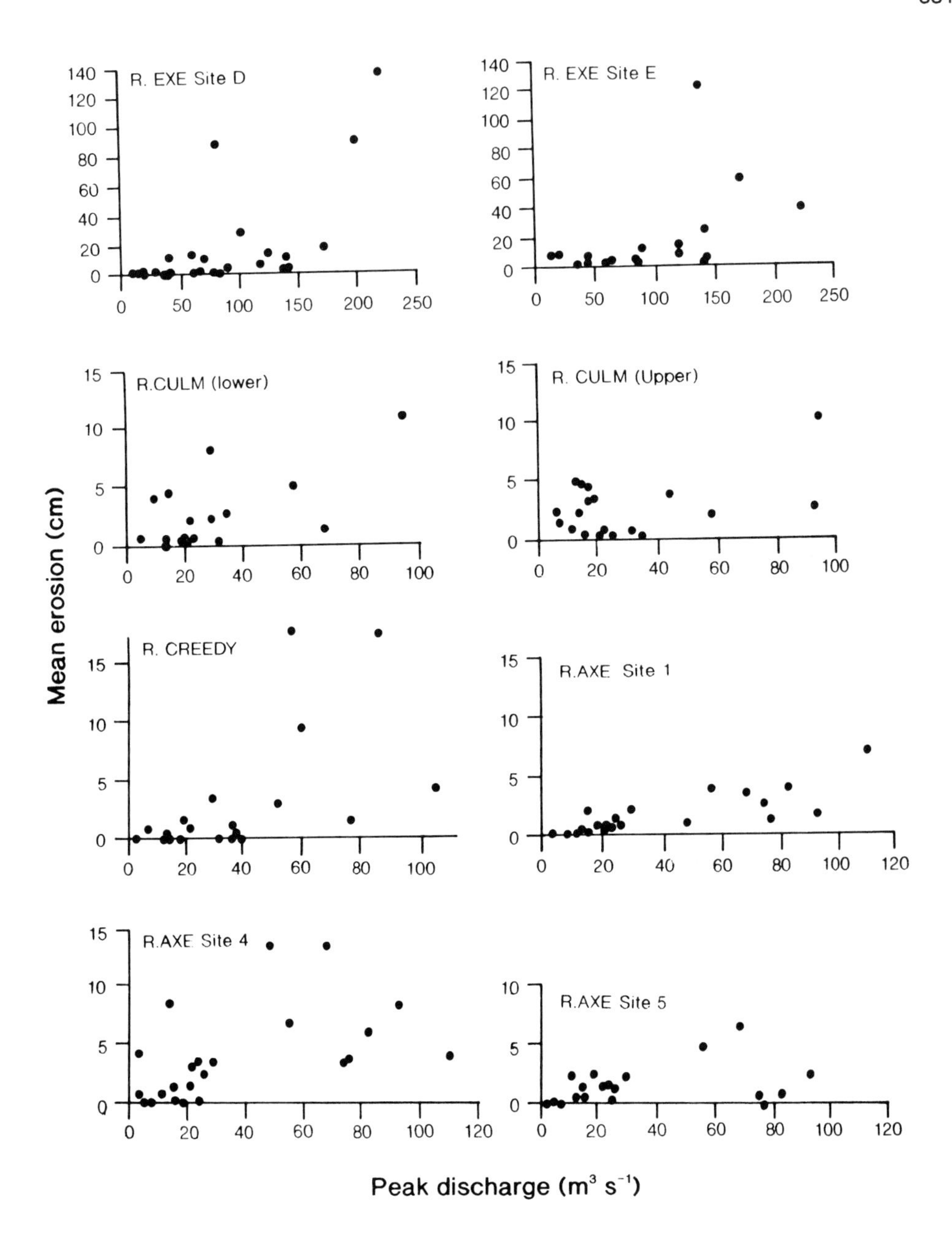

Figure 1. Variability in the erosion-flow relationships for selected rivers in south Devon (13).

hypothetical sequence should be more erosive as it is acting on banks saturated by the first flood (Hypothesis C) (Fig. 2C).

The rigorous testing of such hypotheses allows real progress to be made in the understanding of system dynamics. However, conventional erosion-monitoring techniques would be inadequate here, only revealing the gross impact of the complete event sequence in Fig. 2, and not their crucial separate effects. There are insufficient PEEP field data to subject these hypotheses to a definitive test as yet, but one example serves to demonstrate potential. The diurnal cycles of PEEP output given in Figure 3 show that, at the Arrow site, bank retreat took place after or during the first flow event, with no erosion associated with the second hydrograph. If this pattern were to be repeated often enough, then we would be forced to accept hypothesis B (Fig. 2B) that preparation processes here control the resistance of the bank surface sediment to entrainment processes. Hypotheses A and C (Figs 2A and 2C) could be rejected, and further attention focused on the identification and operation of the preparation processes involved (12).

4. DISCUSSION

Preparation processes have frequently been identified as important on other river banks (and other eroding surfaces). For example, Bello et al. (16) comment on the efficacy of desiccation processes in conditioning a surface prior to sediment removal, Hooke (15) argues that high bank material moisture contents encourage later fluvial entrainment, and Lawler (14 - 17), Hill (18) and Gardiner (19) found that prior freeze-thaw activity increased bank erosion during subsequent flow events. Similar examples can be found in the soil erosion literature (e.g. alternate wetting and drying, surface crust development (6), and seasonal vegetation and crop cover changes (20).

Interestingly, the pattern of bank erosion through the early event(s) in a sequence of closely-spaced storms hypothesized in Fig. 2B and evident in Fig. 3 is entirely consistent with observations of suspended sediment exhaustion effects often noted in such flow sequences (e.g. Walling and Webb (21), their Fig. 5.20). In cases where river banks are the dominant provider of fine sediment to the fluvial system, we might expect a close relationship at the event timescale between amounts of bank retreat and downstream suspended sediment concentrations.
This is depicted schematically in Figure 4, and forms another example hypothesis that can now be tested in future deductive research using combined networks of PEEP sensors and turbidity meters.

Within soil erosion research more widely, similar scope exists for using PEEP sensors at strategic points within conventional measurement frameworks (e.g. erosion pins to detect the spatial picture) to increase understanding of process dynamics, and especially for the testing of hypotheses regarding complex temporal changes in soil erodibility and removal processes. Examples where the PEEP (or similar) system might contribute include :

(i) the influence of changing crop and vegetation cover on erosion intensity. One of the problems that Evans (20, p.200) identifies is that 'eroded fields are visited after erosion has occurred, often a considerable time after' which hinders attempts to determine a vegetation cover threshold for erosion initiation ;

Invariant bank sensitivity

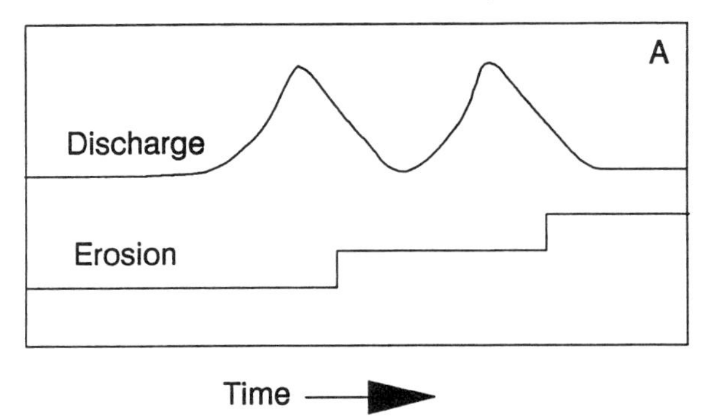

Preparation processes dominant

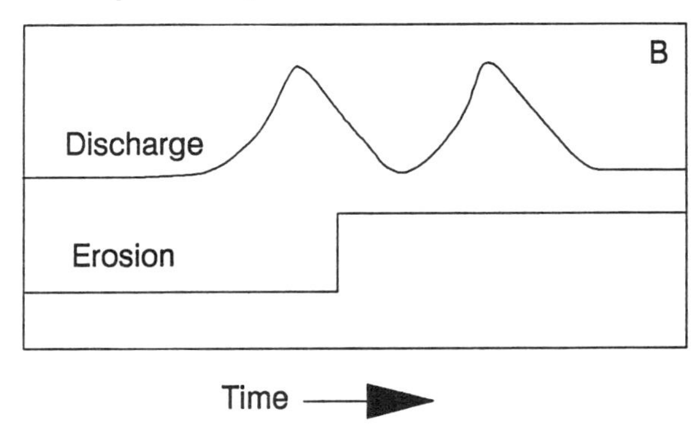

Fluvial pre-wetting dominant

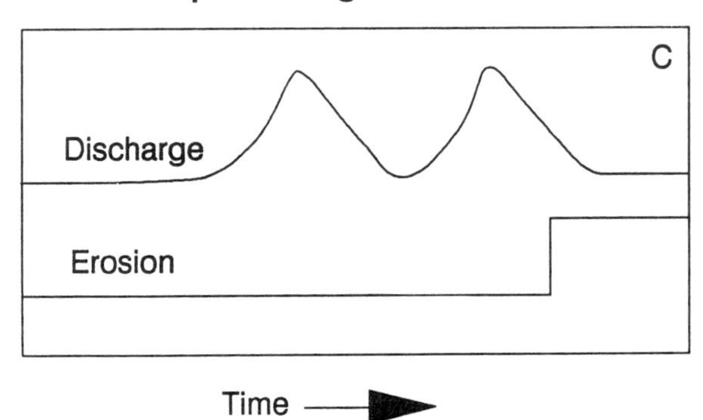

Figure 2. Hypothetical erosional responses to a double-peaked hydrograph event : (A) assuming invariant bank sensitivity ; (B) assuming dominance of preparation processes ; (C) assuming dominance of fluvial pre-wetting processes.

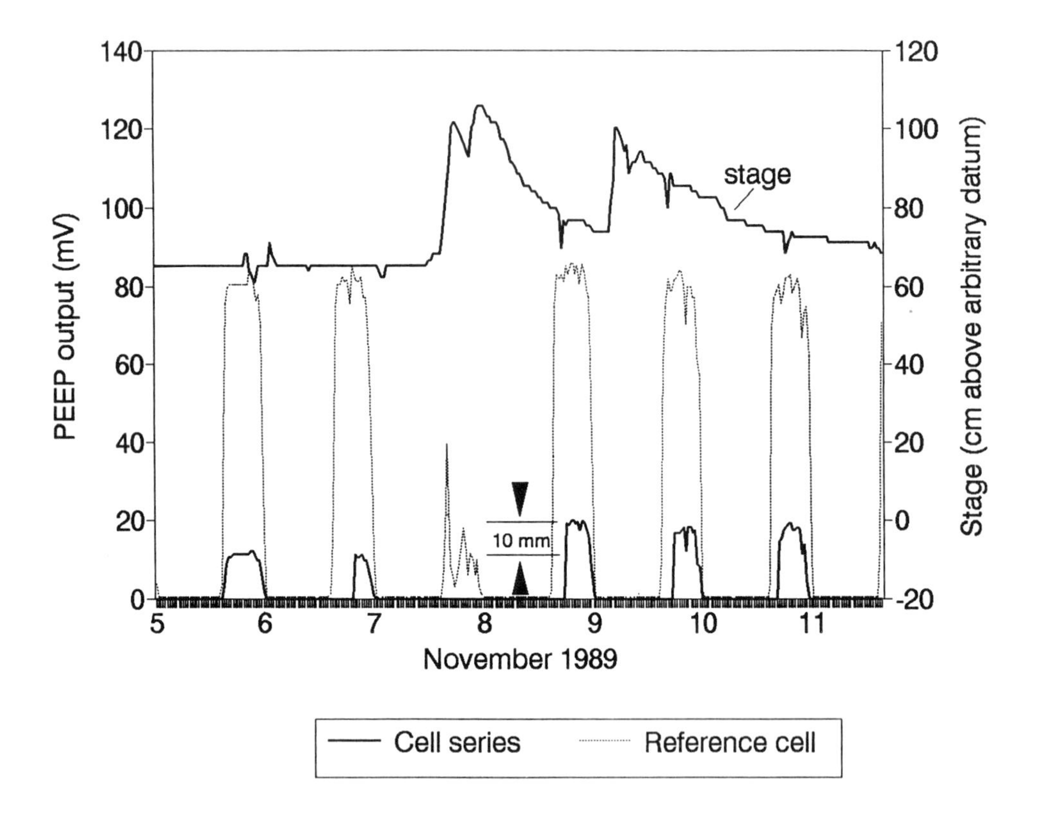

Figure 3. Actual bank erosion response to a double-peaked flow event on the River Arrow in November 1989. All of the erosion (10 mm, as 1 mV approximates to 1 mm of retreat) is effected by the first of the two events, indicating the significance of preparation processes here.

(ii) the fluctuating erosional impact of rainfall intensities (see Fullen, 22). An automatic monitoring technique yields quantitative information on the effects of every rainfall event : this would vastly increase the sample size in such studies and help to identify the presence, nature and importance of any shifts in erosivity thresholds in response to surface changes ;

(iii) gully-head retreat and gully-wall instability processes. Bocco (10, p. 392) argues that 'gully erosion remains a poorly understood process'; continuous data on advance/retreat events, however, may well help to resolve some of the 'confusing and sometimes contradictory results' (p.392) he refers to, and in particular the assumption that 'headcuts migrate with the same form and depth of incision at every erosional event' (p.401), and the precise role of surface- and subsurface flow generation in erosion episodicity ;

Flow, erosion and sediment transport relationships

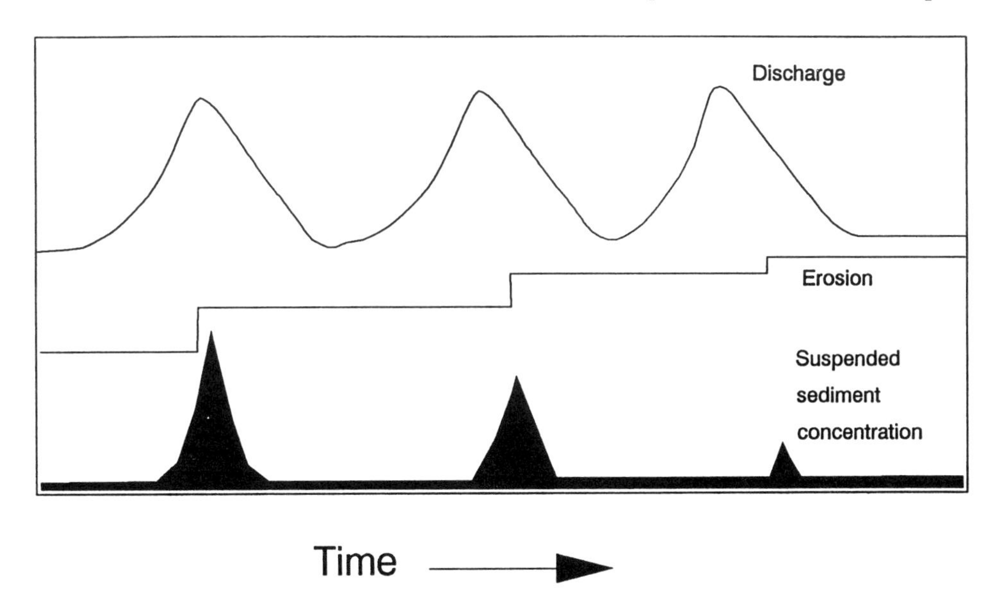

Fig. 4. Schematic links between progressively-decreasing bank erosion through a multi-peaked flow event, as supplies of available prepared surface material are reduced, and resultant suspended sediment concentrations downstream.

(iv) development of drainage ditches associated with, for example, afforestation practices (23 - 24).

(v) complex scour and fill sequences on (semi-arid) hillslopes. Detection and quantification of these is often difficult with traditional methods (25), but potentially straightforward with the PEEP system which records the magnitude and timing of accretion and erosion events with equal resolution.

Work is continuing to refine the PEEP method, develop other techniques, and apply them to a variety of erosional and depositional contexts where continuous information on geomorphological change would help to unravel some of the system complexity. The PEEP reference cell outputs also, of course, provide data on radiation loading of eroding surfaces, which might be useful as an input to desiccation or vegetation-growth models. It is also feasible to connect PEEP sensors to telemetry systems for real-time data acquisition, remote downloading, and erosion-warning capabilities, either for management purposes or to help researchers to time their field visits.

ACKNOWLEDGEMENTS

I am very grateful to Heather Lawler and Robert Brown for assistance, advice and encouragement, Bill Green for access to the R. Arrow site, Mike Dolan and Ali Hasan for help with PEEP installation, Kevin Burkhill for cartographic work and Simon Restorick for photographic processing.

REFERENCES

1. Haigh, M.J. (1977) : The use of erosion pins in the study of slope evolution, British Geomorphological Research Group Technical Bulletin, 18, 31-49.
2. Lam Kin Che, (1977) : Patterns and rates of slopewash on the badlands of Hong Kong, Earth Surface Processes, 2, 319-332.
3. Campbell, I.A. (1981) : Spatial and temporal variations in erosion measurements. In Symposium on Erosion and Sediment Transport Measurement, Florence, International Association of Hydrological Scientists Publication 159, 437-446.
4. Toy, T. J. (1989) : An assessment of surface-mine reclamation based upon sheetwash erosion rates at the Glenrock Coal Company, Glenrock, Wyoming, Earth Surface Processes and Landforms, 14, 289-302.
5. Loughran, R. J. (1989) : The measurement of soil erosion, Progress in Physical Geography, 13, 216-233.
6. Morgan, R. P. C. (1986) : Soil erosion and conservation, Longman, Harlow, 298pp.
7. Lawler, D. M. (1991) : A new technique for the automatic monitoring of erosion and deposition rates, Water Resources Research, 27, 2125-2128.
8. Lawler, D. M. (1992a) : Design and installation of a novel automatic erosion monitoring system, Earth Surface Processes and Landforms, 17 (5).
9. Saunders, I. and Young, A. (1983) : Rates of surface processes on slopes, slope retreat and denudation, Earth Surface Processes and Landforms, 8, 473-501.
10. Bocco, G. (1991) : Gully erosion : processes and models, Progress in Physical Geography, 15, 392-406.
11. Lawler, D. M. and Leeks, G. J. L. (in press) : River bank erosion events on the Upper Severn detected by the Photo-Electronic Erosion Pin (PEEP) system, in : Bogen, J., Day, T. J. and Walling, D. E. (eds) Proc. Oslo Symp., International Association of Hydrological Scientists Publication.
12. Lawler, D. M. (1992b) : Process dominance in bank erosion systems, in Carling, P. A. and Petts, G. E. (eds) Lowland Floodplain Rivers : Geomorphological Perspectives, John Wiley, Chichester, 117-143.
13. Hooke, J. M. (1977) : An analysis of changes in river channel patterns : the example of streams in Devon. Unpub. Ph.D. thesis, University of Exeter, 452 pp.
14. Lawler, D. M. (1986) : River bank erosion and the influence of frost : a statistical examination, Transactions of the Institute of British Geographers, 11, 227-242.
15. Hooke, J. M. (1979) : An analysis of the processes of river bank erosion. Journal of Hydrology, 42, 39-62.
16. Bello, A., Day, D., Douglas, J., Field, J., Lam, K., and Soh, Z. B. H. A. (1978) : Field experiments to analyse runoff, sediment and solute production in the New England region of Australia, Zeitschrift für Geomorphologie N.F. Suppl. Bd., 29, 180-190.

17. Lawler, D. M. (1987) : Bank erosion and frost action : an example from South Wales, In Gardiner, V. (ed.) International Geomorphology 1986, Part 1, John Wiley and Sons Ltd, Chichester, 575-590.
18. Hill, A. R. (1973) : Erosion of river banks composed of glacial till near Belfast, Northern Ireland. Zeitschrift für Geomorphologie, 17, 428-442.
19. Gardiner, T. (1983) : Some factors promoting channel bank erosion, River Lagan, County Down, Journal of Earth Science Royal Dublin Society, 5, 231-239.
20. Evans, R. (1990) : Water erosion in British farmers' fields - some causes, impacts, predictions, Progress in Physical Geography, 14, 199-219.
21. Walling, D. E. and Webb, B. W. (1981) : Water quality, in Lewin, J. (ed.) British Rivers, Allen and Unwin, London, 126-169.
22. Fullen, M. A. (1985) : Compaction, hydrological processes and soil erosion on loamy sands in East Shropshire, England, Soil and Tillage Research, 6, 17-29.
23. Newson, M. D. (1980) : The erosion of drainage ditches and its effect on bed-load yields in mid-Wales : reconnaissance case studies, Earth Surface Processes, 5, 275-290.
24. Leeks, G. J. L. (1992) : Impact of plantation forestry on sediment transport processes, In Billi, P., Hey, R. D., Thorne, C. R. and Tacconi, P. (eds), Dynamics of Gravel Bed Rivers, John Wiley and Sons, Chichester, 641-660.
25. Leopold, L. B. and Dunne, T. (1971) : Field methods for hillslope description, British Geomorphological Research Group Technical Bulletin, 7.

Farm Land Erosion: In Temperate Plains Environment and Hills
S. Wicherek (Editor)
1993 Elsevier Science Publishers B.V.

Use of Caesium-137 in Study of Pedogeomorphic Processes

M.Lehotsky, M. Stankoviansky, V.Linkes

Institute of geography, Slovak Academy of Sciences, Stefanikova 49, 81473 Bratislava, Czechoslovakia.

SUMMARY

Under the conditions of the socialist agricultural activity man's interaction with geosystems displays several specifities, e.g. very large arable land bloks, along with the irrational tactics represented by an agricultural crops production plan dictated "from above" that did not allow th implement the measurer of antierosion and antidegradation soil protection conceived in form of rational crop rotation, of differetiated use of slope section, tolerable fertilization, etc.

In our contribution we are presenting 4 case studies in different landscape types of the investigation of intensity of the erosion-accumulation process and soil degradation state under above mentioned conditions.

For an estimation of the soil loss as well as soil gain, i.e. the intensity of erosion-accumulation process during the period of the last 35 years the content of the radionuclide of Cs 137 in the soil has been investigated. For the purpose of examining the behaviour of degradation processes of the upper solum portion the granulometric composition, humus content, pH value, total exchange capacity of the soil were examined. The localization of sample points was bases on the concepts of elementaryu morphologic units, main morphologic lines and points and their dynamic choric composition.

The result is a delimitation of morpho-pedo units displaying various combinations of denudational, soil-erosion and accumulation processes defined by both the amount (t/ha/year) of soil loss as well as soil gain and specific values of above mentioned soil properties.

RESUME

En raison des types d'activités agricoles qui ont été conduites dans les régions socialistes, des stratégies peu rationnelles choisies, la dégradation des terres est considérable et peu de mesures contre l'érosion ont été prises.

Dans notre communication nous présentons quatre études, correspondant à différents types de milieux, concernant l'intensité du processus d'érosion-accumulation et la dégradation du sol dans les dites conditions.

En vue d'évaluer les pertes de sol ainsi que les gains, c'est-à-dire l'intensité du processus d'érosion-accumulation pendant les dernières 35 années, une étude de la teneur en radionucléide Cs 137 dans le sol a été envisagée. Quant aux processus de dégradation de la partie superficielle du sol, un examen de la composition granulométrique, la teneur en humus, la valeur du pH et de la capacité d'échange totale a été pratiqué. Le choix des points d'échantillonnage était fondé sur des concepts d'unités morphologiques élémentaires.

Le résultat est une délimitation des unités morpho-pédologiques qui mettent en évidence différentes combinaisons de processus d'érosion du sol et d'accumulation, définis par la quantité de perte de sol (t/ha/année), les gains de sol et les valeurs spécifiques des dites propriétés du sol.

INTRODUCTION

During the last four decades an increased study interest in exogenic processes in the geomorphology was aroused. From the broad spectrum of geomorphic processes we shall pay a special attention to the set of erosion-accumulation processes modifying pedo-geomorphic landscape component, i.e. water erosion-accumulation processes. Here belongs the sum of modelling effects of rain and melted snow water flowing down the slope that causes creation of new forms (rills, gullies), thinning of the soil profile, loss of fine soil particles, changes of soil texture and structure and its water regime, fertility decrease, stream and river pollution, water reservoir silting, etc.

For the evaluation of the intensity of erosion accumulation processes we considered useful to use such a method that facilitates determination of the real values of the erosion and accumulation components of the process in the definite time period with respect not only to the activities of rain and snow melting water, but also the beginning of the collectivisation of agricultural land in CSFR. For this purpose the most suitable method seemed the one based on the interpretation of distribution of Cs-137 isotope in the soil.

This method was enriched by the definition of relation between the Cs137 content and other, above all agronomical properties of soil, namely the pH value, humus contents, particle size distribution, total exchange capacity. As examples we presented two model areas-model : Horné Srnie and Bzince.

"Cs-137" METHOD

Caesium-137 represents a by-product of atmospheric nuclear weapons testing and in the European conditions Chernobyl accident. When it reaches the Earth's surface it is adsorbed by clay particles and converted to almost irreversible form. Pioneer approach utilizing Cs-137 as a tracer at the identification and research of soil erosion is represented by works of J.C. Ritchie and J.R. McHenry at the beginning of the seventies (1). Since that time the use of Cs-137 method spread mainly in USA and Canada (2 - 3 - 4), in Australia (5 - 6), in Great Britain (D. E. Walling), in Poland (W. Froehlich), in Italy (7). The first experiments with Cs-137 application in the research of soil erosion were registered in the CIS (8) and Belgium (9). All these works have more or less common aim, but they differ in approaches and utilization of other, supporting methods, which enable more detailed understanding and balance of soil redistribution in model areas. Simultaneously we may observe deviation of the research of slope profiles - catenas to the research of catchments.

FIELD WORK AND LABORATORY METHODS

Field work in two model areas (Horné Srnie-basin, Bzince-slope system) consisted of detailed pedo-geomorphic mapping and subsequent localization of sampling localities and of the proper collection of soil samples. The pedo-geomorphic mapping was based on the analysis of the main morphological lines (valley and ridge lines), main morphological points (contact of valley lines with watersheds), analysis of slope conditions and their properties (vertical and horizontal concavity and convexity) which enable the identification of quasi homogeneous morphological units. Determination of erosive or accumulative soil forms formed a part of work as well. The identification of complex pedo-geomorphic units enables to determine main sources, directions of water and soil movement as well as the identification of suitable localities for sample collection.

These were localized in places with supposed maximum accumulation eventually erosion.

Collection of samples was realized by means of metal cylinder with the diameter of 83 mm namely in soil depth of 0-25 cm and deeper in each 5 cm interval. Total collecting depth was determined according to empirical experience up to soil horizons where there it is not possible to detect Cs-137. So the thickness of solum influenced by water erosion-accumulation processes in positive sense (accumulation) or the negative one (erosion) was identified. Soil sample of one locality represents a sum of cylinder content, driving into each above mentioned depth in 6 points (always 3 and 3 together in the distance of approximately 10 - 15 m). Soil samples were dried either by air, or in the oven. After drying and sieving (<2 mm) samples were analyzed in laboratory. Contents of Cs-137 and Cs-134 in the samples were detected by semiconductor gammaspectrometer system of the firm SILENA working in 4096 regime channel spectra. The evaluation part of the system consisted of a computer PC/AT Olivetti M 290 and evaluation program of gama spectrum "SILENA-M". With regard to undetectable content of Cs-134 in soil samples it was not possible to differentiate the influence of Chernobyl fallout, and that influenced also the method of evaluation. Analysis of other soil properties was realized by current methods used in soil sciences.

MODEL AREA THE HORNE SRNIE

The area under study (scheme 1) is small flowless basin situated on the right side of the Vlára valley. It has a prolonged shape in the direction of inclination. Its upper part presents an expressive amphitheater. The middle part of it is assymetrical as to the inclination and the height (left slope is lower and less inclined) ; the assymetry diminishes in lower part. The basin ends by forested gully. The bottom of dell-like system was in its upper and middle part remodelled by earth-flow. Mixed arable farming is a predominant way of the land-use. There were 4 localities evaluated in this study area. Soils are Dystric Planosol type.

1. Top-ridge infiltration type with non-eroded soil form with predominance of vertical movement of water and mass
1.1. Top-ridge type with flat cross-profile
1.2. Ridge, slightly convex-convex type with accesoric tendency of acceleration and dispersion of water and mass movement
2. Slope type with predominance of horizontal and gravitational water and mass movement
2.1. Slope, erosive-transitory convex-convex type with eroded soil form with tendency of strong acceleration and dispersion of water and mass movement
2.1.1. inclination 10 - 12°
2.1.2. inclination 8 - 9°
2.1.3. inclination 3 - 5°
2.2. Slope dell-like convex-concave, accumulation-transitory type with slightly accumulated soil form, with tendency of relative less strong acceleration and concentration of water and mass movement with inclination about 5 - 8°
3. Bottom type

3.1. Dell-like concave-concave, transit-accumulation type with accumulated soil form with tendency of strong concentration and retardation of water and mass movement with inclination about 4°
3.2. Earth-flow-like, in cross-profile convex-convex transit-accumulation type with different soil forms (from strongly accumulated in depresssions to eroded on edges), with tendency of complicated water and mass movement

Scheme n° 1.
Pedo-geomorphoic types with Dystric Planosols (the model area Horné Srnie)

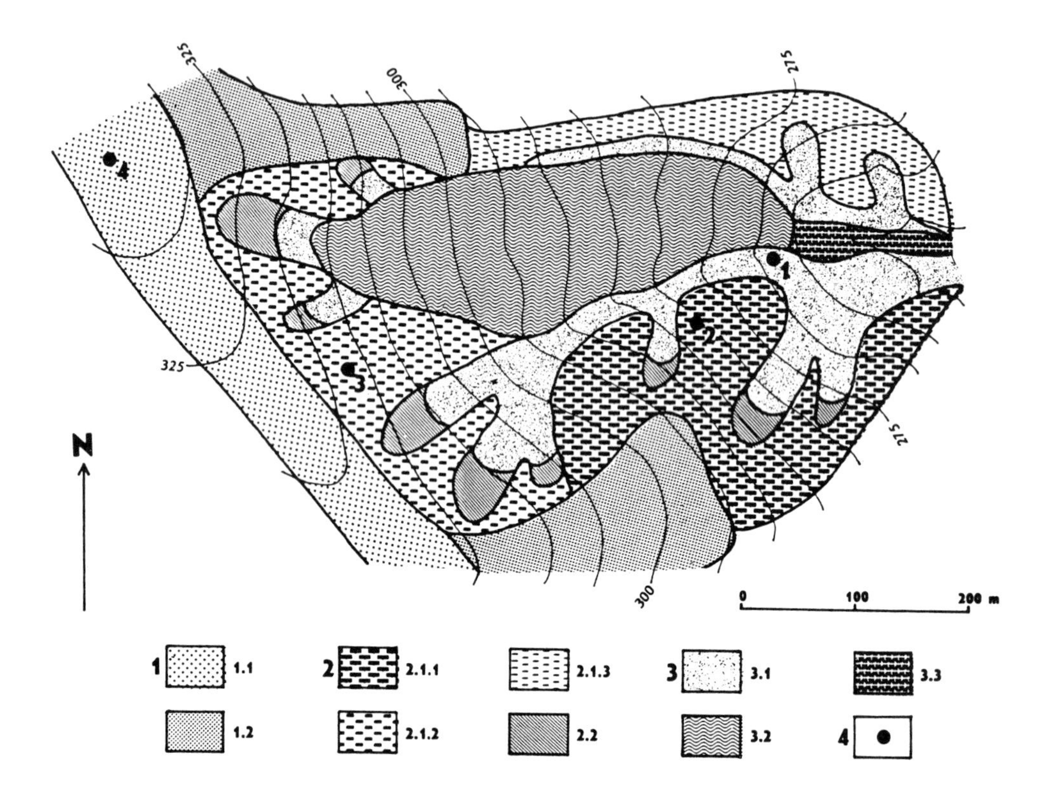

In the referred locality there was the Cs-137 content identifiable to 35 cm depth (Cs-137 content in 0-20 cm depth is 12.3 Bq.kg-1), in localities n° 2 and 3 it was up to the depth of 30 cm. Comparing with the reference locality it indicates a loss of 5 cm soil profile. In average it means 1.4 mm/year and corresponds to about 18-21 ton/ha/year. But from the standpoint of percentual evaluating both localities (n° 2. and 3.) show diverse characteristics. As they are situated in the slope parts of basin there occur not only losses, but the gains of soil substance bound with Cs-137 .

But the gains coming from the upper portions of slopes are identifiable only to the depth of 30 cm. Gain of mentioned material from above at the locality n° 3 up to the depth 25 cm represents 30 % of the corresponding depth of the reference locality, up to 30 cm depth it is 16 % (see graph n° 1). At the locality n° 2 in the same depths it is 4 % or 12 %. In the locality n° 1 Cs-137 is identifiable up to 45 cm depth, what represents an accumulation compared to the reference locality of about 10 cm, this is average 2.8 mm/year, or 36-42 ton/ha/year. From the viewpoint of percentual expression of gain of soil matter bound with Cs-137 as compared to the reference locality we may state the increment of its loss up to the depth of 25 cm about 68 %, to 30 cm about 13 % and to the depth 35 cm about 240 %. Pronounced increase is the result of a depressive, accumulation position in the lowest point of the basin.

Graph. 1. Courbe.

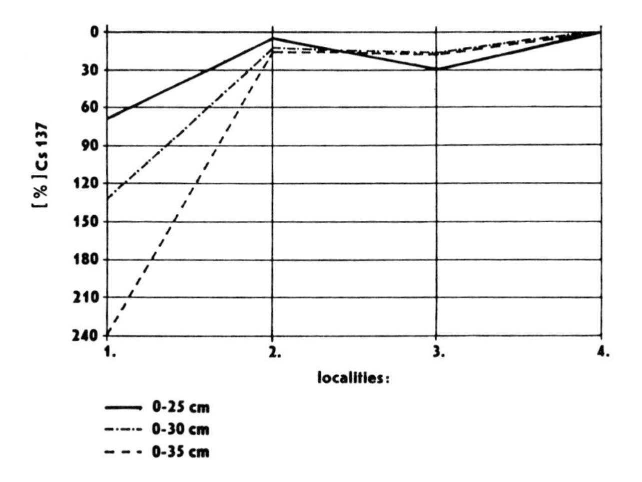

Increased contents of Cs-137 in the localities n° 2 and 3. in the depth up to 25 cm show that these indicate slope position EC (transit-gain in the sense of the mode by Campbell et al. 10), with permanent gain of material from upper portions of slopes. As far as humus content in erosion localities is concerned in the layer 0-25 cm in locality n° 2, it decreased to 31.7 % and to 28,9 % in locality n° 3. Similar course is that of pH values when in locality n° 2 it decreased to 73,9 % and in locality n° 3 to 91,9 %. The accumulation locality shows according to the contents of Cs-137 higher values then the reference one, the humus contents and pH values are lower, though in the upper layers of solum (to 30 cm) only. In deeper layers occurs an increase of all studied properties and that proves an intensive material outwash in the last period. Hence we can see that in the basin the rill and sheet erosion dominate among erosion processes, when into accumulation types enters only a material from surface parts of erosive sections, relatively richer in Cs-137, humus and pH values. The erosive soil forms in comparison with accumulative ones do not show any visible change of the particle size distribution in this study area with Dystric Planosol.

THE MODEL AREA BZINCE

Flat summit-slope-alluvial toposequency is 530 m long with relative height of 29 m (Scheme 2). It is utilized as arable land. Below the foot of slope is possible to identify in the cross-section terrain wave cca 0,5 m high which differentiates underslope system to the proper (younger) system on one side and the allochthonous dell-like part (older) on the other. Five localities were evaluated. Soil type in the slope is Orthic Luvisol. As compared to the Horné Srnie model area we may state that Cs-137 contents in all localities and all sampling depths univocally confirm the hypothetical assumption of its distribution on the erosion-accumulation toposequence. The contents of Cs-137 in the reference locality is 15.6 Bq.kg-1 in the depth 0-25 cm.

Scheme n° 2.
Pedo-geomorphic types with Orthic Luvisols and Eutric Fluvisols (the model area Bzince)

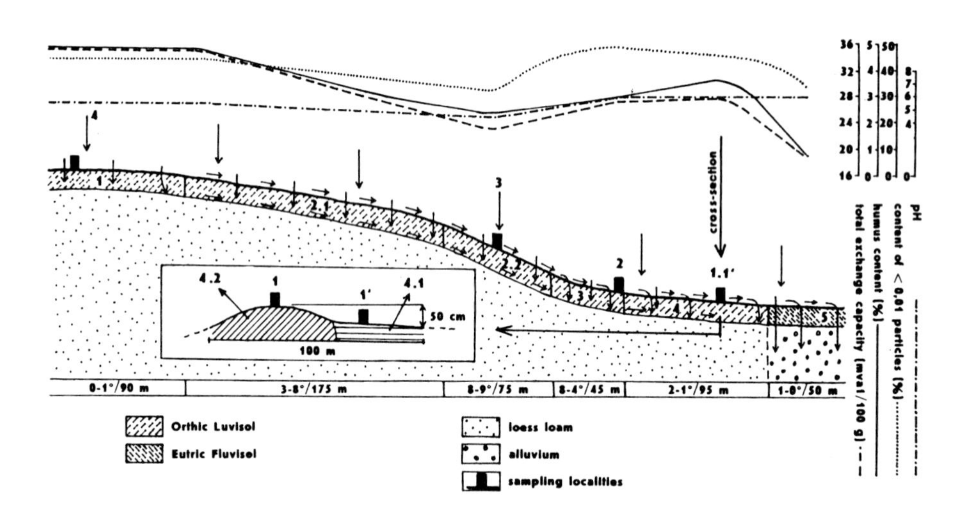

1. Top-ridge infiltration type with non-eroded soil form with predominance of vertical water and mass movement
2. Slope type with predominance of horizontal and gravitational water and mass movement
2.1. Erosive-initiatory type with rectilinear slope and initiation of acceleration of water and mass movement and with slightly eroded soil form
2.2. Erosive-transitory type with rectilinear slope an tendency of strong acceleration of water and mass movement and with strongly eroded soil form
3. Under-slope colluvial accumulation type with accumulated soil form
4. Colluvial accumulation type with convex-convex profile and accumulated and overlayed soil form

4.1. Autochthon-colluvial under-slope type
4.2. Allochthon-colluvial-dell-like type
5. Accumulated fluvial-proluvial type with combined development

Its decrease is presented in the graph N° 2. With regards to the maximum depths where we detected caesium, this is in the erosive-transitory part of catena about 30 cm, in the underslope-colluvial part it is about 45-50 cm. Hence it means that the rate of erosion-accumulation process on toposequence is cca 4.2 mm/year, i.e. 6.0-6.6 t/ha/year. The contents of Cs-137 in localities n° 1 and 1' simultaneously confirm the age diversity and different development of the landscape with hardly determinable landforms. Contrary to the Horné Srnie model area, agro-pedological properties in the landscape type with Orthic Luvisol quasi comfirm the course of values similar to the values of Cs-137. The values of mutually related soil properties (humus contents and T values) show common course. Both properties reach maximum values in the top-ridge type, minimum ones in the erosive-transitory type (humus 56 %, T value 70 %, as compared to the top type). They reach second maxima in the under-slope autochthonous colluvial part. The contents of particles <0,01 show a different course. They reach maximum values in underslope colluvial, expressively accumulation type. Mentioned values decrease in the erosive-transitory type to 75 % of maximum. pH values show similar course with maximum in autochthonous, colluvial, underslope accumulative type. Mentioned values decrease in erosive-transitory type to 74 % of maximum.

Graph. 2

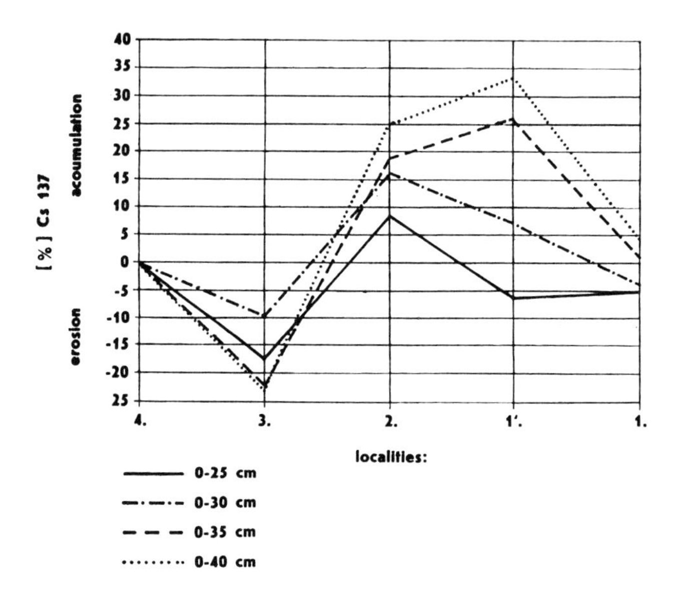

CONCLUSION

This work refers to the utilization of Cs-137 tracer at the identification of erosion-accumulation processes rate.

Preliminary results show that Cs-137 (134) is a suitable tracer for the research of named phenomenon. But in Central European conditions interpretation of the results becomes more complicated. Sometimes it is very difficult to distinguish the situation after the Chernobyl fallout from that before. It seems to be very adequate for comparison of the change of Cs-137 (134) with the change of another, especially agronomical properties of soils under simultaneous elaboration of microgeomorphological mapping.

REFERENCES

1. McHenry, J.R., Ritchie, J.C. (1975) : Redistribution of Cs-137 in southeastern watersheds. In : Howell, F.G., Gentry, J.B., Smith, M.H. (eds.) : Mineral Cycling in Southeastern Ecosystems. Technical Information Center, USA, pp. 452-461.
2. Sutherland, R.A., De Jong, E. (1990) : Quantification of soil redistribution in cultivated fields using caesium-137, Outlook, Saskatchewan, Catena Suppl. 17, pp. 177-193.
3. Sutherland, R. A., De Jong, E. (1990) : Estimation of sediment redistribution within agricultural fields using casium-137, Crystal Springs, Saskatchewan, Canada. Applied Geography, 10, pp. 205-221.
4. Martz, L.W., De Jong, E. (1989) : Natural radionuclides in the soils of small agricultural basin in the Canadian prairies and their association with topography, soil properties and erosion. Catena, vol. 17, pp. 85-96.
5. Loughran, R.J., Campblell, B.L., Walling, D.E. : Soil erosion and sedimentation indicated by caesium-137 : Jackmoor Brook catchment, Devon, England. Catena, vol. 14, pp. 201-212.
6. Elliot, G.L. and al. (1984) : Correlation of erosion and erodibility assessments using Cs-137. Journal of Soil Conservation, 40, 1, pp. 24-29.
7. Bazzofi, P. and al. (1986) : Quantitative evaluation of some processes in experimental areas with different agroclimatic conditions in central Italy. Z. Geomorph. Suppl. Bd. 60, pp. 131-148.
8. Ostrova, I.V. and al. (1990) : Ocenka intensivnosti erozionno-akumulativnych procesov po soderæaniju v poçve cezija-137. Vest, Mosk. Univ., se. Geograf. 5, pp. 79-85.
9. Vanden Berghe, I., Guling, H. (1987) : Fallout Cs-137 as a tracer for soil mobility in the landscape framework of the Belgian loamy region. Pedologie, XXXVII-1, pp. 5-20.
10. Campbell, B.L. and al. (1982) : Caesium-137 as an indicator of geomorphic processes in a drainage basin system. Austral. Geograf. Stud. 20, pp. 49-64.
Kochanoski, R.G. (1987) : Comparison of measured soil Cs-137 losses and erosion rates. Canadian Journal of Soil Science, 67, 1, pp. 199-203.
Rogowski, A.S., Tamura, T. (1970) : Erosional behavior of caesium-137. Helth Physics, 18, pp. 467-477.

Farm Land Erosion: In Temperate Plains Environment and Hills
S. Wicherek (Editor)
1993 Elsevier Science Publishers B.V.

Impact of Modelling in furrow irrigation erosion

Z. Popova and I. Petrova

Institute for Soil Science and Agroecology "N. Poushkarov", 7 Shosse Bankya Str., 1080 Sofia, Bulgaria.

SUMMARY

Surface irrigation is a wide spread practice in Bulgaria. Removal of pollutants and nutrients from irrigated agricultural areas is an unwelcome consequence of this method.

Our study on furrow irrigation erosion is based on two approaches - mathematical modelling and field experiments. As a result of their complex application, quantitative relationships between irrigation erosion indexes (such as quantity of solid and liquid runoff, removal of pollutants, nutrients and humus) on one hand, and furrow irrigation parameters (such as uniformity of water distribution, water permeability of soil surface and application rate) on the other, were established.

Results show that when uniformity of water distribution is optimal, there is from 7 to 26 % runoff of the applied water and together with it from 0,5 to 1,7 t of soil/ha, from 15 to 60 kg of humus/ha, from 1,2 to 5,4 kg of nitrogen/ha and from 2,4 to 6 kg of potassium/ha are removed from silty clay loam soils.

In that case deep percolation is from 4 to 20 % and still from 3 to 4,3 kg of nitrogen/ha are leached.These quantities increase from 30-40 % on clay soils and decrease respectively with the same percentage on sandy loam soils.

To limit these losses, environmentally-sape irrigation methods should be applied such as : surge irrigation, irrigation with runoff flowing to the next set, cut back and reuse systems, which minimize deep percolation and runoff and lead to higher uniformity of water distribution.

RESUME

L'irrigation de surface est pratique courante en Bulgarie. Le déplacement du sol et des substances nutritives dans la région agricole irriguée par sillons résultent de l'impact érosif de l'écoulement de l'eau dans ces derniers. Notre étude sur l'érosion due à l'irrigation par sillons se fonde sur deux approches : la modélisation mathématique et expérimentations sur le terrain. Les quantités mobilisées sont de l'ordre de 0,4 à 1000 kg de sol/ha ; 20 à 40 kg d'humus/ha ; 1,2 à 3 kg d'azote/ha ; 0,6 à 2 kg de nitrates/ha. Le volume de ruissellement atteint 10 à 30 %.

Plusieurs méthodes visant à réduire l'érosion sont envisagées.

INTRODUCTION

Furrow irrigation is a major practice in Bulgaria - it is used on about 60 % of all areas under irrigation. A basic disadvantage of this method are the damages caused to water and soil resources - pollution of surface and ground waters through the soil, humus and nutrients removed by runoff, and pollutants leached by deep percolation. The objective of this study is to apply a model for a quantitative evaluation of the volume of deep percolation and runoff in furrow irrigation (1) for quantitative and qualitative characteristics of furrow irrigation erosion.

MATERIALS AND METHODS

The objective of this study is reached by establishing relationships between the erosion indexes - removal of soil, humus, phosphorus, potassium, etc., and furrow irrigation indexes - relative extention of application time - I, uniformity of water distribution, permeability of soil surface, application rate. Two approaches have been applied - mathematical modelling and field experiments. The experiments were carried out on an area of 14 ha situated near Sofia for production of corn and lysimeters. The soil is deluvial silty clay loam, with average value indexes for the root zone : dry bulk density = 1,36 gm/cc, field capacity - FC = 28 % by weight. The furrow slope was from 2 to 2,5 %, according to the size of the delivered stream - about 0,9 - 1 l/s. Pre-irrigation soil moisture was purposely modelled to range from 60 to 93 % of FC. One, two or three flows per season were delivered respectively on four variants (sets of furrows). For the different cases, water permeability of the soil surface covered practically the whole possible scope - its value, defined as depth of water taken in by the soil for the first hour - K varied from 8 to 46 cm/1st h during the experiments.

A practical furrow irrigation factor - relative extention of application time (I) - was used as an index for the uniformity of water distribution in the experiments. It was defined as a ratio between the additional application time - tad after the stream reached the end of ad the furrow for the time t_1 - and t_1 itself.

$$I = t_{ad} / t_1 \qquad /1/$$

When the parameter 1 (Form. 1) was valued 1 > 0, a flowing away from the end of the furrow began (Fig.1). In that case if the values of the furrow irrigation elements were suitably chosen (size of stream length of furrow), deep percolation occurred at the head of the furrow. When the index I increased together with the runoff and deep percolation, it improved uniformity of water distribution along the furrow length and resulted in higher yield.

In the American irrigation practice, a traditional rule for uniformity is the so called "one quarter time" (2) where the time t (Form. 1) is 1/4 of the application time, i.e. 1 = 3. According to Bulgarian sources (3 - 1) from an economic point of view and for the conditions prevalent in our country, optimal values of I vary from 0,4 to 1.6 for basic furrow irrigated crops.

To make a full balance of water and nutrients on the experimental plot, several experimental methods have been applied. The volume of the delivered water, the distribution of the in-taken water and runoff are defined according to the "inflow-outflow method" (4 - 2) where during the application time, the delivered stream - qd and the runoff stream - qr are measured for every furrow (Fig.1). The volume of the deep percolation for each set of simultaneously irrigated furrows is defined on the basis of the distribution of the water taken in by the soil over the area and the conception for FC of the root zone. The removal and leaching of the pollutants and soil nutrients from the irrigation water is established on the basis of the content of the delivered water, the water infiltrated under the root zone and the solid and the liquid phase of the surface runoff at the end of the plot. Here the method of multi- chemical analysis has been applied. Waters drained under the root zone (below 1,40 m) are examined in lysimeters with the undisturbed structure of the soil and area size of 1,50/4,00 m in conditions identical to those of the experimental plot.

The second approach, applied in this study and adding to the results of the experiments, is mathematical modelling. The simulation is made in relative quantities under a model (1) based on large-scale mathematical description of the spatial distribution of water delivered and its retention in the root zone. The volume of deep percolation and runoff is calculated as a function of the parameters of the model : relative extention of application time - 1 (Form. 1), relative depth of application, coefficient of nonuniformity of stream advance over the surface and indexes of water in-take rate in furrow irrigation. Interface between field data and simulation has been done through parametrization of the model for a variety of conditions. Interpolation and extrapolation of the particular results from the experiments have been possible by simulation.

Figure 1. Changes in available nitrogen content in the runoff dependent on the relative extention of application time 1 (Form.1).

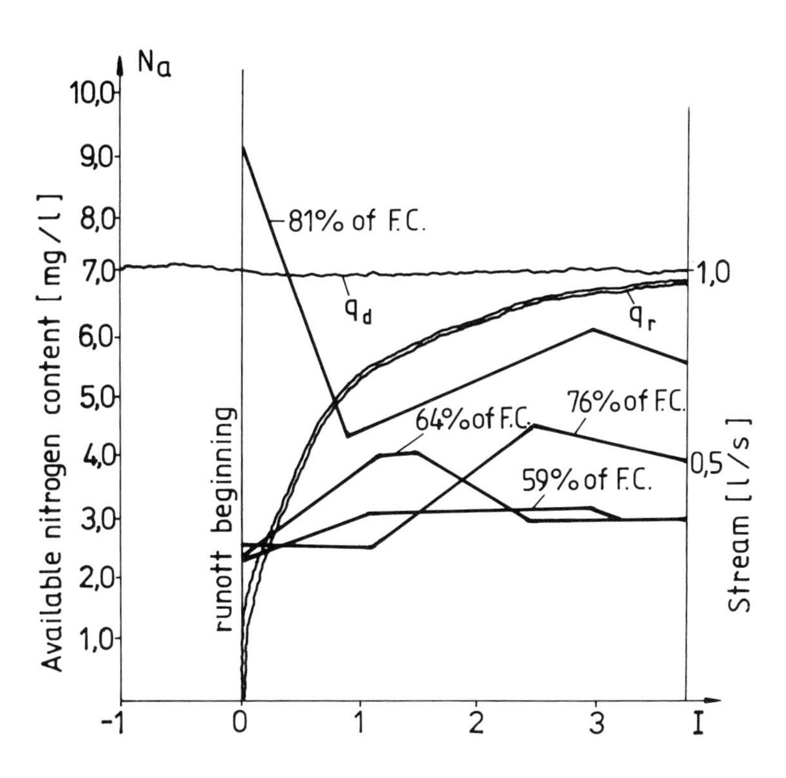

RESULTS

Simulation after the mathematical model shows that within the limits of the economically optimal for Bulgarian conditions uniformity (0,4<1<1,6) and a suitable furrow length, surface runoff is from 7 to 26 % and deep percolation is from 4 to 20% of the delivered volume of water (Fig. 1). The shaded zones, which show the range of the relationships, give the impact of non-uniform stream advance in the simultaneously irrigated non-homogeneous furrows.

The data received after the multi chemical analyses characterize the content of pollutants (nitrogen, phosphorus and soil) and nutrients (potassium and humus) in the surface runoff, and deep percolation under the root zone. Runoff enrichment is comparatively unchangeable - from 2 to 3 g/l - for the concrete experiments, notwithstanding the contrastive conditions of the variants. The removal of humus, total nitrogen and phosphorus with the solid runoff is considerable - correspondingly 3.6, 0.22 and 0.20 % from it.

Analyses of runoff show that only available nitrogen (N_a) concentration varies in time and is considerably influenced by the pre-irrigation level of soil moisture (Fig.1). In most cases an average concentration of available nitrogen Na - between 3 and 6 mg/l - can be accepted with sufficient accuracy, whereas its content in the irrigation water is invariably 1.7 mg/l. It is established that the seasonal flow influences the potassium content in the runoff stream : for the first flow it is l0 mg/l against 3.5 - 4.5 mg/l for the next flows, while concentration of the other elements is comparatively unchanged during the irrigation season - from I to 2 mg/l nitrates, 0.2 mg/l phosphorus.

The analyses of lyzimetric waters show high nitrates content - 47 mg/l - which does not vary significantly during the infiltration in the irrigation season. The content of phosphorus is minimal.

As a complex result of the field experiments, simulations and chemical analyses, many concrete relationships have been established : between the absolute values of the surface runoff and deep percolation, pollutants and nutrients removed by them on one hand, and the indexes of uniformity of water distribution - I (form. 1), water permeability of the soil surface - K and depth of application on the other (Fig. 3, 4).

To illustrate these two contrastive cases concerning water permeability of the soil surface, the values K_1 = 46 cm/h and K_1 = 13 cm/h (shaded in different manners) have been chosen. The appropriate furrow length (80 and 850 m for the particular conditions of the experiments) suits the prerequisite that if I = 0 (form.1) at the head of the furrow, the depth of water taken-in by the soil is equal to the required depth of application - m_r (respectively 1050 and 650 m^3/ha), which cancels the deficit to FC of the root zone. On Fig. 3 the thick line shows the increase of the runoff stream dependent on the extention of I (form. 1) for the particular case of the experiments (FC = 28 %). For the cases of these particular experiments Fig.3 b shows pollutants (nitrogen, phosphorus and soil) removed by surface runoff, while Fig.3 c illustrates nutrients losses (mainly potassium) when irrigation water distribution uniformity improves and relative extention of application time increases. Changes in the volume of deep percolation and nitrogen leached by it under the root zone with achievement of better uniformity of water distribution (1) is shown on Fig. 4.

DISCUSSION

Regardless of the fact that deep percolation and runoff are merely a definite part (%) of the delivered volume of water, when indexes of uniformity of water distribution are fixed (fig.2), their absolute quantity to a unit of area varies with water permeability K_1 and water retention characteristics of the soil (Fig. 3a, 4a). Considerable damage is caused to water and soil resources when soil surface permeability is high (as in the case K_1 = 46 cm/1st 1 h) - Fig. 3, 4. Where silty clay loams of FC = 28 % are concerned (the relationships shown with thick lines), the rate of surface runoff for a single flow is from 150 to 600 m3/ha when the uniformity of water distribution is economically optimal (i.e. 0.4 < 1< 1.6).

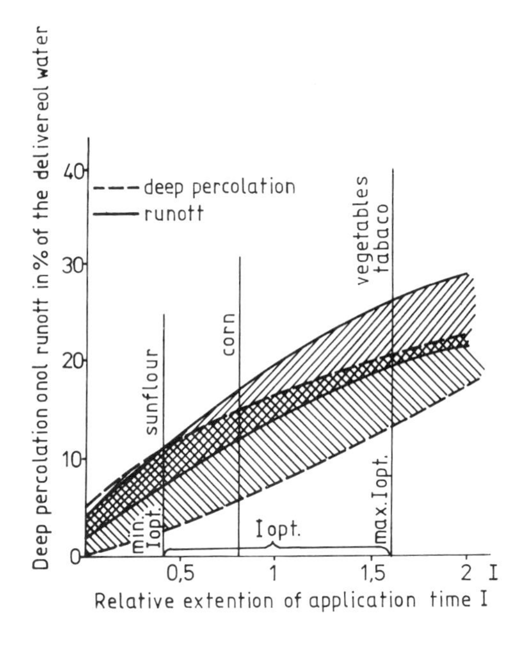

Figure 2. Deep percolation in percentage of applied water calculated as a function of the relative extention of application time 1.

Pollutants removed by runoff in this case are : total nitrogen from 1.2 to 5.4 kg/ha, total phosphorus from 0.7 to 3.5 kg/ha and soil from 0.5 to 1.7 t/ha (Fig. 3b). Potassium losses are from 2.4 to 6 kg/ha (Fig. 3c) and the loss of humus is from 15 to 60 kg/ha. The losses caused by deep percolation should be added to these : with water infiltrated under the root zone (from 10 to 80 m^3/ha), the rate of removed nitrogen is from 3 to 4.3 kg/ha (Fig.4). If these quantities are refered to the rate of fertilization,it means that about 5 % of the active nitrogen, about 3.5 % of the phosphorus and about 6 % of the potassium are removed by deep percolation and runoff for a single flow. If the climatic year has been one of "moderate" precipitation, which for the conditions of these experiments means an irrigation rate of 104 mm

Figure 3. Runoff rate characteristics dependent on furrow irrigation indexes
a) Runoff water losses dependent on the relative extension of application time 1 for different depths of application - m, field capacity FC and permeability of soil surface K1.
b) Rate of removed pollutants (nitrogen, phosphorus and soil) dependent on the relative extension of application time for sitty clay loams.
c) Rate of nutrients losses (potassium) dependent on the relative extention of application time for silty clay loams.

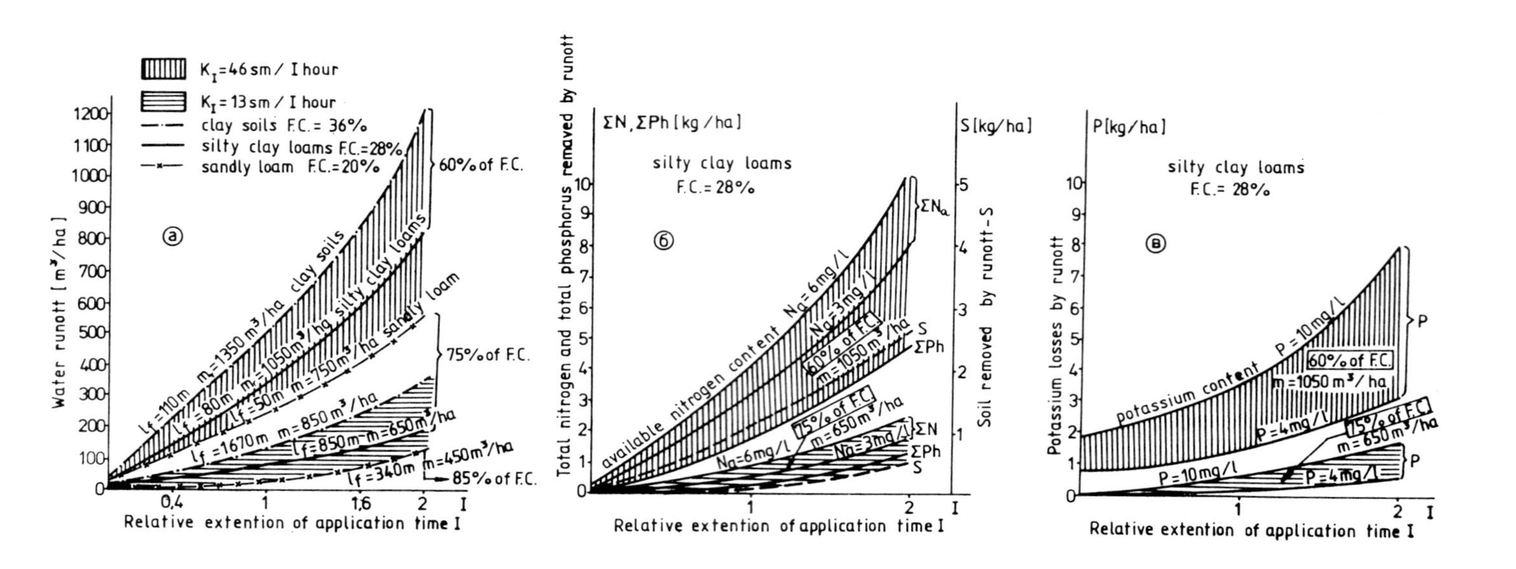

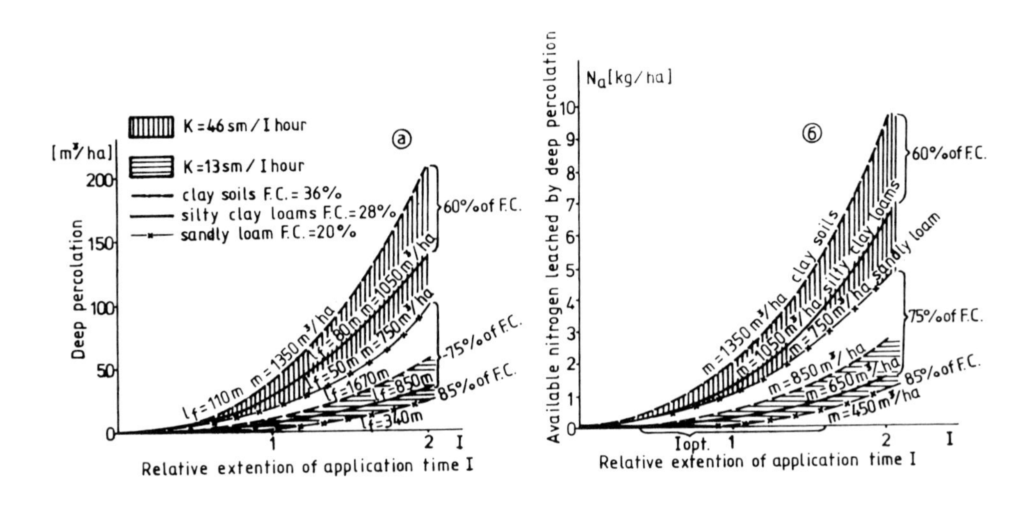

Figure 4. Deep percolation rate characteristics dependent on furrow irrigation indexes and soil type.
a) Deep percolation water losses dependent on the relative extention of application time I for different rates of application - m, field capacity FC and permeability of soil surface K 1.
b) Rate of leached pollutants (nitrogen) by deep percolation dependent on the relative extention of application time 1 for different depths of application - m, field capacity FC and permeability of soil surface K 1.

per year (5), the numbers given above stand for seasonal losses resulting from the irrigation erosion, but when it is a "dry" climatic year (with 6 % probability of irrigation rate) the erosion indexes are two times higher. It should be pointed out that the given rates of losses and damage on water and soil resources (Fig.3, 4) hold true for an ideal case in furrow irrigation practice when stream size and furrow length are suitably chosen, when furrow slope is moderate and flow duration is optimal. That is why in general irrigation practice their values are expected to be much greater.

The next step in the analysis, directed to the generalization of the achieved results of Fig.3, is a discussion of the possibilities for reporting on the geometrical length of an irrigated area. Some revision of the results from Fig. 3 would be necessary when the area under irrigation is irrigated in several sets of furrows (insteas of only one), situated one under the other. In that case we have runoff coming only from the lowest set of furrows, and the accounts from Fig. 3 should refer to the entire area of the irrigation plot by dividing them to the number of the sets in it.

It is interesting that soil type influences the rate of erosion characteristics of the areas under irrigation. When soil conditions change, the quantities of deep percolation and runoff per flow increase from 30 to 40 % on clay soils (FC = 36 %), and decrease respectively with the same percentage on sandy loam soils (FC = 20%) - Fig. 3a, 4a. The increased volume of the deep percolation and runoff on soils of heavier texture presupposes higher quantities of removed substances per flow. Erosion evaluation in these cases should be based on the respective quantities of water (taken from the relationships shown in dotted lines), and the particular content of these waters. Climatic impact should be taken into account by the required number of flows per year as well.

CONCLUSIONS

There are several possibilities of limiting the negative influence of surface irrigation on soil and water natural resources :

1. Irrigation with relatively low water permeability of the soil surface (K_1 should not surpass 10 - 15 cm/1st h) so that the volume of deep percolation and runoff is decreased. It is advisable that soil moisture level is kept over 75 % of FC for loams and over 85 % of FC for clay soils (Fig. 3, 4) during the whole irrigation season through frequent flows.
2. Application of contemporary environmentally-oriented methods of irrigation to minimize deep percolation and runoff. One such method also insuring high uniformity of water distribution of the irrigation water, is surge irrigation - by delivering irrigation water in portions (6). Eliminating surface runoff is possible through cutback irrigation, through irrigation with the runoff flowing to the next set (Varlev, 1975) or reuse systems.

REFERENCES

1. Popova, Z., (1991) : Optimization of Uniformity and Efficiency in Applying Water for Furrow Irrigation, ICID Bulletin, vol. 40, n° 2, pp. 45 - 55.

2. Merriam, J.L., Shearer, M.N., Burt, C.M. (1983) : Evaluation on Irrigation Systems and Practices ; Design and Operation of Form Irrigation on System - ASA, Monograph n° 3, 721 -748.
3. Varlev, I., (1988) : Optimizing the Uniformity of Irrigation and Fertilization, Elsevier Science Publishers, Agricultural Water Management, n° 13, pp. 285 - 196.
4. Cridle, W.D., J. Davis, (1969) : Method for Evaluating Irrigation Systems, Agriculture Handbook n° 82, USDA, Washington.
5. Banov, J., (1988) : Evapotranspiration and Yield of Corn under Irrigation in Different Soil Conditions, PhD dissertation ; "N.Poushkarov" Institute of Soil Science and Agroecology, Sofia.
6. Varlev, I., (1975) : Improvement in Uniformity of Water Distribution in Surface Irrigation, C.K. Academy of Agriculture, Sofia, vol. 8, n° 1, pp. 79 - 84.

Farm Land Erosion: In Temperate Plains Environment and Hills
S. Wicherek (Editor)

Assessing recent rates of soil loss from areas of arable cultivation in the UK

T.A. Quine and D.E. Walling

Dept of Geography, University of Exeter, Amory Building, Exeter, EX4 4RJ, Devon, United Kingdom.

SUMMARY

The growing evidence for soil erosion on agricultural land in the UK has emphasized the need for quantitative assessments of contemporary rates of soil loss. Faced with this need, the authors have employed the caesium-137 technique to investigate rates and patterns of soil erosion on arable land in the UK. The results obtained for 13 fields on a range of soil types are presented and the erosion rates are compared with those obtained by other workers using different approaches. A good general agreement is evident. However, specific comparison of data from two fields reveals substantial differences between caesium-derived and survey-based estimates. Independent evidence is introduced which supports the caesium-derived erosion rate estimates. Finally, the implications of the erosion rate estimates, derived from caesium-137 measurements, are considered in relation to crop productivity, pollution of water bodies and the erosion processes involved.

RESUME

L'érosion des sols a été peu étudiée au R.U. Cependant, une intensification récente de l'agriculture a aggravé les risques affectant les terres arables. En raison de l'absence de données quantitatives sur les pertes en terre dans les secteurs évoqués, les auteurs ont utilisé la méthode du C 137 afin d'évaluer ces pertes sur différents types de sols. Les résultats obtenus pour les 13 sites étudiés indiquent des taux de départ variant de 2,2 à 12,2 t/ha/an et de 0,6 à 10,5 t/ha/an.

On peut estimer que les pertes en terre au cours des 30-35 dernières années ont été interprétées par 10 ou plus. Finalement les estimations des taux d'érosion obtenues par les mesures de Cesium 137 sont en relation avec la productivité végétale, la pollution des eaux et les processus d'érosion concernés.

INTRODUCTION

Faced with increasing concern for the pollution of streams by sediment derived from agricultural land and the greater frequency of visible evidence of soil erosion on cultivated fields, there is a clear need for reliable quantitative assessments of contemporary erosion rates on agricultural land in the UK. Geomorphologists addressing this problem have employed three main approaches ; firstly, detailed erosion plot studies (1 - 2), secondly, field measurement of visible erosion features (3 - 4), and thirdly, the use of fallout radionuclides as sediment tracers (5 - 6). Whilst data obtained using all three approaches have confirmed the existence of an erosion problem, the magnitude of the erosion rate estimates and therefore the resulting assessments of the environmental threat, differ according to the approach employed. This paper will present estimates of erosion rates derived using the caesium-137 technique for fields on a range of soil types in the UK. These data will be compared with rates estimated using other techniques and the significance of the differences will be considered. Independent evidence will be introduced to confirm the caesium-137 derived rates obtained in one study area where differences in the estimates associated with two approaches to erosion rate assessment were most marked. Finally the implications of the high caesium-137 derived rates will be considered.

THE CAESIUM-137 TECHNIQUE

Soil erosion investigations employing caesium-137 have now been undertaken in a wide range of environments and the value of radiocaesium as a sediment tracer has been clearly confirmed (7). The caesium-137 technique for assessing soil erosion rates has been discussed fully elsewhere (8) and will not be described in detail here. In essence, it is based upon assessment of the sediment-associated redistribution of weapons-test caesium-137, most of which was deposited at the ground surface by atmospheric fallout in association with precipitation between 1954 and 1980. In most environments, radiocaesium deposited on the ground surface is rapidly and strongly adsorbed by soil particles (9), and particularly by the finer fraction, and subsequent lateral redistribution takes place in association with eroded, transported and deposited sediment. In the absence of sediment redistribution, the caesium-137 inventory, or total activity per unit area at a sampling point can be expected to be equivalent to the total atmospheric input minus the loss associated with radioactive decay. Negative and positive deviations in caesium-137 inventory from this reference level or inventory may be attributed to the net effect of all processes of erosion and deposition operating during the period since the initiation of caesium-137 deposition (1954). Rates of erosion and deposition based upon caesium-137 measurements therefore represent long-term averages for the period from 1954 to the time of sampling. If the assumptions underlying the technique, which are discussed in detail by Walling and Quine (10), are accepted, then the major control upon the reliability of the erosion and aggradation rate estimates derived using the technique will be the selection of an appropriate calibration relationship. The calibration relationship is necessary to derive quantitative estimates of erosion and aggradation rates from values of the deviation of the caesium-137 inventory from the reference value. Such calibration may be based on either empirically-derived relationships between soil loss and erosion, or application of accounting models (11). In the absence of empirical data, the authors have used a numerical model which represents the movement of caesium-137 in association with eroding and aggrading soil (12 - 11). The results obtained from this model have been verified in a small catchment in Devon (UK) by comparing derived erosion rates with existing data based upon sediment yields and delivery ratios. Further validation of the calibration relationships derived from the model will be discussed below.

ASSESSMENT OF EROSION RATES ON AGRICULTURAL LAND USING CAESIUM-137

In order to assemble quantitative information on long-term erosion rates on agricultural land in the UK and to evaluate the potential of the caesium-137 technique in this environment, 13 fields on a range of soil types representing the major groups under arable cultivation were selected for investigation (Tab. 1).

The sites were chosen on the basis of known propensity to erode and most lay within the study areas of the MAFF/SSEW soil erosion survey (13 - 3). Erosion rates derived for the study fields therefore probably represent upper bound estimates for the soil types rather than average values.

The approach

The same basic approach was used for each field studied. The reference inventory for each locality was estimated by collecting several (ca 5-10) soil cores from a carefully selected site. This was usually undisturbed, uneroded grassland located as close to the study field as possible. Within the study fields a 20 or 25m grid was used to collect 38cm^2 diameter soil cores to a depth of 60 cm. All samples were returned to the laboratory where they were air-dried and sieved to separate the fractions < and > 2mm. The caesium-137 activity of the <2mm fraction was determined by gamma spectrometry and the measured activity for each sample was used to calculate the caesium-137 inventory for the sampling point and this was compared with the reference inventory. The deviations from the reference inventory were used in association with the site-specific calibration relationships referred to above, to estimate the rates of erosion and aggradation associated with the sampling points. UNIRAS software was subsequently used to convert the grid data to a map of the spatial distribution of erosion and aggradation within each study field (cf. Fig. 1).

Estimating erosion rates

Since it is possible to use the calibration relationships to estimate point rates of both erosion and aggradation from caesium-137 measurements, a range of measures of soil redistribution may be calculated for each field. The mean rate of erosion is equal to the total amount of erosion divided by the area subject to erosion. The gross rate of erosion is equal to the total amount of erosion divided by the area of the whole field. The mean and gross aggradation rates may be calculated in similar fashion. The net rate of erosion represents the amount of soil transported beyond or lost from the study field and is equal to the gross rate of erosion minus the gross rate of aggradation. The ratio of net to gross erosion provides a measure of the proportion of eroded sediment transported beyond the cultivated area and is therefore essentially equivalent to the sediment delivery ratio. It should be recognized, however, that in some cases field boundaries may be important in trapping a significant proportion of the transported sediment. These measures of soil redistribution and additional data relating to their spatial distribution are listed for each field in Tab. 2.

If a gross erosion rate of 2 t ha-1 year-1 is considered to be a generous estimate of soil loss tolerance (14), the data in Table 2 suggest that the rates of erosion are excessively high at all the sites investigated, with gross erosion rates ranging from 2.2 to 12.2 t ha-1 year-1. Furthermore, the area of each field subject to rates of soil loss in excess of 2 t ha-1 year-1 exceeds 40 % at all sites except

Keysoe (29 %) and is as high as 81 % at Rufford. These data and the high mean

Table 1.
Soil characteristics and locations of the study fields.

Farm Name/ County	Soil Type	Location
Sandy		
Hole Norfolk	Typical brown sand	52°52′N 1°08′E
Dalicott Shropshire	Typical brown sand	52°33′N 2°20′W
Rufford Nottinghamshire	Typical brown sand	53°07′N 1°05′W
Chalky		
Manor House Norfolk	Brown rendzina	52°54′N 0°45′E
Lewes Sussex	Brown rendzina	50°53′N 0°01′E
Silty		
West Street Kent	Typical brown calcareous earth	51°14′N 1°20′E
Mountfield Somerset	Typical brown earth	50°57′N 2°51′W
Yendacott Devon	Typical brown earth	50°48′N 3°34′W
Wootton Herefordshire	Typical argillic brown earth	52°08′N 2°31′W
Fishpool Gwent	Typical argillic brown earth	51°47′N 2°47′W
Clay		
Higher Dorset	Typical calcareous pelosol	50°38′N 2°33′W
Brook End Bedfordshire	Typical calcareous pelosol	52°15′N 0°25′W
Keysoe Bedfordshire	Typical calcareous pelosol	52°15′N 0°26′W

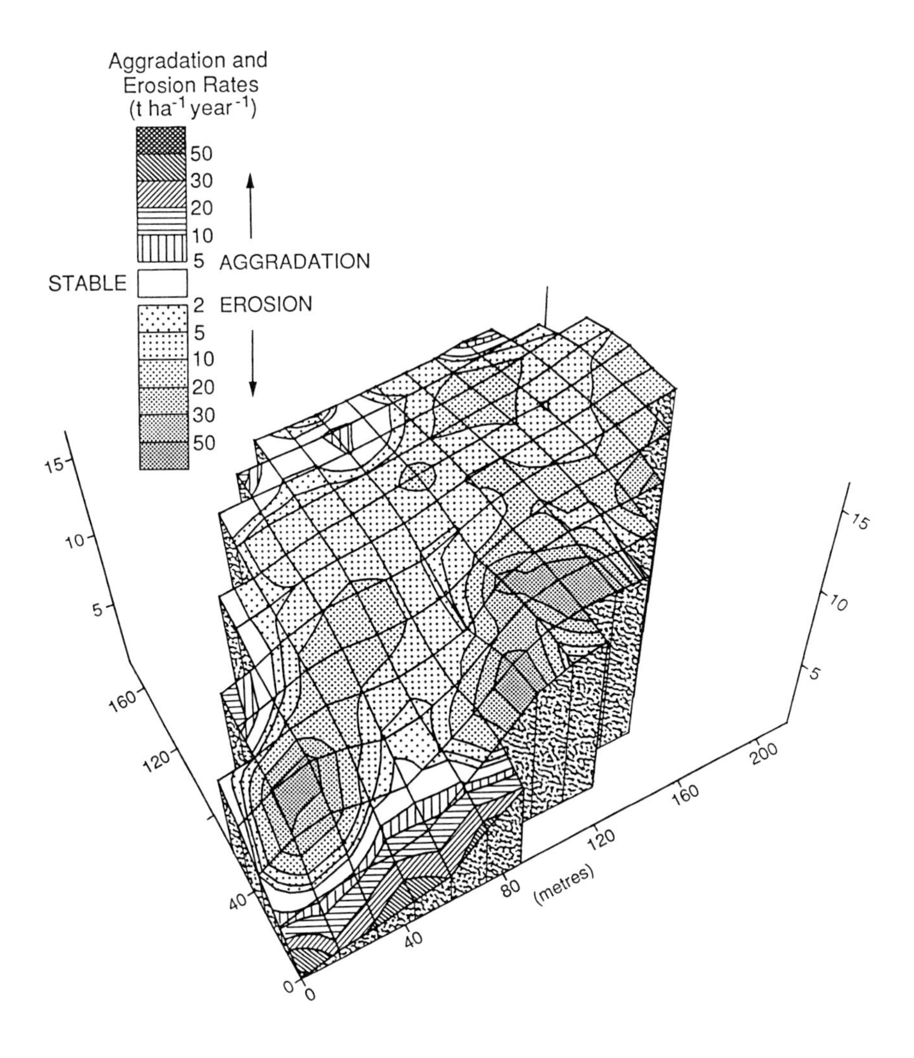

Figure 1. Erosion and aggradation rates within the field at Dalicott Farm, Shropshire, UK.

rates of erosion suggest that a decline in soil depth and productivity may be expected to occur if the problem goes unchecked. This will be considered further in the concluding section.

Table 2.
Rates of soil erosion and aggradation (t ha^{-1} year $^{-1}$), on the study fields, derived from caesium-137 measurements.

Soil Type /Site	Erosion Net Rate	Gross Rate	Mean Rate	% Area >2	>4	% Net/ Gross Erosion	Aggradation Gross Rate	Mean Rate	% Area
Brown Sand									
Hole	3.0	6.3	8.1	56	46	48	3.3	15.6	21
Dalicott	6.5	10.2	12.6	75	63	64	3.7	19.3	19
Rufford	10.5	12.2	13.8	81	72	86	1.8	16.1	11
Brown Rendzina									
Manor Hse.	2.4	6.3	9.4	54	42	38	3.8	11.6	33
Lewes	1.4	4.3	6.2	53	37	33	3.0	9.4	32
Brown Calcareous Earth									
West St.	4.3	7.7	11.1	61	51	56	3.5	11.6	30
Brown Earth									
Mountfield	2.2	4.6	6.1	49	34	48	2.4	9.1	26
Yendacott	1.9	5.3	7.3	57	44	36	3.4	12.5	27
Argillic Brown Earth									
Wootton	2.8	6.4	8.1	43	23	44	3.9	17.7	22
Fishpool	1.9	5.1	8.1	47	34	37	3.2	8.9	36
Calcareous Pelosol									
Higher	3.1	5.2	6.7	61	44	60	2.1	10.1	21
Brook End	1.2	3.6	5.7	40	26	33	2.4	6.5	37
Keysoe	0.6	2.2	4.0	29	20	27	1.5	3.3	46

Comparison with erosion rate estimates derived using other approaches

In any comparison of erosion rates derived using different approaches it is important to recognize the influence of the measurement techniques on the rates obtained. Erosion plots provide measures of net soil loss from a defined area which may be a whole slope transect but is more usually only a segment of the slope. The isolation of small areas may lead to a reduction in the capacity for rill formation and erosion rates may therefore be underestimated. Equally, however, plots which are kept bare of vegetation will be subject to higher rates of erosion than fields where the crop protects the soil surface. The erosion rate data provided by small plots are therefore most readily equated with estimates of possible (vegetated plots) to maximum (bare plots) net erosion due to splash and overland flow (sheet) erosion, and will in most cases underestimate potential rill erosion. In contrast, the rates derived from post-event field surveys will reflect gross erosion associated with large rills and gullys. On some occasions it may be possible to estimate volumes of deposition and thereby derive estimates of net rates of soil loss. However, survey methods are unlikely to identify sediment redistribution associated with sheet erosion or downslope displacement by ploughing. Furthermore, the resolution of the technique is difficult to establish and it is questionable whether rills characterized by a small cross-sectional area which may be widespread, particularly under crops planted on ridges, will be identifiable under even moderate crop cover. Before actually making comparisons of available data for the soil types under study, it can therefore be suggested that net rates of erosion derived using the caesium-137 approach are likely to lie toward the upper limit of the range of values obtained using erosion plots, and that gross rates of erosion derived from caesium-137 measurements are likely to exceed estimates derived using survey methods, particularly in this case in view of the known propensity to erosion of the sites selected for investigation using the caesium-137 technique.

Because erosion rates derived from caesium-137 measurements are long-term averages, the most meaningful comparisons are those with time-averaged data such as long-running plot studies and repeated surveys. Fig. 2 compares the rates obtained using caesium-137 measurements with available plot monitoring estimates (15) and survey-based median rates (3).

All three approaches show a similar pattern of variation with soil texture. The lowest rates in each case were found on the clay soils of Bedfordshire and Cambridgeshire, and both caesium-137 and survey data evidence higher rates on the same textural class in Dorset. No long-term plot data are available for silty soils, but both the caesium-derived and survey based rates suggest that erosion rates are higher than on clay soils. Both data sets also show some geographic variability and it is possible that the higher rates found in Kent may reflect more intensive farming practice, particularly at the site investigated using caesium-137. Boardman and Hazelden (16) have also noted the potential for elevated erosion rates on intensively cultivated silty soils in this area. The caesium-137 estimates and one set those on all other textures except the clay soils. However, both the plot measurements and investigations of extreme erosional events (17) illustrate the potential for severe erosion which may be reflected in the second set of survey data. Finally, both the plot monitoring results and the caesium-137 estimates emphasize that high rates of erosion can occur on sandy soils, a feature further emphasized by the range of 10-45 t ha-1 year-1 measured on a bare plot. Such high rates of erosion on sandy soils are in accord with the data presented by Evans and Cook (13) although the range presented by Evans (3) is somewhat lower.

Overall, the data illustrated in Figure 2 indicate that there is a reasonable degree of agreement between the three approaches considering the differences between the measurement techniques involved. The caesium-derived data lie within the range of rates obtained from long-term plot monitoring studies (15) in all cases except for the clay soils. As might be expected, in view of the known propensity to erosion of the fields studied and the differences between the techniques discussed above, the caesium-137 estimates are higher than or at the upper limit of the range of rates obtained by field survey (3).

Although there would appear to be general agreement between the estimates produced by the different approaches, where survey-derived erosion rate estimates are available for the individual fields studied using caesium-137, greater contrasts are apparent. Boardman (pers. comm.) has surveyed rill erosion in the field on chalky soil near Lewes over a period of 8 years and the resultant erosion rate estimate is only approximately 25 % of the gross erosion rate estimate produced by caesium-137 measurements. This difference may reflect two factors. Firstly, estimates of erosion rates based on survey techniques will exclude the effects of sheet erosion and plough movement. Govers and Poesen (18) refer to two studies in which the rill component of erosion was estimated at less than 25 % of total erosion, but neither of these was on a chalky soil. Secondly, the possible occurrence of low frequency high magnitude erosion events outside the period monitored by Boardman but within the last 35 years must be considered. However, such explanations do not readily account for the differences between the estimates of soil loss obtained using the caesium-137 and survey approaches at Dalicott

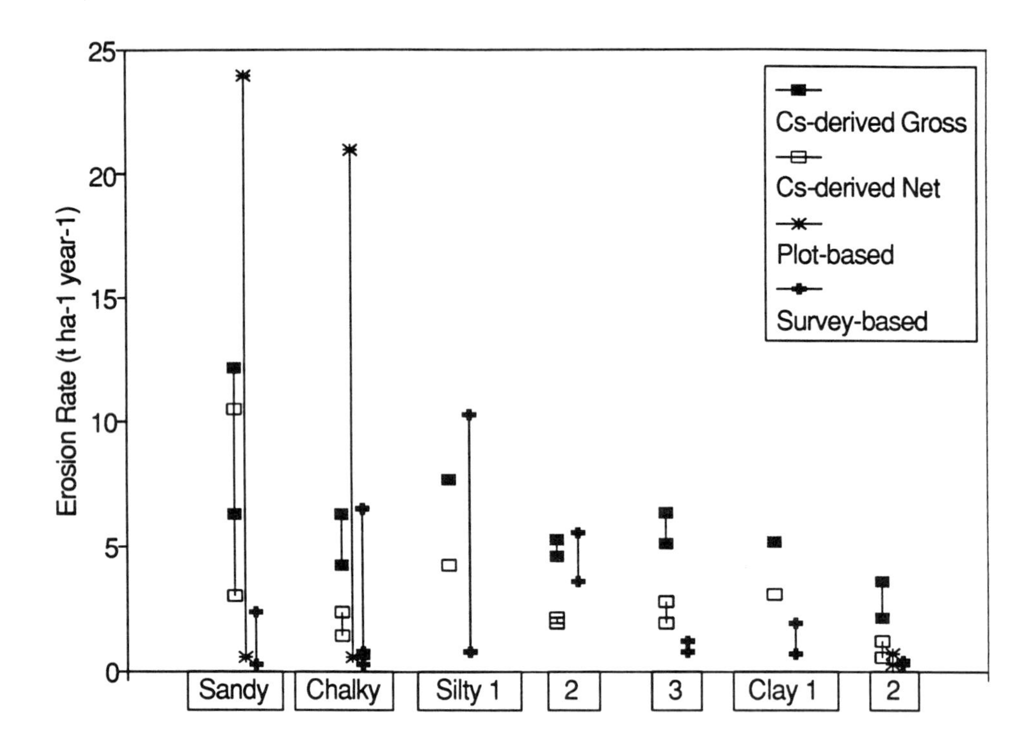

Figure 2.
Comparison of erosion rates derived from caesium-137 measurements with those based on plot studies (15) and field surveys (3).

Farm in Shropshire. At this site, Evans (19) has monitored rills on the field investigated using caesium-137 measurements for 9 years and has estimated the rate of rill erosion at 0.30 t ha-1 year-1 (0.23 m^3 ha^{-1} $year^{-1}$, assuming a bulk density of 1.3 t m^{-3}) compared to caesium-derived net and gross erosion rates of 6.5 and 10.2 t ha-1 year-1, respectively. This difference of more than an order of magnitude has prompted further enquiry.

Verification of erosion rates derived from caesium-137 measurements

If the assumptions of the caesium-137 technique are accepted (10), the major control upon the erosion rate estimates will be the reliability of the calibration relationship. In the present study the calibration relationship was derived using an accounting model which was tested within a small Devon catchment to provide initial verification at the field scale. The validity of the model-based calibration relationships could be further tested by sampling erosion plots for caesium-137 and comparing the calibrated erosion rate estimates for the plots with the direct measurements of soil loss. This has been carried out for erosion plots in Zimbabwe and a good agreement was found between the erosion rate estimates. However, a full test of the model-derived calibration relationships using this approach would require sampling of plots, in a range of locations, which were operational throughout the period of atmospheric fallout, ie from before 1954. Nevertheless, the corroboration currently available suggests that the calibration relationships, derived using the accounting model, provide valid estimates of soil erosion and aggradation rates.

Although more detailed site-specific verification is problematic, it is possible to assess the consistency of the aggradation rates estimated for the Dalicott site. The depth of aggradation estimated from the caesium-137 inventory using the calibration relationship may be compared with the depth indicated by the profile distribution of caesium-137 at the same point. At eroding points concentrations of caesium-137 decline to zero immediately below the plough depth (Fig. 3).

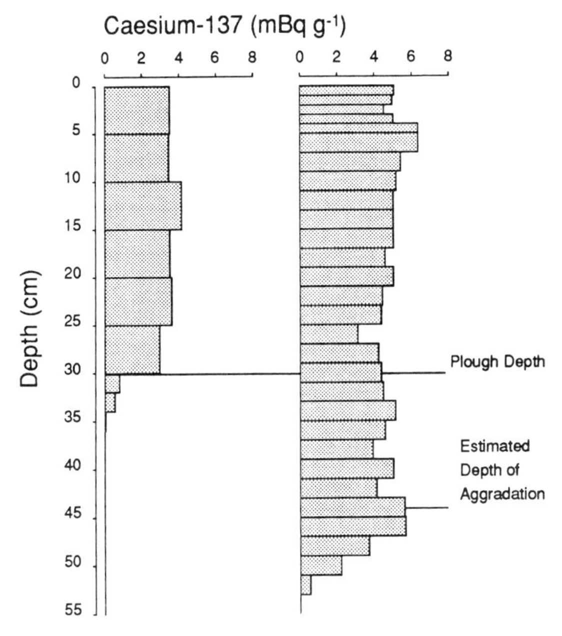

Figure 3.
Profile distributions of caesium-137 from eroding (left) and aggrading (right) sampling points within the field at Dalicott Farm, Shropshire. The depth of aggradation estimated from the caesium-137 inventory using the calibration relationship applied at the site is also shown.

In contrast, where aggradation has occurred significant levels of caesium-137 are found well below the plough depth (Fig. 3). This is due to burial of caesium-137 bearing soil below the lower limit of ploughing by progressive sediment deposition at the surface. The distance from the lower limit of ploughing to the point at which caesium-137 concentrations decline sharply may therefore be taken as equivalent to the depth of aggradation at the sampled point over the period since the initiation of significant fallout. Fig. 3 demonstrates that there is a good agreement between the estimate of aggradation derived using the calibration relationship and the profile evidence for this sample point at Dalicott. This internal consistency supports the validity of the point estimates of aggradation and therefore the gross aggradation rate for the field of 3.7 t ha^{-1} $year^{-1}$. This may in turn be seen as a minimum estimate of the gross erosion rate for the field. While this is only 36 % of the actual estimate of gross erosion it is an order of magnitude greater than the rill based estimate of 0.3 t ha^{-1} $year^{-1}$ (19).

Further validation of the caesium-137 derived rates was obtained by examination of a second field, less than 1 km to the east. At this location, a hedge lies perpendicular to the contour, separating a cultivated field from pasture. The difference in height of the surface of the fields on either side of this field boundary reflects the erosional history of the site (Fig. 4).

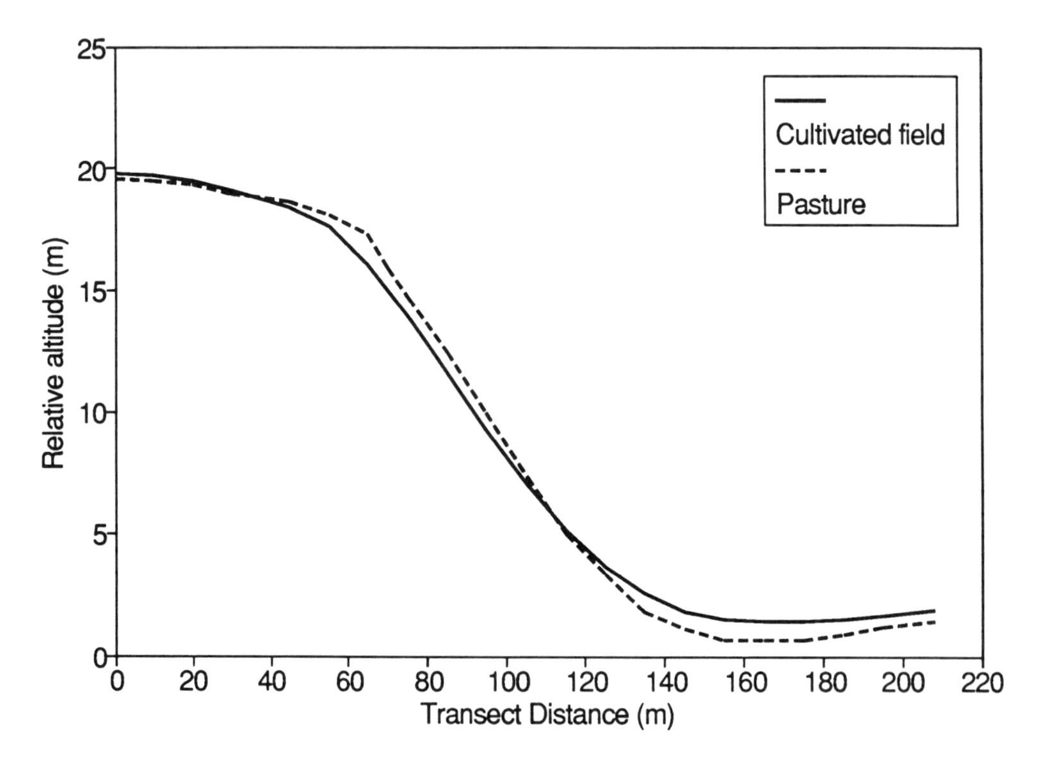

Figure 4. Topographic transects for the cultivated and pasture fields, on either side of a hedge line at Dalicott Farm, Shropshire.

From the slope convexity at the crest to the base of the slope, the level of the cultivated field is considerably lower than that of the hedge and adjacent pasture, reflecting soil loss over a prolonged period. In contrast, on the valley floor sediment deposition adjacent to the field boundary has elevated the surface of the cultivated field above the pasture. If a likely date of enclosure during the late 18 th Century is accepted, it is possible to convert the height difference between the fields to an assessment of erosion rate assuming a bulk density of 1.3 t m^{-3} and negligible soil redistribution within the pasture field. The pattern of erosion and deposition rates calculated in this way (Fig. 5) is one of very high rates of soil loss along the 77 m section from the slope crest to the foot (with a mean of 38 t ha^{-1} $year^{-1}$) and very high rates of aggradation along the 96m section of valley floor (mean of 33 t ha_{-1} $year^{-1}$). The total range of rates, from maximum erosion of 78 t ha^{-1} $year^{-1}$ to maximum aggradation of 51 t ha^{-1} $year^{-1}$, is in close accord with the range derived from caesium-137 measurements for the primary study field (Fig. 1) of 72 t ha^{-1} $year^{-1}$ maximum erosion to 57 t ha^{-1} $year^{-1}$ maximum aggradation. Furthermore, the mean rate of erosion estimated for the transect of 38 t ha^{-1} $year^{-1}$ suggests that the mean rate of erosion for the study field of 12.6 t ha^{-1} $year^{-1}$ is well within the bounds of probability.

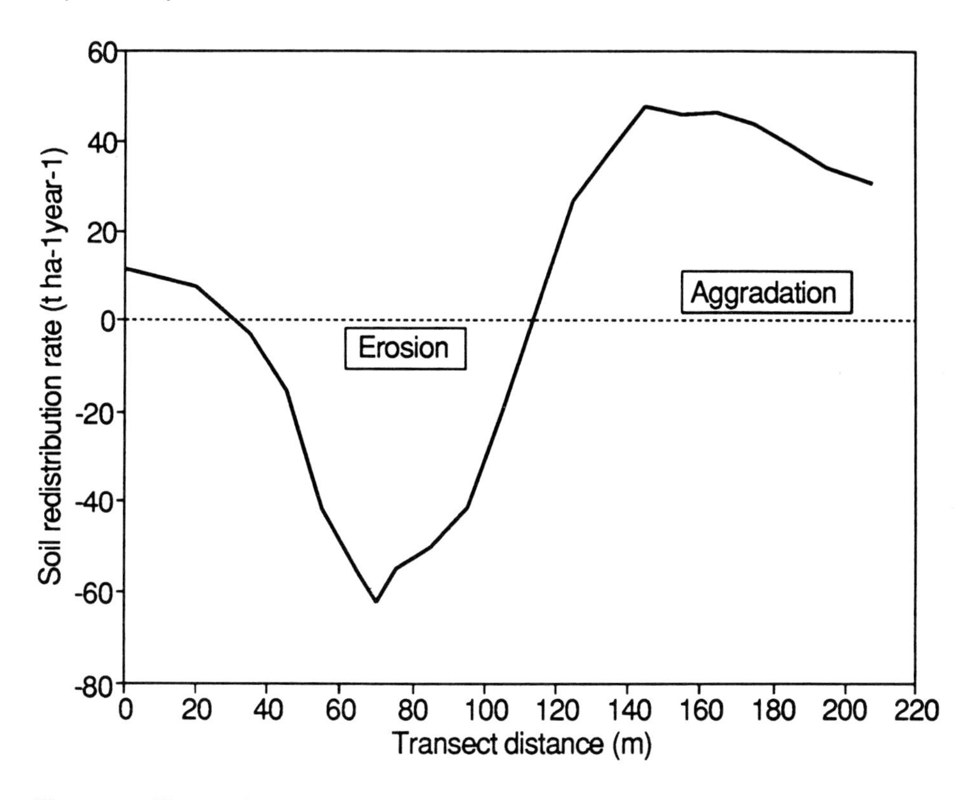

Figure 5. The estimated erosion and aggradation rates in the cultivated field adjacent to the field boundary, based on the difference in height between the cultivated and pasture fields.

The test of the calibration procedure provided by this unusual location may be carried further by comparing erosion and deposition rates estimated for individual cores, collected from the cultivated field along the transect, using caesium-137 measurements with the rates derived from the height measurements. Fig. 6 illustrates the close correspondence between the rates derived using the two approaches. Furthermore, the rates of soil redistribution derived from caesium-137 measurements may be seen to be consistently lower than those derived from the topographic data. When it is recognized that the calculations using the height data assumed a constant rate of soil redistribution over the last 200 years whereas it is generally accepted that there has been a marked increase in rates of soil erosion over the last thirty years, it may be suggested that the estimates derived using caesium-137 are likely to be conservative rather than exaggerated assessments of current rates of soil redistribution.

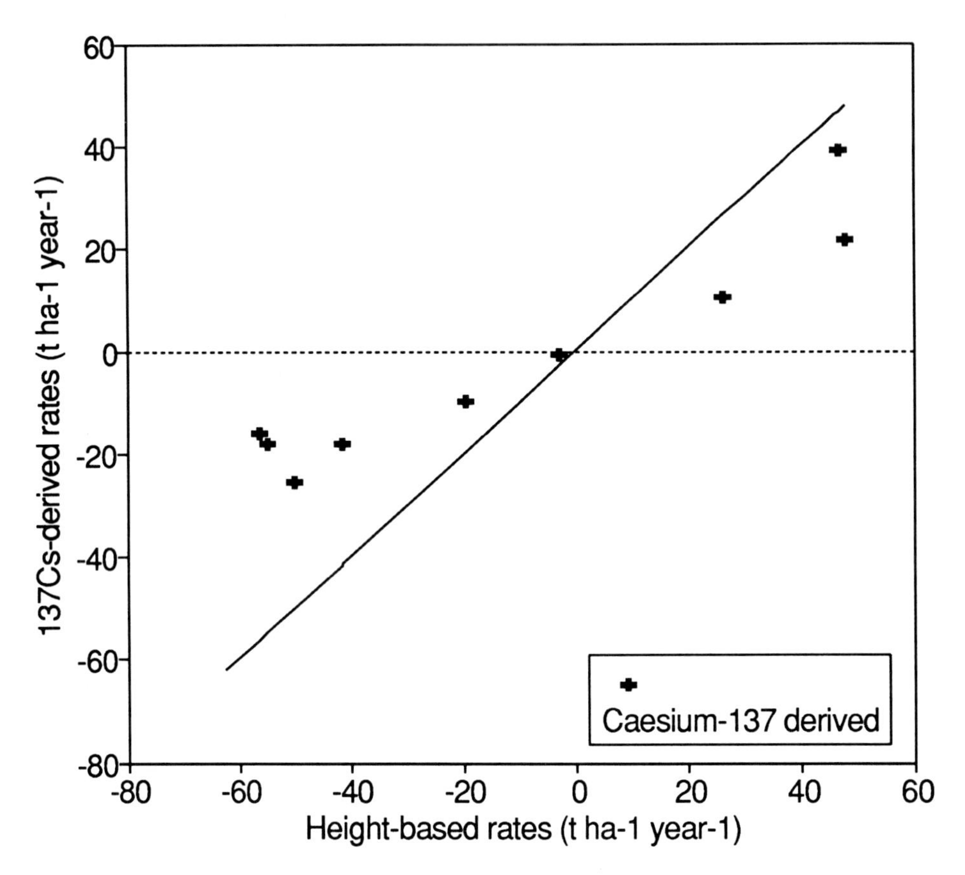

Figure 6. A comparison of the point estimates of erosion and aggradation rates obtained using caesium-137 measurements of individual cores with the estimates based on the contrast in height between the cultivated field and the adjacent pasture.

The evidence for the validity of the estimates of erosion and aggradation derived from caesium-137 measurements necessitates a brief consideration of the reasons for the much lower estimates provided by the rill measurements. Two possible explanations must be considered, firstly, the role of other processes of soil redistribution and secondly, the potential underestimation of rill erosion by small-scale features. The impossibility of assessing soil redistribution by sheet erosion and plough movement through field survey has been recognized by Evans (3) and alluded to above. However, it seems unlikely that rill erosion is responsible for only

3 % of the total soil loss, as suggested by the difference in the values, particularly as Morgan et al (1) identified rill erosion as contributing 20 to 50 % of the total soil loss on bare sandy plots at Silsoe. Underestimation of rilling may therefore be important especially where rill measurements are undertaken after harvesting when only the largest features remain, and this limitation is recognized by Evans (3). This problem is reflected in the estimates of deposition as a percentage of erosion by Evans (3) which occasionally exceed 100 % and in one instance (in Nottinghamshire) is as high as 684 %. In the absence of sediment import from adjacent fields, these data are clearly a result of underestimation of erosion. The potential for underestimation of rilling may be demonstrated by considering the rates of soil loss possible from small scale features. If a field is planted with a root crop on ridges with a density of 2 per metre, and if a rill of 5 cm^2 cross-sectional area forms in each furrow over half the area of the field, this would be the equivalent of a gross erosion rate for the whole field of 6.5 t ha-1 year-1 (1.3 t m^{-3}) in the absence of any other erosion. Features of this size would, however, be virtually indistinguishable.

IMPLICATIONS

In the light of the validation of the caesium-137 derived erosion rate estimates for the Dalicott site it is important to consider the implications of the rates derived for all of the sites using caesium-137 measurements. Three areas of particular significance may be identified, firstly, the impact on crop productivity, secondly, the potential off-site impact, and thirdly, the relative importance of soil redistribution processes. The fields investigated represent the upper end of the spectrum in terms of the magnitude of the erosion problem, but they provide a useful insight into the potential impact of continuing soil loss at current levels. The long-term gross erosion rate exceeds the nominal soil loss tolerance of 2 t ha^{-1} $year^{-1}$ at each site investigated, and mean rates of erosion are 2 to 7 times this figure. Assessing the impact of these rates on crop productivity is problematic, but Evans (20) has provided data which allow an initial estimate of the possible long-term impact. These data provide a first approximation of annual percentage yield reduction of 0.007 % for every tons of soil lost per hectare. If this relationship is used in conjunction with the erosion estimates provided by the caesium-137 measurements, the potential impact of unchecked erosion on future productivity may be estimated. These data suggest that over a 100 year period yield loss for the study fields would vary from 1.6 to 8.9 %, averaging 7 % on the sandy sites and 4.5, 3.9 and 2.7 % on the silty, chalky and clay sites respectively. Yield loss from the eroding areas within the fields (54-89 % of the total areas) would vary from 2.9 to 10.1 % with a mean of 6 %. These estimates are necessarily very tentative and based solely on yield loss due to decrease in soil depth. Decline in yield may also occur in areas of concentrated aggradation, where the deposited soil may be coarse and nutrient deficient. Although some of the reduction in yield as a result of erosion could be offset by increased fertilizer inputs, this would increase both the cost of the crop and the environmental damage caused by pollution associated with fertilizer nutrients.

The off-site impact of elevated erosion rates may be more significant both economically and environmentally than the on-site decline in productivity. Dramatic off-site costs of erosion from agricultural land have been documented by Stammers and Boardman (21), but less attention has been directed to the potential impact on river water quality. The high erosion rates identified in the present study, in association with the evidence of 27-86 % sediment delivery beyond the cultivated areas of the fields, suggests that eroded soil could represent a major non-point source of pollution. The transport of agricultural chemicals in association with this sediment increases the significance of this pollution potential. Walling and Quine (22) have suggested that a sediment delivery ratio to the stream network of 25 % in association with a gross erosion rate of only 4 t ha^{-1} $year^{-1}$ over 5 % of a basin could bring about a doubling of sediment yield compared to essentially 'natural' conditions (with traditional farming practices that do not promote erosion). This gross erosion rate was exceeded at 11 of the 13 fields investigated in the present study and Evans (3) has suggested that 0.1 to 15 % of the agricultural land in Britain is affected by erosion each year. These data, therefore, give cause for concern regarding potential increases in sediment yield for rivers draining agricultural land, particularly in the context of required improvements in river water quality and aquatic habitats.

At the two sites where it has been possible to compare directly rates of rill erosion based on field survey and rates of total erosion derived from caesium-137 measurements, the data suggest that large-scale rilling accounts for 3 and 25% of the total soil loss. This highlights the importance of other processes of soil redistribution in the agricultural landscape of the UK and the need for methods of measurement such as the caesium-137 technique which take account of all erosion processes.

ACKNOWLEDGEMENTS

Much of the research reported in this paper was carried out as part of the NERC/NCC Agriculture and the Environment research programme. We gratefully acknowledge this financial support and the valuable discussions with Dr R.Evans and Dr J.Boardman. We also thank Mr. J. Grapes for assistance with gamma spectrometry, Mr. R.Fry for drawing the diagrams and the various farmers for access to fields for sampling.

REFERENCES

1. Morgan, R.P.C., Martin, L., Noble, C.A. (1987) : Soil erosion in the United Kingdom : a case study from Mid-Bedfordshire. Silsoe College, Cranfield Institute of Technology, Occasional Paper, 14.
2. Fullen, (1991) : A comparison of runoff and erosion rates on bare and grassed loamy sand soils. Soil Use and Management, 7, 136-139.
3. Evans, R. (1988) : Water erosion in England and Wales 1982-1984. Report for Soil Survey and Land Research Centre, Silsoe.
4. Boardman, J. (1990) : Soil erosion on the South Downs : a review. In : Soil erosion on agricultural land (eds. J.Boardman, I.D.L.Foster & J.A.Dearing), 87-105 ; Wiley, Chichester.
5. Walling, D.E. and Bradley, S.B. (1988) : The use of caesium-137 measurements to investigate sediment delivery from cultivated areas in Devon, UK. In Sediment Budgets (ed. by M.P. Bordas and D.E. Walling) IAHS Publication no 174, 325-335.
6. Quine, T.A. and Walling, D.E. (1991) : Rates of soil erosion on arable fields in Britain : quantitative data from caesium-137 measurements. Soil Use and Management, 7, 169-176.

7. Ritchie, J.C. (1987) : Literature relevant to the use of radioactive fallout cesium-137 to measure soil erosion and sediment deposition. Hydrology Laboratory Technical Report, HL-9 ; USDA.
8. Walling, D.E. and Quine, T.A. (1991a) : The use of caesium-137 measurements to investigate soil erosion on arable fields in the UK : potential applications and limitations, Journal of Soil Science, 42, 147-165.
9. Livens, F.R. and Baxter, M.S. (1988) : Chemical associations of artificial radionuclides in Cumbrian soils. Environmental Radioactivity, 7, 75-86.
10. Walling, D.E. and Quine, T.A. (1992) : The use of caesium-137 measurements in soil erosion surveys. In the Proceedings of the Oslo Symposium, IAHS Publication (In press).
11. Walling, D.E. and Quine, T.A. (1990) : Calibration of caesium-137 measurements to provide quantitative erosion rate data. Land Degradation and Rehabilitation, 2, 161-175.
12. Quine, T.A. (1989) : Use of a simple model to estimate rates of soil erosion from caesium-137 data, Journal of Water Resources, 8, 54-81.
13. Evans, R. and Cook, S., (1986) : Soil erosion in Britain. SEESOIL, 3, 28-59.
14. Morgan, R.P.C. (1980) : Soil erosion and conservation in Britain. Progress in Physical Geography, 4, 24-27.
15 Morgan, R.P.C. (1985) : Soil erosion measurement and soil conservation research in cultivated areas of the UK. The Geographical Journal, 151, 11-20.
16. Boardman, J., and Hazelden, J. (1986) : Examples of erosion on brickearth soils in east Kent. Soil Use and Management, 2, 105-108.
17. Boardman, J. and Robinson, D.A. (1982) : Soil erosion, climatic vagary and agricultural change on the Downs around Lewes and Brighton, autumn 1982. Applied Geography, 5, 243-258.
18. Govers, G. and Poesen, J. (1988) : Assessment of the interrill and rill contributions to total soil loss from an upland field plot. Geomorphology, 1, 343-354.
19. Evans, R. (1992) : Erosion at Dalicott Farm, Shropshire - extent, frequency and rates. First International European Society for Soil Conservation Congress. Post-Congress Tour Guide.
20. Evans, R. (1981) : Assessments of soil erosion and peat wastage for parts of East Anglia, England. A field visit. In Soil Conservation : Problems and Prospects (ed. R.P.C. Morgan), 521-530 ; Wiley, Chichester.
21. Stammers, R. and Boardman, J. (1984) : Soil erosion and flooding on downland areas. Surveyor, 164, 8-11.
22. Walling, D.E. and Quine, T.A. (1991b) : Recent rates of soil loss from areas of arable cultivation in the UK. In Sediment and Stream Water Quality in a Changing Environment, IAHS publication, n° 203, 123-131.

Farm Land Erosion: In Temperate Plains Environment and Hills
S. Wicherek (Editor)
1993 Elsevier Science Publishers B.V.

Rainfall Simulation Tests for Parameter Determination of a Soil Erosion Model

M. Schramm and D. Prinz

Institut für Wasserbau und Kulturtechnik, Universität Karlsruhe, Kaiserstr. 12, D-7500 Karlsruhe 1, Germany.

SUMMARY

Rainfall simulation experiments carried out on 10 different erosion plots in SW-Germany on loess soils and in NW-Algeria on marl soils demonstrated that overland flow tends to concentrate along a few flow paths. Detachment and transport of soil particles are significantly affected by flow concentration. For this reason a model concept was introduced which allows one to calculate the hydraulic parameters and the erosion rate for rill flow rather than broad shallow flow. Only partial areas contribute to flow and erosion in the initial stage of runoff. The rainfall simulation data were used to calculate the flow effective areas for different initial conditions. The results of the parameter determination for the different locations support the model concept.

RESUME

Des expériences de simulation de pluies menées dans 10 parcelles d'érosion dans le S.O. de l'Allemagne sur des sols loessiques et dans le N.O. de l'Algérie sur sols marneux prouvent que les ruissellements pluviaux en nappe se concentrent le long de quelques chenaux d'écoulements. Le détachement et le transport des particules de sols sont affectés de façon significative par la concentration de l'écoulement; Pour cela un modèle a été décrit qui permet de calculer les paramètres hydrauliques et le taux d'érosion pour les écoulements en rigoles plutôt que pour les écoulements étendus et peu profonds. Dans le premier stade du ruissellement seules quelques zones contribuent à l'écoulement et à l'érosion. Les données de pluies simulées furent utilisées pour estimer les zones effectives d'écoulements selon différentes conditions initiales. Les résultats obtenus sur différents sites confirment le modèle étudié.

1. INTRODUCTION

Many models have been developed to predict infiltration and its variation with time. With these models it is now possible to determine the amount of rainfall that produces runoff, the excess rainfall. Routing procedures allow one to calculate the runoff from hillslopes, based on the assumption of broad sheet flow over the entire slope segment. Flow equations such as the Manning-Strickler equation are used for these conditions, but roughness coefficients of the classical open channel hydraulics may not be applicable for flow of water in thin sheets over rough ground.

On the steeper slopes, overland flow frequently occurs as concentrated flow in rill areas and sheetflow in interrill areas. The erosion process is significantly affected by the concentrated flow in the rills which transports the sediments from upland areas. Only limited information is available for the partitioning of total runoff into rill flow and interrill flow. An approach has been incorporated in this study to account for concentration of both flow and erosion in the rills. The hydraulic parameters are calculated for the rill flow rather than the sheetflow.

In order to obtain sufficient data for the determination of the necessary parameters a new type of rainfall simulator was incorporated into this study. To represent the complete process of rill and interrill erosion, the experimental plot and thus the rainfall simulator had to be large enough. Design and construction of such a modular rainfall simulator were realized by ERTI, Munich.

2. THE RAINFALL SIMULATOR

The modular rainfall simulator is designed for plots 4.0 m wide and a maximum length of 22.5 m (Photo 1).

Photo 1. The rainfall simulator for large experimental plots.

To keep the weight low the whole assembly was made of aluminum. The simulator consists of 10 equal units being combined in pairs 2.0 m apart. The support assembly of a unit is 4.5 m long and equipped with 3 sprinkler modules. The distance between the modules is 1.5 m. A module consists of a water storage tank with floater valve, nozzle, a gear motor to oscillate the nozzle and gutters to return the excess water that accumulates when the nozzle reaches the turning point. The nozzles are placed 3.0 m above the soil surface. Pressure at the nozzles is controlled by adjusting the air pressure in the storage tanks. Thus the pressure at the nozzles is independent of changes in elevation on sloping terrain. The gear motors are controlled with an electronic remote control unit that regulates the rainfall intensity by varying the length of the periods of no flow between spray applications.

Because of the modular design of the system the simulator can be used on plots of various sizes. The units and sprinkler modules can be used separately, e. g. as sprinkler infiltrometer or in the laboratory.

The distribution of intensity was determined during extensive tests carried out in the laboratory. Drop size distribution, drop impact velocity and the kinetic energy of the rain produced by the Veejet 80/100 nozzles were determined photographically on a smaller simulator at the Institute of Hydrology and Water Resources Research.

Uniformity of simulated rainfall and a level of kinetic energy comparable to natural rainfall are important in hydrological and erosion studies. For the Karlsruhe area the kinetic energy of natural rainfall was obtained as a function of intensity from 15 years of observations. The calculation of the energy is based on an adaptation of the Marshall-Palmer drop size distribution (1 - 2). The values for the Karlsruhe area are about 20 % lower than those calculated with the equation derived by Wischmeier and Smith (3). An equation derived by Brandt (4) corresponds closely with the observations (Fig. 1).

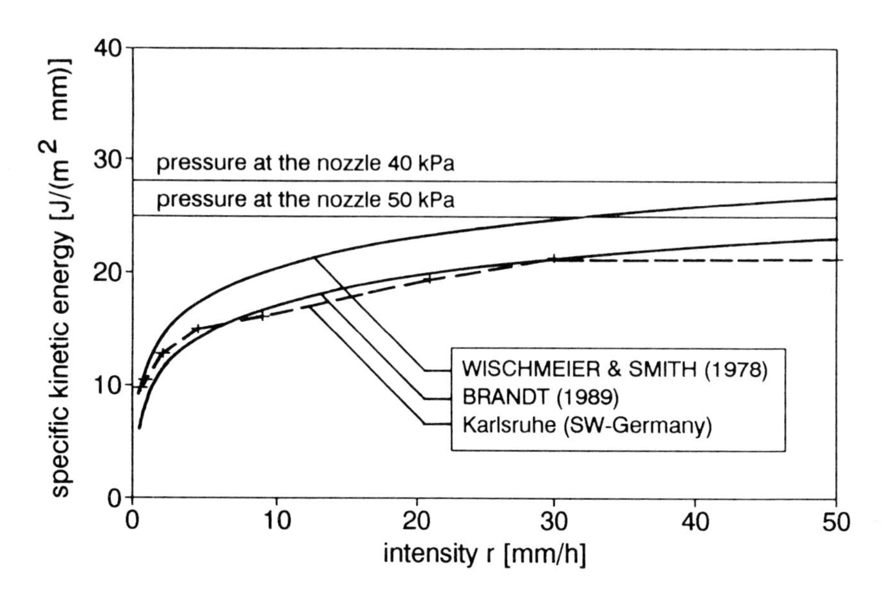

Figure 1. Relationship between kinetic energy of rainfall and intensity.

The kinetic energy of simulated rainfall for a pressure at the nozzle of 50 kPa corresponds to the results obtained by Wischmeier & Smith for intensities of more than 30 mm/h. A better adaption to the Karlsruhe conditions could probably be obtained if a higher pressure at the nozzle is used. A higher pressure would produce smaller drops. However, the resulting increasing intensity would require longer off-periods between spray applications.

3. MODEL CONCEPT

The basic concept of the erosion model introduced in this chapter is similar to that on which the mathematical simulation model EROSION2D is based (5). The external forces acting on the soil particles are the momentum fluxes produced by raindrops and by overland flow. In order to detach soil particles, the shear stress exerted by overland flow and raindrops has to overcome the resistance of the soil to erosion caused by internal friction, cohesion and gravity.

The momentum flux iq of overland flow can be defined as

$$i_q = q\,\rho_q\,\Delta z\,v_q \qquad (1)$$

q = discharge per unit width [$m^3/m/s$], ρ_q = fluid density [kg/m^3], Δz = width of slope segment [m], v_q = mean flow velocity [m/s]

The mean flow velocity vq can be calculated from the Manning-Strickler equation :

$$v_q = \frac{1}{n}\,\Delta y^{\frac{2}{3}}\,S^{\frac{1}{2}} \qquad (2)$$

n = Manning's roughness coeff. [$s/m^{1/3}$], Δy = flow depth [m], S = slope gradient [-]

The momentum flux of the rainfall is the sum of the momentum fluxes of all raindrops falling on a unit area. Under the assumption of a drop size distribution after Marshall and Palmer the momentum flux of the drop diameter class j can be calculated from :

$$\begin{aligned} i_{r,j} &= r_j\,\cos\alpha\,A\,\rho\,v_{r,j}\,(1-C) \\ &= N_0\,\exp\left(-4.1\,r^{-0.21}\,1000\,d_j\right)\,v_{r,j}^2\,d_j^3\,\frac{\pi}{6}\rho\,(1-C)\,\cos\alpha \end{aligned} \qquad (3)$$

r_j = intensity of drop size class j [m/s], α = slope angle, A = area [m^2], C = soil cover [-], ρ = density of water [kg/m^3], $v_{r,j}$ = terminal velocity of class j [m/s], N_0 = 2000, d_j = mean drop diameter of class j [m], r = rain intensity [mm/h]

The mean terminal velocity of raindrops of class j can be calculated according to the semi-empirical approach proposed by Beard (6) :

$$\begin{aligned} Re_j &= v_j\,\frac{d_j\,\rho_a}{\mu} = P^{1/6}\,e^{Y_j} \\ \text{in which}\quad P &= \sigma^3\,\frac{\rho_a^2}{\mu^4\,g\,(\rho-\rho_a)} \\ Y_j &= \sum_{n=0}^{5} C_n\,x_j^{\,n} \\ x_j &= \ln\left(\frac{16}{3}B_j\,P^{1/6}\right) \\ B_j &= g\,\frac{\rho-\rho_a}{\sigma}\,\frac{d_j^2}{4} \end{aligned} \qquad (4)$$

Re_j = Reynolds number of drop motion at terminal velocity, v_j = terminal velocity of drop size class j [m/s], g = acceleration of gravity [m/s^2], C_n = correction factor, μ = dyn. viscosity of air [kg/m/s], σ = surface tension [N/m], ρ_a = density of air [kg/m^3]

The gross momentum flux of the rainfall is given by

$$i_r = \sum_{j=1}^{k} i_{r,j} \tag{5}$$

$k=$ *number of drop size classes*

Because of the heterogeneity of the soil, the transfer of the momentum of raindrops to individual soil particles cannot be described in detail. Only a portion of the momentum is effective for detachment and transport. For simplification only the component that acts downslope is taken into account :

$$i_{r,\alpha} = i_r \sin\alpha \tag{6}$$

Erosion occurs if the momentum flux caused by the raindrops and the overland flow overcomes the resistance of a particular soil to erosion icrit. Its value can be defined as the momentum flux necessary to initiate erosion

$$i_{crit} = q_{crit}\, \rho_q\, \Delta z\, v_q \tag{7}$$

$q_{crit} =$ *flow rate at initial erosion*

The erosion coefficient

$$E = \frac{i_q + i_{r,\alpha}}{i_{crit}} \tag{8}$$

can be correlated with the sediment discharge empirically (7) by means of the equation

$$q_S = (E-1)\ 1{,}75 \cdot 10^{-4} \quad \left[\frac{kg}{m\ s}\right] \tag{9}$$

with q_s = O for E $\leq$ O. Because of the physical basis of the governing parameters I_r, i_q, and i_{crit} the model concept is applicable to different resolutions in space and time.

4. PARAMETER DETERMINATION

Values for Manning's n can be taken from reference works, but the calibration factor I_{crit} has to be determined from experiments. In this study, some 30 rainfall simulation experiments were carried out on 10 different plots - 5 in an experimental basin in SW-Germany and 5 in a marl region of NW-Algeria - to determine I_{crit}. The durations of the simulator runs were between 30 and 100 minutes and the rainfall intensities were between 25 mm/h and 46 mm/h. Representative results from the rainfall simulation experiments are presented in this chapter. Topographic and soil data for these experimental plots are listed in Tab. 1.

Table 1.
Slope and soil type of the experimental plots.

plot no.	slope [%]	soil type (German classification)
1	26.2	Keuper-Regosol
2	16.8	Pararendzina (silty loess)
3	10.8	Pararendzina/ Kolluvisol (silty loess)
4	21.0	Syrosem (badlands)
5	18.0	Vertisol (marl, clay content >50 %)

4.1. Rill parameter

As shown by the rainfall simulation experiments, overland flow is more likely to concentrate along a few flow paths (Fig. 2). In most models hydraulic parameters such as Manning's n or flow depth Δz are calculated under the assumption of shallow flow in a wide channel (e. g. Creams : (8), Erosion2D : (5)). Only the recent erosion model Wepp (9) distinguishes between broad shallow flow in interrill areas and concentrated flow in rill areas, a method that is more realistic. Hence, the momentum flux for overland flow in this investigation was calculated as the discharge in a certain number of rills. The rill density of the study sites was in the range of 0.8 - 1.3 rills per meter of width and thus a mean rill density of 1 rill per meter of width was used. These values are in accordance with those from the rainfall simulation runs that were conducted during the development of Wepp (10).

For the determination of I_{crit}, the flow rate was assumed to be the same in each rill. This simplification was necessary because no rill discharges were measured during the experiments. A stochastic approach to partition flow between rills was developed by Gilley et al. (10), but it has not yet been integrated into the modified Erosion 2D model.

The cross section of the rills was taken to be a triangle with a ratio of width to depth of 3:1. The advantages of the use of a triangular cross section are the ease with which one can calculate the hydraulic radius and the variation of width with flow rate. This pattern appeared to represent satisfactorily the observations during the simulation experiments. However, the effect of different rill shapes on the calculated results has still to be investigated.

4.2. Hydraulic parameter

In this study a dye tracer was used in measuring the flow velocity in the rills so as to calculate Manning's n from Eq. 2. Values for n calculated from these

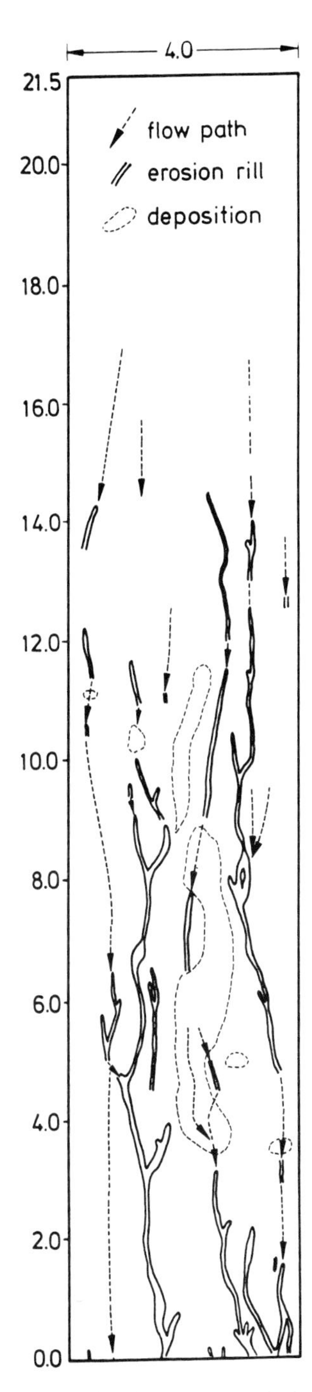

Figure 2. Rill pattern of plot n° 1 after simulation run 1.91.

observations ranged from 0.038 to 0.060 for the experiments in SW-Germany. Because of the small number of experiments it was not possible to relate a value of n to a specific land use or soil management. For simplification a mean value of 0.05 was chosen for the German plots and 0.08 for the Algerian plots.

4.3. Erosion resistance

During the simulation tests, runoff samples were taken at intervals of 3 or 5 minutes to determine the variation of the erosion rate with time. For each time step the erosion resistance parameter I_{crit} was adjusted until equal values of measured and calculated sediment discharge qs were achieved. The results from the determinations of the erosion resistance i_{crit} are presented in Fig. 3 to 6.

The higher values of I_{crit} for all curves are representative of conditions when runoff first occurs. In most cases I_{crit} then decreases and reaches a constant minimum value after a certain time. This final value is about 0.6 mN for seedbed condition and 1.5 - 2.0 mN for non-cultivated fallow at experimental plot n° 1 (Fig. 3). In simulation runs 3.90 and 2.91, the value for seedbed condition occurred initially, but the value then increased as the simulation run progressed. The final value in these cases is that for non-cultivated soil. The depth of the rill erosion during these experiments was great enough that the plough layer was reached during the tests and the erosion resistance I^{crit} then increased.

The results from plot n° 2 (Fig. 4) are in accordance with the results from plot n° 1. The final erosion resistance on the relatively flat plot n° 3 was only reached during the third simulation run (Fig. 5). The results from four rainfall simulation experiments on two different sites in Algeria are shown in Fig. 6. The variation of I_{rit} is similar to those from the German plots n° 1 and 2.

4.4. Contributing areas

If the erosion resistance is assumed to depend only on the type of soil and the management practice, but not on the runoff, initial moisture content, and other time related parameters, the variation with time of I_{crit} is a consequence of the incorrect determination of qs from Eq. 9. Effects like surface sealing or cracking are not significant because the graphs of I_{crit} are similar for non-cultivated and cultivated soils under dry and wet conditions. However, this is not the case for infiltration capacity and runoff generation, thus different erosion rates were found for different initial conditions.

Since only partial areas contributed to the flow in the initial stages of runoff, soil particles could only be removed from these areas. As the simulation runs progressed, more and more areas contributed to both runoff and soil detachment until erosion occurred on nearly the entire plot area.

The determination of I_{crit} was based on a sediment discharge rate qS related to the entire plot size. Thus Eq. 9 has to be modified as following

$$\left(\frac{q_S}{A_P}\right)_{measured} = \frac{1}{A(t)} (E-1)\ 1.75 \cdot 10^{-4} \tag{10}$$

which yields

$$q_S = \frac{A_P}{A(t)} (E-1)\ 1,75 \cdot 10^{-4} = \frac{1}{\beta} (E-1)\ 1,75 \cdot 10^{-4} \tag{11}$$

A_P = *plot size* $[m^2]$, $A(t)$ = *effective area* $[m^2]$,
β = *effective area ratio* [-]

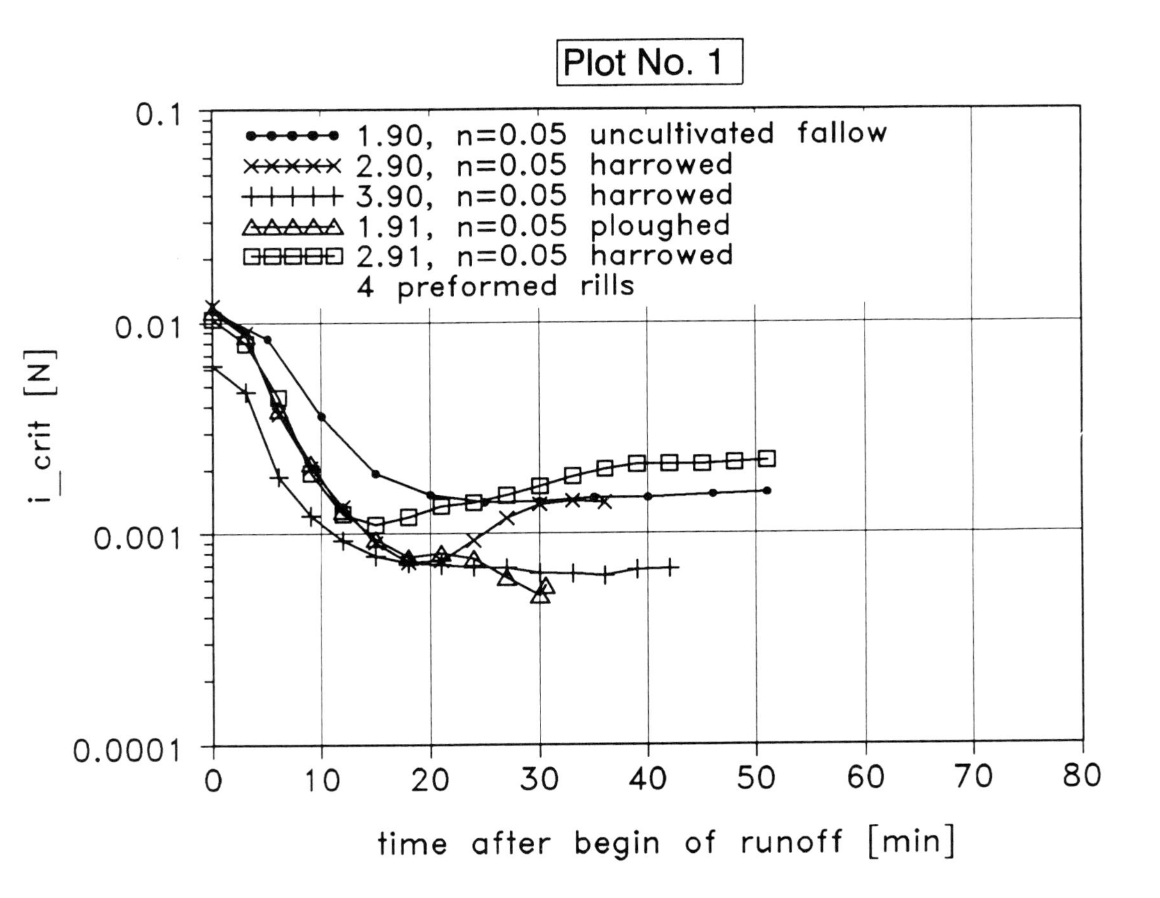

Figure 3.
Variation of the erosion resistance i_{crit} with time for plot n° 1.

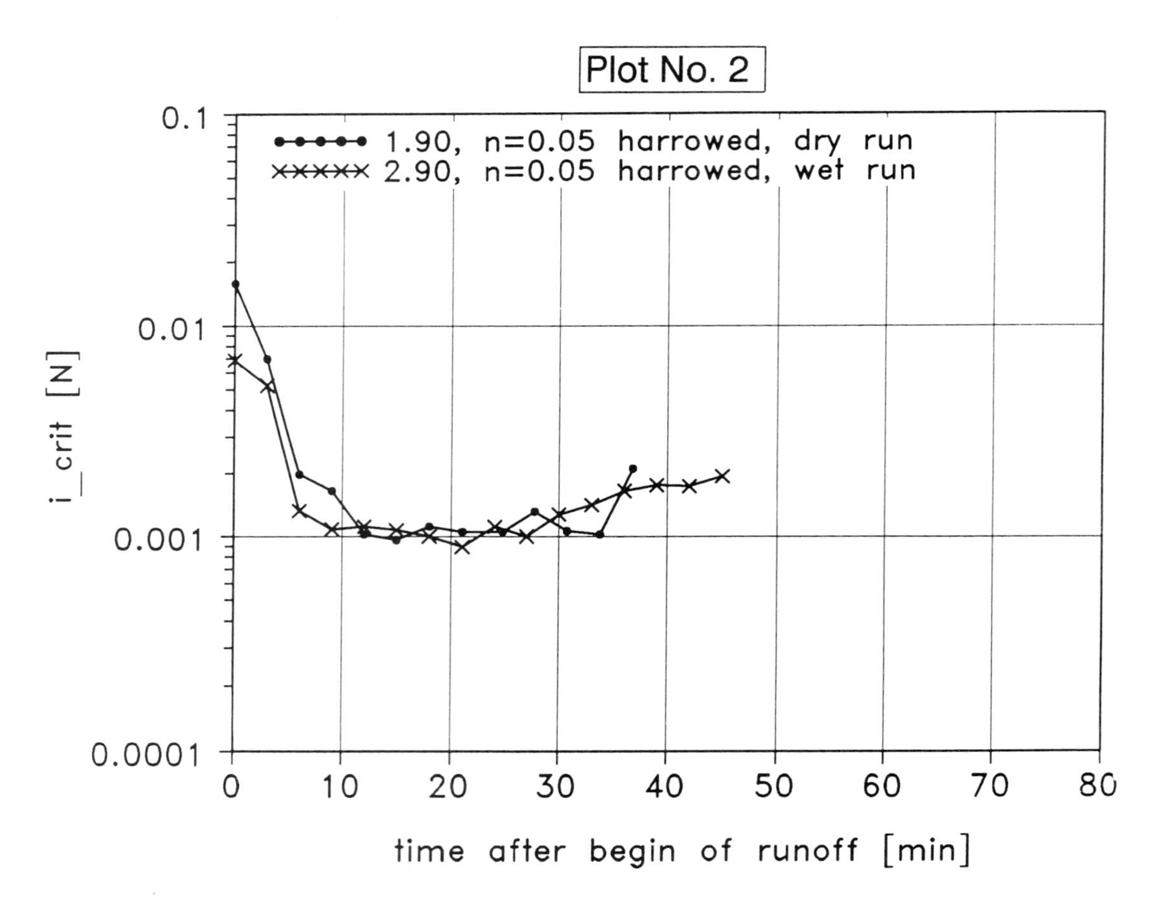

Figure 4.
Variation of the erosion resistance i_{crit} with time for plot n° 2.

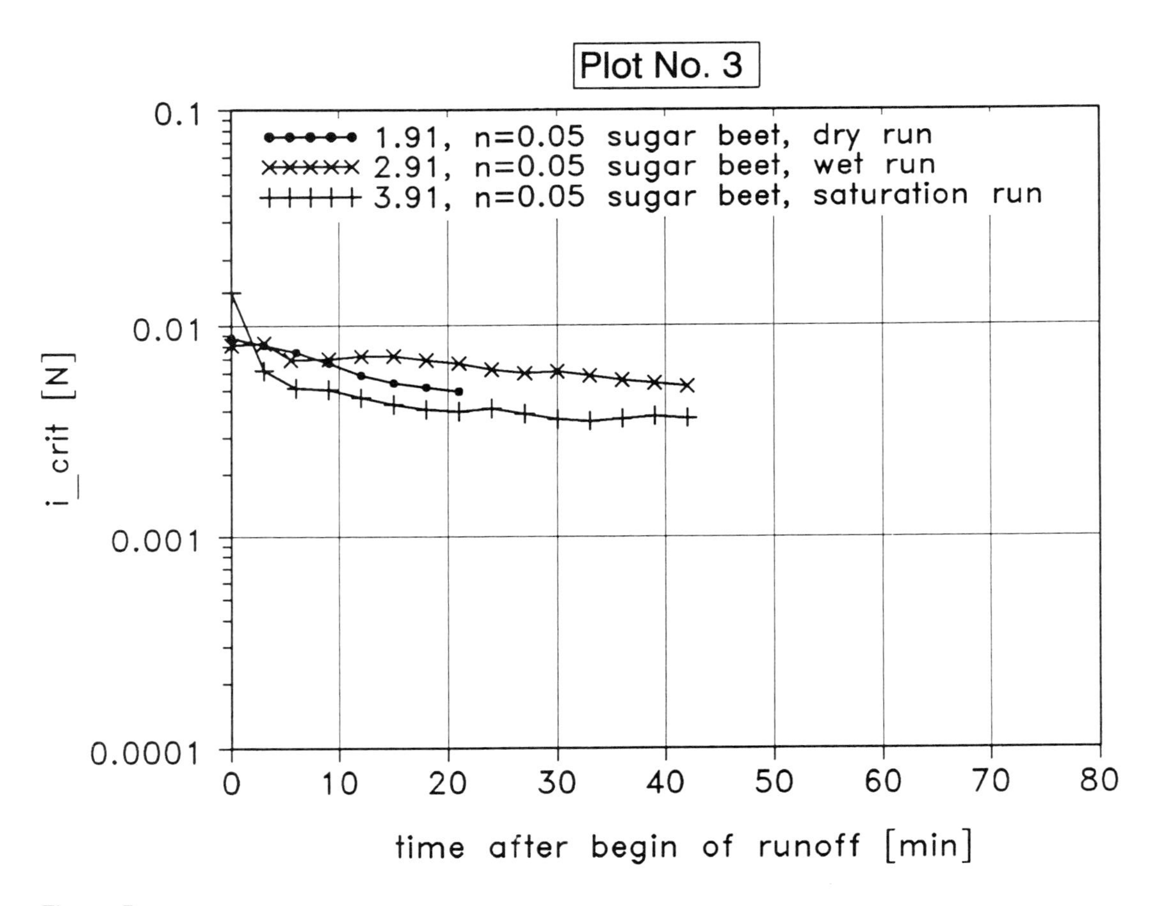

Figure 5.
Variation of the erosion resistance i_{crit} with time for plot n° 3.

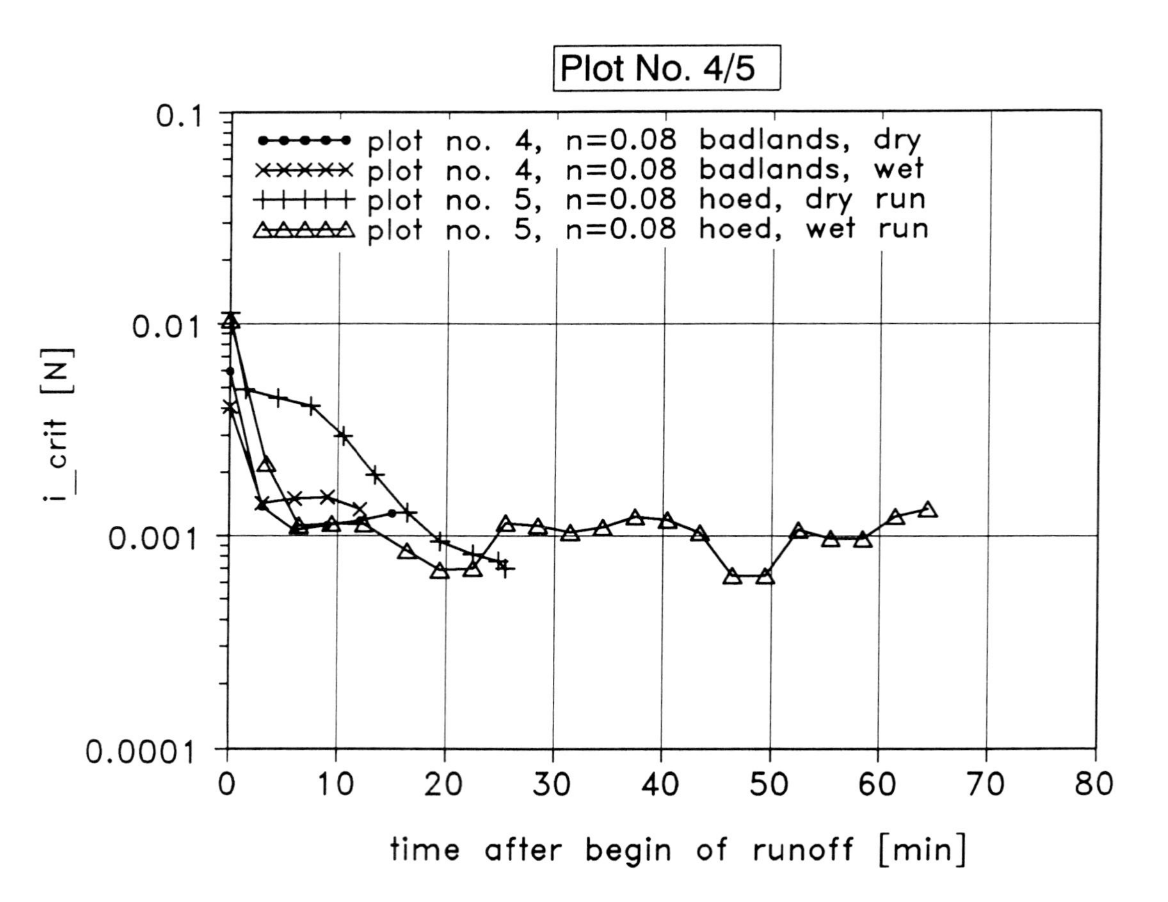

Figure 6.
Variation of the rosion resistance i_{crit} with time for plot n° 4 and 5.

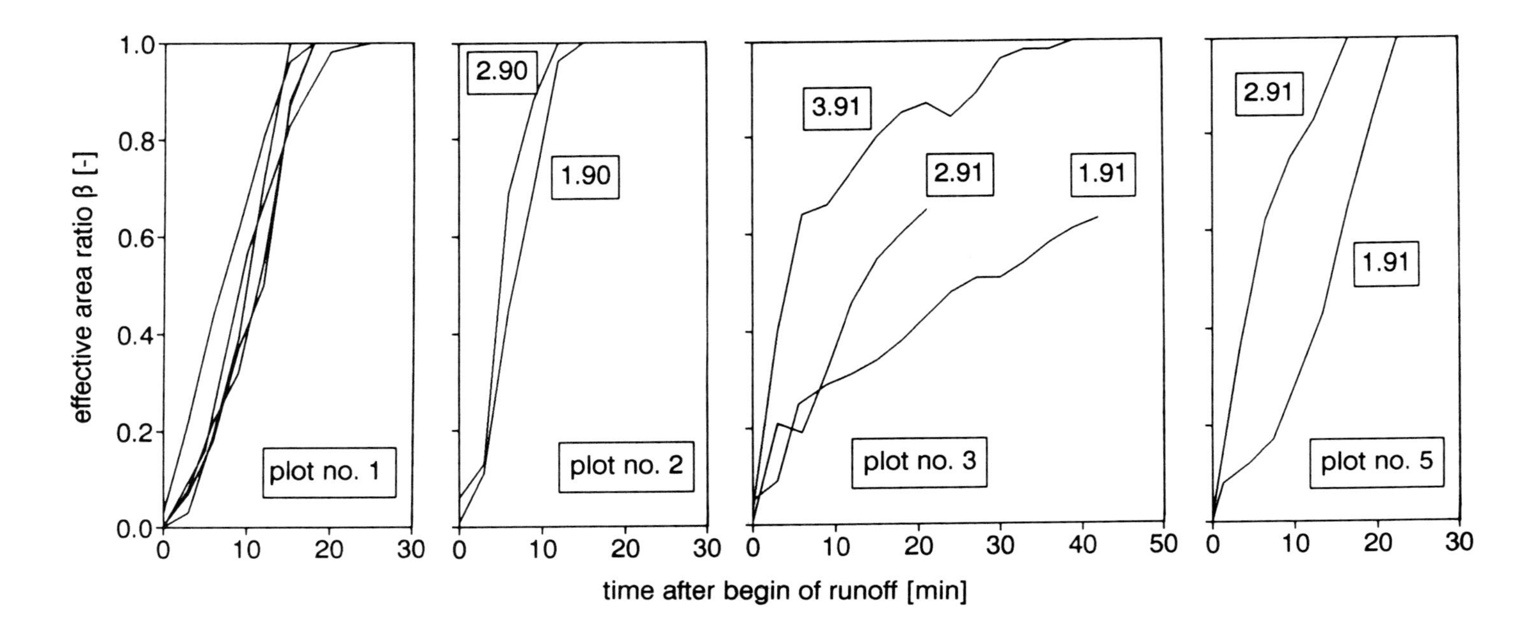

Figure 7. Variation of the effective area ratio ß with time.

The graphs for plots n° 1 and 2 show no significant effect of the initial moisture conditions on steeper slopes. However, for the less steep plot no. 3 the effective area ratio increases faster as the initial moisture content of the soil increases. A similar result was obtained for plot n° 5. Here, swelling of the marl soils (more than 50 % clay) probably affected the generation of runoff. Because only few experiments were made, the effect of plant cover on the increase with time of the area ratio ß could not be determined. However, covered areas probably contribute later to runoff and soil erosion than do uncovered areas. This finding could help to explain why the effective area ratio for the covered plot n° 3 increased less than did those for the uncovered plots.

5. CONCLUSION

Rainfall simulation experiments on large erosion plots demonstrated the effect of concentration of flow on the hydraulic parameters and on the erosion process. The experiments showed that calculations based on the assumption of broad shallow flow over the entire plot area are not realistic. Thus an approach was introduced to predict soil erosion as a function of the momentum fluxes of rainfall and of flow concentrated in rills. The areas that generate runoff are limited, and only in these areas are soil particles detached and transported. For this reason the contributing area concept was incorporated into the analysis of both processes and results. Data from a loess region in SW-Germany with silty soils and from a marl region in NW-Algeria with clay soils support the model concept and provide typical results.

Further research is needed to obtain more complete information about the formation of rills, the rill density and the effect of the flow profile on the determination of such hydraulic parameters as Manning's n, flow depth or flow velocity. At this moment little information is available about the effect of soil characteristics (clay content, macropores, water content, etc.), topography and plantcover on the extent of the contributing area.

ACKNOWLEDGEMENTS

The research project on which this paper is based was financed by the Bundesminister für Forschung und Technologie, grant n° 02WA89030.

REFERENCES

1. Strantz, R. (1971) : Typen der Regentropfenspektren I. Meteor. Rdsch., 24. Jg., Heft 1.
2. Diem, M., Strantz, R. (1971) : Typen der Regentropfenspektren II. Meteor. Rdsch., 24. Jg., Heft 1.
3. Wischmeier, W. H., Smith, D. D. (1978) : Predicting rainfall erosion losses - a guide to conservation planning. U. S. Dept. of Agriculture, Agriculture Handbook n° 537. Washington, DC.
4. Brandt, J. (1989) : Simulation of kinetic energy of rainfall under vegetation. Paper presented to British Geomorph. Res. Group symp. on vegetation and geomorphology, Bristol, UK.

5. Schmidt, J. (1992) : Modelling long-term soil loss and landform change. Hydraulics and Erosion Mechanics of Overland Flow, ed. A. D. Abrahams and A. J. Parsons, University College London Press (in press).
6. Beard, K. V. (1976) : Terminal velocity and shape of cloud and precipitation drops aloft, J. Atmos. Sci., 33, 851-864.
7. Schmidt, J. (1988) : Wasserhaushalt und Feststofftransport an geneigten, landwirtschaftlich bearbeiteten Nutzflächen. Diss. FU Berlin.
8. Knisel, W. G. (1980) : Creams : A field scale model for chemicals, runoff, and erosion from agricultural managed systems. NSERL Report No. 2, USDA-ARS Nat. Soil Erosion Res. Lab.
9. Foster, G. R., Lane, L. J. (1990) : USDA - Water Erosion Prediction Project : Hillslope profile version documentation. NSERL Rep. No. 2, USDA-ARS, W. Lafayette, Indiana.
10. Gilley, J. E., Kottwitz, E. R., Simantonl, J. R. (1990) : Hydraulic characteristics of rills. Transactions of the ASAE, Vol. 33 (6), 1900-1906.

List of mathematical symbols

A	area [m^2]
$A(t)$	effective area [m^2]
A_P	plot size [m^2]
C	soil cover [-],
C_n	correction factor [-]
d_j	mean drop diameter of class j [m]
E	erosion coefficient [-]
g	acceleration of gravity [m/s^2]
i_{crit}	erosion resistance coefficient [N]
i_q	momentum flux of overland flow [N]
i_r	momentum flux of the rainfall [N]
$i_{r,j}$	momentum flux of drop size class j [N]
k	number of drop size classes
n	Manning's roughness coeff. [$s/m^{1/3}$]
N_0	a factor, 2000
q	discharge per unit width [$m^3/m/s$]
q_{crit}	flow rate at initial erosion [m^2/s]
q_S	sediment discharge [kg/m/s]
r	rainfall intensity [mm/h]
r_j	intensity of drop size class j [m/s]
Re_j	Reynolds number of rain drop motion at terminal velocity [-]
S	slope gradient [-]
v_j	terminal velocity of drop size class j [m/s]
v_q	mean flow velocity [m/s]
$v_{r,j}$	terminal fall velocity of drop size class j [m/s]
α	slope angle
β	effective area ratio [-]
Δy	flow depth [m]
Δz	width of slope segment [m]
μ	dyn. viscosity of air [kg/m/s]
ρ	density of water [kg/m^3]
ρ_a	density of air [kg/m^3]
ρ_q	fluid density [kg/m^3]
σ	surface tension [N/m]

Farm Land Erosion: In Temperate Plains Environment and Hills
S. Wicherek (Editor)
1993 Elsevier Science Publishers B.V.

Soil erosion and badlands areas

P. Ballerini [a], M. Brunori [a], S. Moretti [a], G. Rodolfi [b].

[a] Dipartimento di Scienze della Terra, Universita di Firenze, Italy.

[b] Dipartimento di Scienza del Suolo e Nutrizione della Pianta, Universita di Firenze, Italy.

SUMMARY

This work comes by a three years research project developed by the Earth Sciences Department of the Florence University financially supported by ENEA (National Agency for New Technologies, Energy and Environment - Rome). The task of this project consist mainly in evaluating soil erosion in a badlands sample area of central Italy. Here, in a sector of central Apennines facing the Adriatic coast, the geology is characterised by marine plio-pleistocene sediments constituted of a conglomerate, sand and clay sequence. A high rated erosive phenomena takes place producing wide and deep badlands landscape (calanchi). The work is developed on a series of plots, for the sediment yield interception, having the same shape and surface but different steepness, exposure and vegetation cover. The sediment produced by the rainfall and runoff is channelled towards a flowmeter and after, towards a case for settling and collection.

A further control of the detached material along the slope is done using the WEPP (Water Erosion Prediction Project) model with which it is possible to analyse the different soil (particles) distribution due to a simulated or a real rainfall along a profile. Using such a model several characteristics of the considered soil have to be analysed.

A multitemporal geomorphologic analysis helps to distinguish the different areas in which the badlands evolution has been more or less important in the past. By means of this work it is possible to characterise the landscape evolution as a function of the parameters that influence the soil loss, especially in areas where badlands are highly developed as recorded in several places of Italy, from the piedmont to Calabria and Sicily.

RESUME

Cet article résulte de la collaboration du Département des Sciences de la Terre (Florence) et l'ENEA (Agence Nationale pour la Technologie Nouvelle, l'Energie et l'Environnement - Rome).

Le projet vise à évaluer l'érosion des sols sur des parcelles expérimentales dans une région de bad-lands. La région d'échantillonnage se situe en Apennin central dans un secteur constitué de sédiments marins sableux et argileux pléistocènes.

L'érosion est très active sur la fraction argileuse ce qui donne naissance à des bad-lands larges et profonds. Le travail sur le terrain tente d'intercepter les sédiments à l'aval de parcelles qui ont à peu près toutes la même forme et la même surface ; leur pente, leur exposition et leur couverture végétale varient.

Les sédiments mis en mouvements sont canalisés et vont se déposer dans des bacs. Une station météorologique a été installée afin d'évaluer directement l'influence de la pluie sur les phénomènes érosifs. Le but de cette expérimentation toujours en cour est de caractériser l'évolution du milieu en fonction de paramètres agissant sur la perte en sol dans des secteurs où les bad-lands sont très développés. Les relations entre paramètres physiques et degré d'érosion sont envisagées particulièrement dans des zones où les bad-lands sont très développés et enregistrés dans plusieurs sites d'Italie, du pied mont à la Calabre et la Sicile.

INTRODUCTION

A considerable amount of researches have been undertaken on the problem of soil loss and the mechanics of water erosion is rather well understood and documented. To the other hand the basic processes of soil erosion, including the principles of its control and prevention, are commonly and schematically identified as follow : 1) detachment of particles by raindrops falling on a bare soil; 2) and runoff development on a slope, carring on the particles previously or by water shear stress detached and outwashed from the slope. In taking in account these consideration the basic factors affecting soil loss by water are : erodibility of the soil, climatic aggressiveness (rain erosivity), morphology, land use and management.

In such a context this study is collocated, either to obtain quantitative data on soil erosion of clayey-silty soils, and to understand the geomorphologic evolution on respect of the existing relations between the most significative factors which operate in central Italy and in particular on the badlands landscape ("calanchi").

The word "calanco", that may be considered a particular form within the badlands landscape, generally defines a rather small hydrographic unit with a tributary system in which each channel is separated from the others by means of a more or less sharp ridge with a slope depending on the physical and mechanical characteristics of the bedrock.

To a general level it is possible to consider that the "calanco' as a landscape developed under a certain morphoclimatic period, obviously different from the one actually present (generally dryer), which helps a more uniform distribution of the badlands areas on the same lithological unit (1). It was also not strongly connected to the slope exposure and controlled mainly by water erosion processes and locally by structural elements (2).

A climatic change toward a moist period but with marked seasonal contrast most similar to the actual situation, could induce a marked differentiation in respect of the degradational processes along the slopes depending on the insolation intensity. Going on with such climatic condition in one hand the badlands formation has a regression phase on the northward exposure on the other hand there is a conservation of the same forms on the south-facing slopes. This could be explained by the maintenance of a certain slope because of the physical characteristics of the materials (water content, internal friction angle) mainly dependent by the solar radiation. The exposure factors it is influent, in the badlands genesis, only to permit the maintenance of those characteristics generated in a different morphoclimatic period under the action of a different process. Under this point of view the badlands ("calanco") could be consider as a residual landform surviving in the actual climatic period (1).

Some of these considerations, briefly resumed in Tab. 1, are quite difficult to transfer in the practical research if not supported by historical data and models which permit to simulate the processes in well known climatic situation as well morphological condition. The opportunity to use models such as USLE or WEPP is important also considering their limitations and constraints.

Table 1.
Badland landforms affecting factors after different authors.

	Dominant factor	Authors	Year	Site
Lithology	Texture	Sheideegger	1968	USA
	"	Vittorini	1977	Valdera
	Clay mineralogy	Sfalanga & Vannucci	1975	Valdera
Structure	Monoclinal structure	Castiglioni	1933-1935	Romagna
	Structural bonds & presence of caprock	Lulli	1974	Valdera
	"	Guasparri	1978	Valdorcia
Landform dynamics	Neotectonics and mass movement	Guerricchio & Melidoro	1979-1982	Basilicata
	"	Dramis et al.	1984	Marche
	Basal undercutting	Rodolfi & Frascati	1979	Valdera
	"	Alexander	1980	Basilicata
	Shallows slides	Lulli & Ronchetti	1973	Valdera
		Vittorini	1971-1979	Valdera
Climate	Slope aspect	Passerini	1937	Romagna
	"	Panicucci	1972	Valdera
	Morphoclimatic heritage	Rodolfi & Frasacati	1979	Valdera
Man's influence	Deforestation	Dramis et al.	1984	Marche
	Intensive agriculture	Alexander	1980	Basilicata

After Mazzanti & Rodolfi (11) and Rodolfi (12) updated.

GEOLOGICAL OUTLINES

The sample area chosen for this research is located in the Teramo province and includes the Atri municipality (Fig. 1).

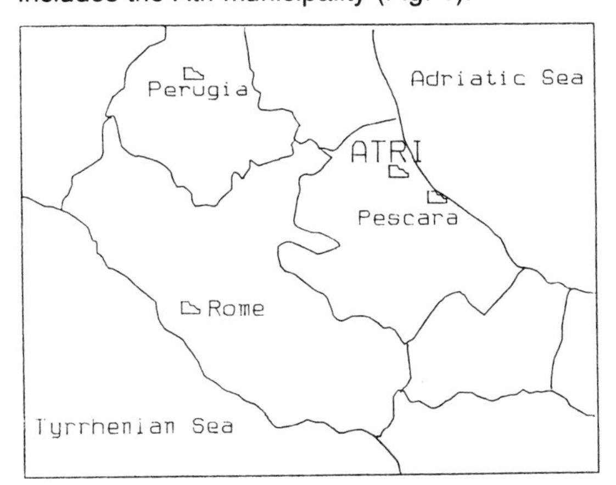

Figure 1. Location of the sample area.

The landscape is characteristic of the badlands areas that occur on the marine Pliocene deposits which a prevalently fine texture (3). It is characterised by a series of sub-parallel rivers mainly west-est oriented. These rivers, cut down the deposits so that it is possible to identify quite well the stratigraphy of the area (figure 2) :
- middle Pliocene clay and marls; on these materials the most part of the badlands is formed ; such deposits pass in concordance into sand and sandy-clay of the upper Pliocene ;
- lower Calabrian deposits, mainly sandy-clayey with marls and fine sand interbedded (4 - 5) ;
- sand and clay with some coastal conglomerate levels dated to the lower Calabrian ;
- continental conglomerate, forming very steep escarpments bordering remnant plateau, at the top.

The morphology of the area is strongly influenced by this lithological succession. In fact there are several structural surfaces which indicate the resistant materials covering the more erodible underlying deposits.
In the areas where such a cover has been removed by the erosive processes there is a completely different morphology, in which badlands and rolling areas alternate, of course not in the same proportion.

In Italy a similar geolithological situation is distributed from the inner part of the Alpine chain to the southern part of the country, along the Apennine ridge, wedging into the Mediterranean sea. The climatic condition, in this case, becomes extremely different from the north to the south, gradually increasing the aggressiveness. In the Atri area such climatic condition is comparable to the Mediterranean areas characterised by a dry period during summer and by a concentration of rainfall during autumn (with an absolute maximum) and spring (secondary maximum). The rainfall are not frequent and short in time but with a strong intensity (Tab. 2).

Table 2.
An example of some rainfall events during 1990 winter and their associate erosional figures.

RAINFALL OF 1990						
Date	**Time** *(h)*	**Intensity** *(mm/h)*	**Max int.** *(mm/h)*	**Hight** *(mm)*	**Runoff** *(mm)*	**Soil loss** *(kg)*
10/04/90	5.25	7.89	72.00	41.43	25.40	21.60
19/05/90	5.98	4.04	59.24	24.16	11.50	10.08
15/06/90	9.73	1.49	30.02	14.52	2.00	1.44
18/07/90	3.57	4.00	13.78	14.27	0.00	0.00
08/08/90	8.90	3.23	23.95	28.76	0.00	0.00
11/09/90	0.75	17.95	38.81	13.46	5.20	3.96
16/11/90	17.22	4.68	22.01	80.63	37.80	15.12
01/12/90	11.30	2.08	12.73	23.48	0.00	0.00
05/12/90	9.15	1.89	5.64	17.26	0.00	0.00
06/12/90	11.20	1.03	3.43	11.55	0.00	0.00
12/12/90	11.60	2.54	36.00	29.44	13.40	8.64
13/12/90	22.23	1.98	30.00	43.99	21.70	12.60
17/12/90	23.17	2.74	30.00	63.48	35.80	19.08

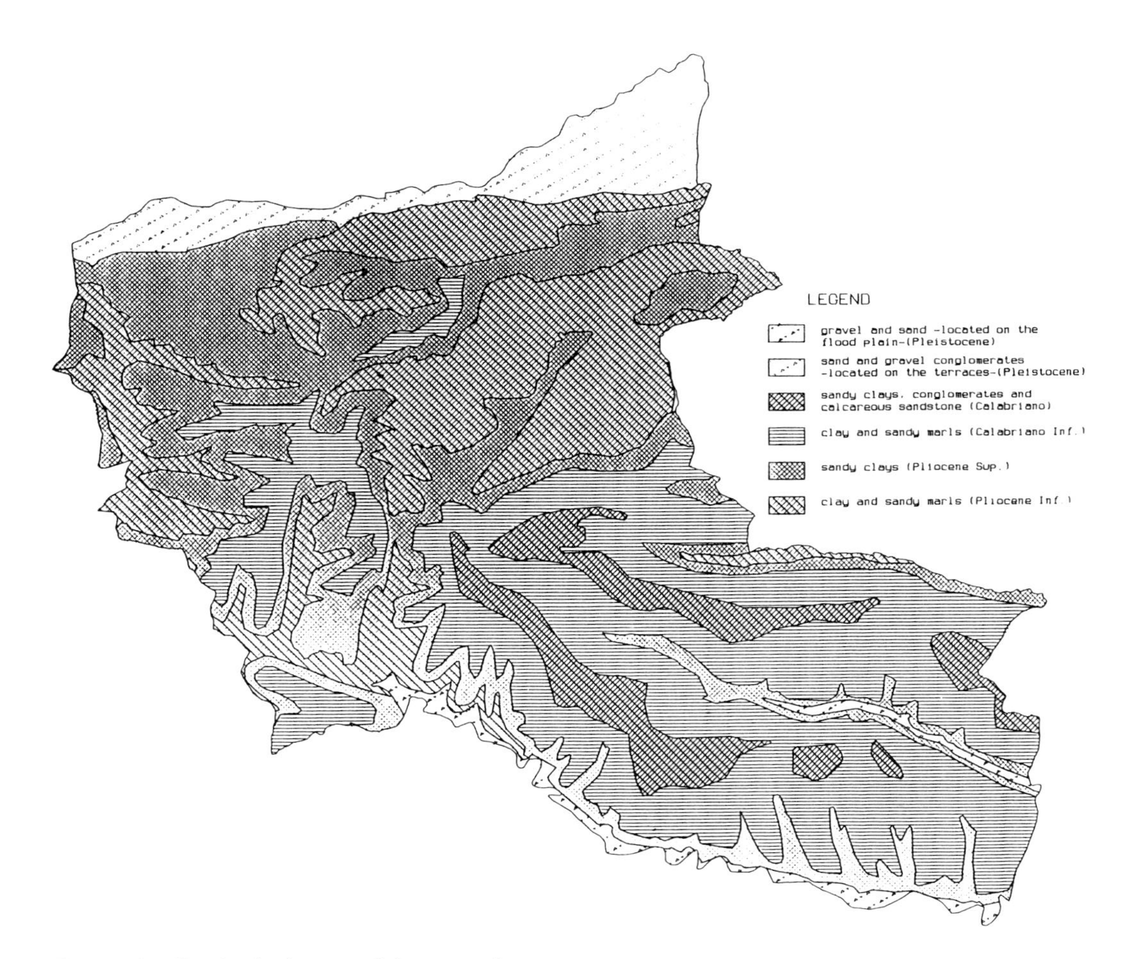

Figure 2 - Geological map of the sample area

Due to such condition, in the areas where these clayey Pliocene deposits outcrop, there exist a concurrence of different natural factors (1) ; on the other hand an antropic component, not properly trascurable, is present too.
All these factors can be resumed as follows :
- soils and substrata with a high suitability in respect of the weathering processes, because of their poor physical characteristics (soil erodibility) ;
- relevant aggressiveness of climatic conditions (rain erosivity) ;
- strong agricultural activity for some hundred years, or recently associated with a not conservative management ;
- rather intense geological (tectonics) activity with common catastrophic phenomena (earthquakes).
Deriving from this situation a strongly dissected landscape often results, moreover a logical consequence of a fast landscape evolution controlled by the two main processes which influence the slope morphology : water erosion and mass movement.

MATERIALS AND METHODS

A series of experimental plots was realised on a large divide between two badlands sub-watershed. A meteorological station was associated to the plots themselves. Furthermore the two sub-watershed were kept under photographic control to detect the landform evolution during the rainfalls, and in particular devoted to record the surficial mass movements developing on the rather shallow clayey soils during and after rainfall of strong intensity. Up to now, limiting to a sample area, the badlands facing distribution over 30 years (1954-1985) was computed (Fig. 3) comparing the difference between the two periods. The observation was done by means of photointerpretation analysis trying to separate the active badlands areas from the partially remodelled and not active ones.

The plots have the purpose of measuring the surface water erosion with the general methodology used for estimating soul loss by Wischmeier and Smith (6). The data analysis follows the normal USLE (Universal Soil Loss Equation) procedure associated to the new WEPP (Water Erosion Prediction Project) profile model (7) together with field rainfall simulations. WEPP model was associated to several sample rainfall during 1989 and 1990 and some of them are represented in figure 4.

RESULTS AND DISCUSSION

The data recorded in the bare fallow plots confirms that erodibility (mean value of 0.38 by Wishmeier method) is not constant throughout the year but varies as the cosinusoidal function in different period of the year (8 - 9).

Regarding the WEPP simulation associated to real rainfalls there are some correspondence between the data obtained from the field plots, also if a longer period of observation and associated simulation is needed to have a confirmation on representativeness of the WEPP soil loss forecast (Tab. 2).

Figure 4 (a,b,c) represents some examples of the real and simulated rainfalls (by WEPP model), in which the runoff has been computed by the model. For these events the soil loss is computed too.

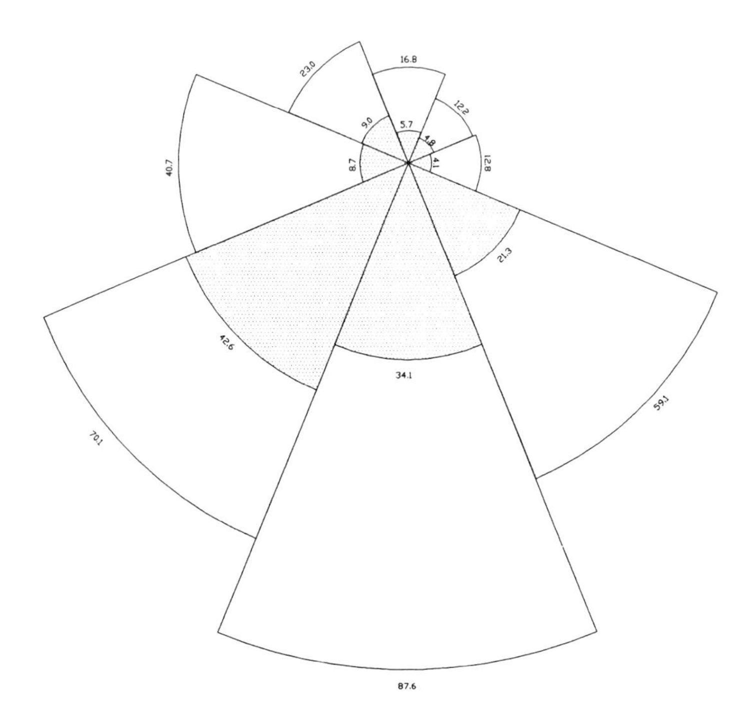

Figure 3. Badlands surface comparison (in ha) between 1954 and 1985 by polar representation.

The soil eroded from the experimental plots was examined in weight and in its physical characteristics (figure 5) and associated with the rainfall over a sample period determined by rainfall amount of at least 12.5 mm, the distribution of water discharge was computed too ; the measure was done by a continuous water discharge instrument.

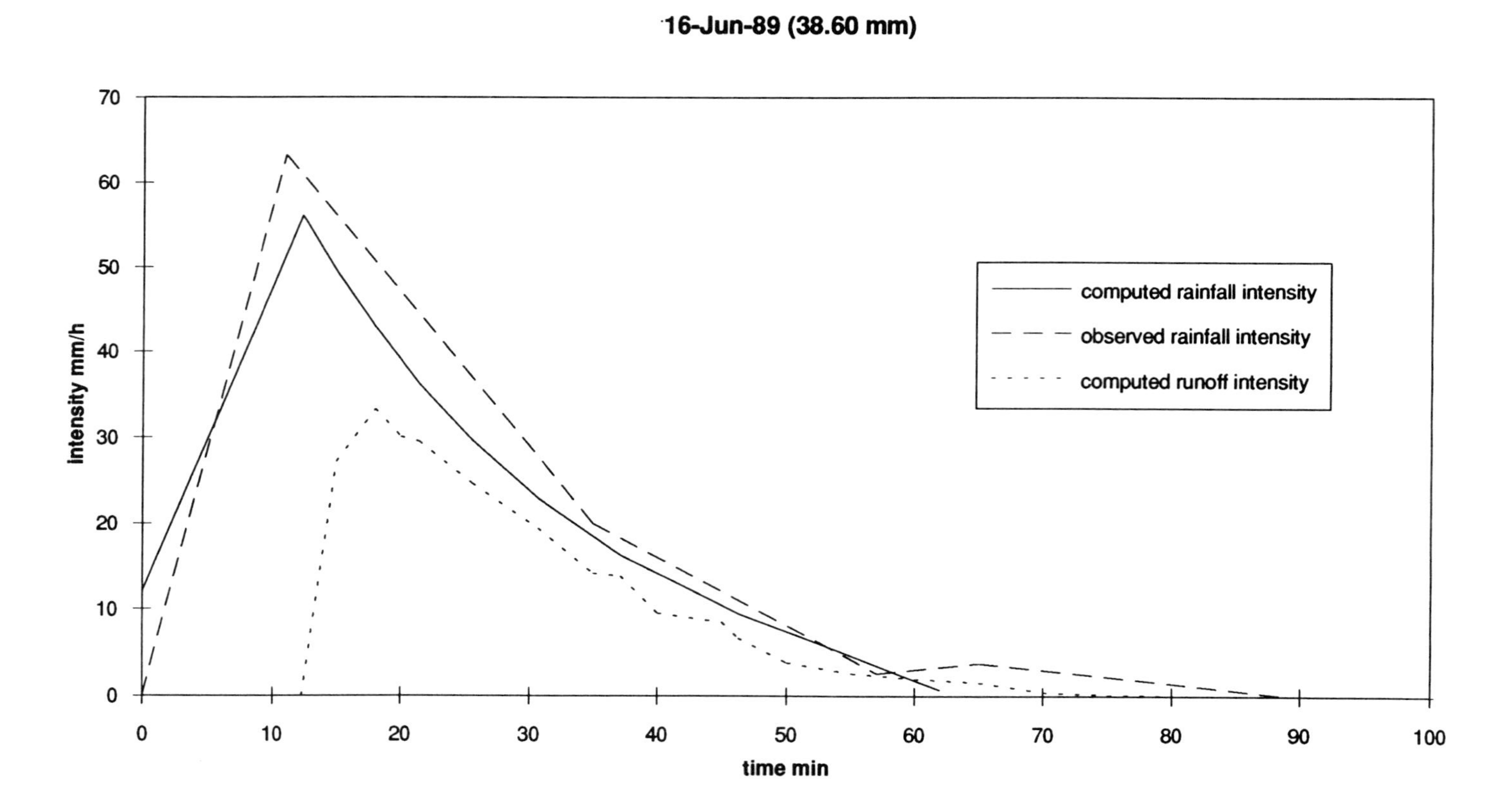

Figure 4a. Computed and observed rainfall data on 16 June 1989 rainfall.

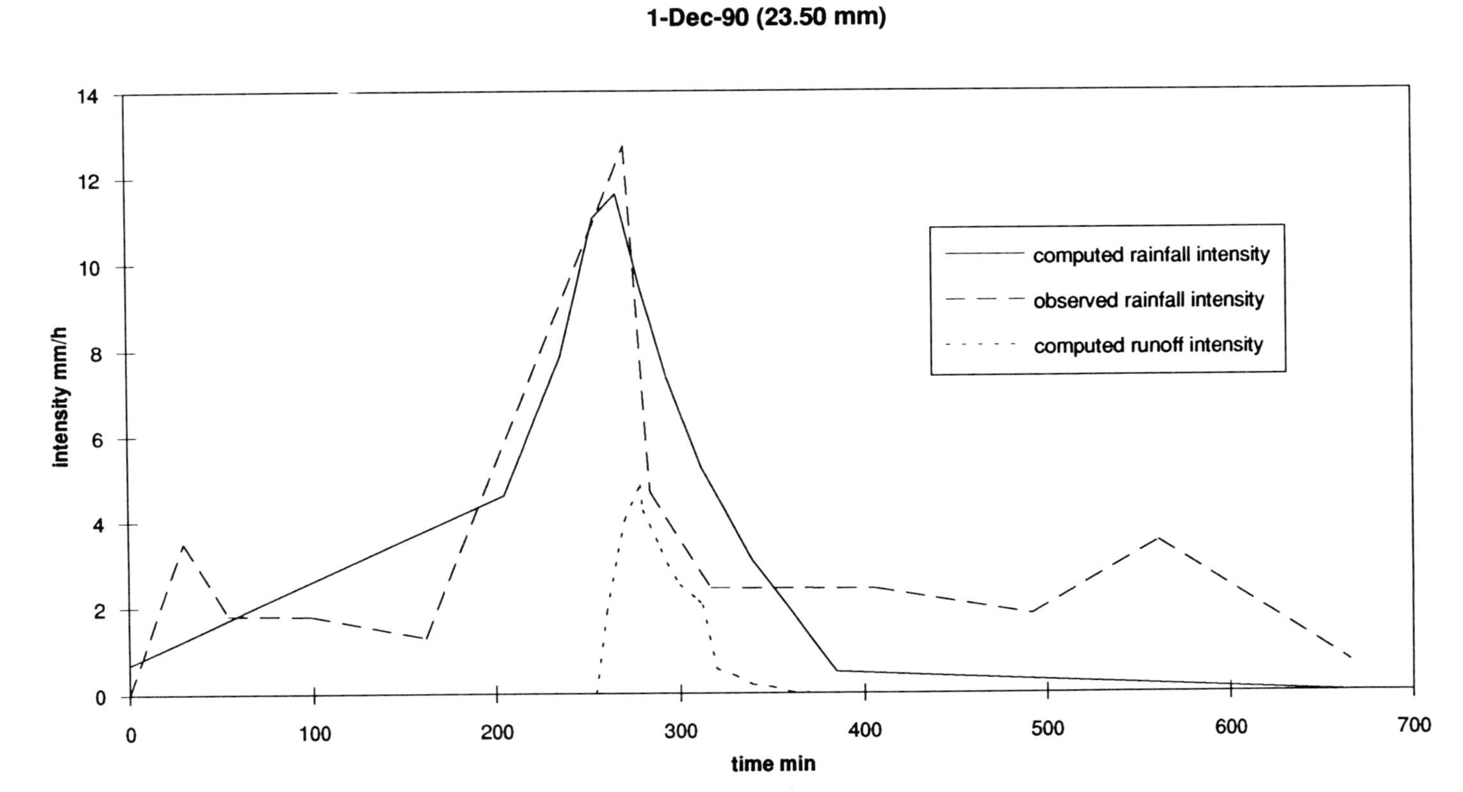

Figure 4b. Computed and observed rainfall data on 1 december 1990 rainfall.

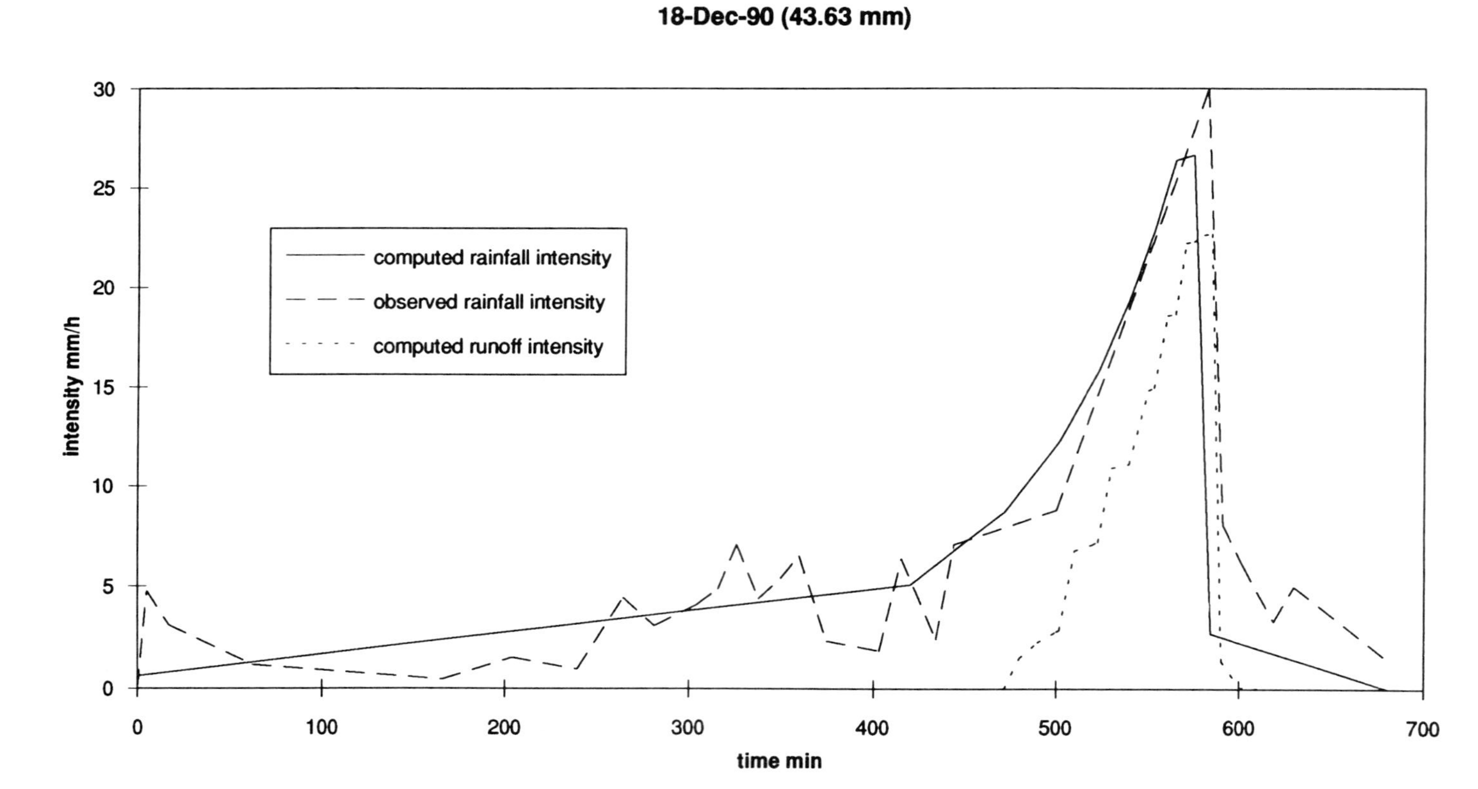

Figure 4c. Computed and observed rainfall data on 18 december 1990 rainfall.

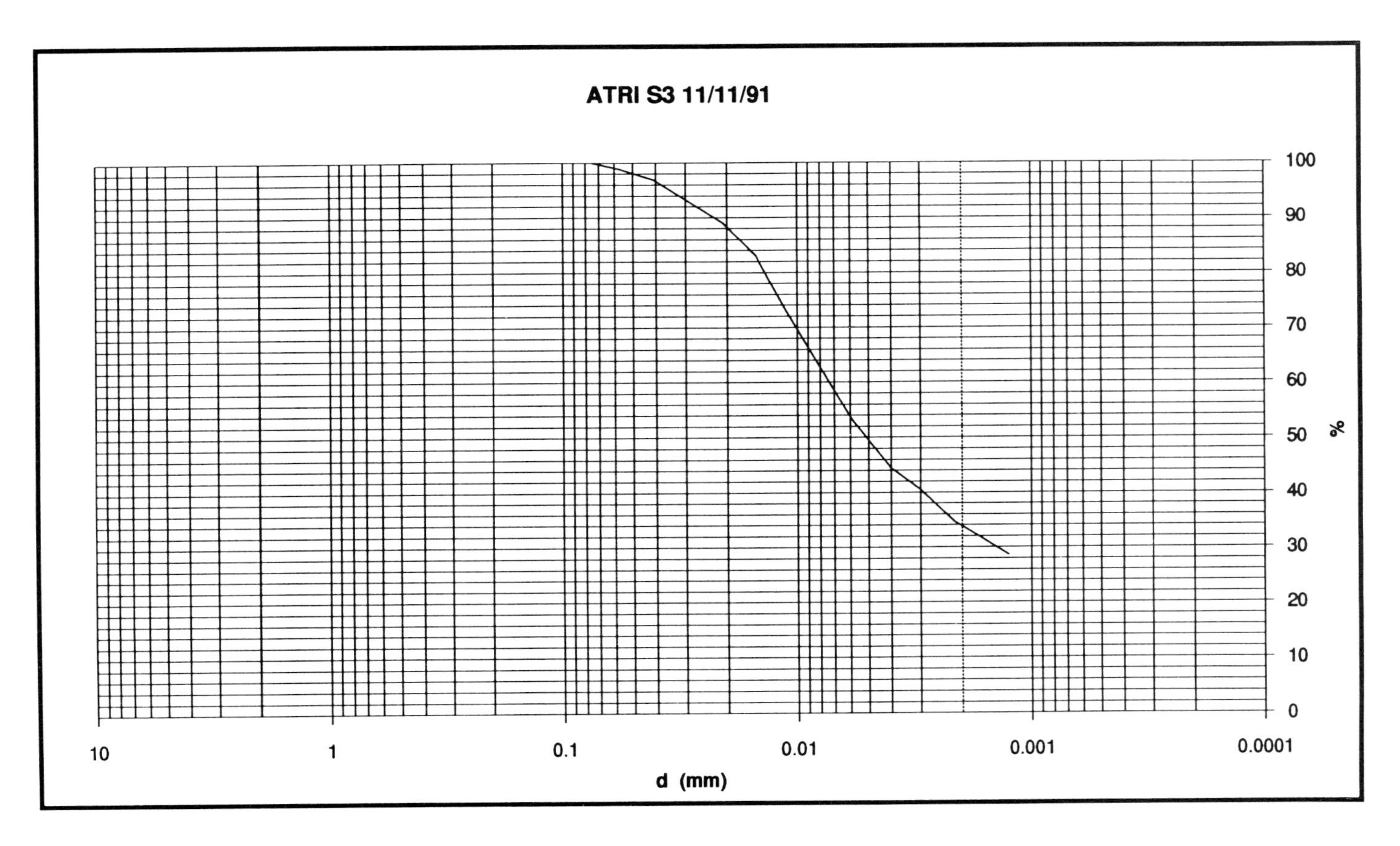

Figure 5. Representative texture curve of the clay deposits.

Concerning the resulting data of soil loss obtained from field plots, it seems possible to forecast soil loss on a monthly basis both with the Universal Soil Loss Equation and the WEPP profile model, however the second has to be related with more precise knowledge of rainfall distribution to know exactly the antecedent soil moisture content. It is, in fact, very important for these clayey soils to estimate the possible presence of cracks, which can change very much the infiltration and runoff condition. Furthermore the WEPP profile model it is rather useful to estimate the along slope changes of particles movement.

Regarding the study of the mass movements evolution, it has been noted that the seasonal changes have a strong influence on their evolution. The shallow soils (40-50 cm depth at maximum) forms an isolated body on the underlying bedrock creating a surface on which the water can infiltrate and flow changing the mechanical properties and the previous equilibrium. This fact creates the condition to a shallow but very rapid mass movement comparable as a mudflow.

In the Atri area the process of badlands evolution is mainly influenced by a succession of mass movements evolving during and after intense rainfall periods. The most part of the processes has been observed (during 1989, 1990 and 1991) along the slopes and the channels where the movement is concentrating its action this confirms one of the assumption of table 1 (10), under such condition the evolution by means of water surface erosion was limited at the upper parts of the slopes or on the larger divides. Furthermore the actual situation is evolving in a conservative way as shown in figure 3, this is probably due to the increasing of man activity which is mainly directed in remodelling the landscape for an easier management and exploitement of the land.

REFERENCES

1. Rodolfi G., Frascati. (1979) : Cartografia di base per la programmazione degli interventi in aree marginali (area rappresentativa dell'Alta Valdera) : carta geomorfologica. Annali Ist. Sper. studio e Difesa Suolo Vol. IX.
2. Guerricchio A., Melidoro G. (1982) : New views on the origins of badlands in the plio-pleistocenic clays of Italy. Proceeding IV Congr. IAEG, 2.
3. Brunori F., Moretti S. (1990) : G.I.S. in geology : an application for engineering geology maps. VI Int. IAEG congress. Amsterdam.
4. Valloni R., Parea G.C. (1984) : Late quaternary marine terraces in central Italy. V European Congress of Sedimentology, Marseille.
5. Parea G.C. (1984) : Uplifted strandlines along the front of the apenninic chain. V European Congress of Sedimentology, Marseille.
6. Wischmeier W.H., Smith D.D. (1978) : Predictin rainfall erosion losses - a guide to conservation planning U.S.D.A. - Agricultural handbook n.537.
7. Laflen J.M. (1990) : Water Erosion Prediction Project - Hillslope Profile Model, second edition. NSERL report n° 4, National Soil Erosion Research Laboratory USDA - Agricultural Research Service, West Lafayette.
8. Zanchi C., Bazzoffi P., D'egidio G., Nistri L. (1983) : A new rainfall simulator, with improved characteristics, for field erosion studies - Ann. Ist. Sper. Studio e Difesa del Suolo XIV. Work supported by ENEA - ROMA.
9. Bazzoffi P., Canuti P., Moretti S., Rodolfi G., Zanchi C. (1986) : Quantitative evaluation of some slope processes (surface erosion, mass movements) in experimental areas with different agro climatic conditions in central Italy. Z. Geomorph. N.F. Suppl.-Bd. 60.

10. Lulli L., Ronchetti G. (1973) : Prime osservazioni sulle crepacciature dei suoli delle argille plioceniche marine nella Valle dell'Era, Volterra (Pisa). Ann. Ist. Sper. Studio e Difesa Suolo, Vol. 4
11. Mazzanti R., Rodolfi G. (1988) : Evoluzione del rilievo nei sedimenti argillosi e sabbiosi dei cicli neogenici e quaternari italiani. in Canuti P. Pranzini E. (A cura di) : La gestione delle aree franose. Ed. delle Autonomie, Roma.
12. Rodolfi G. (1991) : Forme di erosione nei sedimenti neogenici e quaternari. In Mazzanti R. (A cura di) La gestione delle aree collinari argtillose e sabbiose. Ed. delle Autonomie, Roma.

Farm Land Erosion: In Temperate Plains Environment and Hills
S. Wicherek (Editor)

Agriculture on the Brazilian basaltic plateau. Cultivation practices and soil conservation : First results of the Potiribu project

N. M. dos R. Castro, P. Chevallier, A.L.L. Silveira da

Institute of Hydraulic Research, Convention IPH-ORSTOM-CNPq (Brazil-France), Federal University of Rio Grande do Sul, C.P. 530, 90001-970 Porto Alegre RS, Brazil.

SUMMARY

Over the last twenty years soya bean semi-monoculture has developed on the Brazilian basaltic plateau (states of Paraná, Santa Catarina and Rio Grande do Sul), causing serious soil degradation problems : linear erosion, compaction, increased solid loads in watercourses and deposition in reservoirs, besides changes in hydrological regimes. After a brief description of the environment, the authors present the first hydrosedimentological results of a research project to measure agricultural impact on the environment in this region. Observations performed in 1 m≤ plots, under simulated and natural rainfall and hydrological events for two catchment sizes (110 ha and 19.4 km≤) are analysed.

RESUME

Une quasi-monoculture du soja s'est installée sur le plateau basaltique brésilien (Etats du Paraná, de Santa Catarina et du Rio Grande do Sul) depuis une vingtaine d'années, entrainant de sérieux problèmes de dégradation des sols : érosion linéaire, compaction, augmentation des charges solides dans les cours d'eau et du dépôt dans les retenues, altération des régimes hydrologiques. Après une description sommaire du milieu naturel, les auteurs présentent les premiers résultats hydro-sédimentologiques d'un projet de recherche dont l'objectif est de mesurer l'impact de l'agriculture sur l'environnement de cette région. Des observations réalisées sur des parcelles de 1 m≤ sous pluies simulées et naturelles, ainsi que des évènements hydrologiques pour deux tailles de bassins versants (110 ha et 19,4 km≤) sont analysés.

INTRODUCTION

The South American basaltic plateau covers a large part of three states in the South of Brazil (Paraná, Santa Catarina and Rio Grande do Sul). These regions have been used for agriculture for several decades, following immigration (mainly German and Italian) to a region formerly used for intermittent cattle breeding.

Complete forest felling, farming without attention to good cultivation practices, the systematic and abusive use of chemicals in agriculture soon led to serious soil and water conservation and quality problems. Recently, however, authorities, cooperatives and farmers have become aware of these problems and have attempted to protect this extremely cultivated region which makes a major contribution to the Brazilian economy.

In 1989, a research project began to study the impact of agriculture on the environment, with cooperation between the Institute of Hydraulic Research of the Federal University of Rio Grande do Sul (IPH/UFRGS) in Porto Alegre and the French Institute of Scientific Research for Development in Cooperation (ORSTOM).

THE ENVIRONMENT

The geographical environment

Borges and Bordas (1) study a large 230,000 km≤ area in Brazil, covering the eastern half of the states of Paraná and Santa Catarina and the northern half of the state of Rio Grande do Sul, located approximately between the 24 th and 30 th parallels South and between the 50 th and 55 th meridians West (Fig. 1). The authors distinguish homogeneous physical-climatic zones in this region based on three criteria (rain erosivity, soil erodibility and relief) and two zones critical for soil conservation : a marked relief (mean slope greater than 50 %), on the border of the plateau, and another with moderate relief (mean slope between 10 % and 20 %).

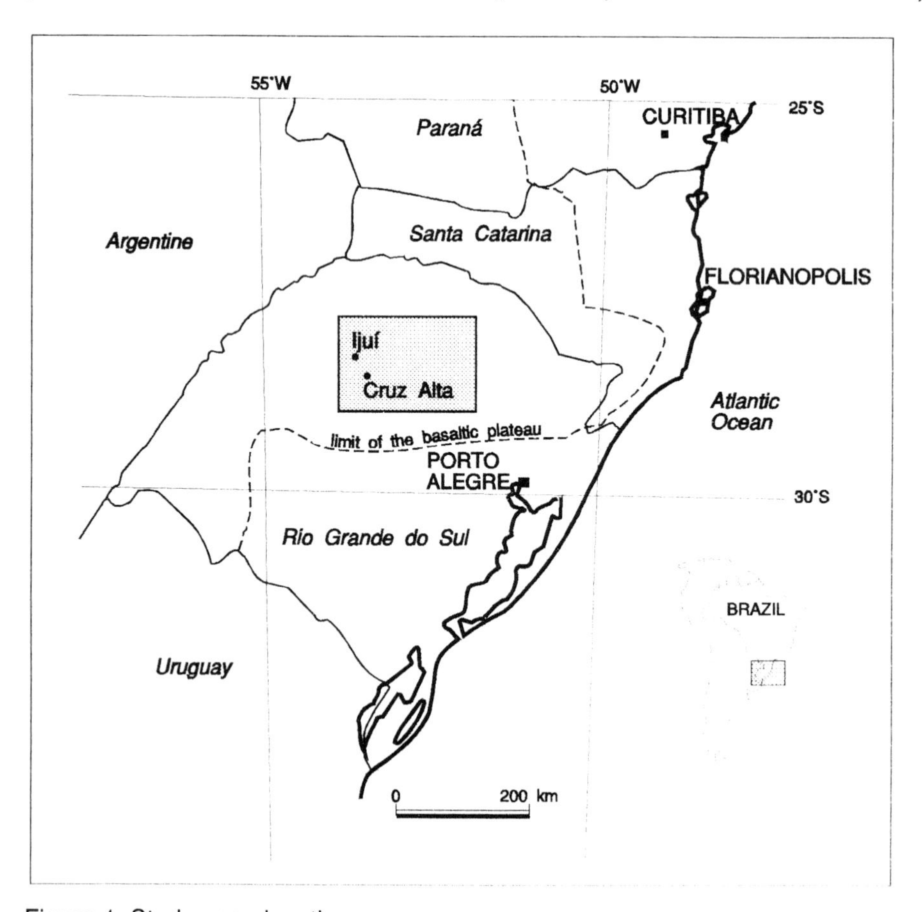

Figure 1. Study zone location.

A group of catchments belonging to the first zone had been studied at the end of the seventies and had shown the important role played by rainfall in soil losses (2). Agriculture in this area consists mainly of cattle breeding and small-scale farming, often on steep slopes (mainly maize, soya beans, black beans, vineyards). Properties are small, several dozen hectares at most.

The catchment of the River Potiribu was chosen to study the second area, which is extremely planted. It is a tributary of the River Uruguay and enters the Atlantic Ocean through the Plata Estuary together with the River Paraná. Two medium-size towns (population approximately 100,000 each), Ijuí and Cruz Alta, are nearby. The research project benefits from the support of powerful agricultural cooperatives in the region (Cotrijuí and Cotricruz), local municipal authorities (especially in the county of Pejuỹara) and the University of Ijuí.

Geology and soils

The substratum consists of continental basaltic flows in approximately horizontal layers (*trapp*), about fifteen meters thick, separated sometimes by *inter-trapp* sand or sandstone lenses (3). These formations are 130 to 140 million years old (Lower Cretaceous, Upper Jurassic).

Soils in the Potiribu catchment are mainly *dark red* or *violet latosols* and *structured violet soils* (*latossolos vermelhos escuros ou roxos, terra roxa estruturada*), according to the Brazilian classification. Rich in clay, they are always well structured and are not hydromorphic (4). Table 1 shows mean granulometry. Soils are often deep, up to 6 meters and more. In cultivated areas they form *structural crusts* (according to the name give by Casenave and Valentin, (5)), with small aggregates which adhere to the lumps resulting from ploughing which breack up with time.

Table 1.
Mean granulometry and location of the main soil types.

Name	depth (cm)	medium sand (%)	fine sand (%)	silt (%)	clay (%)	location
latossolo roxo	0-80	10	8	22	60	Top and mid-hillslope
latossolo vermelho escuro	0-80	15	15	20	50	Top of the hillslope
terra roxa estruturada	0-20 20-120	13 5	16 7	21 16	50 72	Mid-hillslope

Climate

Nimer (6) describes the climate in this region as *humid temperate.* There is no dry season.

According to IPAGRO (7), mean minimum monthly temperature at Cruz Alta is 14°C during the coldest month (July) and the mean maximum is above 24°C during the warmest month (January), but the extreme temperatures are often below 0°C or above 35°C. Relative humidity throughout the year ranges from 65 % in December to 80% in June. Mean evaporation is between 2.3 mm/day in June and 4.2 mm/day in December for an annual total close to 1,200 mm.

The rainfall regime is regulated by the confrontation between the cold, dry polar anticyclone and the warm, humid South Atlantic anticyclone. Frontal discontinuity, which is active all year long, is regularly shifted from south to north, causing generally long storms of moderate intensity. This special circulation creates a local rainfall regime with practically uniform distribution throughout the year ([6]).

Mean annual rainfall in the region is 1,700 mm, with monthly means between 120 and 150 mm (8). However, marked irregularities occur both between and within years, as shown by observations made since 1989 (Fig. 2). Daily precipitation with an annual return period is 88.9 mm.

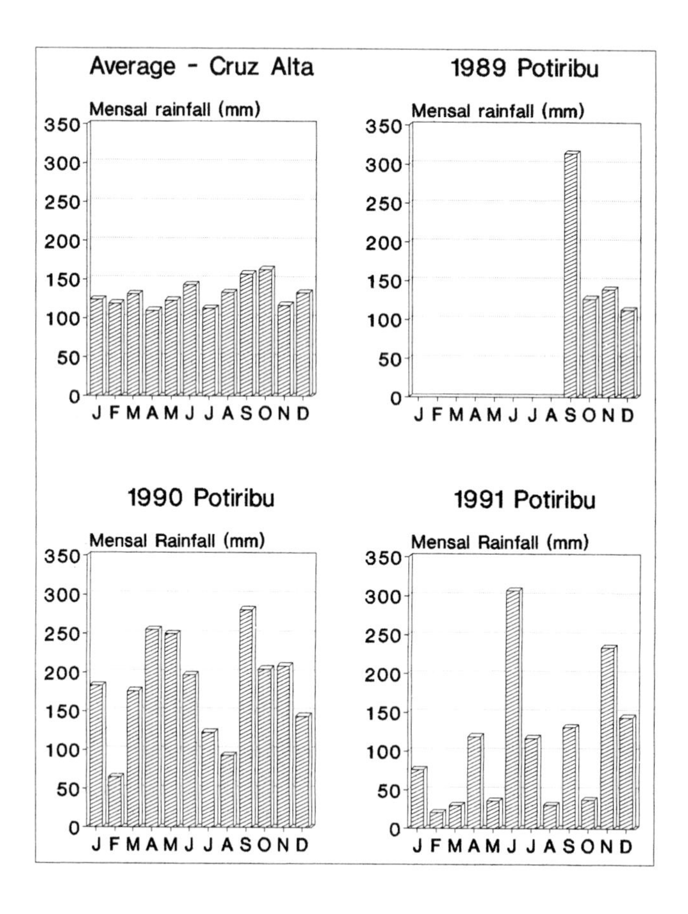

Figure 2. Mean monthly rainfall (mm).

The detailed study of storm events observed for 16 years at Ijuí shows an intensity of 115 mm/h in 5 minutes and 68 mm/h in 30 minutes with an annual return period. The Wischmeier erosivity index has a mean value of 468 I.S.u. and a value of 62 I.S.u. for a single event which occurs with annual frequency.

Agricultural activity and conservationist measures

Most of the area is used for crops and cattles. A few some small stands of the original forest remain, usually close to roads or on property limits. Riparian forests remain where streams are deeply marked ; they disappear when the talweg widens or has a large flood zone.

Cattle breeding is practiced in areas inappropriate for mechanized agriculture : steep slopes, boggy areas, shallow or excessively eroded soils.

Soya beans have been grown almost in monoculture for the last twenty years. Brazil is the second world producer and soya bean exports play an important role in repaying the foreign debt. The crop is sown in November and harvested in April-May. It undergoes several agrochemical treatments (fertilizer, fungicides, pesticides) during growth, and heavy machinery is required for soil preparation and harvesting. The present tendency among some producers, related to economic conditions for soya bean trading, is eventually to replace this crop with maize, which has a similar crop cycle. Twice-yearly cropping is less underspread, and depends more on the farmers' finances than on agricultural factors. The second crop is usually wheat or oats, and the latter is often ploughed back into the soil as green fertilizer. Sowing take place between May and July and harvesting just before soya bean planting in November.

The crop stands are medium to large ; rarely less than 10 ha, often over several hundred hectares. All agricultural work is performed by machines. Fertilising by air is done only by the farmers with the largest cultivated stands. Irrigation is not practiced and the use of crop rotations and fallow periods is disappearing.

As this heavy agriculture developed on agriculturally rich and mechanically fragile soils, many problems of erosion arose as shown by the onset of deep gullies (*voÿorocas*) in the path of concentrated surface flows and by the soil compaction which reduces infiltration capacity. Since the intensive development of soya bean crops on the basaltic plateau, a significant increase in the solid load in rivers, accelerated sedimentation in reservoirs and more severe floods or low flows have been observed.

Farmers have adopted two measures to counteract these effects :

- Construction of strips separed by banks (*terraÿos*) whenever the slope is greater than 5 %. These follow contour lines and their width, 10 to 30 m, depends on the slope. The strips are separated by an embankment, 50 to 80 cm high (*murundus*) and approximately 2 m wide. A furrow, to help infiltration on the one hand and drain excess surface runoff on the other, is dug immediatly above the embankment. Often a row of maize is planted on the crest to ensure stability. These structures are made anew each year by farmers using special machinery. If they breack stet unexpectedly during the crop cycle, they are immediately repaired.
- reduction in the number of operations required to work the soil and the use of more appropriate machinery (disk plough, rotavator,...). The cooperatives try to promote the *direct sowing* technique for soya beans after the spring harvest, without reworking the soil.

These anti-erosion measures are accompanied by increased agro-chemical treatment, to counter what is believed to be a loss in yield.

Finally, the cooperatives try to encourage greater diversification in the production of oleaginous plants (particularly colza and sunflower), but this has not been very successfull.

MATERIALS AND METHODS

The device used to estimate runoff and soil erosion related to agricultural practices, yields estimates approppiate for different space and time scales affected. Two such estimates may be distinguished : for small plots, giving attention mainly to production processes, and those carried out on basic geomorphological units, i.e. small catchments. The construction of the device has been described elsewhere ; in this paper we discurs only some observations which we consider significant.

Rainfall simulation and natural rainfall micro-plot

A detailed study of surface hydrodynamics and the tendency to erosion is performed using the small rainfall simulator built by ORSTOM (9). This device, which is now well-known, simulates rainfall events of intensities ranging between 30 and 150 mm/h on a 1 m≤ plot. For our study it will be used on an agricultural plot which has been banked against erosion, with a mean slope of 10 to 12 %, strictly following the crop calendar described above (10).

A group of three 1 m≤ plot (A, B and C) receives a serie of four storms at different seasons of the agricultural year. Two campaigns have already been carried out : at the beginning and end of the spring cycle, on oats, respectively one week (5 % cover) and fourteen weeks (80 % cover) after sowing. The simulated storm is based on the typical hyetogram with one-year return period : a total of 90 mm in 75 minutes, maximum intensity 100 mm/h during 10 minutes. It is repeated four times during four consecutive days.

Soil humidity is measured up to 80 cm depth by sampling before each storm. Surface runoff intensity is recorded continuously at the outflow from the plot. Runoff is systematicallysampled on one (B plot) of the three plots every two or three minutes to analyse the solid load. The solid particles deposited in the collection system are gathered and weighed at the end of each experiment.

A fourth 1 m≤ plot (D plot) installed at the same site is connected to a 60 liter tank : it is examined after each natural storm, and thus the runoff production and solid load under natural conditions as a function of the type of storm, and of soil status and plant development, may be evaluated.

Catchments

The River Potiribu catchment (Fig. 3), the natural frame of our study, is controlled a few kilometers east of Ijuí by a natural waterfall with a small hydropower plant above which the drainage area is 563 km≤.

During this stage of the study, two embedded sub-basins are studied, the Turcato catchment (19.4 km≤) and the Donato catchment (110 ha). The small Donato catchment is completely cultivated and banked. About 60 % of the Turcato catchment is cultivated ; of the remainder, 25 % is grazing land or saturated flood zones (25 %) and 15 % is the town of Pejuỹara (population approximately 2,500). These catchments are equipped with limnimetric stations which allow continuous measurement of water levels. Flows and solid discharges are often measured, if possible during floods. The River Potiribu catchment is equipped with a network of 9 raingauges and 11 rainrecorders to measure precipitation.

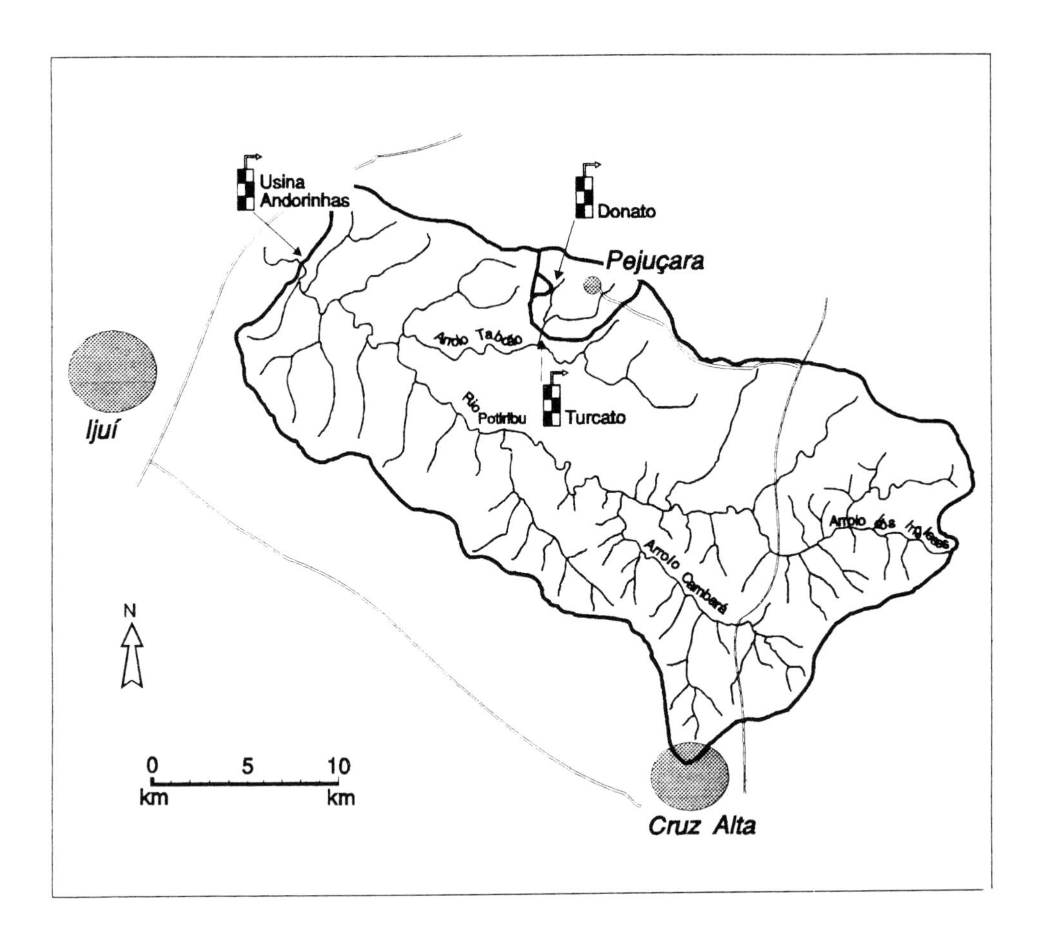

Figure 3. River Potiribu catchment at Andorinhas power plant.

FIRST RESULTS

Rainfall simulation and natural rainfall micro-plot

Table 2 shows the main results obtained for the eight storm events simulated on plot B during the two campaigns carried out between May 29 and June 1, 1991, and between August 26 and September 2, 1991. The mean slope of the plot is 11%.

Table 2.
Simulated rainfall results on plot B (1 m^2)

Campaign/ storm	Rainfall (mm)	Humidity for 80 cm depth (mm)	Variation for 80 cm depth (mm)	Runoff volume (mm)	Runoff coeffi-cient (%)	Conc. (suspe n-sion) (g/l)	Total conc. (g/l)	Soil loss (suspension) (g/mm of rain)	Total soil loss (g/mm of rain)
1 / 1	78,3	276	-	0,4	0,5	-	-	-	-
1 / 2	86,8	315	39	18,2	21	0,83	1,54	0,17	0,32
1 / 3	89,9	318	3	33,4	37	1,43	2,26	0,53	0,58
1 / 4	88,9	326	8	40,4	45	1,28	1,96	0,58	0,89
2 / 1	82,2	236	-	28,9	35	1,01	1,64	0,36	0,58
2 / 2	88,2	308	71	52,6	60	0,67	1,48	0,40	0,88
2 / 3	88,7	303	-5	56,6	64	0,68	0,91	0,43	0,58
2 / 4	85,3	300	-3	57,8	68	1,61	2,37	1,09	1,60

During the first period the soil had just been ploughed with a disk plough, the oats had been sown one week before, and the average height of vegetation was only a few centimeters with average cover of approximately 5 % of the plot area. The previous natural rains had occurred 15 mm, 8 days before, and 19 mm, 17 days before.

During the second series of measurements, 14 weeks later, the oats were between 50 and 80 cm high, covering 80% of the area. The soil had not been worked mechanically between the two periods. The last heavy rain had occurred 22 days before (28 mm).

This Table calls for some comments :

1. Continuous increase in runoff is observed during a campaign on the one hand, and higher mean value of this runoff during the second campaign, on the other. This shows the important role played by soil, to the detriment of plants, since greater plant cover should logically lead the opposite results. Two effects are combined : the vertical linear structure of oats, with little interception of rain, and progressive compaction of the soil which is initially cultivated, due to the effect of rain and its own weight.
2. Internal vertical drainage is very rapid : after the first rainfall in each of the campaign, when a dry water profile is refilled, variations in the 80 cm depth soil stocks are very small between two storms. In other words, rainfall is drained off in less than 24 hours, at a depth greater than 80 cm, since the effect of renewed evapotranspiration (2 to 4 mm/day) is small. Several small inconsistencies in the water balance will be noted in Table 2 due to different methods for measuring variables.
3. Soil losses captured in the suspended load at the outflow of the plot and related to the rainfall volume increase during the first campaign : they are almost stable, except in the case of the last storm which is suspicious, during the second campaign, since it has a lower value than the maximum value of the first campaign. This observation is consistent with the process mentioned in the first comment regarding progressive soil compaction during the crop cycle.

Table 3 shows the measurements performed in plot D, which is identical to the rain simulation plots and was observed under several natural rainfall events between June and September, 1991, i.e., during one period of the oats crop cycle after that observed under simulated rainfall. This type of experiment is subject to errors in measurement, and only orders of magnitude of the results should be taken into account. It is especially noted that observed values, both in terms of liquid and solid flow, are slightly higher than those recorded under simulated rainfalls, but are still of the same order of magnitude. This is explained by the fact that, under the conditions in the region studied, it is likely that the kinetic energy of natural rainfall is higher than that supplied by the rainfall simulator : this hypothesis is suggested by a partial study performed in the laboratory (11), but is yet to be confirmed under natural conditions.

Table 3.
Natural rainfall results on plot D (1 m^2)

Date	Rainfall (mm)	Erosivity (S.I.unit)	Runoff volume (mm)	Runoff coefficient (%)	Total concentr. (g/l)	Total soil loss (g/mm of rainfall)
06,17,91	35,1	4,68	11	31	2,79	0,86
06,18,91	57,6	10,7	41	71	2,60	1,85
06,27,91	45,6	20,4	43	94	2,77	2,60
07,10,91	15,8	2,09	6	38	2,57	0,98
07,21,91	28,5	1,31	7	25	0,87	0,22
09,18,91	36,9	4,55	13	36	0,65	0,23

Catchments

We will comment on the results observed in the Turcato and Donato catchments during two rainfall events which occured on June, 4 and 5, 1991 (Fig. 4), and June, 18 and 19, 1991 (Fig. 5). This period is immediately after the rainfall simulation and partially corresponds to observation of the plot under natural rainfall. Some of the features of these two events are shown in Table 4.

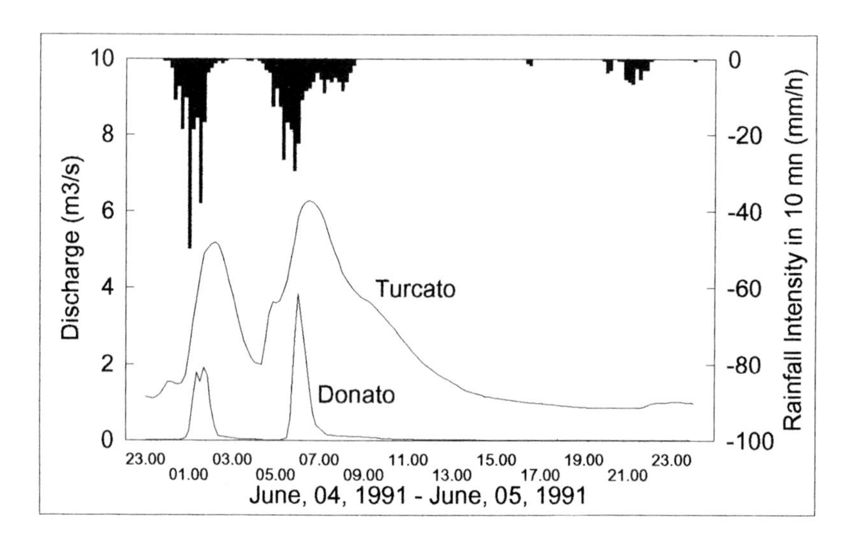

Figure 4. Event of June 4 and 5, 1991.

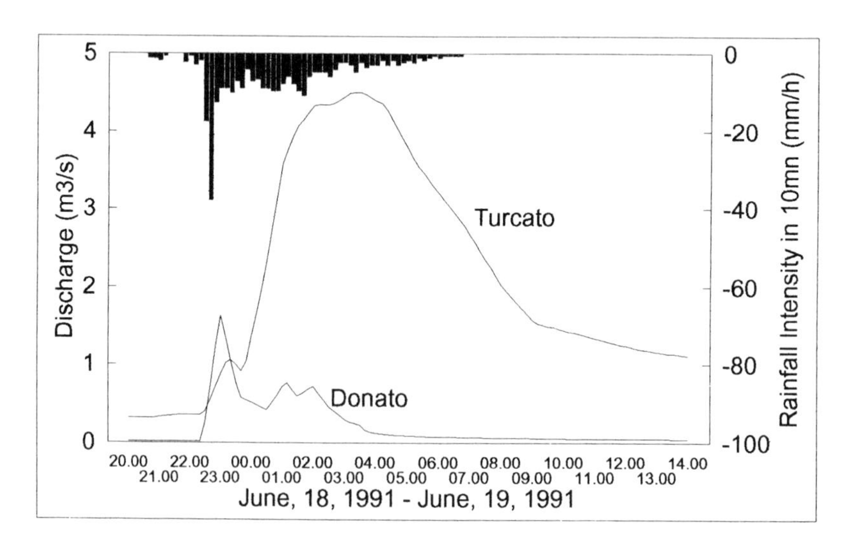

Figure 5. Event of June 18 and 19, 1991.

Table 4.
Results of two storms in the Donato and Turcato catchments

Date /catchment	Rainfall (mm)	Erosivity (S.I.unit)	Runoff volume (mm)	Runoff coefficient (%)	Mean concentr. (g/l)	Total soil loss (g/m≤. mm rain)	Total soil loss (t/ha)
06,04-05,91 /Donato	74,6	22,2	12,4	17	-	-	-
06,04-05,91 /Turcato	74,6	22,2	5,91	8	2,15	0,20	0,15
06,18-19,91 /Donato	50,3	10,7	7,19	14	5,28	0,76	0,38
06,18-19,91 /Turcato	50,3	10,7	3,46	7	-	-	-

This Table allows several observations :

1. For relatively heavy storms, of the order of magnitude of simulated rains, the rapid runoff coefficients are moderate, and distinctly lower for the Turcato catchment.
2. Mean concentrations of suspended material observed in the water course, on the other hand, are very high, greater than those observed in plots. This means that the solid load continues to increase initially with diffuse flow on the hillslope and then concentrated in gullies and water courses. It is particularly visible in the Donato catchment which has no saturated areas and is only farmed with banking.
3. Soil loss values are also very high. For purposes of comparison, Bordas al. ([12]) present a mean value of 0.95 t/ha.year, observed in large catchments (between 1,000 and 10,000 km≤) cultivated in this region. The mean concentrations (of the order of 0.2 to 0.4 g/l for a very average-sized flood) observed in the River Potiribu catchment at Andorinhas (563 km≤) might mean that the value proposed based on a mean concentration of 0.1 g/l by Bordas al. ([12]) may have been underestimated.

CONCLUSION

The study discribed in this article is still in progress and many observations are not yet available or cannot yet be used, especially those concerning the granulometry of moving particles and their physico-chemical qualities. The main goal here is to call attention to soil conservation problems in an intensively farmed environment, using modern methods, on fragile soils and under very humid climate conditions considering that is a temperate region. It seems to us particularly useful to study this situation, on the one hand to try to solve locally generated problems, and on the other because it may soon and speedily run into situations of large crop environments a sfound in less critical regions such as Europe.

AKNOWLEDGMENTS

This paper was translated into English by Professor Hedy L. Hoffmann and revised by Professor Robin T. Clarke of the Hydraulic Research Institute of the Federal University of Rio Grande do Sul in Porto Alegre (Brazil).

REFERENCES

1. Borges A.L. de O., Bordas M.P. (1988) : Choix de bassins représentatifs et expérimentaux pour l'étude de l'érosion sur le plateau basaltique sudaméricain. Sediment Budgets (Proceedings of the Porto Alegre Symposium, December 1988). IAHS Publ. no. 174.
2. Silveira G.L. da (1982) : Erosão do solo na encosta do planalto basáltico no Rio Grande do Sul. Representatividade dos parâmetros envolvidos na produÿão de sedimentos. Dissertaÿão de mestrado. IPH-UFRGS.
3. Ibge (1986) : Levantamento de recursos naturais, volume 33. Folha SH22 Porto Alegre e parte das folhas SH21 Uruguaiana e SI22 Lagoa Mirim. Projeto Radam Brasil. 792 p. + maps.
4. Carvalho A.P. de, Abrão P.U.R., Fasolo P.J., Pötter R.O. (1990) : Levantamento semidetalhado dos solos da bacia do Arroio Taboão (Pejuÿara/Ijuí-RS). Multigr., 41 p. + map 1:25 000).
5. Casenave A., Valentin C. (1988) : Les états de surfaces de la zone sahélienne. Influence sur l'infiltration. ORSTOM, coll. Didactiques, 230 p.
6. Nimer E. (1988). Climatologia do Brasil. 2a ediÿão. IBGE, Rio de Janeiro. pp. 195-264.
7. Ipagro (1989) : Atlas agroclimático. Rede meteorológica. Estado do Rio Grande do Sul. Porto Alegre, 3 volumes.
8. Chevallier P., Castro N.M. dos R. (1991) : As precipitaÿões na região de Cruz Alta e Ijuí (RS - Brasil). Anais do IX Simpósio Brasileiro de Recursos Hídricos (Rio de Janeiro, 10 a 14 nov 1991). Vol. III, pp. 183-192.
9. Asseline J., Valentin C. (1978) : Construction et mise au point d'un infiltromètre à aspersion. Cah. ORSTOM, sér. Hydrol., vol XV, n° 4 : 321-350.
10. Silveira A.L.L. da, Chevallier P. (1991) : Primeiros resultados sobre infiltraÿão em solo cultivado usando simulaÿão de chuvas (Bacia do Rio Potiribu - RS). Anais do IX Simpósio Brasileiro de Recursos Hídricos (Rio de Janeiro, 10 a 14 nov 1991). Vol. I, pp. 213-221.
11. Semmelmann F.R., Chevallier P., Alcântara W. de, Silveira A.L.L. da (1991) : Determinaÿão da energia cinética de chuvas simuladas. Anais do IX Simpósio Brasileiro de Recursos Hídricos (Rio de Janeiro, 10 a 14 nov 1991). Vol. I, pp. 89-97.
12. Bordas M.P., Lanna A.E., Semmelmann F.R. (1988) : Evaluation des risques d'érosion et de sédimentation au Brésil à partir de bilans sédimentologiques élémentaires. Sediment Budgets (Proceedings of the Porto Alegre Symposium, December 1988). IAHS Publ. n° 174.

Farm Land Erosion: In Temperate Plains Environment and Hills
S. Wicherek (Editor)
1993 Elsevier Science Publishers B.V.

Evaluation of Erosion Processes on an Area of the Karla basin in Larissa, Greece

S. A. Floras and I. Sgouras

IXTEL, Theofrass Str. 1, 413-35 Larissa, Greece.

SUMMARY

The study is related to the evaluation of the erosion processes which are taking place nowadays. We made a detailed study of the soils across a soil sequence (Cross section), which is traversing the surveyed area from east-northeast to west-southwest. Twelve (12) soil profiles (from which we selected six (6) for the purpose of the study) were described and studied in detail along this soil sequence and lots of laboratory data were received. From geological and geophysical studies the isodepths of the sediments were received and the volume of the sedimentation which came from the eroded material of the surrounding area was calculated. Experimental devices were also used to collect and estimate the eroded material. Finally a SPOT-1 satellite image was used to interpret the erosional processes of the study area.

RESUME

La région que nous avons étudiée est située au sud du bassin de Karla. Cette étude se propose d'évaluer les processus d'érosion actuels et de comparer diverses méthodes utilisées. Ces méthodes ont été de trois types : Travail sur le terrain et mesures, télédétection, traitement d'images par ordinateur.

Nous avons étudié dans le détail les sols d'une région qui prend en écharpe le bassin de Karla de l'est/nord-est à l'ouest/sud-ouest. Cette région comprend une grande partie de l'ancien lac Karla et le terrain accidenté autour des villages de Néo Monastiri et Agnaderi. Douze types de sols qui ont été étudiés en détail ont été classés en suivant la typologie de la classification des sols.

En outre, nous avons étudié des photographies aériennes et des images satellitaires, grâce à ces outils les processus ont pu être examinés de façon apprfondie. Dans la dernière étape de notre étude, nous avons utilisé des techniques de traitement SPOT 1 par ordinateur et des méthodes d'analyses statistiques.

INTRODUCTION

The area under study (Fig. 1) comprises part of the former lake Karla and the hilly area west-southwest of the lake, close to the villages of Neo Monastiri and Agnaderi. The former lake Karla is a tectonic depression of the recent geological epoch (Pleistocene) (1 - 2), inside which the two rivers Pinios and Titarisios were depositing their loads, consisted mainly of silt, clay, marly and coarser materials. The opening of the Tempi valley during the middle of the pleistocene epoch and the outlet of the pinios river through the narrow opening to the Aegean Sea, affected the rate of the sedimentation in the lake Karla, having as a result the confinement of the lake surface which finally occupied the southeast part of the Prefecture of Larissa and Magnesia. Finally the lake Karla was accepting the waters of the surface runoff, only of the surrounding area and its acreage was following the climatic fluctuations. In the southeastern area of the former lake Karla we find fossils inside the Quaternary deposits, which are of lacustrine and marine origin. The clay of the alluvial cover is containing lots of salts and the acreage coresponds to the limits of older lake Karla. The drainage of the lake Karla to Pagasitikos gulf was made possible through an artificial tunnel which was constructed in 1964 under the lower parts of Maurovouni and Pilion mountains.

Today lake Karla has been fully drained and its soil is under cultivation. Recently a small reservoir of approximately 3000 stremmas (Fig. 1) has been constructed for the environmental improvement and the irrigation needs of the crops. A number of small reservoirs in the area are under planning to cover the same needs as above. The limits of the lake towards the mountainous area, are covered by coarser colluvial materials. The eastern peripheral border of the lake is comprised of gneisses, gneiss-schists and Karstic marbles, while very often are found schists inside the marbles (2). The area west of the lake is consisted of river and torrential holocene deposits containing clay, silt and sand together with gravels and stones of various mineralogical composition (1 - 2). In the southwest area of the lake we find Tertiary marls.

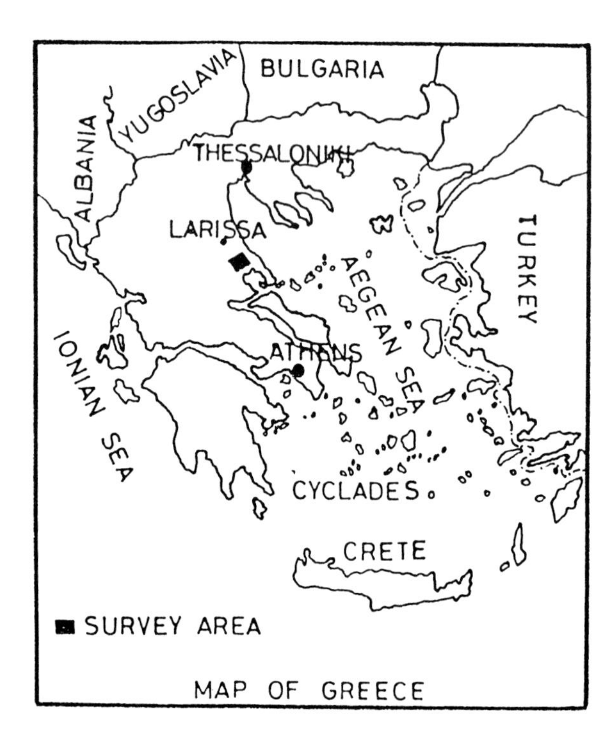

Figure 1. Map of Greece showing the study area.

MATERIALS AND METHODS

1. Field Observations

Natural and artificial soil profiles are able to help in the determination of the degree of erosion in plain and sloping areas. Across the cross section A-A' (3 - 4) (Fig. 2 and 4) of the former lake Karla are shown the derived soil profiles. On the southwest hilly area the depth of the soil horizons and the appearance of the parent material on the soil surface, give the clue of the degree of erosion, while on the less sloping areas the heterogeneity of the deposited materials which is distinguished macroscopically in several profiles from the soil color, the difference in soil texture, the different kind of rocks and the presence or not of fossils, give the rate of the sedimentation in specific chronological periods and in conclusion the rate and the degree of erosion (Profile P1, Photo 1) (Floras, S. Sgouras, I, 1992).

2. Chemical and Mineralogical Analyses

On areas with little slope we may estimate the heterogeneity of the soil horizons and the geological strata, whichin relation to their depth they show the rate of erosion and deposition in the surrounding area and on the bottom of the former lake Karla. Thus in the profile P1 (Table 1) after the depth of 48 cm there is a change in the chemical and mineralogical composition of the material. In the profiles P7 and P9, (Photo 2) differences are observed in various depths, while in the profile P4 which is located within the lake there is an homogeneity in the surface horizons, which means that the materials were deposited in short time during the epoch of holocene deposition as the products of intense erosion of the surrounding area. From the results of chemical analysis (8) of table 1 it is possible to estimate the chemical erosion, which is inferred from the rate of the translocation of the chemical elements in various depths of the profiles. There are also differences in heavy and light minerals in horizons and layers across the cross-section A-A' (Fig. 2).

3. Geological, Geophysical and Paleontological Measurements

The deposits in the former lake Karla are mainly of the pleistocene and holocene epochs having slight slopes to the center of the lake, while we find rather thin pliocene deposits to the southwest of the lake and plio-pleistocene deposits to the west borders of the Karla lake. The pleistocene sediments were distinguished from the numerous fossils that were found in various layers after drilling. The tectonic activity and the erosion of the hills and low mountains which surround the area, formed the today relief and topography of the study area. The erosion products contain mainly clayey, silty and sandy material, together with gravels and stones, which mineralogically are comprised of quartz, feldspars and calcium carbonate. The clay minerals are the montmorillonite and vermiculite. In the broader area of the former Karla lake there have been made a lot of drillings and geophysical studies in order to determine the thickness of the plio-quarternary sediments. These studies were conducted by the Electro-watt and Sogreah companies. From the combination of the lithological profiles, the drillings and the geophysical and geological studies, were found the isodepth curves of the pleistocene and holocene deposits. After the drawing of the isodepth curves and the estimation of the occupied area, the two diagrams of the isodepth sediments were compiled, which refer to the pleistocene and holocene sediments (Fig. 3).

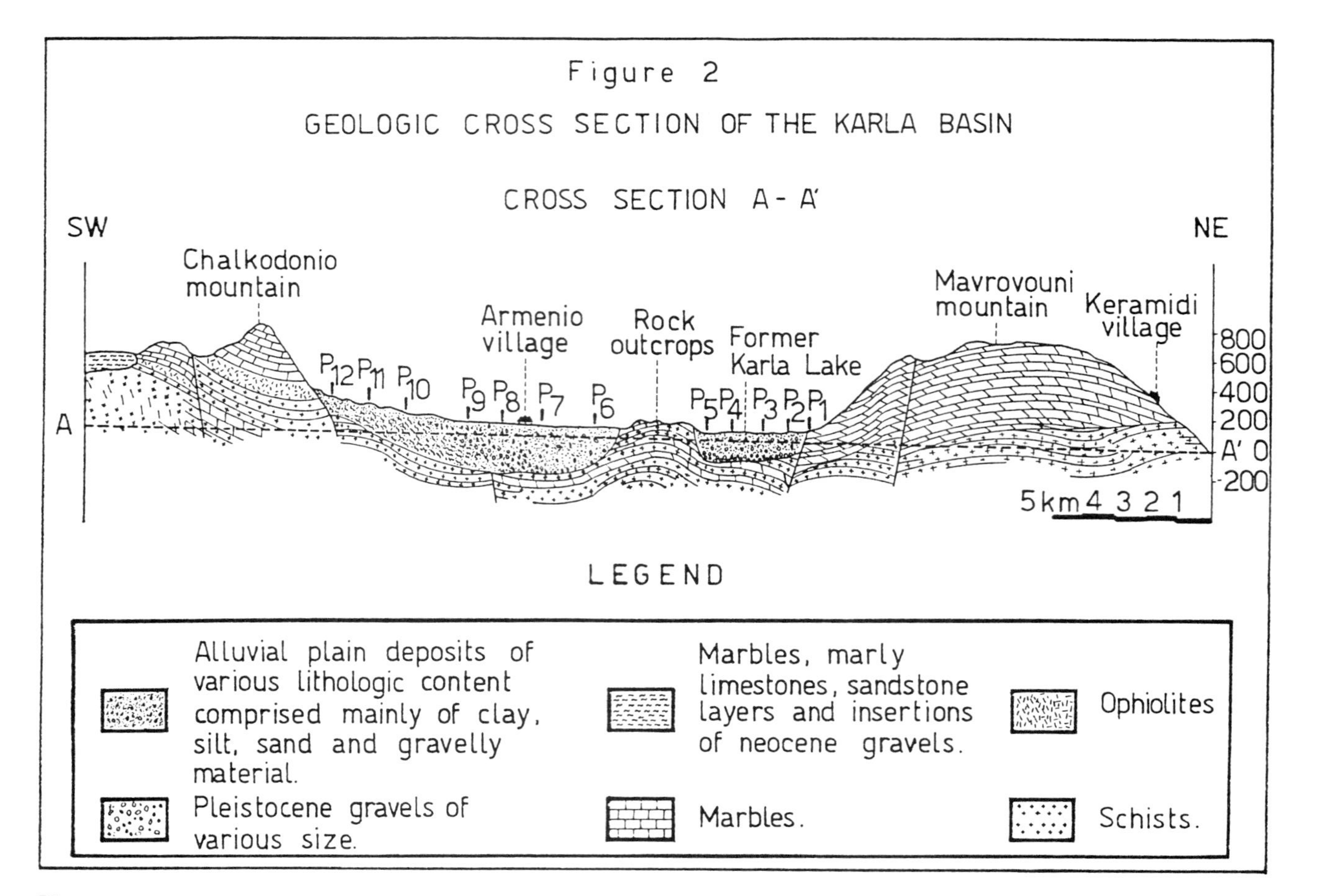

Figure 2. Cross section of the surveyed area.

Table 1.
Analytical data of Soil Profiles.

LOCATION	No of PRO-FILE	DEPTH of HORIZON	NAME of HORI ZON	TEXTURE				CLASSIFI-CATION IN USDA AND/OR FAO SYSTEM
				Sand	Silt	Clay	Char	
Edge of drained lake Karla	P1	0-30	A	64	22	14	SL	USDA: Typic Xerofluvent or FAO: Jc, Calcaric Fluvisol
		30-40	C1	72	16	12	SL	
		40-48	C2	79	13	8	LS	
		48-70	2Ab	33	33	34	CL	
		70-110	2Ck1	29	25	46	C	
		110-150	2C2	29	33	38	CL	
Drained lake Karla	P4	0-25	Ap	17	23	60	C	USDA: Vertic Calcixeroll FAO: Zm or Mollic Solonchak
		25-45	ACk	17	26	57	C	
		45-76	Ck1	17	24	59	C	
		76-104	Ck2	15	21	64	C	
		104-129	Ckn3	19	15	66	C	
		129-150	Ckn4	13	19	68	C	
Stefano-vikio (edge of d. lake)	P7	0-25	Ap	39	24	37	CL	USDA: Typic Xerofluvent FAO:Calcaric Fluvisol(Jc)
		25-58	ACk	43	22	35	CL	
		58-95	Ck1	59	14	27	SCL	
		95-120	Ck2	55	16	29	SCL	
Stefano-vikio	P8	0-23	Ap	29	38	33	CL	USDA: Calci-xerollic Xerochrept FAO: Calcic Cambisol(Bk)
		23-50	Bw1	31	28	41	C	
		50-75	Bwk2	27	26	47	C	
		75-105	BCk1	31	28	41	C	
		105-150	BCk2	41	21	38	CL	
Stefano-vikio	P9	0-20	Ap	30	34	36	CL	USDA: Calci-xerollic Xerochrept FAO: Bk or Calcic Cambisol
		20-40	ABk	29	33	38	CL	
		40-66	Bwk1	36	30	34	CL	
		66-102	Bw2	49	23	28	SCL	
		102-124	Ck1	65	15	20	SCL	
		124-150	Ck2	51	25	24	SCL	
Megalo Monasti-ri village	P10	0-35	Ap	35	19	46	C	USDA: Typic Xerofluvent FAO:Calcaric Fluvisol(Jc)
		35-65	Ck1	43	19	38	CL	
		65-93	Ck2	43	26	31	CL	
		93-130	Ck3	51	26	23	SCL	

Table 1. (continuation)
Analytical data of Soil Profiles.

No of PRO-FILE	ORGANIC MATTER gr/ 100gr	P OLSEN ppm	K_2O mgr/ 100g Soil	$CaCO_3$ CONT-ENT %	pH 1:1	EC_e X 10^3	TOTAL SALTS %
P1	1,57	30,5	3,0	6,51	7,8	<3	-
	0,73	24,5	1,7	7,14	8,0	<3	-
	0,50	33,5	1,5	6,09	8,1	<3	-
	1,47	36,0	1,4	10,71	8,1	<3	-
	1,27		2,5	18,06	8,4	<3	-
	1,40		2,6	13,02	8,5	<3	-
P4	3.18	72.5	10.0	23.94	8.1	<3	-
	2.91	33.5	3.0	22.47	7.8	<3	-
	2.54	22.5	3.1	28.14	8.2	<3	-
	1.07		5.4	34.44	8.4	7.0	0.51
	0.46		4.6	35.49	8.4	8.7	0.72
	0.63		5.1	31.5	8.4	8.7	0.72
P7	2.17	33.5	8.1	16.38	7.8	<3	-
	0.73	22.5	1.1	18.27	8.1	<3	-
	0.23		0.7	18.48	8.2	<3	-
	0.36		0.5	21.84	8.2	<3	-
P8	2.04	39.0	4.0	20.16	7.9	<3	-
	1.34	11.0	1.5	14.28	8.0	<3	-
	1.27	17.0	0.9	20.16	8.1	<3	-
	0.56		1.0	22.68	8.2	<3	-
	0.46		1.0	18.06	8.2	<3	-
P9	1.61	22.5	3.6	18.69	7.7	<3	-
	0.97	17.0	1.6	17.64	7.9	3.5	0.13
	1.07	14.0	1.3	16.8	8.0	3.3	0.13
	0.60		1.4	14.49	8.1	3.1	0.12
	0.47		1.1	18.9	8.3	4.0	0.09
	0.40		0.7	25.41	8.3	3.2	0.10
P10	0.804		0.6	1.47	7.3	4.0	0.16
	0.335		0.2	37.8	8.0	<3	-
	0.100		0.4	28.56	8.2	<3	-
	0.033		0.6	21	8.2	<3	-

Table 1. (continuation).
Analytical data of Soil Profiles.

No of PRO-FILE	EXCHANGEABLE CATIONS					C.E.C meq/ 100gr	ESP %
	Na meq	Ca meq	Mg meq	K meq	Ca/Mg		
P1	0,15	8,93	6,17	0,33	1,45	15,95	-
	0,15	6,18	6,17	0,23	1,00	12,65	-
	0,15	5,87	4,13	0,16	1,42	10,45	-
	0,25	9,55	6,93	0,35	1,38	21,45	-
	0,45	15,93	12,87	0,59	1,24	30,25	-
	1,25	14,93	9,67	0,57	1,54	26,40	-
P4	0.35	19.05	15.15	1.42	1.26	36.85	-
	0.7	18.98	19.72	0.78	0.96	40.70	-
	1.4	15.30	14.20	0.73	1.08	31.90	-
	4.35	15.67	10.13	0.95	1.55	31.90	13.64
	6.2	2.74	14.26	0.97	0.19	24.20	25.62
	6.95	3.44	15.51	1.07	0.22	27.50	25.27
P7	0.2	16.48	2.74	1.15	6.01	21.45	-
	0.25	11.68	6.17	0.38	1.89	18.70	-
	0.35	6.61	4.87	0.21	1.36	12.65	-
	0.35	6.18	7.30	0.16	0.85	14.30	-
P8	0.25	12.10	9.00	0.71	1.34	22.55	-
	0.35	11.12	17.28	0.59	0.64	29.70	-
	0.3	18.42	8.08	0.35	2.28	27.50	-
	0.3	12.36	9.41	0.33	1.31	22.55	-
	0.25	11.05	8.75	0.33	1.26	20.90	-
P9	0.35	15.85	8.75	0.63	1.81	26.95	-
	0.40	14.42	10.58	0.42	1.36	26.40	-
	0.35	17.86	7.94	0.33	2.25	26.95	-
	0.30	9.67	7.88	0.23	1.23	18.15	-
	0.25	6.05	4.20	0.16	1.44	11.00	-
	0.25	8.79	3.81	0.16	2.31	13.20	-
P10	0.30			0.78		31.35	-
	0.25			0.40		24.75	-
	0.25			0.40		19.80	-
	0.30			0.40		21.45	-

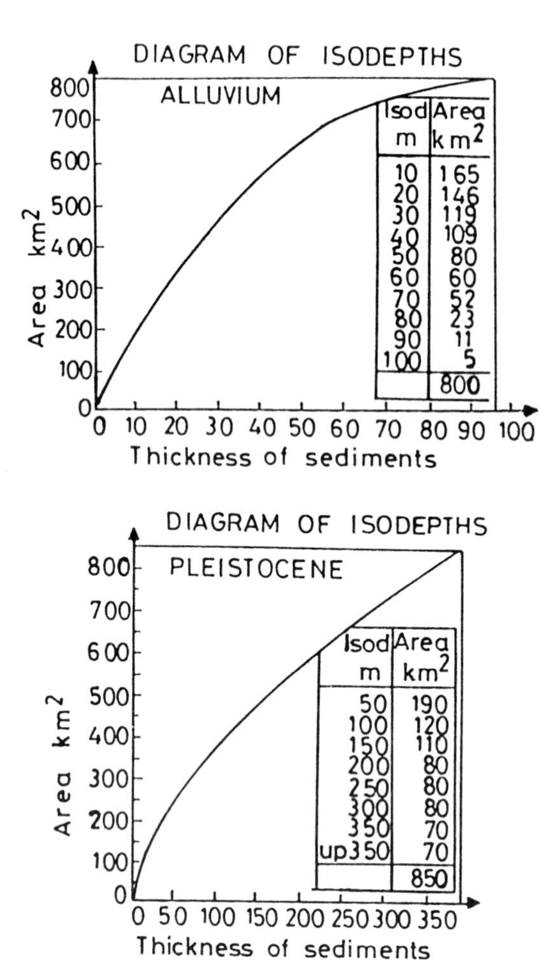

Fig. 3. Isodepth diagrams indicating the thickness of the sediments.

From the isodepth diagrams the total volume of the sedimentation was estimated for each epoch, which are calculated for the holocene deposits as Va = 21,730 x 10^6 m^3 and the pleistocene deposits as Vp = 130,150 x 10^6 m^3.

The mean annual sedimentation rate for each epoch of the quarternary period is estimated as follows : Total duration of the holocene sedimentation is estimated to 30,000 years. Total Volume of sedimentation : Va = 21,730 x 10^6 m^3

Photo 1.
Profiles of the study area across the cross section.

Photo 2. Profiles of the study area across the cross section.

Total yearly sedimentation : 724,333 m^3
Sedimentation area : 800 x 10^6 m^2
Height of yearly sedimentation : 0.000905 m or 0.905 mm
Total duration of the pleistocene sedimentation is estimated to 1,500,000 years.
Total Volume of sedimentation : Vp = 130,150 x 10^6 m^3
Total yearly sedimentation : 86,767 m^3
Sedimentation area : 850 x 10^6 m^2
Height of yearly sedimentation: 0.000102 m or 0.102 mm

It is observed that the erosion processes are more active during the holocene epoch, but if we consider the glacial periods of the pleistocene epoch, when essentially there is no erosion, then the difference should not be considered as high.

4. Experimental measurements

In the hilly area which is located westwards of the former lake Karla and which is composed of plio-pleistocene materials rich in calcium carbonate and marl, the rate of erosion was estimated in relation to the slope, from experimental devices specific for the collection and measurement of the volume of erosional material which were induced from the rain and irrigation of the agricultural fields.

These results and the measurements from empirical data that is, estimation of the erosion rate from the soil type, measurements in specific places in the countryside etc., gave the following mathematical equation $\Delta = 4.07 \times S^{1.52}$ (J. Sgouras, 1991)

Where Δ = the estimated erosion
S = the slope

5. Use of satellite imagery

A SPOT-1 satellite image KJ 88/271 of 5/4/1989 (Fig. 4) was used for the interpretation of the erosional processes on the area which was studied. The erosional processes are made perceptible on satellite images from the physiographic elements such as : relief, rocks, surface soil, hydrographic or natural drainage network, drainage conditions, land use and the general spectral signature of the satellite image (5 - 6 - 2 - 7). According to some research findings (4) the spectral bands 3 and 5 of TM of Landsat-5 shows very good discrimination of erosional surfaces with mean radiometric values of 65 and 140 in the grey scale of 256 (where 0 is the total absorption and 256 total reflectance). As we can see from the SPOT-1 image on the southwest side of the study area the relief is quite conspicuous because the area is hilly and the drainage network very intricate. The highly reflecting areas on this part of the image are the more susceptible to erosion and the erosion risk is much more greater there. Channel 3 of TM corresponds to channel 2 of SPOT-1 XS which is more applicable generally to soil studies. As it has been referred previously the south-west part of the study area, that is the area around the villages of Agnaderi and Neo Monastiri is consisted mainly of Tertiary marls. This marly material contains lots of clay and calcium carbonate which is very susceptible to erosional processes, because it is easily transported by the rain water in sunspension during overland flow and the profile of erosion surfaces are usually vertical and the formed slopes with compound gradients. From the village of Neo Monastiri going down to the village of Stefanovikio there is a transitional zone of very low hills and slight sloping terrain. This transitional zone covers the area from Neo Monastiri till the middle of the distance between Neo Monastiri and Stefanovikio. Grey tone, relief and land use help to conclude that erosion is much less than of the previous hilly marly area. Going further down to the flat plain which is adjacent to the recent Karla lake, grey tone and land use justify the conclusion that this area is not beting eroded, but on the contrary is accepting eroded material from the hilly area. The northeast part of the surveyed area comprises the recently dried (1964) lake Karla which on the SPOT-1 image is very well visible. As we can see the land use and the grey tone of that part of the image is quite different from the rest of the area. This is the lowest part of the whole area and is still accepting all the eroded material from the surrounding low mountains and hills. On the outer northeast part of the image there is a very well formed alluviocolluvial fan which is the limit between former lake Karla and the low Maurovouni mountain consisted mainly of gneiss or gneiss-schist. Again the image grey tone, land use and relief are quite obvious elements in concluding that the erosion processes are still very active here. The SPOT-1 satellite image is a very useful tool suitable for erosional studies because of its high spectral and especially ground resolution which together with its stereoscopic capabilities is becoming unique.

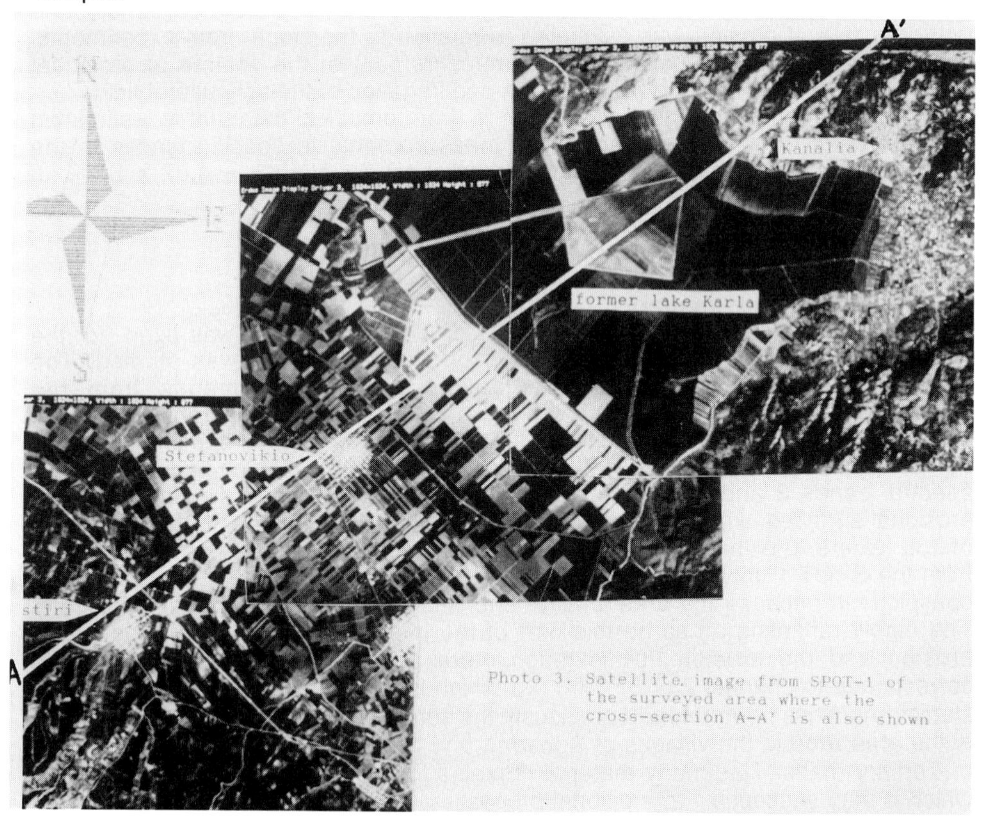

Figure 4. SPOT-1 satellite image of the study area.

RESULTS AND DISCUSSION

From the macroscopic observation and description of the natural and artificial soil profiles, useful conclusions were derived, about the composition of the soils and the sediments and we formed a general idea of the processes which prevailed on the area studied. The chemical and mineralogical analyses helped us on the definition of the composition of the soils and sediments and gave us information about the continuity or discontinuity of the strata and the erosion of the surface soil horizons. With the geological, geophysical and paleontological measurements we derived information about the thickness, extend and volume of the sediments and to some extent a view of the general erosion of that region. With the help of the experimental devices we drew conclusions on the rate of the erosional processes. This method is time consuming and quite laborious but gives good results. Finally the use of satellite imagery to similar studies is very contributing and can supplement the conventional methods. The advantages from the use of the satellite images and especially of the SPOT-1 and SPOT-2 images are the following to name but a few : overall coverage of the area, stereoscopic capability, digital image processing and image vision. In short the satellite imagery is becoming from day to day an indispensable tool for the environmental scientist in order to manage and protect the natural resources.

BIBLIOGRAPHY

1. Floras, S. (1990) : The process of Erosion in the Agricultural Soils of Greece and Especially of Thessaly.
2. Sgouras, J. (1991) : Relations of Rich in $CaC0_3$ Geologic Substrata with the Properties of the Overlying Soils
3. Constantinidis, D. (1976) : Etude Hydrogéologique du Bassin de la Thessalie Orientale, Phd thesis, Grenoble.
4. Floras, S. (1992) : Analysis and Classification of Spectral Signatures, Structure and Texture of the Physiographic Elements with Remote Sensing Techniques, Phd Thesis, Aristotelian University of Thessaloniki.
5. Astaras, Th., Labrinos, N. Soulakelis, N. (1990) : A Drainage System Analysis Evaluation of, and Comparison Between Landsat-3 RBV, Landsat-5 TM and SPOT-1 PA Images Covering The Central Macedonia District, Greece.
6. Bergsma, E. (1980) : Interpretation Exercises. Department of Soils, ITC The Netherlands.
7. Silleos, N. (1990) : Mapping and Evaluation of the Agricultural Land and Soils.
8. Alexiades, K.A. (1967) : Physical and Chemical Analysis of Soil, Aristotelian University of Thessaloniki.
9. Curran, P.J. (1989) : Principles of Remote Sensing.
10. Zachar, Dusan, (1982) : Soil Erosion, Developments in Soil Science 10, Elsevier Science Publishing Company,New York.

Farm Land Erosion: In Temperate Plains Environment and Hills
S. Wicherek (Editor)

Terrace degradation and soil erosion on Naxos Island, Greece

R. Lehmann

Geographisches Institut, Universität Basel, Basel, Switzerland

SUMMARY

As a typical cycladic island, Naxos represents the historical changing of the natural landscape into a strongly anthropogenic influenced landscape. To cultivate the mountenous island, the steep slopes have been terraced since several hundreds or some thousand years. Destruction of the natural vegetation and soil movement during the terrace construction led to the first soil erosion, the dimension of which cannot be traced back. Today the abandonment of agricultural use and maintenance, overgrazing and fire cause a general terrace destruction and thereby severe soil erosion on an irreversible scale.

Main type of terraces is a parallel level bench terrace, i.e. the stone wall terrace and scarp terrace which is reinforced with vegetation. Mean terrace hight is 2 m. Comparison between field box measurements shows, that apart from slope angle, vegetation is here the most important control factor of erosion. Practically no erosion can be found under a 100 % cover of the Phrygana or Maquis vegetation. The protection applies to the soil and the terrace walls and scarps. Starting with soil moving down from one terrace to the other, the erosion leads finally to the exposure of the bare rock surface. The estimated average soil loss was in that case between 40 and 50 t/ha/yr. Those bedrock areas are the oldest given up terrace slopes with an age of about 200 to 250 years. Data of measuring bridges correspond with the estimated soil decrease of 0.3 cm/a. The high mediterranean variability of rainfall leads to much higher erosion values in case of heavy rainstorms.

Autumn to early winter is the time with the highest erosion, while the surface of the sparse Phrygana is bare of herbs and grasses and rainfall is heavyest. Livestock increases soil erosion due to overgrazing, destruction of stone walls and walking on steep terrace scarps.

To get an evaluation of the landscape, a key for mapping and classification of mediterranean agricultural terraces has been developed. It is an important tool to obtain a basis for the conservation of soil and environment. Five stages indicate the condition and the decay of the terraces. The key classifies relief, soil, erosion, vegetation and stone wall or scarp condition. An easy practicability allows the application without spezialized knowledge. The method is valid on areas from 0.25 ha up and on slope angles between 10° and 35°. But more gentle or steeper slopes may also be classified. A recommended use is given for each terrace stage, comprising integrative soil-, water- and vegetation conservation. The estimation of the total soil loss is possible on the basis of a nomograph.

RESUME

Ile typique des Cyclades, Naxos illustre la transformation d'un sol sous l'influence anthropique. Afin de cultiver cette île montagneuse, les pentes raides ont été aménagées en terrasses depuis plusieurs centaines voire milliers d'années. La destruction de la couverture végétale et la construction des gradins ont provoqué les premières manifestations érosives dont l'ampleur aujourd'hui mérite d'être soulignée. De nos jours, l'occupation du sol et la gestion agricole ont été abandonnées ; le surpâturage et les incendies provoquent la destruction générale des terrasses et une érosion du sol irréversible.

Les terrasses sont limitées par des murs de pierres, la partie raide étant renforcée par de la végétation. La hauteur moyenne des gradins est de 2 m. On ne trouve aucune érosion là où existe une couverture végétale fermée de type Phrygana ou maquis. Cette protection s'applique au sol, aux murs.

Là ou l'érosion intervient, elle agit généralement de façon régressive du bas du versant vers le haut, d'une terrasse à l'autre ; elle aboutit à la dénudation du versant. Dans ce cas, la perte en terre peut être estimée à 40-50 t/ha/an. Les variations de hauteur des précipitations en région méditerranéenne peuvent encore aboutir à une aggravation des phénomènes évoqués.

L'érosion la plus forte se situe en automne et au début de l'hiver, période où la terre n'est pas protégée et où les pluies sont les plus intenses. Le surpâturage a pour effet d'augmenter l'érosion du sol, le bétail en marchant sur les murs contribue à les détruire.

Afin d'évaluer le milieu, on a développé une méthode de cartographie et de classification des terrasses cultivées en Méditerranée. Cette méthode constitue un outil important pour la conservation du sol et de l'environnement. Cinq étapes indiquent la condition et le degré de dégradation des gradins. La méthode en question, de par sa simplicité, peut être utilisée sans connaissance spécialisée : elle s'applique aux surfaces d'au moins 0,25 ha et aux pentes de 10° à 25°. Des pentes d'un angle plus faible ou plus fort peuvent toutefois être prises en compte. On suggère une utilisation adaptée au milieu, y compris la conservation intégrale du sol, de l'eau et de la végétation. Une estimation de la perte de sol totale est possible sur la base d'un nomographe.

1. INTRODUCTION

1.1 Historic background and causes of erosion

Like many parts of the Mediterranean, Naxos (Fig. 1) is characterised by early and continuous settlement and acquisition of land. The typical natural woods of Greece were diminished by a strong human pressure since 2000 B.C. (1 - 2 - 3). The exploitation of building timber, fire wood and resin, but also simple burning to expand pasture land and the effects of war have to this day gradually destroyed the woods (see also 4). Besides deforestation and destruction of the natural vegetation, construction of stone- walled terraces led to the first soil erosion, the dimension of which cannot be traced back (see also 5). Rackman & Moody (6) suggest six main reasons for construction but they emphasize that backgrounds cannot clearly be proved. This usual measure to cultivate the steep slopes since several hundred to some thousand years was applied all over the mountainous lands. The walls which were built to stabelize the soil and to gain ground-water recharge collapsed and were rebuild several times during the centuries according to human pressure. Reasons for recent abandonment of agriculture and maintenance on terraced slopes are economic alternatives. Since the 60`s until today massive migration from agriculture to industry (Athens) and tourism reduced the cultivation of the mountainous areas due to lack of labour force. Farmers left their terraced fields to shepherds and transhumance, and that leads to overgrazing and burning. Comparison with correlations in Yemen show interesting similarities (7 - 8).

Figure 1. Location of Naxos.

1.2 Area and environment

On Naxos Island agricultural terraces were constructed from sea level up to the highest mountains (800-1000 m NN). Steep slopes are common (15°-35°). Terraces are a characteristic element of the environment. Overuse as pasture and fire cause a general terrace destruction and thereby severe soil erosion on an irreversible scale. Cultivation and maintenance occurs today only in the vicinity of the small villages. Main type of terraces is a parallel level bench terrace (stepped terrace), i.e. the stonewall terrace and scarp terrace which is reinforced with vegetation (earthen terrace). As an archaeologist Rackham (6, with Moody) supports the experience, that historic earthen terraces are unusual. Mean terrace height is 2 m.

The Prygana vegetation covers most of the abandoned lands only 30 to 40 % in parts and is characterised as a 30 to 100 cm high dwarfshrub-steppe (9, p.101). Poterium spinosum and Thymus capitatus are typical representatives. Wood or Maquis covers only some small areas and are highly endangered due to burning. Burned areas remain nearly without vegetation for 2 years in minimum. Herbs, grasses and small sprouts of shrubs which survive fire are pioneer plants.

The soil of the terrace slopes is a highly anthropogenic soil, displaced by human activity and shifted by soil erosion. It shows hardly any horizontation except a 5 to 10 cm thick, darker humusly layer on top of the mainly undisturbed terraces. This soil type is a sandy to loamy terrace colluvium. Soil types of slopes without terracing and rocky landscapes are mainly rankers, rendzinas and lithosols/

regosols (see also 10). The colour (variations of brown) depends on the bedrock underneath as well as the stonyness (10-50 %). Parent rocks are marbles, schist, granodiorite and conglomerates. Average thickness of the terrace colluvium ranges from 2 m at the terrace wall to 0.2 m towards the slope. Potsherds can be found in every depth of the profiles.

Naxos is part of the dryest region of Greece and is characterised by a semiarid climate. Humid phases with surplus of rain last only for weeks. Annual amount of precipitation (end of September to beginning of May) are fluctuating between 139 and 721 mm, the average value is 378 mm/a. Precipitation measurements of 35 years show a maximum of 65 mm in 24 hours (2, p. 110 f.).

Environment and historical development give reasons for today`s ecological and economical problems. Degradation of terraces and vegetation lead to soil erosion and reduction of retention. Therefore the environment is suffering under severe ecological problems concerning microclimate, water balance and soil.

2. METHODS AND PROCEDURES

2.1 Classification

To evaluate the landscape a key for mapping and classification of mediterranean agricultural terraces has been developed. (Appendix, see after References). It describes the condition, and recommended use helps to realize landscape planning and protection. It may also be regarded as a long-term damage mapping and analysis. The method is an important tool to obtain a basis for the conservation of soil and environment. It starts with mapping of the areas to be investigated. The key includes condition of the stone-walled terraces and the scarps of those held up by earth banks (Tab. 1). The dip of the terrace surface and the cross-section are classified in tab.2, the erosion forms and thickness of soil in tab. 3 and 4, and also contemporary use and vegetation (Tab. 5) as well as stone cover (Tab. 6). The classification in the tables results in points. The number of the points (6-30) reveals a terrace stage between one and five in tab. 7. Stage one means a very good, five a very bad stage and therefore high total soil loss.

The assessment of the parameters in tab. 1 to 6 takes place in the field, one is ascertained per table. The method is valid on areas from 0.25 ha up and on slope angles between 10° and 35°. But more gentle or steeper slopes may also be classified. With the key it is possible to map small slopes as well as wide areas. The classification allows to recommend the future use concerning agriculture, pasture, maintenance, regeneration and afforestation (Tab. 8). The recommended use is given for each terrace stage, comprising integrative soil-, water- and vegetation conservation. A simple model of a terrace slope allows to estimate the total soil loss.

2.2 Further methods

Contemporary soil erosion is measured in 5 typical environments with different terrace stages and vegetation cover. Three methods are used for quantification : measuring bridge (11), field box or sediment trough (12) and damage mapping (13). Results of contemporary erosion will be presented in Lehmann (14) and in a short note 3.4.

3. RESULTS

3.1 Processes

The erosion on terraced slopes starts with stones falling from the top of the wall or local caving-in and slumping. Using these spots as trails, walking livestock increases soil erosion, what may be called "livestock erosion" (on Naxos an average density of about 230 animals per km^2 - sheep and goats - is existing). Surface water is running through these small gullies, which become larger and get in contact after some years of erosion. Starting with soil moving down from one terrace to the other, the erosion leads finally to the exposure of the bare rock surface. Soil islands may be held by dwarfshrubs and rests of stone walls are common. Fig. 2 shows five terrace stages, a distinction of which is very simple. Soil loss is going along with flattening of scarps and steepening of terrace surfaces, an adaptation to the general slope angle. While soil thickness and infiltration is decreasing, surface run-off increases. Most erosion forms, like small gullies or fans underneath, occur in the mean stages of degradation. Temporal course : after five or six arid summer months, the sparse Phrygana is bare of herbs and grasses. Overgrazing and local or widespread burning reduces vegetation density in general. Livestock movement has loosened the soil surface on steep trails and terrace scarps. Stone walls may have suffered under walking, creating new drainage furrows for running surface water. Soil life is sparsely active and infiltration is reduced, creating higher runoff. The first rainfalls at the end of September or in October, often thunderstorms, may cause the highest erosion. Erosion values decrease during winter, when annual vegetation, herbs and grasses start to grow on the bare soil surface between dwarfshrubs.

3.2 Example for using the classification

Investigating a terrace slope, stone wall or scarp, terrace surface, erosion forms, soil thickness, vegetation, using and stone cover have to be taken into consideration. First case: a stone walled terrace is marked by single slumping, but most of the wall is in good order, that comes to 2 points ; the angle of the terrace surface is about 2 degrees, which amounts to 1 point; some small gullies and fans underneath characterise the erosion forms, resulting in another 2 points ; soil thickness is 1.5 m at the stone wall, but decreases in areas of few gullies, that comes to 2 points ; there is no agricultural use except pasture and maybe a few olive trees, so that results in 3 points ; stone cover is high with about 10 percent, coming to 3 points. The result of 13 points stands for a good terrace condition and a terrace stage of 2 (Photo 1).

In the second example some rests of terrace scarps mark the former terrace locations (5 points), while the angle of the small terrace surfaces exceeds 12° (5 points). Erosion forms like rills expand over the whole area, but few gullies and accumulations remain (4 points). Soil thickness is irregular and does not reach 0.5 m in most parts of the area. Bedrock is visible on some spots (5 points). Land is used as pasture, some burned vegetation may occur within the Phrygana or low Maquis (5 points). At least stone cover may not be very high, for example 8% because of granular disintegration (3 points). The result of 27 points shows a very bad terrace condition and a terrace stage of 5 (Photo 2).

Photo 1. Rests of maintained terraces (stage 1 and 2) in the mountains of Naxos about 600 m above sea level (Skeponi area). The whole area was terraced, showing today terrace stage 3 to 5. The large light slope in the background was burned about 2 years ago. The dark ash was transported by water and wind. Bare soil rests and rocks are lighter than the vegetation cover of the areas beside.

Figure 2. Degradation of terraces, showing five stages of decay and the effects on infiltration, ground water recharge and surface run-off.

Photo 2. Close up of terrace stage 5 and a field box. Small rests of stone walls, terrace surfaces and bare rock surfaces (Granodiorite) indicate a very poor condition. Characteristic representative of the Phrygana vegetation is here Poterium spinosum.

Every terrace stage indicates a practical measure to stabelise or improve the environment.

In the first case full agricultural use is recommended. If there is no possibility for that, cultivation for feed and supervised pasture at times on terrace surfaces should be organized. Maintenance of stone walls or vegetation cover on scarps should be obligatory. Transhumance is to be excluded, even if there is no use at all. That means the area has to be fenced.

The second case, showing terrace stage 5, is more difficult to handle. Once a farmer has left his land, interest on the degraded terrace slopes is low, except for shepherds and transhumance. Cultivation is not possible any more and pasture is to be excluded. In that case, fencing is recommended as well as afforestation where still possible (see also 15). The fast setting-in and the high quantity of surface flow lead to the necessity to conserve the water in ditches or basins for oozing away to help in recharge of ground water, or to collect it in tanks for irrigation.

3.3 Estimating absolute soil loss

The calculation of the absolute soil loss (ASL) of abandoned terrace slopes is based on a model of terrace degradation (Fig. 3), and of a terraced slope (Fig. 4), which is structured on the classification.

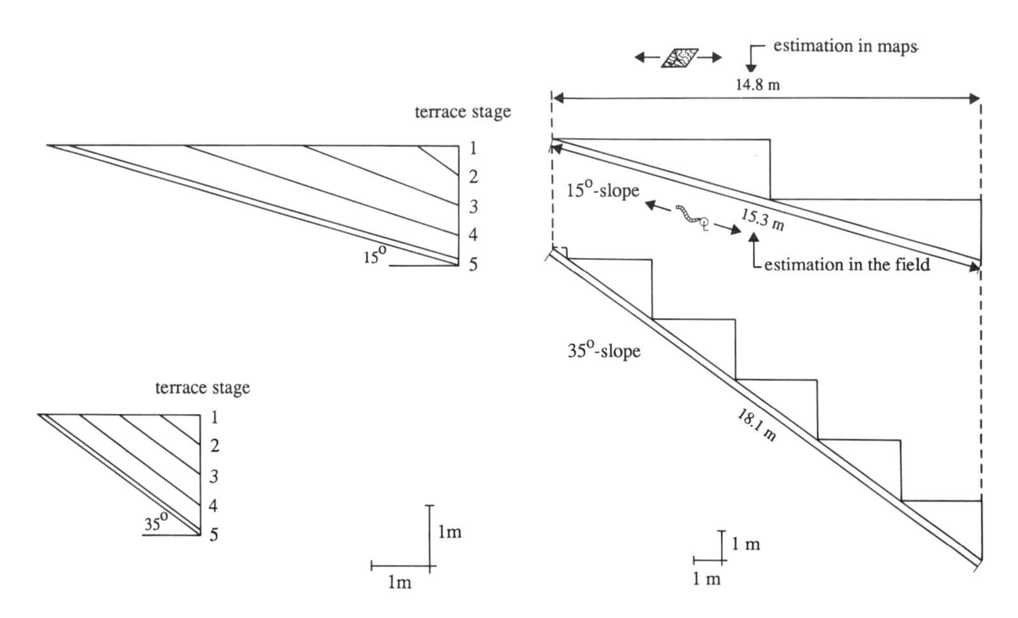

Figure 3. Model of the terrace degradation and terrace stages, 15°/35°-slope. Basis for the estimation of the absolute soil loss. Interaccumulations are disregarded.

Figure 4. Model slope for the computation of area and the basis of the nomograph.

The erosion will be estimated since the last construction. From experience mean terrace height is 2 m and most are constructed on slopes between 15° and 35°. The model shows a cross section of terrace degradation, terrace stages and soil loss on a slope with 15° and 35° (Fig. 3, Tab. 1). Development of erosion is not exactly the same in both cases. Small interaccumulations on top of the terrace were disregarded.

Table 1
Estimation of the absolute soil loss, example for one terrace, 15°/35°-slope [m^3/m^2 and %]

	15°-slope		35°-slope	
terrace stage	ASL one terrace m^3/m^2	%	ASL one terrace m^3/m^2	%
1	0	0	0	0
2	0.2	3	0.2	7
3	1.4	20	0.7	25
4	3.6	50	1.6	57
5	6.6	92	2.5	89

Computation of area (m^3/m^2) is made with the help of a model slope (Fig. 4, Tab. 2). The calculations of a single terrace (Fig. 3) are applied to the slope (Fig. 4). Soil thickness and erosion are almost equal concerning slope angle (Fig. 5). Starting at the edges of the terraces, the 35°-slope with more terraces shows higher erosion values (m^3/m^2) from stage 2 to stage 3. At stage 4 and 5 the higher value of the 15°-slope can be put down to the higher mean soil thickness (15° : 0.94 m, 35° : 0.79 m).

The estimation of soil loss is possible in maps (= vertical evaluation) and in the field. On the map slope length is shorter, that is why soil thickness respectively erosion per m^2 seem to be higher (Tab. 2). A correction factor is necessary for slopes steeper than 20° (Tab. 3) to come to accurate erosion values (m^3/ m^2).

After mapping and classifying a terrace slope, the absolute soil loss is directly readable in the nomograph, based on the model (Fig. 6). The nomograph is built up on the model of the 35°-slope. The absolute erosion value of slopes more gentle than 30° has tobe corrected for terrace stage 2 and 5 with a factor entered inthe nomograph. Results are based on estimation in the field. If the calculation of area was carried out in maps, the result of the nomograph (ALS) has to be corrected with a third factor : 20° - 30° ASL x 1.1 ; > 30° ASL x 1.2

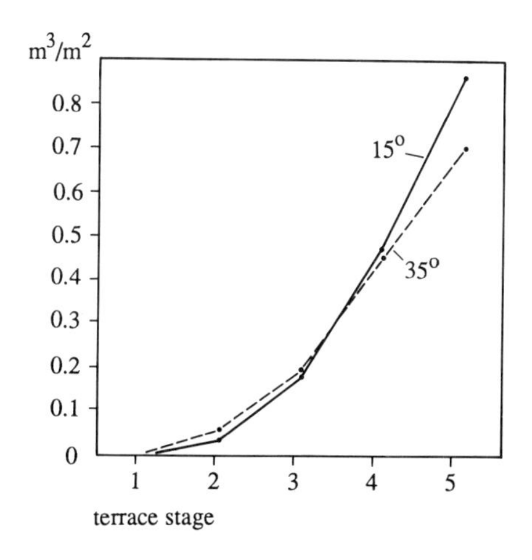

Figure 5. Absolute soil loss, 15°/35°-slope.

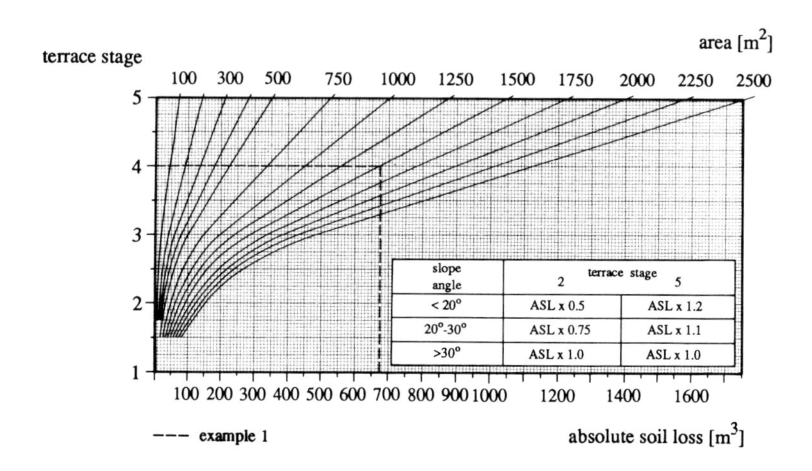

slope angle	terrace stage 2	terrace stage 5
< 20°	ASL x 0.5	ASL x 1.2
20°-30°	ASL x 0.75	ASL x 1.1
>30°	ASL x 1.0	ASL x 1.0

Figure 6. Nomograph for the computation of area, absolute soil loss. 35°-slope, with correction factors for slope angles <30°, terrace stage 2 and 5.

Table 2
Estimation of the absolute soil loss, example for a terrace slope, 15°/35°-slope [m^3/m^2]

	ASL m^2/m^2			
terrace stage	field evaluation slope lengh 15.3/18.1 m		map evaluation slope lengh 14.8 m	
	15°	35°	15°	35°
2	0.03	0.06 *	0.03	0.07
3	0.18	0.19	0.20	0.24
4	0.47	0.45	0.49	0.55
5	0.86	0.70 *	0.89	0.85

* values < 30° see correction factor in the nomograph.

Table 3
Estimation of the absolute soil loss in maps, correction factor for slopes steeper than 20°. ASL in m^3/m^2

slope angle	correction factor
<20°	ASL x 1.0
20°-30°	ASL x 0.9
>30°	ASL x 0.8

Examples for estimating ASL with the nomograph :

Example 1
Terrace stage 4 (Tab. 2)

675 m^3 absolute soil loss (ASL)

Area 30 x 50 m = 1500 m^2

Slope angle 32° : no correction factor

Example 2
Terrace stage 2

105 m^3 ASL

Area 35 x 50 m = 1750 m2

x 0.75 = 79 m^3 ASL

Slope angle 25° : correction factor (nomograph)

In comparison with the ASL an estimation of mean soil losses between 40 and 50 t/ha/yr leads to an age of the abandonment of the terraces of 200 to 250 years. These erosion values are assumed because of obviously higher rainfall in the past. Old large-scale gullies and valley forms which contemporary do not underlie any formation lead to this conclusion. The data of most measuring bridges at localities of small gullies on terrace edges correspond with the estimated soil decrease of 0.3 cm/a. It also has to be taken into consideration that the high mediterranean variability of rainfall leads to much higher erosion values in case of heavy rainstorms.

3.4 Short notes on contemporary erosion

Apart from the erosion at spots of terrace edges (as mentioned above), values are rather low. But most slopes with degraded Phrygana vegetation suffered even in the last dry years without heavy rainfalls under erosion values between 1 and 3 t/ha/yr. This amount lies in all probability over soil formation, estimated to $\leq$0.5 t/ha/yr, and therefore under the limit of tolerance. Inclined extensive terraces (> 10°) under agricultural use show erosion between 20 and 26 t/ha*a. Basically the comparison between field box measurements shows that apart from slope angle, vegetation is here the most important control factor of erosion. Practically no erosion can be found under a 100 % cover of the Phrygana or Maquis vegetation, even though small surface flow can occur. The protection applies to the soil and the terrace walls and scarps. In contrast to Central Europe a well developed vegetation with high shrubs and trees protects the stonewall against collapsing (verbal note, J. Riedmiller, Heidelberg : filling-in of drainage pores with roots causes bowels and slumping of the wall in Central Europe). Experiences from Naxos show no construction with drainage fillings behind the wall or the necessity for it.

4. CONCLUSIONS

The classification represents a first step and practicable method to investigate terraced areas for erosion damages where limited data are available. The mapping and classifying allows rapid evaluation of the condition of little examined areas and gives recommendations for future use. It is the long-term aim to achieve the conservation of water, soil and vegetation, the stabelisation of the ecological system, and to give basis for a management of devastated landscapes. The method has significance for convertation of scientific results into practical measures. Experiences of the mapping show in some cases a need of generalisation especially while mapping in small scales. The recommended uses are intended for areas with some km^2 in size. In very large and remote landscapes most practical measures cannot be carried out. But the recommendations concerning grazing, burning and fencing are definitely applicable. Results of the calculation of the absolute soil loss (ASL) do not give exact values, because of the modelling. But the approximation is a valuable help for a good estimation. An adaptation to terraced landscapes in further regions is possible. As a long-term damage mapping it gives no evidence about current erosion. Limits of application are determined by size of area or the necessity of precise data.

ACKNOWLEDGMENTS

The author is greatly indepted to Prof.Dr.H. Leser for initiating the projekt and Dr.D. Schaub, both Basel, for his constructive and critical remarks. For drafting the illustrations he wishes to thank Mrs. L. Baumann. This study was possible through the financial support of the Swiss National Foundation (SNF).

REFERENCES

1. Beuermann, A. 1956 : Die Waldverhältnisse im Peloponnes unter besonderer Berücksichtigung der Entwaldung und Aufforstung.- In : Erdkunde, Bonn, Band X, Heft 1/4, p. 122-136.
2. Philippson, A. 1948 : Das Klima Griechenlands, Bonn, 233 p.
3. Von Trotta-Treynen, H. 1916 : Die Entwaldung in den Mittelmeerländern.- In : Dr. A. Petermanns Mitteilungen, 62. Jahrgang, Gotha, p. 248-253 und 286-292.
4. Papamichos, N. 1990 : Soil erosion after forest fires in mountain areas of Greece.- E.S.S.C. Newsletter, 2-3, p. 16-21.
5. Brückner, H. & Hoffmann, G.1989 : Man Induced Erosion Processes in Mediterranean Countries - Evidences from Geoarcheology and Geomorphology.- Geoöko plus, Vol. 2 : Symposium "Mediterranean Erosion", Program and Abstracts of Papers, ICG 1989, p. 7, Bensheim.
6. Rackham, O. & Moody, J. 1990 : Terraces.- Paper read at the Seventh International Symposium on Agriculture in Ancient Greece at the Swedish Institute at Athens, 16-17 May, 1990.
7. Alkämper, J., Haffner, W., Matter, H.E. & Weise, O.R. 1979 : Erosion Control and Afforestation in Haraz, Yemen Arab Republic.- Giessener Beiträge zur Entwiklungsforschung, Reihe II, Band 2, Giessen.
8. Vogel, H. 1987 : Terrace farming in Yemen.- Journal of Soil and Water Conservation, Vol. 42, N° 1, p. 18-21.
9. Rechinger, K.H. 1951 : Phytogeographica Aegaea.- Österr. Akademie der Wissenschaften, Denkschriften, Wien, Bd. 105, 2.Halbband, 2. Abt.
10. Louis, A. & De Roubaix E. 1962 : Les sols de l'île de Naxos.- Pedologie, Gent, XII,$_1$, p. 15-49.
11. Coelho, C.A., Shakesby, R.A. & Walsh, R.P.D. 1989 : Effects of Land Use Management Practices in Water Quality and Soil erosion, Agueda Basin, Portugal.- Geoöko plus, Vol. 2 : Symposium "Mediterranean Erosion", Program and Abstracts of Papers, ICG 1989, p. 9-10, Bensheim.
12. Herweg, K. 1988 : Bodenerosion und Bodenkonservierung in der Toskana, Italien.- Physiogeographica, Basler Beiträge zur Physiogeographie, Bd. 9, 175 p., Basel.
13. Rohr, W., Mosimann, TH., Bono, R., Rüttimann, M. & Prasuhn, V. 1990 : Kartieranleitung zur Aufnahme von Boden-erosionsformen und -schäden auf Ackerflächen.- Materialien zur Physiogeographie, Basel, H. 14, 56 p.
14. Lehmann R. 1993 (in prep.) : Landschaftsdegradierung, Boden-erosion und -konservierung auf der Insel Naxos, Griechenland.- Physiogeographica, Basler Beiträge zur Physiogeographie, Basel.

15. Kahl L. & Jacob, M. 1990 : Aufforstung arider und semiarider Regionen nach dem Kallidendron-Verfahren unter Verwendung gering belasteter Klärschlämme.- Holz-Zentralblatt, Stuttgart, Bd. 116, Nr. 128, p. 1988-1992.
Thornes, J.B. (ed.) 1990 : Vegetation and erosion.- 518 p., Wiley.
Goudie, A.S. (ed.) 1990 : Techniques for desert reclamation. Wiley, 271 p.

TERRACE CLASSIFICATION

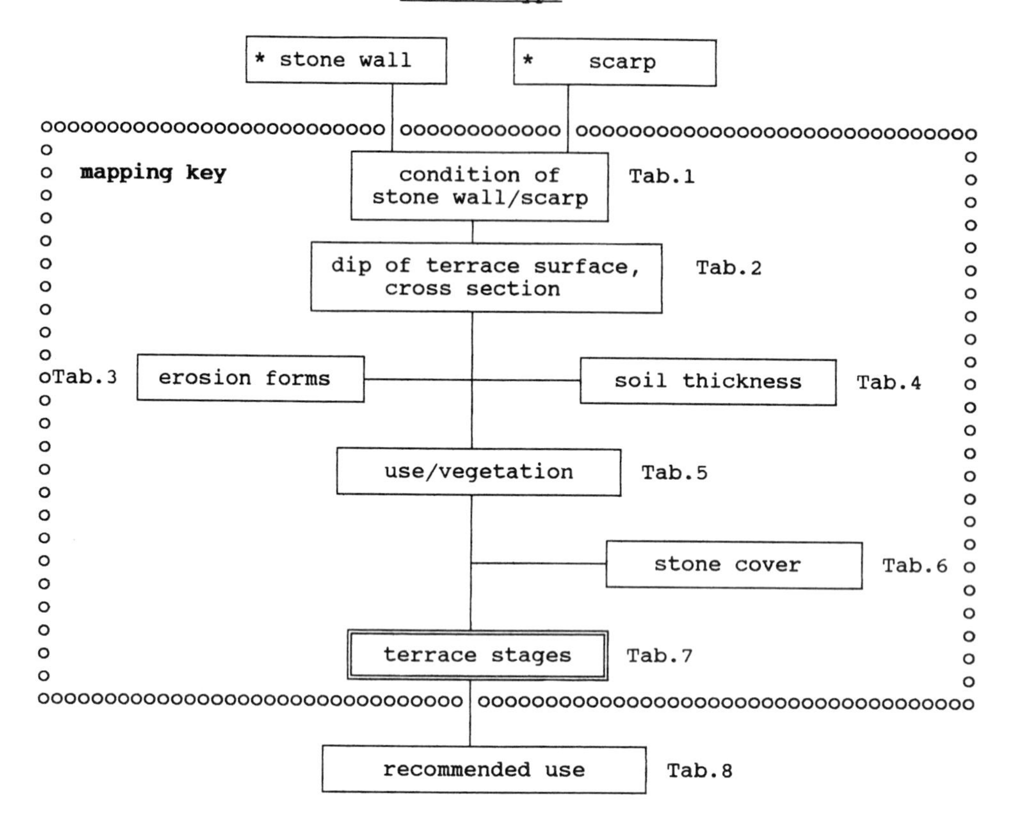

* stone wall terrace: significant construction of the terrace with stone walls. Marking with a "T" in the map.

  scarp terrace: the scarp may show rests of stone walls. If the share of the stone wall is high or inbetween an area scarp and stone wall terraces are existing, the area may marked with a ("T") in the map.

Short mapping regulation

Plots with an area from about 0.25 ha up may be mapped (scale 1:5000). A survey and assessment of much smaller areas from about 100 m² up is possible. Parameters in tab.1 to tab.6 are to investigate in the field. Per table one assessment is to be carried out. The sum of the points result in the terrace stage (Tab.7). Recommended use is given in tab.8.

stone wall	scarp	points
stone wall intact	scarp intact, vegetation cover 100%	1
single collapsing or slumping, beginning "livestock erosion"	single trails of live-stock lead over the scarp beginning "livestock erosion"	2
numerous caving-ins at the stone wall, livestock and water erosion form small gullies and fans underneath	numerous trails of live-stock lead over the scarp decreasing vegetation cover (30-70%)	3
Rests of stone wall mark the former terrace course	eroded and flattened coved edges mark the former terrace course	4
small isolated rests of eroded stone wall, former terrace course only incomplete to recognize	small isolated rests of eroded coved edges and scarps, adaptation to the general slope angle	5

Tab.1. Condition of stone wall or scarp.

cross-section	slope angle [°]	points
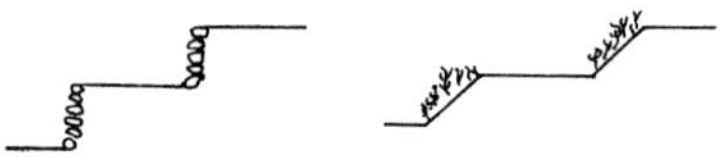	0 - 2	1
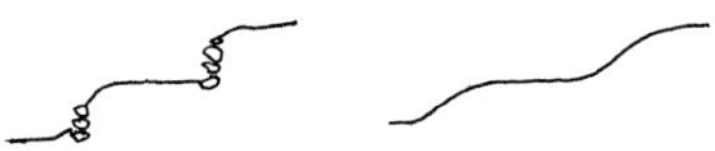	>2 - 5	2
	>5 - 7	3
	>7 - 12	4
	>12	5

Tab.2. Dip of the terrace surface and cross-section.

stone wall	scarp	points
no forms; selective and little rill erosion at terrace transitions possible		1
initial forms: collapsing of stone walls, slumping	*	2
linear on short sections: gully-like short rills and furrows, underneath small interaccumulations		
linear across several terraces: rills/furrows (trails of livestock) and interaccumulations (fans); widened gully-forms, eroded/leveled edges *		3
linear to extensive: network of rills and furrow systems (livestock trails); small interaccumulations; advanced flattening of the edge; decrease of furrows, gully-forms overlap #		4
extensive to linear: braided surface flow forms some rills; rests of furrows may existing; no interaccumulations #		5

* increase of erosion forms
# decrease of erosion forms

Tab.3. Erosion forms.

* new mechanical construction

Thickness	Description	Class
0,5 - 2	stone wall: extreme up to 3 scarp: extreme up to 8 (*)	1
0,5 - 2	short sections with reduced thickness	2
0,5 - 1	partly up to 2 possible	3
0,2 - 1	partly rests up to >1 possible	4
0 - 0,2	partly rests up to >0.5 possible	5

Tab.4. Soil thickness at the terrace edge [m].

Use/vegetation	Class
agricultural use: intensive/extensive, supervised pasture at times scarp: 100% vegetation cover (Phrygana)	1
agricultural use extensive no use: growing up of Phrygana vegetation scarp: degradet vegetation as a result of repeated burning and overuse as pasture *	2
extensive or no agriultural use; partly rests of terrace surfaces used for feed or olive trees; vegetation: Phrygana; frequent pasture *	3
no agricultural use; partly small scale use for feed, traditional cultivation, no mechanisation; vegetation: Phrygana; frequent pasture *	4
no agricultural use; degraded, open Phrygana, frequent pasture *	5

* usually transhumance; in most cases overgrazing and frequent burning prevent dense Phrygana vegetation on abandoned areas; decreasing of varieties and individuals.

Tab.5. Use/vegetation.

low	1 - 2	1
medium	>2 - 5	2
high	>5 - 10	3
very high	>10 - 30	4
extreme high	> 30	5

The degree of weathering has to be taken into account: highly weathered bedrock means also little stone cover. Example: granular desintegration of granite.

Tab.6. Stone cover [%].

sum of points	terrace stage	terrace condition
6 - 10	1	very good
11 - 15	2	good
16 - 20	3	moderate
21 - 25	4	poor
26 - 30	5	very poor

Tab.7. Terrace stages.

terrace stage	recommended use
1	a. full **agricultural use**, intensive/ extensive b. **maintenance:** stone wall or vegetation cover on scarps scarp: in general no need of care
2	a. full **agricultural use**, extensive b. **pasture:** cultivation of feed and supervised pasture at times on terrace surfaces c. **maintenance:** repair of stone wall move back of eroded soil; scarp: planting/regeneration of vegetation
3	a. full **agricultural use** (extensive) only possible with labour intensive restauration; use of rests of terrace surfaces for cultivation of feed; b. **pasture:** supervised pasture and rotation c. **abandonment:** afforestation/natural regeneration of vegetation; seeding
4	a. **agricultural use:** extensive cultivation of feed on small rests of terrace surfaces b. **pasture:** supervised pasture and rotation c. **abandonment:** afforestation/natural regeneration of vegetation; seeding
5	a. **agricultural use:** not possible b. **pasture:** exclusion c. **abandonment:** afforestation, where possible d. **special measures:** conservation of surface run-off

Fencing-in is obligatory.
Pasture: correct and adapted stocking rate and grazing system (eg. three-herd, four-pasture).

Tab.8. Terrace stages and recommended use for terrace slopes.

Farm Land Erosion: In Temperate Plains Environment and Hills
S. Wicherek (Editor)

Soil erosion evaluation on hillslopes in Taiwan

K.F.A. Lo, S.H. Chiang and B.W. Tsai

Department of Natural Resources, Chinese Culture Univ. ; Department of Geography, National Taiwan University, Taipei, Taiwan.

SUMMARY

Watershed data are being collected in Taiwan for testing the AGNPS model, a micro-computer program to model the erosion process and simulate the water erosion and transport of sediment, nutrients and flow for watersheds ranging from 1 to 50,000 ha size. With proper modifications of the Universal Soil Loss Equation factors, excellent agreements between the simulated and measured sediment yields were obtained for the Tsengwen Reservoir watershed.

The model input data are collected initially by the technique of remote sensing and geographic information system, and processed with the ARC/INFO Geographic Information System computer software. The predicted sedimentation depth for the entire reservoir watershed averages about 5.9 mm/yr, which is not significantly different from the observed rate, but exceeds what is observed in the U.S. and the Yellow River Basin. It is, therefore, necessary to prescribe appropriate soil and water conservation practices to control the sedimentation problem in reservoir watersheds in Taiwan. The model is also capable of identifying areas within the watershed with high erosion and sediment yield. This will provide a guide in formulating policies and developing plans to counteract erosion effects, to optimize farm output, and to stablize economic development.

RESUME

Les techniques d'occupation et de la gestion du sol employées dans le passé deviennent aujourd'hui dangereuses en raison de l'accroissement démographique qui parfois engendre des dégats irréversibles. Ces changements de pratique dans l'exploitation des terres se traduisent souvent par une augmentation des taux d'érosion sur site et par une diminution de la fertilité. Il est également urgent d'évaluer les dégâts hors-site que causent spécifiquement l'érosion en nappes et par rigoles, l'érosion éolienne, ainsi que l'érosion par ravinement. Le Service de Recherche Agricole du Ministère de l'Agriculture américain a élaboré, en collaboration avec l'Agence de Contrôle de la Pollution du Minnesota et le Service de la Conservation du Sol, un programme informatique visant à simuler le processus d'érosion, l'érosion hydrique, ainsi que le transport des sédiments et des nutriments et l'écoulement dans des bassins de 1 à 50 000 ha. Le modèle qui en résulte - appelé AGNPS, pour Agricultural Nonpoint Source Model - est capable de traiter un très grand nombre de conditions rencontrées dans les bassins. Il permet de décrire : 1) l'ampleur des phénomènes : la quantité de sédiments et de nutriments générée et transportée à l'intérieur d'un bassin, 2) l'origine de ces phénomènes sources et/ou des drainages déversant des quantités substan-tielles de polluents dans les voies d'eau, 3) les conséquences potentielles pour l'environnement des nouvelles pratiques d'exploitation des terres, de traitements utilisés pour la conservation. Les scénarios de gestion impliquant les récoltes et les labours ainsi que diverses techniques de gestion et les mesures de contrôle sont aussi envisagés.

La comparaison des données recueillies sur le Bassin Tsengwen à Taïwan avec celles du modèle AGNPS dont on voulait tester la performance a permis de mettre en évidence, après modification des facteurs de l'Equation Universelle de Perte de Sol, de très grands points de convergence entre la production de sédiments simulée et celle mesurée sur le terrain.

Les données utilisées par le modèle, recueillies grâce au système d'information géographique et de repérage à distance, sont traitées par ordinateur (logiciel ARC/INFO). La perte de sol totale ainsi estimée pour l'ensemble du Bassin atteint environ 5 mm/an. Ce résultat correspond étroitement aux estimations fondées sur de précédentes données concernant la sédimentation dans les bassins. Cela prouve encore que le modèle AGNPS est capable de fournir une évaluation quantitative des dégâts sur site/hors-site, d'estimer la performance du bassin et de planifier des stratégies de conservation au niveau local et régional, voire à celui du pays.

INTRODUCTION

Because of its high densities of population and economic activity, Taiwan has accumulated a large backlog of environmental problems. In recent years, population in addition to socioeconomic pressure and abusive cultivation practices have aggravated the problem of soil erosion and sediment yield. The major concern about erosion was the damage done to the soil itself - the loss of soil and of soil productivity. However, when eroded soil is carried off the farm by runoff, it may end up in local stream, rivers, canals, or irrigation and hydroelectric reservoirs. The loss of topsoil that reduces land productivity may also reduce irrigation, electrical generation, and the navigability of waterways. In Taiwan, most of the large reservoirs have already experienced serious siltation problems resulting from natural landslides and improper land use in the upper reaches (1). The longevity of these reservoirs probably will all be shorter than planned.

Further downstream, sedimentation can increase flood hazards, deteriorate water quality, and harm coastal resources, including corals, mangroves, and fisheries. Another problem of more recent concern involves the contaminants that are carried into the streams. Dissolved fertilizer and pesticides may be carried into streams and rivers affecting plants and aquatic life, and even land animals and men (2). Hence, there is a critical need to identify site specific off-site damages caused by sheet and rill, wind, ephemeral qully and gully erosion.

Recently, many erosion and sedimentation models have been developed for predicting fluvial soil erosion and sediment yield. The "AGNPS" (Agricultural Non-Point Source Pollution model) model has the capabilities of evaluating nonpoint source pollution at any point and predicting site-specific sedimentation within a catchment. This paper describes the adaptation of this model in evaluating and predicting soil erosion in the Tsengwen Reservoir Watershed, Taiwan.

AGNPS MODEL

The AGNPS model was developed by the USDA's Agricultural Research Service in cooperation with the Minnesota Pollution Control Agency and the Soil Conservation Service (3). It is a computer simulation model designed to analyze nonpoint source (NPS) pollution in watersheds. The primary objectives of AGNPS development were to : 1. Obtain uniform and accurate estimates of runoff quality with primary emphasis on sediment and nutrients, 2. Analyze potential impacts of conservation alternatives (and land uses) for watershed management, and 3. Provide a flexible and easy-to-use model.

AGNPS also provides a means for assessing watershed conditions and objectively evaluating storm-related generation and transport of NPS pollution within watersheds.

The model subdivides the catchment watershed into uniform square areas called "cells". Potential pollutants are routed through cells in a stepwise manner, proceeding from the headwaters of the watershed to the outlet. This allows flow as well as water quality parameters at any point be examined. All characteristics of the watershed are expressed at the cell level.

AGNPS is a single-event based model. Basic model components include hydrology, erosion, and sediment and chemical transport. In the hydrology portion, calculations are made for the runoff volume and peak concentrated flow. The erosion portion computes total upland erosion, total channel erosion, and a breakdown of these sources into five particle size classes (clay, silt, sand, small aggregates, and large aggregates). Upland erosion is routed through the watershed. Sediment transport is also calculated in the five particle classes as well as the total. The chemical transport portion is separated into one part handling soluble pollutants and another part handling sediment-attached pollutants. Transport of nitrogen, phosphorus, and chemical oxygen demand (COD) are estimated throughout the watershed. The model also treats sediment from gullies and inputs of water, sediment, nutrients, and COD from animal feedlots as point sources and routes them with contributions from nonpoint sources. Water impoundments, such as tile-outlet terraces, are considered as depositional areas for sediment and sediment-attached chemicals.

Various output options are available with the model. The outputs can be examined for a single cell or for the entire watershed. A preliminary output includes watershed and cell area, storm precipitation and erosivity, runoff volume and peak rate, and a detailed analysis of the sediment and nutrient yields. Also given are estimates of sediment delivery ratio, sediment enrichment ratio, sediment concentration and sediment yield for each of the five particle size classes. The detailed nutrient analysis includes the unit area amount of N, P, and COD in runoff and N and P for sediment adsorbed nutrients ; and the N, P, and COD concentration in the runoff.

GENERAL DESCRIPTION OF STUDY AREA

The Tsengwen Reservoir watershed is located in the northeastern corner of the Chiayi County, bordering the Tainan County in the south. Yu Shan (Jade Mountain) is situated in the east. Towards the northeast lies the Alishan Mountain Range. A large floodplain is located in the west. The central portion is undulating hills and ridges. Therefore, the prevailing slope of the Tsengwen Reservoir watershed is towards the west (Figure 1).

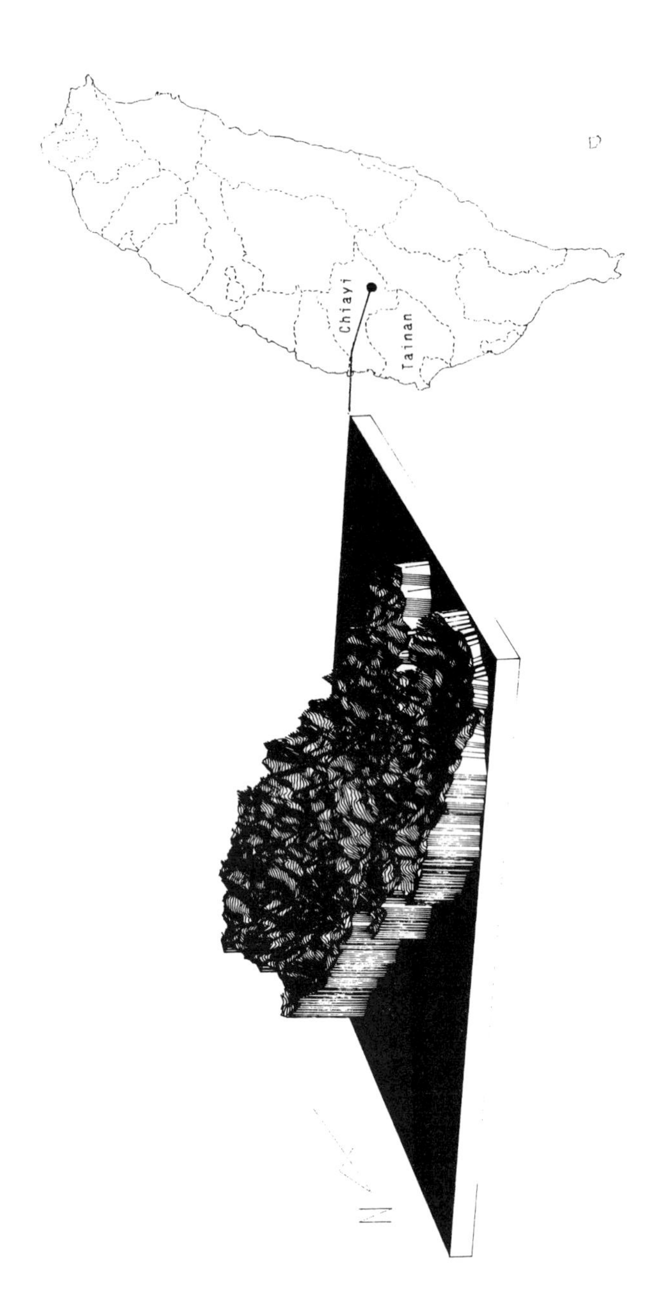

Figure 1. Location map of Tsengwen Reservoir Watershed.

The major river system in this watershed is the Tsengwen River. The main dam is constructed at Tapu to store water for the Tsengwen Reservoir. The entire length of Tsengwen River is about 138 km. The average inclination ratio is about 1:57. It originates from Suishan (2,609 m elevation), passes through Nanhsi, Shanshang, and Matou before reaching the sea.

According to 30 year rainfall records (1946-1985), there is a definite trend of increasing rainfall from the west towards the east. Between 1,000 and 2,500 m elevation, annual rainfall often exceeds 3,000 mm. However, heavy rain is concentrated during summer from May to September with monthly rainfall of above 100 mm and comprises 83 % of the yearly rainfall. The average annual temperature is about 24°C. From May to October, average monthly temperature often exceeds 25°C. Evaporation rate is quite high, usually averaging 100 mm/month and reaching 1,693 mm/year.

There are two major groups of soil in the study area. The first group belongs to the Pale Colluvial soils formed from sandstone or shale, or both. It is found on undulating to hilly slopes of 7 % to 35 %. It is usually well-drained. Soil depth ranges between shallow (<20 cm) to deep (>120 cm). Soil structure consists of crumb or weak subangular blocks and is friable to firm. The color of this soil is dark yellowish brown to yellowish brown, or dark brown to strong brown. Soil texture is between sandy loam to silt clay. The pH is rather low, ranging from 4.0 to 5.5. According to the U.S. Taxanomy, it is often classified as Hapludalfs or Haplustalfs (4).

The other group of soil is located on rolling to hilly slopes of 25 % to 40 %. It belongs to the local colluvium formed from sandstone or shale, or both. It is also well-drained with very shallow (<20 cm) to deep (>120 cm) soil depth. This soil usually is very dark brown, very dark gray, or very dark grayish brown colored. Soil texture ranges from sandy loam to silty clay loam. Its structure can be friable to very friable with crumb, weak subangular block, and many fragments of parent rock. The pH for this soil is low, between 4.2 and 5.5. The eqivalent Great Group Soil Taxonomy classification for this soil is Hapludalfs or Haplumbrepts (4).

The entire watershed is situated between elevation of about 100 to 1,000 m. Most areas are covered with natural forest, conservative forest and experimental forest. A small area belongs to the reservation land of the hilltribes. Currently, land development in this watershed is still scattered and limited. Because of abundant rainfall on hillslopes, any large scale land development would induce runoff and cause severe soil erosion and massive landslides (5).

STUDY METHODS

The procedures in conducting this study is displayed in Figure 2. The ARC/INFO Geographic Information System (GIS) computer software is selected to process and analyze the necessary input data. Database for the entire study area is structured using the Integrated Terrain Unit Mapping (ITUM). Data layers consist of : 1) The Ministry of the Interior 1/25,000 topographic maps, 2) The Central Geologic Survey Institute 1/50,000 geologic maps, and 3) The Soil and Water Conservation Bureau 1/25,000 soil maps.

The available natural resources theme maps for the study area include climate, hydrology, geology, soil, elevation, slope steepness, aspect, land use, and vegetation.

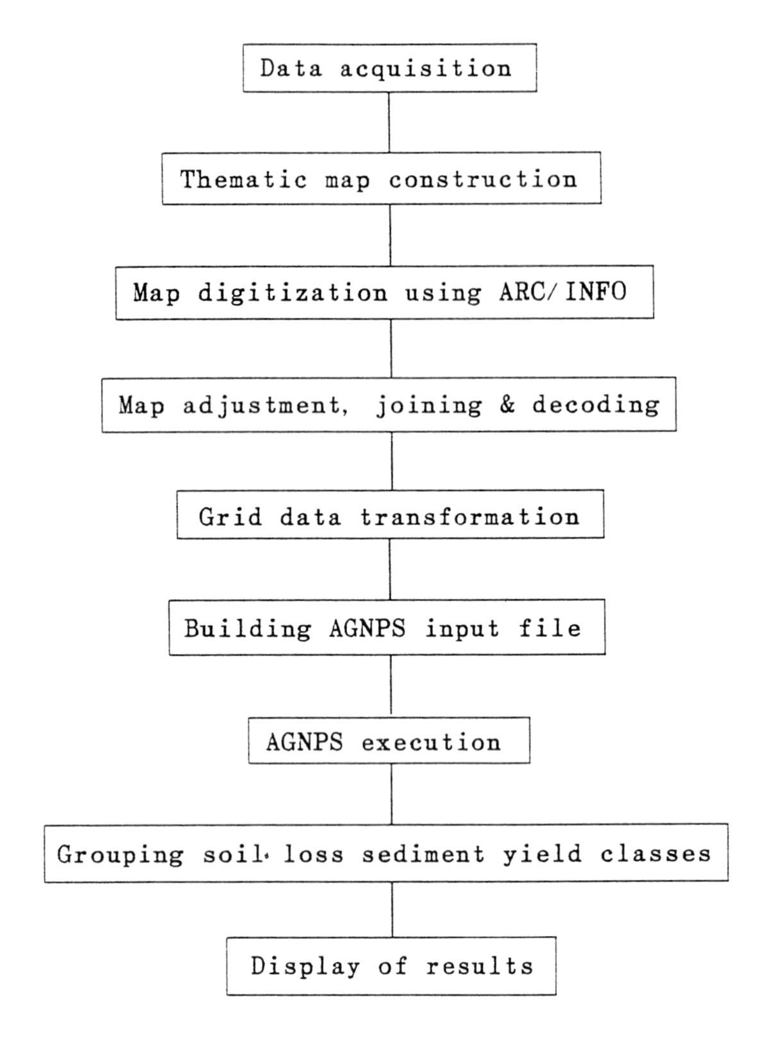

Figure 2. Study procedures to integrate GIS and AGNPS model.

Table 1 summarizes the input data file parameters for the AGNPS model. The rainfall erosivity factor (R) in the USLE (Universal Soil Loss Equation) is obtained from the iso-erodent map published by Huang (6). Wann and Hwang (1989) estimated the soil erodibility factor (K) in Taiwan using Wischmeier and Smith (8) nomograph method. Their values are used in estimating the K factors. Slope steepness and slope length factor (LS) are derived from the topographic maps. The crop management factor (C) are based on a collection of representative C values obtained from field experiments under natural conditions. The supporting practice factor (P) is assumed the maximum value of unity to represent the worst case when none of the conservation control measures are used.

Table 1.
AGNPS model input data file parameters.

No.	Catchment Input
1	Catchment identification
2	Cell area
3	Total number of cells
4	Precipitation amount
5	RAinfall erosivity value

No.	Code	Cell Parameter
1	CE	Cell number
2	RC	Number of the cell into which it drains
3	CN	SCS curve number
4	LS	Average land slope (%)
5	SF	Slope shape factor (uniform, convex, or concave)
6	SL	Average field slope length
7	CS	Average channel slope (%)
8	CSS	Average channel side slope (%)
9	N	Manning's roughness coefficient for the channel
10	K	Soil eroodibility factor (K) from USLE
11	C	Cropping factor (C) from USLE
12	P	Conservation practice factor (P) from USLE
13	SCC	Surface condition constant (factor based on land use)
14	A	Aspect (one of 8 possible directions indicating the principal drainage direction from the cell)
15	T	Soil texture (sand, silt, clay, peat)
16	F	Fertilization level (zero, low, medium, high)
17	AF	Incorporation factor (% fertilization left in top cm of soil)
18	PS	Point source indicator (indicates existence of a point source input within a cell)
19	GS	Gully source level (estimate of amount of gully erosion in a cell)
20	COD	Chemical oxygen demand factor
21	IF	Impoundment factor (a factor indicating presence of impoundment terrace system within the cell)

Parameter values such as curve number, Manning's roughness coefficient, surface condition, are based on actual land use conditions and the corresponding table values provided in the model users' guide (3). To form the input data file needed to operate the model, the watershed was first divided into 1,900 uniform cells. Each cell occupied approximately 25 ha in area. The drainage direction of each cell, defined as the direction of flow leaving the cell, was estimated based on its shape and topographic setting to establish the watershed drainage pattern.

Knowing the drainage direction, the runoff receiving cell can be easily determined. Fertilization rate is assumed to be not high and is incorporated into the soil. No point source, gully source and impoundment are included. The COD factor usually varies according to different land use. A value of 50 mg/l is used, corresponding to the "pasture and open" land use type (3). A listing of the input data file compiled for this watershed is shown in Table 2.

Table 2.
Partial listing of the input data file.

CE	RC	CN	LS	SF	SL	CS	CSS	N	K	C	P	SCC	A	T	F	AF	PS	GS	COD	IF
1	5	70	0.0	1	500	0.0	10.0	0.08	0.4	.05	1.00	.59	4	2	0	10	0	0	50	0
2	4	70	40.0	1	500	20.0	10.0	0.08	0.4	.05	1.00	.59	6	2	0	10	0	0	50	0
3	12	70	17.0	1	500	9.0	10.0	0.08	0.4	.05	1.00	.59	6	2	0	10	0	0	50	0
4	15	70	35.0	1	500	18.0	10.0	0.08	0.4	.05	1.00	.59	4	2	0	10	0	0	50	0
5	14	70	36.0	1	500	18.0	10.0	0.08	0.4	.05	1.00	.59	6	2	0	10	0	0	50	0
6	7	71	0.0	1	500	0.0	10.0	0.08	0.4	.10	1.00	.59	3	2	0	10	0	0	50	0
7	6	70	0.0	1	500	0.0	10.0	0.08	0.4	.05	1.00	.59	7	2	0	10	0	0	50	0
8	19	70	42.0	1	500	21.0	10.0	0.08	0.4	.05	1.00	.59	5	2	0	10	0	0	50	0
9	20	70	31.0	1	500	16.0	10.0	0.08	0.4	.05	1.00	.59	5	2	0	10	0	0	50	0
10	20	70	43.0	1	500	22.0	10.0	0.08	0.4	.05	1.00	.59	6	2	0	10	0	0	50	0
11	21	70	12.0	1	500	6.0	10.0	0.08	0.4	.05	1.00	.59	6	2	0	10	0	0	50	0
12	23	70	52.0	1	500	26.0	10.0	0.08	0.4	.05	1.00	.59	5	2	0	10	0	0	50	0
13	23	70	54.0	1	500	27.0	10.0	0.08	0.4	.05	1.00	.59	6	2	0	10	0	0	50	0
14	26	70	63.0	1	500	32.0	10.0	0.08	0.4	.05	1.00	.59	4	2	0	10	0	0	50	0
15	14	70	26.0	1	500	13.0	10.0	0.08	0.4	.05	1.00	.59	7	2	0	10	0	0	50	0
16	27	70	50.0	1	500	25.0	10.0	0.08	0.4	.05	1.00	.59	6	2	0	10	0	0	50	0
17	18	70	43.0	1	500	22.0	10.0	0.08	0.4	.05	1.00	.59	3	2	0	10	0	0	50	0
18	34	70	60.0	1	500	60.0	10.0	0.08	0.4	.05	1.00	.59	6	2	0	10	0	0	50	0
19	36	70	56.0	1	500	28.0	10.0	0.08	0.4	.05	1.00	.59	5	2	0	10	0	0	50	0
20	37	70	65.0	1	500	33.0	10.0	0.08	0.4	.05	1.00	.59	5	2	0	10	0	0	50	0

RESULTS AND DISCUSSIONS

Erosion factors affecting the tropical soils are significantly different than those for the temperate soils. The climate in Taiwan is humid and tropical to subtropical. Based on limited data source, the field measured soil loss in Taiwan is estimated to be approximately 15 percent of the USLE predicted soil loss (9 - 10). Table 3 shows a summary output from the AGNPS model with rainfall amount of 2,000 mm and rainfall erosivity index of 2,500. The total average annual soil loss for the entire Tsengwen Reservoir watershed is estimated to be about 903 t/ha. The total sediment yield is approximately 2.71 million tons.

Table 3.
Summary output of the AGNPS model.

Watershed Summary

Watershed Studied	TSENG-WEN RESERVOIR WATERSHED	
The area of the watershed is	117230	acres
The area of each cell is	61.70	acres
The characteristic storm precipitation is	80.00	inches
The storm energy-intensity value is	2500	

Values at the Watershed Outlet

Cell number	1889	000
Runoff volume	75.4	inches
Peak runoff rate	433163	cfs
Total Nitrogen in sediment	42.19	lbs/acre
Total soluble Nitrogen in runoff	13.72	lbs/acre
Soluble Nitrogen concentration in runoff	0.80	ppm
Total Phosphorus in sediment	21.10	lbs/acre
Total soluble Phosphorus in runoff	0.85	lbs/acre
Soluble Phosphorus concentration in runoff	0.05	ppm
Total soluble chemical oxygen demand	855.07	lbs/acre
Soluble chemical oxygen demand concentration in runoff	50	ppm

Sediment Analysis

Particle type	Area Weighted Erosion Upland (t/a)	Area Weighted Erosion Channel (t/a)	Delivery Ratio (%)	Enrichment Ratio	Mean Concentration (ppm)	Area Weighted Yield (t/a)	Yield (tons)
CLAY	20.26	0.00	93	15	2200.21	18.79	2202992.0
SILT	32.15	0.00	18	3	683.94	5.84	684807.7
SAGG	201.48	0.00	0	0	95.77	0.82	95894.0
LAGG	124.77	0.45	0	0	3.07	0.03	3073.1
SAND	24.66	0.21	0	0	0.58	0.00	577.6
TOTAL	403.33	0.01	6	1	2983.58	25.48	2987344.0

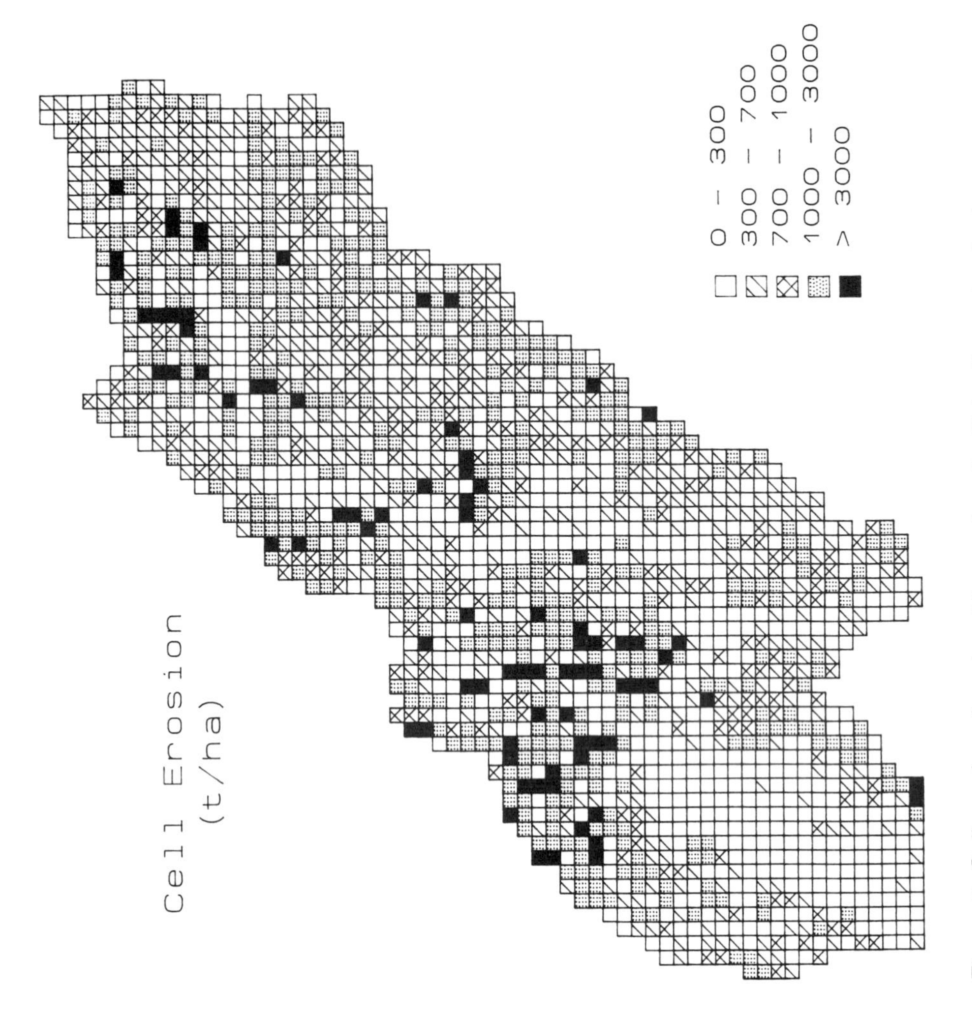

Figure 3. Soil erosion class level for each cell within the watershed.

In recent years, the government has placed strong emphasis on the erosion problems surrounding reservoir watersheds. Dedicated agencies have been established to manage and perform reservoir watershed conservation works. As a result, significant reduction in reservoir sedimentation problems has been observed. During 1973 to 1983, the average sedimentation rate at Tsengwen Reservoir was about 8.8 mm/yr. For the next 4 years, the rate has declined to about 7.7 mm/yr. This trend is still continuing at a fast pace (2 - 11).

Based on the model estimate of about 2.71 million tons of sediment yield per year and assuming an average bulk density of 1.0 g/cm3 (or t/m^3) and a catchment area of 460 km^2, the estimated sedimentation rate is about 5.9 mm/yr. This rate is not too different from the observed rate, further emphasizing the applicability of AGNPS in estimating erosion problems in large watersheds in Taiwan. This estimated sedimentation rate, however, far exceeds that observed in the United States. The average annual sedimentation rate recorded for 48 reservoirs managed by the Tennessee Valley Authority is only about 0.25 mm (12). The Yellow River is world famous for its high erosion rate. Its turbidity measurements often exceed 37.6 kg/m^3. However, the highest sedimentation rate within the entire river basin never exceeds 7 mm/yr (12).

Although significant climatic and geographic differences exist between these regions, it is clear that the soil conservation programs need to be intensified in Taiwan in order to preserve the precious soil resource which has direct influence on the stability of agricultural development.

Figure 3 subdivides amount of soil loss for each cell into five level groups. Out of the entire watershed area, over 33 % loses less than 300 t/ha soil. About 23% loses between 300 to 700 t/ha ; 14 % between 700 to 1,000 t/h ; and over 3 % exceeding 1,000 t/ha. Based on the cell output, high erosional areas may be easily identified within the entire watershed. Subsequent land development should avoid such areas because they need to be adequately protected with appropriate conservation plans.

CONCLUSIONS

The AGNPS model is a computer simulation model designed to analyze nonpoint source pollution in watersheds. It has the capabilities of evaluating nonpoint source pollution at any point within a watershed. It also allows a more realistic view of the sediment and nutrient yields on an areal basis. With adequate modification and adaptation, the model has been shown to produce reasonable estimates of sediment yield at the Tsengwen Reservoir watershed.

The predicted sedimentation depth for the reservoir watershed averages about 5.9 mm/yr, which is not significantly different from the observed rate, but exceeds what is observed in the U.S. and the Yellow River Basin. It is, therefore, necessary to prescribe appropriate soil and water conservation practices to control the sedimentation problem in reservoir watersheds in Taiwan.

The model is also capable of identifying areas within the watershed with high erosion and sediment yield. This will provide a guide in formulating policies and developing plans to counteract erosion effects, to optimize farm output, and to stablize economic development.

REFERENCES

1. Chang, C. Y. D., P. C. Chiang, Y. P. Chu, H. H. M. Hsiao and Severinghaus L. L. (1989) : Taiwan 2000 : Balancing economic growth and environmental protection. Instit. of Ethnolongy, Academia Sinica, Nankang, Taipei, Taiwan, ROC. 748 pp.
2. Lee, S. W. (1985) : Watershed management and protection in Taiwan. Construction World, Taipei, Taiwan, (in Chinese).
3. Young, R. A., Onstad, C. A., Bosch D. D. and Anderson W. P. (1987) : AGNPS : An agricultural nonpoint source pollution model : A watershed analysis tool. USDA Conservation Research Report 35, 150 pp.
4. MARDB (Mountain Agricultural Resources Development Bureau) (1986) : Slopeland's Soil Survey Report : Chiayi and Yunlin Counties, 154 pp. (in Chinese).
5. Ho, C. W. and Chiang S. H. (1990) : Disasters from Sloping Lands. National Science Council, Taipei, Taiwan, R. O. C., 28 pp. (in Chinese).
6. Huang, C. T. (1979) : Study on the rainfall erosion index in Taiwan. J. Chinese Soil and Water Conservation. 10 (1) : 127-144. (in Chinese).
7. Wann, S. S. and Hwang, J.I. (1989) : Soil erosion on hillslopes of Taiwan. J. Chinese Soil and Water Conservation 20 (2) : 17-45. (in Chinese).
8. Wischmeier, W.H. and Smith, D. D. (1978) : Predicting rainfall erosion losses : A guide to conservation planning. U S D A Agric. Handbk. 537, Washington, D. C., 58 pp.
9. Lo, K. F. A. (1989) : Applicability of a process-based model for assessing soil erodibility in Taiwan. J. Chinese Soil and Water Conservation 20 (1) : 31-38.
10. Liao, S. H. (1990) : Application of geographic information system in estimating soil erosion potential. M. S. Thesis, National Chung Hsin University, 99 pp. (in Chinese).
11. Lee, S. W. (1991) : Reservoir watershed conservation and management. Science Monthly 22(4) : 272-276. (in Chinese).
12. Chiang, S. H. (1991) : The reservoir and water resources in Taiwan. Science Monthly 22(4) : 251-258. (in Chinese).

Farm Land Erosion: In Temperate Plains Environment and Hills
S. Wicherek (Editor)
1993 Elsevier Science Publishers B.V.

"Possibilities for an environmentally-sound restructuring of agriculture in the new Bundesländer"

C. AHL

European Parliament, DG IV/STOA, L - 2929 Luxembourg

SUMMARY

After a brief introduction into the STOA programme of the EP and the research methodology some aspects of the former GDR agriculture are outlined. Different definitions for sustainable agriculture lead to the possibilities for an environmentally-sound restructuring of agriculture, which will require a new Common Agricultural Policy of the Commission of the European Community.

RESUME

Après une brève introduction dans le programme STOA, dans cet article certains aspects de l'agriculture de l'ex. R.D.A. sont soulignés. Différentes définitions pour une agriculture à encourager conduisent à la restructuration de l'agriculture vis à vis de l'environnement ; ceci demandera une nouvelle politique d'Agriculture commune de la Commission des Communautés Européennes.

FOREWORD

STOA is the Scientific and Technological Options Assessment programme of the European Parliament. STOA was established in 1987, administratively connected to the Directorate General for Research, DG IV. In brief, STOA works as follows : a Member of the European Parliament is concerned about the future development of science and technology in a specific field and puts a question to the STOA Panel. The Panel decides upon and launches a project on the appropriate scale. External research institutions, universities, consultancies, etc., both inside and outside of the EC, are contracted to deliver their findings and results according to the rules laid down for the particular project.

1. INTRODUCTION

In late 1990, a Member of the EP made the proposal to the STOA programme to undertake a study of the five new Länder, titled : "Possibilities for an environmentally-sound restructuring of agriculture in the new Bundesländer". Since it is necessary to restructure agriculture in this part of Germany anyway, and because it therefore offers possibilities for investigating different sustainable scenarios, the new Länder seem to offer a useful test-bed for the introduction of sustainable agricultural systems on a large-scale basis. The aims of this project are not only to focus on the re-establishment of a certain kind and scale of agriculture, but also to discuss the proposed Common Agricultural Policy (CAP) (COM (91) 258 Final 1991) (1) reform and its impact to future land-use systems. Possible policy options to support sustainable agriculture will be outlined, too. According to the European Community's Programme of Policy and Action in relation to the Environment and Sustainable Develoment "Towards sustainability" (COM (92) 23-Vol.II) [2], environmetally friendly development has to be an integral part of the definition and implementation of other Community policies. Future CAP cannot be reduced to the question of prices for agricultural goods, milk-quotas and pension funds for early retirement, as the environmental issue now has a legal basis in the Treaty of Maastricht.

2. RESEARCH METHODOLOGY

A conventional inductive research approach would be :
- identifying some specific problems,
- proposing the hypothesis,
- testing the hypothesis, and
- evaluating it.

But this type of generalization, from manageable pieces of scientific problems, implies that results from few variables could be applied over a broad area, and which could falsify the whole system [3].

An ideal Technology Assessment (TA) study is characterized by the following items [4] :
- the ethical, legal, economic, social and ecological dimensions,
- including besides the short - and medium term intended and unintended effects, the long term and synergized effects, and results in a scenario development for the proposed new technology (in this case CAP reform and sound-restructuring of agriculture in the new Länder). Potential beneficial and adverse impacts have to be assessed ; the analyses must be followed by evaluation.

The problem of soil erosion and land degradation is explained by both the technical and scientific arguments - texture, surface sealing, rainfall intensity etc.- and by the practises of the farm manager. He may be a farm manager from a cooperative or a private farm, circled by a decision-making matrix :

Information	Economic	Production Enterprise
Social Environment	FARM MANAGER	Soil (site)
Legal Environment	Pest & Waste	Climate (region)

Figure 1. The decision-Making Unit (after Frescoe & Westphal 1988) [19]

Similarly, a policy manager is embedded in a decision-making matrix. A TA study with an agricultural background should enable policy makers on the european level, as well as farm managers on the regional level, to enlighten their decisions in the view of future benefits and potential disbenefits according to the input in the decision-making matrix.

HISTORICAL BACKGROUND - FORMER GDR

Due to the political system in the former GDR, the total farming system : crop rotation planning, livestock density as well as the working technology, was more or less ordered and planned by the centralized administrative and party staff. In most cases the natural conditions of the agriculturally used area were neglected. Schmidt [5] estimated the degraded and anthropogenic deteriorated soils to cover up to 2.4 millions ha, i.e. 40 % of all agricultural land. Although the use of fertilizer and agro-chemicals was low or relatively comparable to use in EC countries, the potential misuse resulted from the distribution behaviour starting on the regional level down to the field level. For 1989, quantities of chemical nitrogen applied were: in the FRG about 135 kg/ha ; in the GDR about 134 kg/ha (according to the quantities of nitrogen delivered to the agricultural sector in the former GDR, the amount applied would have been less). About 4.6 kg/ha of chemicals were used on arable land in both countries in 1989.

Livestock density, falling sharply over the last two years, was around 100 large animal units (LAU)/100 ha (80 LAU/100 ha EC(12)), but storage capacity for slurry was only sufficient for 60 days. Slurry was applied to less than 11 % of agricultural land, so overdosages frequently occurred on the regional level (Stat. Amt of former GDR).

Soil erosion and soil compaction due to the specific conditions (large fields neglecting natural borders for soils and slopes, heavy machines etc.) has already been outlined by Frielinghaus at this conference (cf. infra).

THEORETICAL BACKGROUND - SUSTAINABLE AGRICULTURE

Although the concept of sustainability is sometimes regarded to be a philosophical one, the two key concepts are strictly focused on "Our Common Future" [6] :

- the concept of "needs", particularly the essential need of the world's poor,
- the idea of limitations imposed by the state of technology and social organization on the environment's capacity for meeting present and future needs.

Presently, the European Community is a long way from sustainability in many areas [7], and for agricultural development only a vague definition of sustainability is given in the Fifth Environmental Action Programme : "Traditionally, the farmer has been the guardian of the soil and the countryside. By careful husbandry, including integrated crop and livestock farming and waste management, farmlands were passed in sound conditions from one generation to the next".

A more detailed definition is given by Harwood [8] : "Sustainable agriculture is a system that utilizes an understanding of natural processes along with the latest scientific advances to create integrated, resource-conserving farming systems. These systems will reduce environmental degradation, are economically viable, maintain a stable rural community, and provide a productive agriculture in both short and long term."

In the 's-Hertogenbosch conference of the FAO, the Sustainable Agriculture and Rural Development Programme was launched and the conference adopted the following definition of the SARD programme [9] : "... (is) the management and conservation of the natural resources base, and the orientation of technological and institutional change in such a manner as to ensure the attainment and continued satisfaction of human needs for present and future generations. Such sustainable development (in the agriculture, forestry and fisheries sectors) conserves land, water, plant and animal genetic resources, is environmentally non-degrading, technically appropriate, economically viable and socially acceptable."

In Harwood's and FAO's definition included is the aim to maintain a stable rural community, which leads to the social component of sustainable development of agriculture. FAO focused also on genetical resources, which for example is especially a treasure for developing countries for resistance-breeding.

As migration of inhabitants from the countryside to the fast growing towns has been regarded as a problem for LDC's, in many parts of Europe, especially in the new Länder, zones of land are depleted off the peasant society lack to any income possibility.

POSSIBILITIES FOR RESTRUCTURING

A reconstruction of agriculture and land-use is to be defined by the aims and goals of what the society wants to support. Two recent TA studies have outlined the possible ways.

The Office of Technology Assessment has forecasted in its study (10) a two track agriculture for the future, one with a high intensified cash crop agriculture and another branch with all the other possibilities, part time farming, ecological farming, etc. Less favoured areas will be abandoned and nature will recover these areas. For the Deutsche Bundestag, the Enquete-Commission for Technology Assessment has indentified four possible pathways for the future development of agriculture (11) :

1 - world-market oriented agricultural policy creates hi-tec agriculture which is highly competitive due to the international division of labour ; the agricultural production and land use has been shifted to the best sites, subsidies are reduced sharply ;
2 - agricultural policy is oriented to the internal EC market, continues with trade barriers and levies on imported goods, less reduction of subsidies, development of regions with a high share of cash crops and regions with mostly direct marketed goods with consumer oriented quality ;
3 - introduction of a system of quotas, large-scale extensification of the production, high administrative surveillance efforts ;
4 - regionalization of agricultural policy into small areas with a site matched agri-culture under the rules of ecological/integrated farming.

Assuming the present development of traditional agricultural production and a non-increased consumption level in the future, it is calculated that the agricultural area has to be reduced by 30 % to avoid any further surplus production. Grain production alone will have already created by 1996 a stock surplus of 45 million tons. Having an average yield of 4.5 tons/ha, 10 million hectares have to be set aside if a business-as-usual scenario is to be pursued.

As the set aside scheme was not as successful as expected (1991 : 1.1 million ha (EC 12)) and also rejected by scientific testing as being not environ-mentally-sound (12), land-use scenarios for the future must include a mixture of production schemes which will cover the whole landscape.

The suitability of the landscape for a certain production-scheme should be based on a Geographical Information System (GIS), as already in use for fertilizer application models (13).

Geographical Information System				
Soils	Climate	Geomor-phology	Geology	Administrative Boundaries

Figure 2. Geographical Information System Set up.

Referring to these conditions, arable land use, grassland use (and nature preservation) will show regionally sharp contrasts. For the farming society, the net income will differ according to the region the farm is situated. In less favoured areas, or restrictive land-use zones (such as in water catchment areas or nature reservation resorts), environmental quality as a public asset is part of the output of farming and is to be financed by the society. This compensation should be correlated to the farmers role in the protection of the rural environment. The landscape preservation should be more fully recognized and renumerated accordingly [1], but payments on environmental grounds should not be restricted to farms of a certain size as it is still uncertain why small farms should be environmentally-sound at all [14]. As shown in Fig. 3, the range of land use will be spread from extensive farming to landscape conservation, restruction and preservation. Most of the intensive farming methods to get a maximum yield level will be abolished in future, the individual farmer of today, acting "rationally" under the rules of the present agricultural policy, makes use of common property resources to increase his farm income, but eventually exploiting, degrading and destroying them. Social costs of the farmers' methods are externalized, like pollution of groundwater, destruction of a desirable landscape [15], or diminishing of flora and fauna, especially due to large plots [16]. On the other hand, extensive farming and preservation and conservation of the landscape do require action of the legislative body, the European Parliament. The financial incentives for the extensification scheme, launched in 1990/1991, are expected to be revised though participation on the european level was little with 77.000 ha in the past [17]. But, payments for environmental management should not be a compensation for not polluting or not destroying wildlife habitat or landscape, they should be tied directly to ecological work (extensification, hedging, creation of ponds, receiving of rare animals, warding of nature zones)[18] - [15].

The process of measurable benefits starts with the application of satisfactory yardsticks, giving them relative values, which will be valid also for future generations. An overestimation of subjectively valued goods like colourful rare birds or landscapes with an outstanding beauty do not stand for scientifically and ethically spoken equal status of all species, the aim is to preserve the generality of a landscape with natural flora and fauna [15].

AGRI-ECOLOGICAL ENVIRONMENT						
conventional land use	efficient and environmentally-sound land use		extensive land use			
intensive farming		restrictive farming	alternative farming	no use of arable area	extensive farming and landscape conservation	
maximum yields	integrated	rules for protected areas	several acknowledged methods	fallow, change of use of arable land afforestation	subsidized production for landscape conservation	landscape conservation without production
parameters						
use of latest findings for high yield and profit	use of latest findings in ecological and economic knowledge	according legal rules for farming in protected areas	acc. to rules for alternative farming (IFOAM)	Keeping of conservation rules	use of "natural fertility"; conservation plan	landscape conservation plan
maximum yield	high yield	reduced yield	natural yield	no crop/yield	small yield	no crop
narrow CR[1)]	balanced CR[1)]	restricted CR[1)]	varied CR[1)]	green fallow	natural grassland	natural vegetation
high fertilization (over dosage)	balanced fertilization	restricted fertilization	no chemical, only organic fertilizer and stone meals	prohibition of chemical fertilizers, seewage sludge, town wastes and others	small chemical fertilizers, organic fertilizers	
intensive soil working	adjusted and conservative soil tillage	adjusted and conservative soil tillage	shallow turning and deep loosing soil tillage	soil tillage only for conservation	without soil tillage	
prophylactic plant protection	plant protection "upon view" control, improving selectivity	restricted plant protection	plant protection without chemically synthetised substances	no plant protective agents	mechanical plant protection	
high income due to externalisation of costs	balanced income	reduced income with subsidies	sustainable income with higher prices for the products	subsidies for fallow land	low income plus subsidies	renumerated conservation

CR = crop rotation

Figure 3. Land Use Systems (after Breitschuh 1992 (20).

An environmental payment strategy should increase the income of all farming conditions to a maximum value to allow every kind of farming to be successful, in less favoured areas as well as in highly favoured regions, under the conditions towards a sustainable development (Fig. 4).

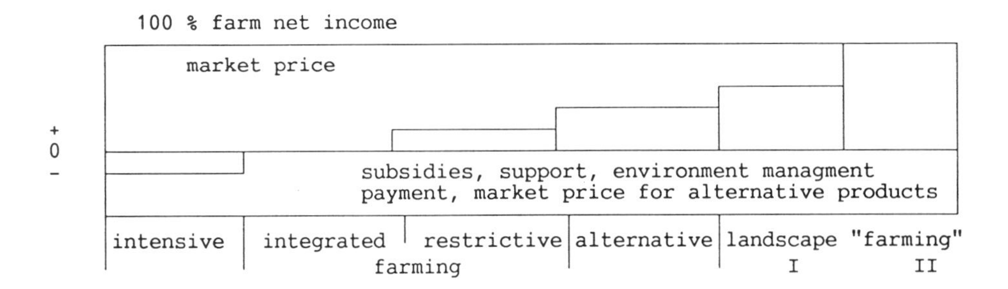

Figure 4 : Environmental Management Payment Strategy

CONCLUSION

In the past european farmers and consumers have received economic benefits from a production scheme according to the Treaty of Rome (Article 39) which has promoted and supported agricultural production. The Treaty of Maastricht of which one part is the "Towards Sustainability" programme will set goals for environmentally-sound development. In accordance with this, the CAP reform is a first step. Although a principle aim of the CAP reform is to reduce the surpluses food, and, secondly as "accompanying measures" the agri-environmental action programme is incorporated. Future reform cannot neglect the demands of sustainability. The project of "Possibilities for an evironmentally-sound restructuring of agriculture in the new Bundesländer" will create scenario studies for future land use to guide the CAP reform to a sustainable agricultural reform.

ACKNOWLEDGEMENTS

The author whish to thank grateful Mr. Patrick Keen, M. sc. and STOA Scholar, for revising the article upon correct english.

REFERENCES

1.COM (91) 258 (1991) : The development and future of the Common Agricultural Policy. Commission of the European Communities, Brussels.
2. COM (92) 23 - Vol. II (1992) : Vers un developpement soutenable ©
Programme Communautaire de Politique et d'Action pour l'Environnement et le Développement Durable et Respectueux de l'Environnement. Commission of the European Communities, Brussels.
3. Oberle, S.L. & Keeney, D.R. (1991) : A case for agricultural systems research. J. Environ. Quality. 20, 4-7.

4. Beusmann, V. (1989) : Agrarökonomische Fragen der Technikfolgenabschätzung. in : S. Albrecht (Hg.) : Die Zukunft der Nutzpflanzen. Campus Verlag, Frankfurt/New York.
5. Schmidt, R. (1991) : Anthropogene Veränderung und Degradation landwirtschaftlich genutzter Böden in den neuen Bundesländern Deutschlands. Z. f. Kulturtechnik und Landentwicklung 32, 282-290.
6. W.C.E.D. (1987) : Our Common Future. [Brundtland Report]. World Commission on Environment and Development, Oxford/New York, Oxford University Press.
7. Kraemer, R.A. (1992) : The European Community's Fifth Action Programme on the Environment (First Draft of an Analysis). Workshop Proc. 30/31 March1992, LUFA Thüringen, in press.
8. Harwood, R.R. (1989) : History of sustainable agriculture : U.S. and international perspective. In : Sustainable Agricultural Systems, eds. C.A. Edwards, R. Lal, P. Madden, R.H. Miller, and G. House, Soil & Water Conservation Society, Ankeny, Iowa, U.S.A.
9. FAO (1991) : The den Bosch declaration and agenda for action on sustainable agriculture and rural development - report of the conference. FAO/Netherlands Conference on Agriculture and the Environment, 'S-Hertogenbosch, The Netherlands, 15-19 April 1991.
10. OTA (1986) : Technology, Public Policy and the Changing Structure of American Agriculture. U.S. Congress, Office of Technology Assessment, OTA-F 285, Washington D.C.
11. Deutscher Bundestag (1990) : Bericht der Enquete-Kommission "Gestaltung der technischen Entwicklung ; Technikfolgenabschätzung und - Bewertung - Landwirtschaftliche Entwicklungspfade. Deutscher Bundestag, Drucksache 11/7991.
12. DBG (1991) : Ist Flächenstillegung von Ackern aus bodenkundlicher Sicht sinnvoll? Arbeitsgruppe Bodenschutz der Deutschen Bodenkundlichen Gesellschaft, Oldenburg 1991.
13. Andres (1991) : Soil fertility data banks as a tool for site-specific K-recommendations. Proc. 22nd Colloquium Int. Potash Institute, Bern.
14. CPRE (1991) : CAP in hand: submission from the Council for the Protection of Rural England to the Houses of Lords on the development and future of the CAP (91) 100. Council for the Protection of Rural England, Warwick House, Buckingham Palace Road, London, U.K.
15. Jenkins, T.N. (1990) : Future Harvest : The economics of farming and the environment : proposals for action. Council for the Protection of Rural England, Warwick House, Buckingham Palace Road, London, U.K.
16. Reschke, K. (1991) : Okologischer Wiederaufbau geschädigter Agrarlandschaften-Gedanken zu den fünf neuen Bundesländern. Z. f. Kulturtechnik und Landesentwicklung 32, 291-300.
17. COM of the EC (1992) : The Agricultural Situation in the Community -1991 Report. Commission of the European Communities, Brussels, Luxembourg 1992.
18. von Weizsacker, E.U. (1988) : Not a miracle solution but steps towards an ecological reform of the Common Agricultural Policy. Institute of European Environmental Policy, Bonn, London, Paris.
19. Frescoe, L.O. & Westphal, E. (1988) : A classification of farm systems. Expl. Agric. 24, 399-419.
20. Breitschuh, G. (1992) : Preliminary Report : Possibilities for an environmentally-sound restructuring of agriculture in the new Bundesländer. STOA/Lufa Thüringen, Luxembourg, Jena 1992.

Farm Land Erosion: In Temperate Plains Environment and Hills
S. Wicherek (Editor)

Field experiments on the reduction of sediment yield from arable land to receiving watercourses (N-Kraichgau, SW-Germany)

J. Baade, D. Barsch, R. Mäusbacher and G. Schukraft

Geographisches Institut der Universität Heidelberg Im Neuenheimer Feld 348, D-6900 Heidelberg, Germany.

SUMMMARY

Soil erosion represents a serious problem for agriculture in the hilly, loess covered regions of central Europe. Beside the well-known on-farm damages soil erosion is followed by considerable off-side damages in the hydrological system (e.g. siltation of waterways, input of nutrients and pollutants, etc.). Most of the sediment mobilised by soil erosion is derived from small drainageways on the fields where overland flow and interflow or returnflow concentrates and triggers linear erosion or thalweg gullying, respectively.

In our field experiments a small drainageway of a zero-order basin was set aside and stabilized with a permanent plant cover and fascines running across the thalweg. This 'grassed waterway' can be crossed by agricultural machinery. First results show that these measures reduce the linear erosion by approximately 50 %.

In addition, a retention pond has been constructed at the outlet of a first-order catchment to allow sedimentation of soil particles from other sources (e.g. rill-interrill erosion) in the watershed. Depending on the characteristics of the runoff events this measure prevents 40 % to 90 % of the suspended matter from entering the receiving watercourse.

Applying this system of conservation measures reduces the total sediment yield from the investigation area to the receiving watercourse on the average by 65%.

RESUME

A côté des dégâts - d'ailleurs bien connus - sur les terres agricoles proprement dits (érosion linéaire en rigoles - ravinement), des dégâts importants surviennent aussi dans le réseau hydrographique. (apport de substances nutritives et polluantes, envasement...). La plupart des sédiments mobilisés par l'érosion des sols sont issus de petits vallons qui se traduisent, sur la surface les dominant, par des phénomènes d'érosion linéaires et de ravinements.

Nos enquêtes sur le terrain, ont montré que les parties des champs les plus touchées étaient retirées de la production et plantées d'une couverture végétale que pouvaient traverser les machines agricoles. Les premiers résultats montrent que ces mesures réduisent jusqu'à 50 % l'érosion linéaire. Les parcelles couvertes d'herbe épaisse jouent le rôle de filtre retenant les particules de sol qui, sans cela, seraient emportées jusqu'aux vallons.

De plus, une zone de rétention a été construite pour permettre la sédimentation des particules du sol en provenance d'autres sources dans le bassin d'alimentation. Selon l'intensité des orages, elle retient entre 40 % et 90 % de la matière en suspension, évitant par là même, son rejet dans les ruisseaux.

L'application de ce système de mesures de conservation réduit la masse totale de sédiments produits dans la zone d'investigation d'environ 65 %.

1. INTRODUCTION

During rainstorms soil erosion by water causes considerable on-farm damages. In the short term the downwash of seeds, plants and nutrients causes crop damage and financial losses for the farmers (1 - 2). In the longer term soil erosion is followed by an irreversible degradation of the soil profil. Despite the intensive research on soil erosion and soil protection "the spectre of erosion has continued to enlarge" (3, p. 92). According to Auerswald (4, p. 663) soil erosion increased in Germany by 24 % between 1975 and 1985 caused by changes in the tillage system and land use during the past decades. Especially for loess covered areas one has to keep in mind that the Pleistocene sediment 'loess' is a nonrenewable resource. Therefore, Nortcliff's statement that "once the soil is removed by erosional processes reclamation is impossible" holds specially for these areas (5, p. 249).

Beside the well-known on-farm damages soil erosion is followed by considerable off-side damages in the hydrological system (6). The siltation of waterways increases the flood risk by aggradation of the streambeds and the maintenance costs of waterways. In addition, the input of nutrients and pollutants are other resultant problems to mention. Meanwhile comparative studies in the U.S.A. even indicate that the numerous off-site damages exceed on-site damages (7).

2. SIGNIFICANCE OF EPHEMERAL GULLY EROSION IN THALWEGS

Traditionally, soil erosion has mainly been seen as a problem of sheet erosion and rill-interrill erosion. The USLE (8) represents this tradition. Recently the role of ephemeral gully erosion has been emphasized (9 - 10 - 11). The concentration of overland flow and interflow in natural drainageways creates ephemeral gullies. Poesen (12, p. 39) describes these erosion features as "continuous temporary channels, erased by normal tillage, but recurring in the same place during subsequent runoff events". He furthermore observes that the depth of these gullies in deep loess loam soils is essentially controlled by the plough depth, which is usually 25 cm).

Lately, De Ploey (10, p. 150) drew attention to the fact that incision in the upper reaches of small valleys of zero-order basins often starts already at relatively low rainfall intensities. From his studies in Central Belgium and the loess areas of the Paris Basin he calculated critical rainfall intensities between 8.5 and less than 1 mm/hr.

There is strong evidence that in certain areas soil loss from ephemeral gullies is comparable to rill and interrill erosion (13). Because of the direct connection between intermittent thalwegs and the perennial drainage system gully erosion might even be responsible for the most part of the sediment yield from arable land to receiving streams. During two subsequent rainstorms in February 1990 up to 38 % of the sediment yield (17.6 t) from our investigation area in the loess belt of SW-Germany were derived from the thalweg of a zero-order basin (0.039 km^2) which drains less than 10 % of the arable land (0.445 km^2) in the catchment area (0.621 km^2).

3. APPLIED PREVENTION METHODS

In order to reduce the sediment yield from arable land to receiving streams the following conservation measures have been applied (14) :

1. Establishment of a grassed waterway in a zero-order basin to reduce ephemeral gully erosion.

Based on the mapping of ephemeral gullies in the investigation area in March 1990 the most affected thalweg was selected for the survey. This 260 m long thalweg drains a zero-order basin of 0.039 km^2 with a local relief of 25 m. The entire catchment is used as arable land and conventional tillage with cross slope farming is applied. Winter wheat, maize and sugar beets are planted ; a rotation typical for this loess region in SW-Germany. Mean slope gradient of the valley bottom is 4.4° but in some places 6.8° is reached, details are given in Fig. 2.

In the valley bottom a 3 to 5 meters wide and 220 m long grassed waterway was established (Photo 1). Grass and other weeds were used for the plantation. In addition every 15 m fascines (8 to 10 meters long, 30 to 40 cm high) made out of organic geotextiles (coco fibre net) were positioned across the thalweg (Fig. 2). In total an area of 1800 m^2 (= 4.5 % of the arable land in this zero-order basin) was used for this measure.

Photo 1.
Stabilization of the thalweg with a grassed waterway and fascines. Photo shows situation one year after the establishment (Photo : Jussi Baade).

2. Establishment of a sediment retention area at the outlet of a first-order catchment to diminish sediment yield to the receiving watercourse.

At the outlet of the first-order catchment (0.621 km^2) an area of 4000 m^2 is used for the construction of a sediment retention area. A natural succession of plants covers this former field. In addition fascines, similar to the ones used in the thalweg, were set up across the slightly inclined slope (slope gradient 1.9°) (Fig. 1). Compared to the area of arable land in the catchment (0.445 km^2) less than 1 % is used for this measure.

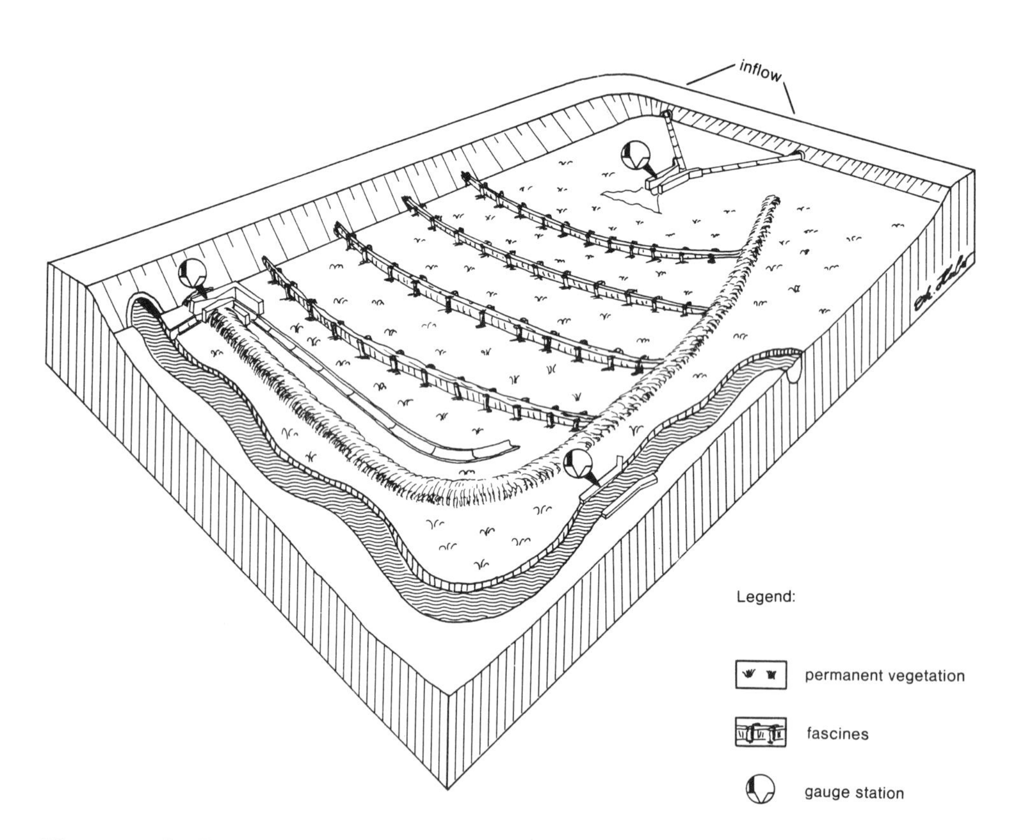

Figure 1. Outline of the sediment retention area at the outlet of the first-order catchment

The runoff and the suspended sediments enters the retention area through an input gauge station and spreads laterally over the area. At the lower end of the field a small channel takes the runoff to an output gauge station. Continuous measuring of discharge and an event triggered sampling at the input and output gauge stations allows the calculation of in- and outflow of water and sediments as well as the evaluation of the sediment trap efficiency (15).

4. RESULTS

4.1 Grassed waterway in the thalweg

Volumetric mapping of gully erosion in the thalweg before and after the construction of the grassed waterway was used to quantify the protective effects of this measure. During the surveys in winter 1989/90 and winter 1990/91, the width and the maximum depth of the gully were measured every 2 metres. Assuming V-shaped cross-sections the volume of the gully was computed. Govers (16) and Schmidt (17) use a similar method in their surveys. Soil loss is calculated from the volume of the ephemeral gullies based on a soil density of 1.5 g/cm³ for the plough layer.

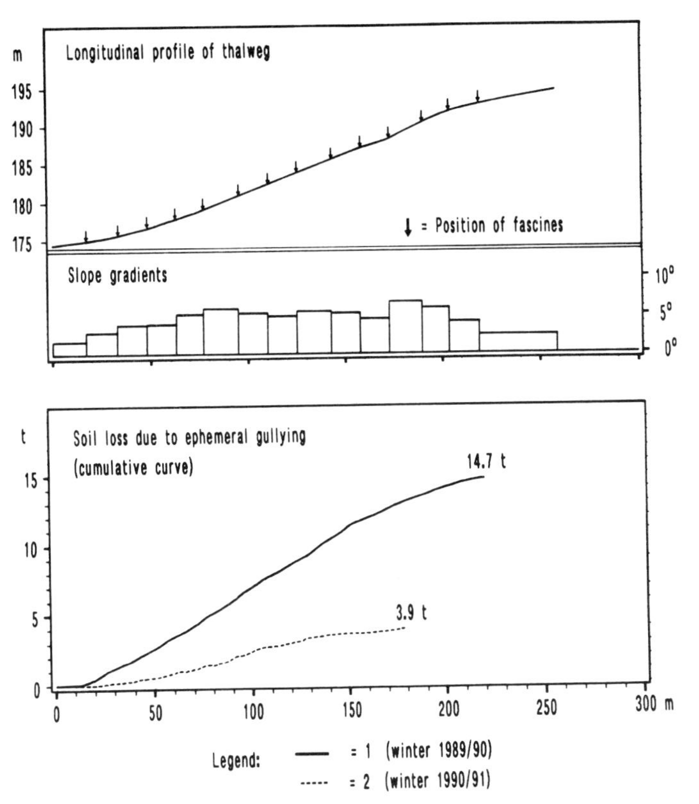

Figure 2. Cumulative curves of soil loss in the stabilized thalweg due to ephemeral gullying in the two periods : winter 1989/90 and winter 1990/91.

Fig. 2 shows clearly that the total soil loss by ephemeral gully erosion in the thalweg was considerably smaller after the introduction of the protective measures. Because of the known large variability of soil erosion during subsequent time periods the difference in total soil loss is not only attributable to the protective effects of the grassed waterway. Comparative volumetric mapping of an unstabilized gully was carried out to allow a comparison of the two periods. Soil loss in this gully amounted to 4.6 t in winter 1989/90 and 2.2 t in winter 1990/91, respectively. The comparison of these data indicates that the intensity of erosion in winter 1990/91 was less than half (47.8 %) of that in the winter before. From this proportion the potential soil loss for the stabilized thalweg in winter 1990/91 was calculated to be 7 t. Compared to the actual soil loss of 3.9 t during this period the reduction of soil loss amounts to 44 %.

4.1.1 Effects of grass cover and fascines

Observations of the grassed waterway during runoff events show that a permanent vegetation cover is needed to protect the surface against ephemeral gully erosion. From the selected plants only grass guarantees a year-round vegetation cover. The other tested weeds had the disadvatage of regression during the winter. Actually the starting-point of the ephemeral gully which developed during the winter 1990/91 (see Fig. 2) coincide with the boundary between the plots with grass cover (upslope) and the downslope plots where other weeds had been planted. The later plots were reseeded with grass in spring 1991.

The set-backs in the development of a permanent vegetation cover during the first winter gave the opportunity to observe the effects of the fascines more closely. Runoff with relatively high content of suspended matter partly blocks the holes of the coco fibre net, creates a 10 to 20 cm deep backwater and consequently allows sedimentation. Estimations of the volume of these small alluvial depositions indicate that 10 % of the soil detached by ephemeral gully erosion in winter 1990/91 has been trapped by this process. Furthermore, the fascines fan out the concentrated runoff and therefore reduce the risk of concentrated flow erosion immediately downhill of the fascines. This demonstrates that fascines play an important part in the initial stage of the development of the grass cover in the thalweg of zero-order basins which are susceptible to erosion throughout the year. Recent research focuses on the question, whether the fascines can partly be removed after a few years.

In any conservation measure, the interests of the farmer have to be taken into account. Discussions with the farmer revealed that lifting planting equipment when crossing the grassed waterway is not felt as an unacceptable obstacle to the work [18]. Crossing the grassed waterway during dry periods did not leave any marks (e.g. during the harvest of wheat in August 1990, photo 2). But during wet soil conditions it might lead to considerable damage of the grass cover increasing the risk of gully incision and/or maintenance costs.

Photo 2.Crossing the grassed waterway during wheat harvest in August 1990 (Photo : Jussi Baade).

4.2 Sediment retention area

From January 1990 to October 1991, 8 major runoff events are registered and sampled (Tab. 1). The maximum discharge (max Q) measured at the input gauge station range from 20 l/s during a snowmelt runoff in February 1991 to 335 l/s following a thunderstorm in June 1990. Maximum concentration of suspended sediment (max Cs) in the runoff at the input gauge station varied from 2.3 g/l to 23.2 g/l. The computed sediment inflow from the investigation area to the retention area extended from 1 to 8.5 tons per event. Total sediment yield for the period mentioned above adds to 35.15 t or an average sediment yield of 0.43 t/ha/yr, respectively. This rather low value might be attributed to the unusual dry summer of 1991. A calculation based only on the results of 1990 issues an average sediment yield of 0.68 t/ha/yr.

Table 1
Sediment trap efficiency of the sediment retention area (Jan. 1990 - Oct. 1991)

date	max Q (l/s)	max Cs (g/l)	sediment inflow	sediment outflow	E[1]
15.02.90	133	6.5	8.45 t	4.88 t	42 %
27.02.90	150	>2.3	6.50 t[2]	3.38 t	48.0 %
30.06.90	335	23.2	8.60 t	3.28 t[2]	61.9 %
22.09.90	130	11.1	3.67 t	0.87 t	76.3 %
18.11.90	43	9.5	1.06 t	0.64 t	39.6 %
20.11.90	78	2.6	2.09 t	1.15 t	44.9 %
26.02.91	21	12.7	2.26 t	0.25 t	88.9 %
20.03.91	32	13.8	1.94 t	0.19 t	90.2 %
Total			34.57 t	14.64 t	57.7 %

Remarks : 1) sediment trap efficiency $E = (S_{IN} - S_{OUT})/S_{IN} * 100$
2) result is partly based on extrapolation

The efficiency ranges from 40 % to over 90 %. Except of the quite low efficiency in November 1990, the results display an increasing tendency. Observations during runoff events revealed that sedimentation on the retention area is basically achieved by the reduction of flow velocity due to the lateral spreading of the runoff and the roughness of the ground cover. The fascines showed no visible effects. Therefore, the increasing efficiency is interpreted as a consequence of the development of the permanent vegetation cover. The high effectivness during the February and March events in 1991 is attributed to a high amount of organic litter on the ground after the winter.

Assuming a uniform sedimentation on the retention area a mean aggradation of 2 mm/yr is calculated. From this, a life time of 50 years can be estimated. Because the mean annual sediment yield from the catchment was quite low during the period under investigation, this estimation may be too high.

V. CONCLUSION

The establishment of a grassed waterway in a small natural drainageway on arable land reduces sediment loss due to ephemeral gullying by approximately 50%. A permanent grass cover is necessary for a year-round protection of the soil surface. Crossing of grassed waterways with agricultural machinery should be avoided during periods of high soil moisture content. Otherwise considerable damages to the grass cover followed by further incision of gullies and/or higher maintenace costs might occur.

The construction of a sediment retention area at the outlet of this small agricultural catchment facilitates the reduction of sediment yield to the receiving stream by 40 to 90 %. Taking into account the very low proportion of arable land used and the low maintenace costs for this measure, this represents a very efficient way to reduce sediment yield from arable land to the receiving watercourses.

The overall reduction of sediment yield from the investigation area to the receiving watercourse by applying the introduced system of conservation measures is calculated to be 65 %.

ACKNOWLEDGEMENT

The work was funded by the federal state Baden-Württemberg, FRG, as a part of the 'Projekt Wasser-Abfall-Boden' (PW 88 070) at the Kernforschungszentrum Karlsruhe GmbH (KfK).

REFERENCES

1. Crosson, P. (1985) : Impact of erosion on land productivity and water quality in the United States. In : El-SWAIFY, S.A., and al. (ed.) : Soil Erosion and Conservation. Ankeny, Iowa, 217-236.
2. Nowak, P.J. (1988) : The costs of excessive soil erosion. Journal of Soil and Water Conservation 43, 307-310.
3. Higgitt, D.L. (1991) : Soil erosion and soil problems. Progress in Physical Geography 15, 91-100.
4. Auerswald, K. (1989) : Prognose des P-Eintrags durch Bodenerosion in die Oberflächengewässer der BRD. Mitteilungen der Deutschen Bodenkundlichen Gesellschaft 59, 661-664.
5. Nortcliff, S. (1986) : Soil loss estimation. Progress in Physical Geography 10, 249-255.
6. Clark, E.H.I. (1985) : The off-site costs of soil erosion. Journal of Soil and Water Conservation 40, 19-22.
7. Prato, T. and H.Q. Shi (1990) : A comparison of erosion and water pollution control strategies for an agricultural watershed. Water Resources Research 26, 199-205.
8. Wischmeier, W.H., and D.D. Smith (1978) : Predicting rainfall erosion losses. - A guide to conservation planning. USDA, Agric. Handbook No. 537, Washington.
9. Auzet, A.V., J. Boiffin, F. Papy, J. Maucorps and J.F. Ouvry (1990) : An approach to the assessment of erosion forms and erosion risk on agricultural land in the Northern Paris Basin, France. In : Boardman, J., et al. (ed.) : Soil erosion on agricultural land. Chicester, 383-400.
10. De Ploey, J. (1990) : Threshold conditions for thalweg gullying with special reference to loess areas. CATENA Suppl. 17, 147-151.
11. Poesen, J.W.A. and G. Govers (1990) : Gully erosion in the Loam Belt of Belgium : Typology and control measures. In : Boardman, J., et al. (ed.) : Soil erosion on agricultural land. Chichester, 513-530.
12. Poesen, J.W.A. (1989) : Conditions for gully formation in the Belgian loam belt and some ways to control them. In : Schwertmann, et. al. (ed.) : Soil erosion protection measures in Europe. Cremlingen-Destedt, 39-52.
13. Thorne, C.R. and L.W. Zevenbergen(1990) : Prediction of ephemeral gully erosion on cropland in the south-eastern United States. In : Boardman, J., et al. (ed.) : Soil erosion on agricultural land. Chicester, 447-460.

14. Baade, J., D. Barsch, R. Mäusbacher and G. Schukraft (1990) : Geländeexperiment zur Verminderung des Sedimenteintrags von landwirts-chaftlichen Nutzflächen in kleine Vorfluter (Gut Langenzell, N-Kraichgau, SW-Deutschland) (PW 88 070). KfK-PWAB 5, 49-64.
15. Heinemann, H.G. (1984) : Reservoir trap efficiency. In : Hadley, R.F. and D.E. Walling (ed.) : Erosion and sediment yield: some methods of measurement and modelling. Cambridge, 201-218.
16. Govers, G. (1987) : Spatial and temporal variability in rill developement processes at the Huldenberg experimental site. CATENA Suppl. 8, 17-34.
17. Schmidt, R.-G. (1979) : Probleme der Erfassung und Quantifizierung von Ausmaß und Prozessen der aktuellen Bodenerosion (Abspülung) auf Ackerflächen. Methoden und ihre Anwendung in der Rheinschlinge zwischen Rheinfelden und Wallbach (Schweiz). Physiogeographica 1, Basel.
18. Moldenhauer, W.C. and C.A. Onstad (1977) : Engineering practices to control erosion. In : Greenland, D.J., and R. Lal (ed.) : Soil conservation and management in the humid tropics. Chichester, 87-92.

Farm Land Erosion: In Temperate Plains Environment and Hills
S. Wicherek (Editor)
1993 Elsevier Science Publishers B.V.

Analysis of catastrophic erosion in Czechoslovakia : The reflections of the structure of agriculural land and the physical conditions of soils

M. Kundrata and J. Ungerman

Institute of Geographical Sciences, Mendlovo Nám 1, 662 82 Brno, Czechoslovakia.

SUMMARY

Changes in the structure of agricultural land over the past 40 years have seriously contributed to the erosion of the soil. Indeed, it is ploughed land covering a large surface area on too steep or variable slopes that is especially affected. As the choice of crops is very limited (wheat, silage maize, sugarbeet, potatoes), the crop cover is not guaranteed during the danger periods of intense rainfall (end of spring, beginning of autumn).

The catastrophic erosion caused by intense rainfall calls for a detailed study on two accounts :

- These exceptional events demonstrate in dramatic fashion the damage caused to soil and crops during a very short period of time,
- The consequences of these phenomena ought to bring about a change in agricultural methods so that steps against erosion can be taken. Especial attention needs to be given to the process of soil erosion caused by particularly intense rainfall. Indeed, the average thickness of the soil stripped by erosion may be as great as several centimetres, and in extreme cases run into tens of centimetres.

In this paper we give the results of our analysis of the periodicity and localization of this intense rainfall. Several cases of exceptional erosion in South Moravia will be commented on with regard to the physical condition of the soil (for example, its content in clayey substances and humus, its air retention capacity, the composition of the lower strata of arable soil).

In several instances we present the projects for restructuring the agricultural land in order to assure the protection of the soil against erosion, and to improve the ecological balance in agricultural areas.

RESUME

L'évolution de la structure foncière agricole au cours des 40 dernières années a contribué considérablement à l'aggravation des processus d'érosion des sols. En effet, ce sont les terres labourées, les vastes parcelles à pente forte, qui sont surtout atteintes. Le choix des cultures étant très simplifié (blé, maïs, betterave à sucre, pommes de terre), la couverture végétale n'est pas suffisamment assurée pendant les périodes affectées par des précipitations de forte intensité (fin de printemps, début d'automne).

Les cas d'érosion catastrophique causée par de telles pluies méritent une étude détaillée pour deux raisons :

- ces faits exceptionnels offrent une démonstration spectaculaire des degâts provoqués sur les sols et les cultures agricoles en un très court laps de temps,

- les conséquences de ces phénomènes devraient entraîner un changement des méthodes utilisées au sein du milieu agricole afin de prendre des mesures contre l'érosion. Il convient d'attribuer une attention toute particulière au processus d'érosion du sol provoqué par des pluies d'intensité extrême. En effet, la couche moyenne de sol, enlevée par l'érosion, peut atteindre alors quelques centimètres, voire même, quelques dizaines de centimètres.

Les résultats de l'analyse de la périodicité et de la localisation des précipitations de haute intensité vont être soulignés. Plusieurs cas d'érosion exceptionnelle sur le territoire de la Moravie du Sud seront commentés en relation avec l'état physique des sols (contenu des substances argileuses et d'humus, capacité de rétention d'air, tassement des couches inférieures des sols arables). En quelques exemples nous allons présenter les projets de restructuration du fonds foncier agricole pour assurer la protection des sols contre l'érosion et pour améliorer l'équilibre écologique dans les paysages agricoles.

In this article we analyse the following theses which are connected with the steep development of devastating erosion events over the last few decades.

1) Thesis concerning anthropogenic factors which influence the intensity of erosion more than natural factors in the country (e.g erodibility of soils).
2) Thesis looking at the transcending of the threshold, the stability between the soil formation and erosion processes, and the retension capacity of soils.
3) Thesis examining the strenthening of variabilities in the quality of soils in one field due to enhanced erosion processes.

DISCUSSION OF THESIS N°1

The process of industrialisation of agriculture in the ÇSFR has developed more sporadically in comparison with other western countries. There were no big differences in the trends of the growth of energy and material inputs, but the soil degradation processes have been enhanced by large scale land amalgamation, and the disrespect for natural conditions neither in the structure of fields nor in the choice of crops and fertilizers used.

Extremely large fields (up to 100 ha even in mountainous areas at 600 m), and the growing of monocultures vulnerable to erosion (esp. maize for silage, beetroot or potatoes) have occurred. Additionally the largescale use of heavy mechanisation has caused significant soil compaction. Extensive attempts to plough higher positions (400-700 m a.b.s.) has caused the same effect as "inversion" in all types of relief of the crystalline Bohemian Massif, the Karst and Carpathian flyche : upper positions on watersheds have been more intensively ploughed and managed then stabilised enclaves of small woods grasslands and orchards, which have remained on steeper slopes along the river valleys.

In the hilly and lowland regions a high percentage of the land was ploughed (even 99 %) and fields up to 200 hectares were amalgamated. In comparison with the average size of fields before collectivisation, the size of fields rose by the order of 100 times. Especially in the dissected relief of some hilly lands (e.g. central Moravian Carpathians) the erosion on the tops of slopes is very apparent, with the exposure of the bedrock. Moreover, the accumulation of soil on the foothills has started to cover the fertile chernozemic soils by eroded deposits from the upper slopes.

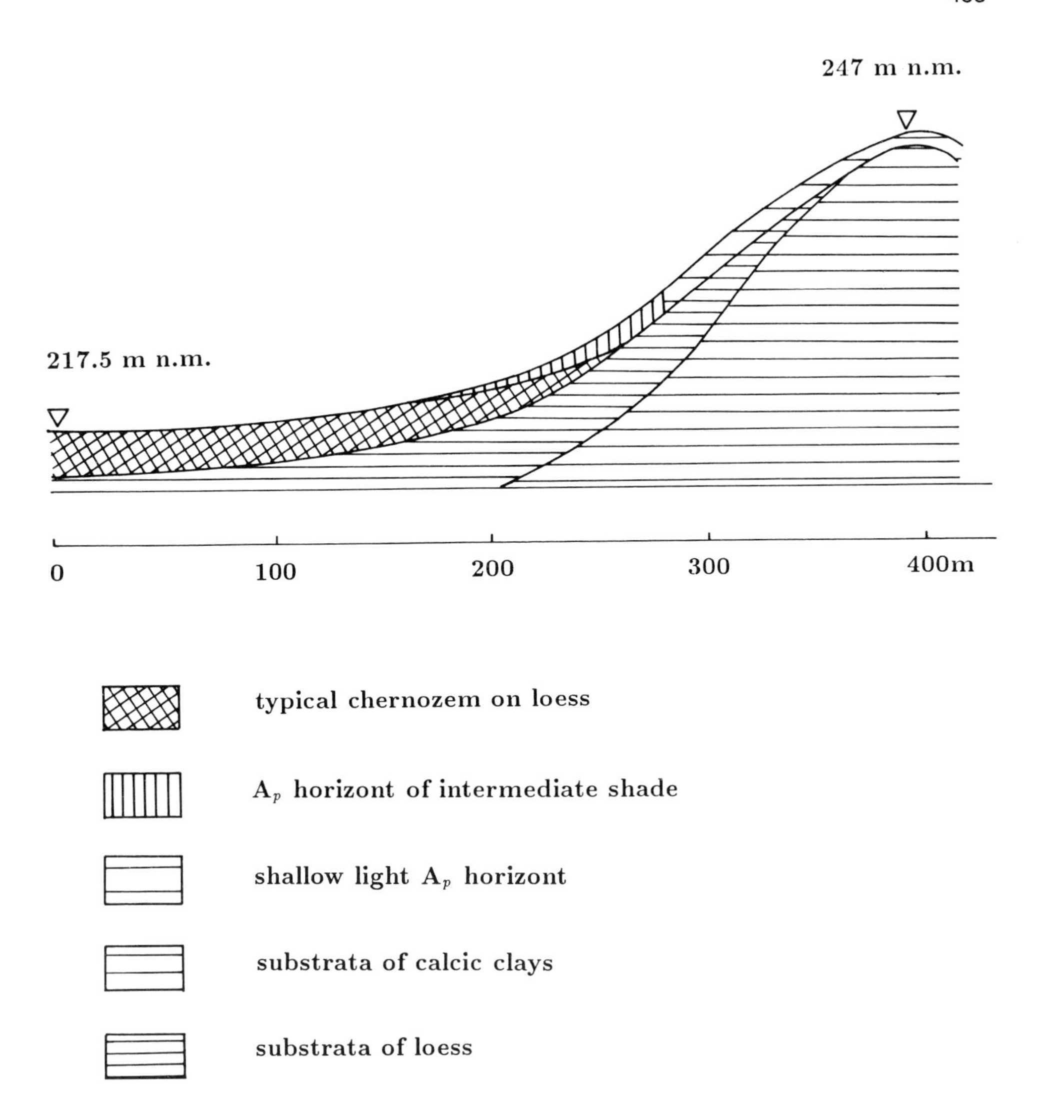

Figure 1. Slope profile in the hilly near Slavkov (Austerlitz).

Due to these structural changes in land-use, and in the physical conditions ot soils, there is no difference in the intensity of erosion in different natural conditions. If heavy precipitation occurs in small watersheds of a high percentage of ploughed land with no vegetation (risk season April-June), the same values of soil loss and the same erosion forms have been observed on the deep soils of gentle slopes of the lowlands and on steeper terrain of the highlands with shallow stony soils. Especially after heavy rains of the intensity between 30-100 mm in two hours the influence of soil type, bedrock and even the angle of slope is minimised. Most significant is the influence of the micro-relief and the structure of the runoff channels. The creation of catastrophic runoff erosion and floods, influence small watersheds by the following factors :

1) Percentage of soil without vegetation or with the seedlings of maize, sugarbeet, potatoes and other vegetables and also orchards and vineyards. Much lower effects of erosion after heavy rainfall has been observed on fields with winter wheat in comparison with neighbouring maize fields (Fig. 2).

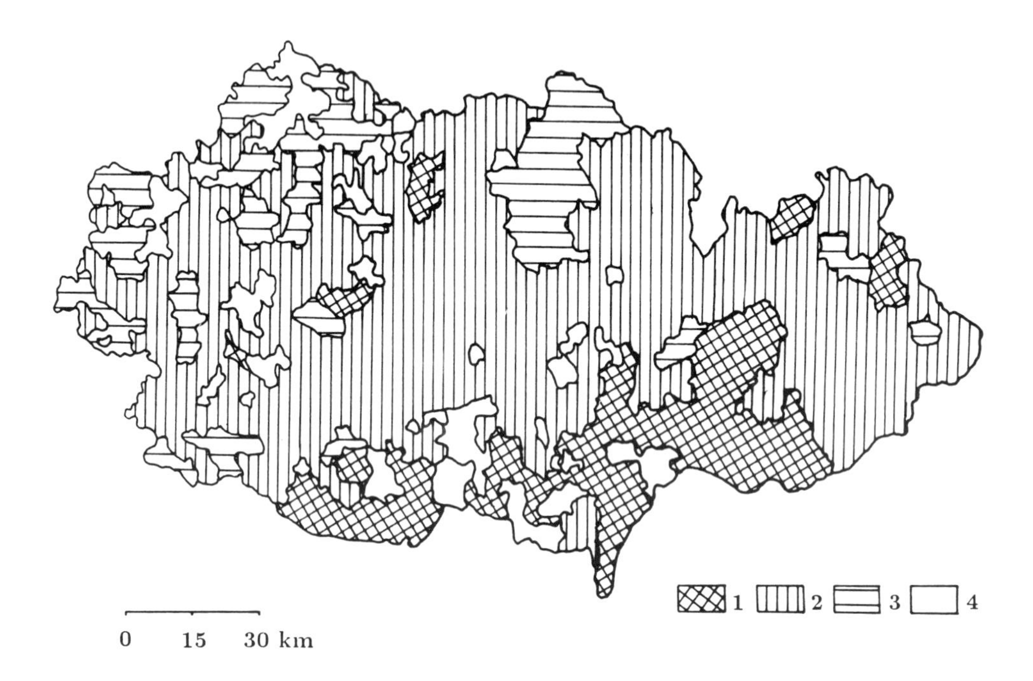

Figure 2. The erosion potential of agricultural land of South Moravia calculated from relative share of crops which do not protect soil from erosion.
The Key : 1. high and very high erosion potential ; 2. average to high erosion potential ; 3. low and average erosion potential ; 4. not evaluated areas.

2) Long uninterrupted slopes covered by monoculture stands enables the runoff to concentrate.
3) The extreme decrease in the porosity and the infiltration capacity of soils, in comparison with the period before the last wave of mechanisation and collectivisation in 1968. The comparison of the infiltration capacity of soils under different crops discussed by Kasprzak (1), and this trend is also described in other sources. e.g Cihlar (2) describes the rise of factor C in USLE from 0.1 in 1945 to 0.4 in 1980 (New Brunswick Canada).

EXAMPLES OF CATASTROPHIC EROSION EVENTS IN DIFFERENT TYPES OF THE MORAVIAN LANDSCAPE IN THE 1980s

1) *Lowlands and terraces of flat relief* and gentle slopes up to 2°. The villages Pravlov and Nêmçiçky near Dolní Kounice (ca. 13 km SW. from Brno) were effected by frontal precipitation with hail in June 1990, when 60 mm fell in 2 hours.

Photo 1.
Large scale (8 steps-floors) terraces near Divaky village (25 km S.E. from Brno) were heavily eroded after the August storm in 1989 (up to 100 mm in 2 hours). One of the arguments to constract this terrace was to protect the land from erosion.

The effected watershed had a flat terrain of average slope angle of 1° and fields of sizes between 50-100 hectares had been planted with maize (seedlings of 5 cm). Most of the watershed is formed by brown soils on loess tertiary clays and terraces of the River Jihlava. The precipitation started with hail which destroyed the maize seedlings and compacted the soils. Even from the flat parts of the watershed, the topsoil from 10-30 mm thick was washed out and the seeds of maize appeared at the surface. The village Pravlov was flooded and covered by mud including the local church. On the vineyards, located on gentle slope of sandy river terrace, 2 metre deep gullies appeared.

2) *Dissected Hilly Lands*
The village Vàzany above Litava (3 km SW from Slakov - Austerlitz).

Photo 2.
Village Vazany after flood in June 1990 (3 km from Austerlitz).

Photo 3.
Eroded maize fields up to 1 km long upstream Vazany small basin after intensive rain in June 1990 (3 km from Austerlitz).

The small agricultural watershed is created by a shallow valley of a small stream going through the village before it meets the small River Litava. The size of the basin is 4 km^2, it is 3 km long, and the average slope is 2°. The topsoil is created by chernozemic and brown soils on flychic bedrock of heavy tertiary clays with layers of sandstone. During the critical period (June 1990) the whole watershed was ploughed and 95 % of the land was amalgamated into cooperative fields. More than 80 % of the agricultural land was planted with maize and sugarbeet. The precipitation came in two waves during two frontal storms. First the intensity of the first rains was about 30 mm in two hours causing strong erosion and flooding of the village streets. The second more intensive precipitation came two weeks later and its intensity was estimated at around 60-80 mm in two hours. Minimally 10mm thick layer of topsoil (ca 40,000 m^3 of soil) was lost from the watershed with the exception of the 20 % planted as winter wheat. The whole village was flooded by a muddy flow above the line of windows and the army had to help with the rescuing and rehousing .

Intensive erosion started in the flat watersheds and the concentrated run off created wide gullies in the depressions. Due to the lack of exact measurements, the devastation created enormous discussion. Hydrologists built their explanations on waterflow measurements on the River Litava and they predicted much more evaporation and infiltration then the experts which researched the runoff. The modelling of the whole flood wave was complicated by the sediment saturation of water and the barrier of the village which both modified the runoff.

A basin of similar sized slopes and land use was effected by catastrophic erosion above the village of Lesany (6 km west from Prostêjov) in May 1987. On the 19th May 1987 25-35 mm of rain fell in 30 minutes. This precipitation caused strong runoff erosion, but the catastrophy came the next day when 11 mm fell compacting the soil. After the method of Wischmeier's USLE (3), the maximum soil loss was 9.77 tonnes $ha^{-1}yr^{-1}$, but during the erosion event 10 mm of topsoil was washed out. Gullies and rills were also observed. The only parts of the watershed not effected by strong erosion were those areas planted with winter wheat.

A more extensive area (50 km^2) in the hills S.E. of Brno, was effected by similar erosion after storm rains in August 1989. At the peak of the storm up to 100 mm fell in 2 hours. Especially vineyards, orchards, and largescale terraces near the village of Diváky were damaged, but similar erosion occurring on the fields with maize and sunflowers initiated gullies in several places.

The Highlands

Catastrophical erosion events occurred quite frequently also in small watersheds of the highlands both in crystalline and the Carpathian part of the country without any difference in the intensity of erosion. The example of a catastrophy in the village Luka above the River Jihlava is described by Kundrata and Ungerman (4). This small watershed makes it possible to identify the reasons and factors of extreme runoff and erosion. In the small basin above the city Brumov (E. Moravia) of the size 4 km^2 (1/3 of forests, 1/3 of orchards and grassland, and 1/3 of arable land) all the ploughed land was planted with maize in 1985. The slope of the fields ranging from 7-10° were from 300-500 metres long. In May 1985 a local storm with 30 mm of rain in $1_{1/2}$ hours created not only intensive erosion, but also flooding to the centre of the city up to a height of 1 metre (Photo 4).

The influence of natural conditions on the creation of extreme runoff and erosion, grows with the size and diversity of land use in the basin. But the soil and bedrock properties are not so important as the micro-relief, shape of basin, and the intensity of rainfall. Catastrophic floods connected with erosion, occur also in the basins of sizes around 50 km^2. The examples of smaller areas, the River Besének in 1985 or the small River Balinka in 1986 which meets the Svratka watershed could be mentioned. Even though the basins are forested, minimally 1/3 of the retention function of the forests is not sufficient for the following reasons :

1) Forests remain mostly on the slopes of valleys, but below the decisive runoff accumulation points.

Photo 4.
Especially shallow cambisols are endangered by total degradation due to erosion. Luka near Jihlavon, after catastrophic rain in April 1988 (80 km west from Brno).

2) Forests are spruce monocultures with a smooth ground surface covered with dead needles, which do not capture the runoff. The proof of these effects are the frequent recent alluvial fans along the foot of the slopes of the forested slopes of the valleys (Fig. 3).

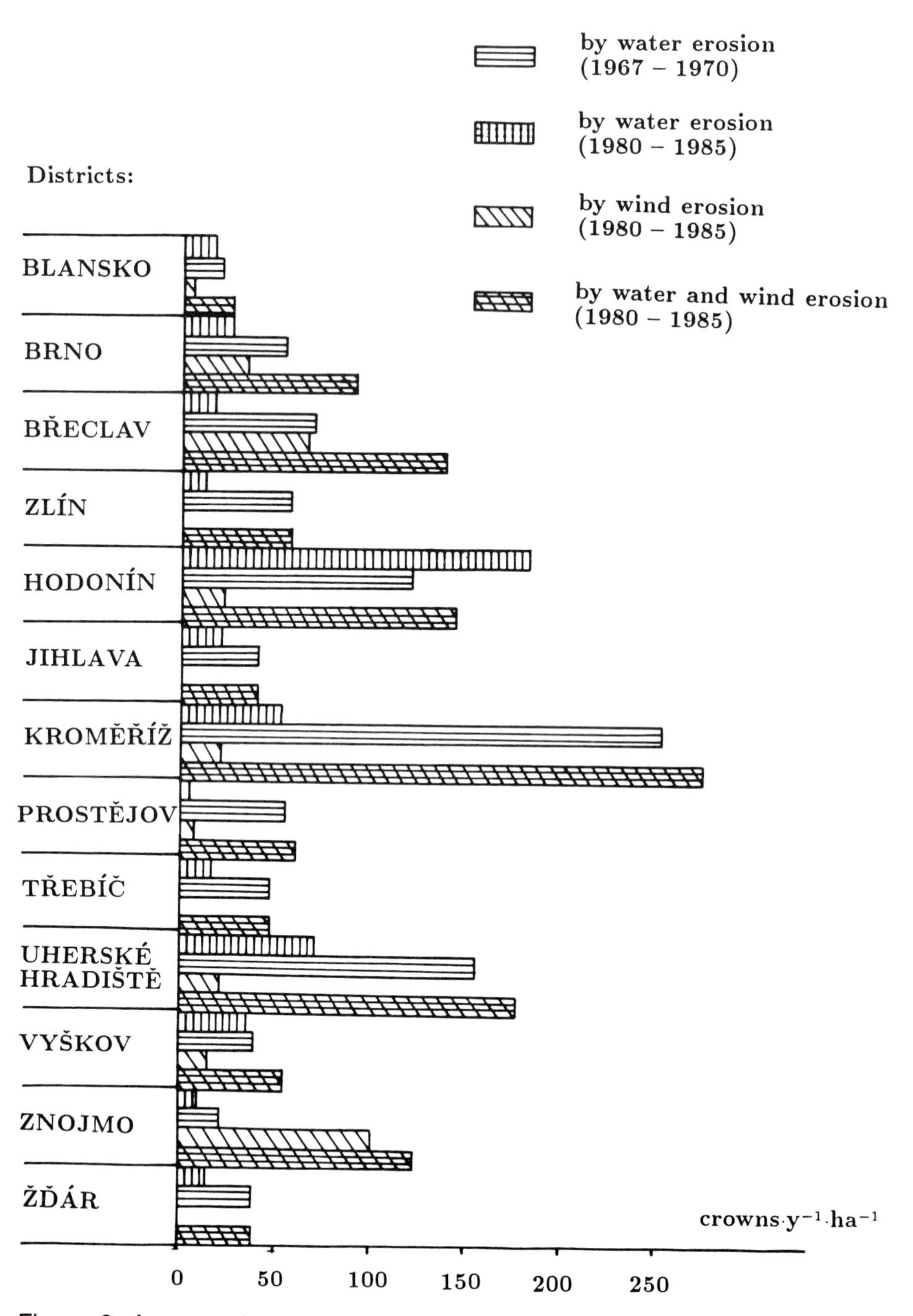

Figure 3. Average demages on crops in Czech crowns per year and hectar of arable land in South Moravia.

Catastrophic erosion events create the biggest concern due to visible damage in small areas. But more important factors of soil loss are caused by average rain, snowmelt and wind erosion. This is proved also by research in different countries (5 - 6 - 7). The importance of winter and early spring season for the annual sediment yield due to erosion, is proved also by the measuring of sediments in two small rivers provided by the Czech Hydrological and Meteorological Institute (8).

Table 1.
The sediment characteristics of the basins of the Rivers Olsava and Louçka and the calculation of the intensity of erosion in the basins based on the principle of sediment delivery ratio (9). The River Olsava flows through the flychic Carpathians and the River Louçka flows through the crystalline Bohemian massif.

Parameter		Units	R.Olsava	R.Louçka
shape of basin			fan	long
medium altitude		(m a.s.l.)	367.2	532
precipitation		(mm yr^{-1})	706	630
medium specific - runoff		(l.s-1 km^{-2})	5.49	5.42
medium flow		(m3 s^{-1})	2.20	2.09
size of basin		(km^2)	401.23	386.20
ploughed land		(km^2)	162.91	124.0
ploughed land		(%)	45	58
forested land		(%)	30	20
sediment flow	1985	($t.yr^{-1}$)	18,948.3	5,303.5
	1986	($t.yr^{-1}$)	11,160.0	2,933.0
	1987	($t.yr^{-1}$)	55,318.2	26,790.0
average	1985-87	($t.yr^{-1}$)	28,475	11,675.5
sediments in	1985	(%)	15.2	36.4
winter	1986	(%)	55.2	44.8
(D,J,F,M)	1987	(%)	62.2	57.9
maximum sediment flow		($t.day^{-1}$)	13,355 (87)	10,519 (87)
SDR - amplitude		(%)	4-12	4-12
SDR - average		(%)	8	8
river bed erosion		(%)	6-16	6-16
out of ploughed lan		(%)	10	10
av. erosion in basi		($t.ha.yr^{-1}$)	6.4=0.4 mm	3.4=0.2 mm
maximum		($t.ha.yr^{-1}$)	32.4=2.2 mm	16.3=1.1 mm
minimum		($t.ha.yr^{-1}$)	1.9=0.13 mm	0.5=0.04 mm
av.erosion from ploughed land		($t.ha.yr^{-1}$)	13.4=0.9 mm	9.4=0.6 mm
maximum		($t.ha.yr^{-1}$)	71.3=4.8 mm	45.3=3.0 mm
minimum		($t.ha.yr^{-1}$)	4.3 =0.3 mm	1.5=0.1 mm
pollution of river		(class)	severe (4th)	low (2nd)

DISCUSSION OF THESIS 2 - EXCEEDING THE THRESHOLDS

The intensity of erosion is 2 orders higher on 1/3 of arable land in the ÇSFR than the soil formation rate which is from 0.03 to 0.1 mm yr-1 in central Europe (depending on the bedrock). It is apparent from table 1 that in both river basins the average erosion from the arable land is 10 times higher than the ability of soil formation. The intensity of erosion is closely connected with other soil degradation processes like humous loss, soil compaction, destruction of soil aggregates due to over mechanisation and so on. The increasing amount of catastrophic erosion events, in small water basins especially in 1980s, are warning indicators of exceeding the threshold of the infiltration and retention capacities of the soil. The soil body has not been protected yet and the all attention is concentrated on crop damage, communities, and property. The expenses of agricultural insurance companies have increased so much that the system of insurance had to be changed last year.

The rapid degradation of the soil quality has not been visible on the yields up to now because the lands have been over fertilized even in less favourable conditions, which is also indicated by the pollution of water courses. Nowadays when subsidies to agriculture are strictly limited, the use of fertilizers will decrease and after the exhaustion of the support for fertilisers we can estimate the real effects of soil degradation on its production capacity.

Decreasing the production capacity is very real especially on shallow stony soils of high altitudes, on crystalline bedrocks where the top soil has been eroded. A large part of the eroded slopes has to be converted to grasslands. The soils on the more favourable substrate of loess or calcic clays are endangered by the extensive decrease in production, but more important factors of these soils is the increase in the differences between the layers and quality of the soil profile discussed in thesis n° 3. The threshold condition of the soil balance are closely connected with the entire disturbance of the stability of the landscape. The renovation of the retention soil capacity is one of the basic prerequisites for a more sustainable development of agriculture.

DISCUSSION OF THESIS 3 - GROWTH OF THE VARIABILITY IN THE QUALITY OF SOILS IN FIELDS

Observing the erosion processes on a detailed scale led us to hypothesize that soil loss and sedimentation are primarily influenced by micro-relief, and that the differentiation inside single fields or slopes is very varied.

From the year 1986 the correlation between the colour of soil, its quality and shape of relief, has been observed in the central Moravian region of the Carpathians. The correlation between the light eroded patches and convex divergent shapes of relief on the one hand, and the dark sedimentation patches and concave and convergent shapes are apparent at least in 85 % of observations. The light bedrock of clays, loess and sand soils sharply contrast with the prevailing black soils and enables us to distinguish the changing colours by remote sensing methods. Polygenetic origins of soils and historical changes in the land-use does not allow complete interpretation, but the present intensity and longterm agricultural use of the area prove the importance of the erosion processes on the differentiation of soil quality.

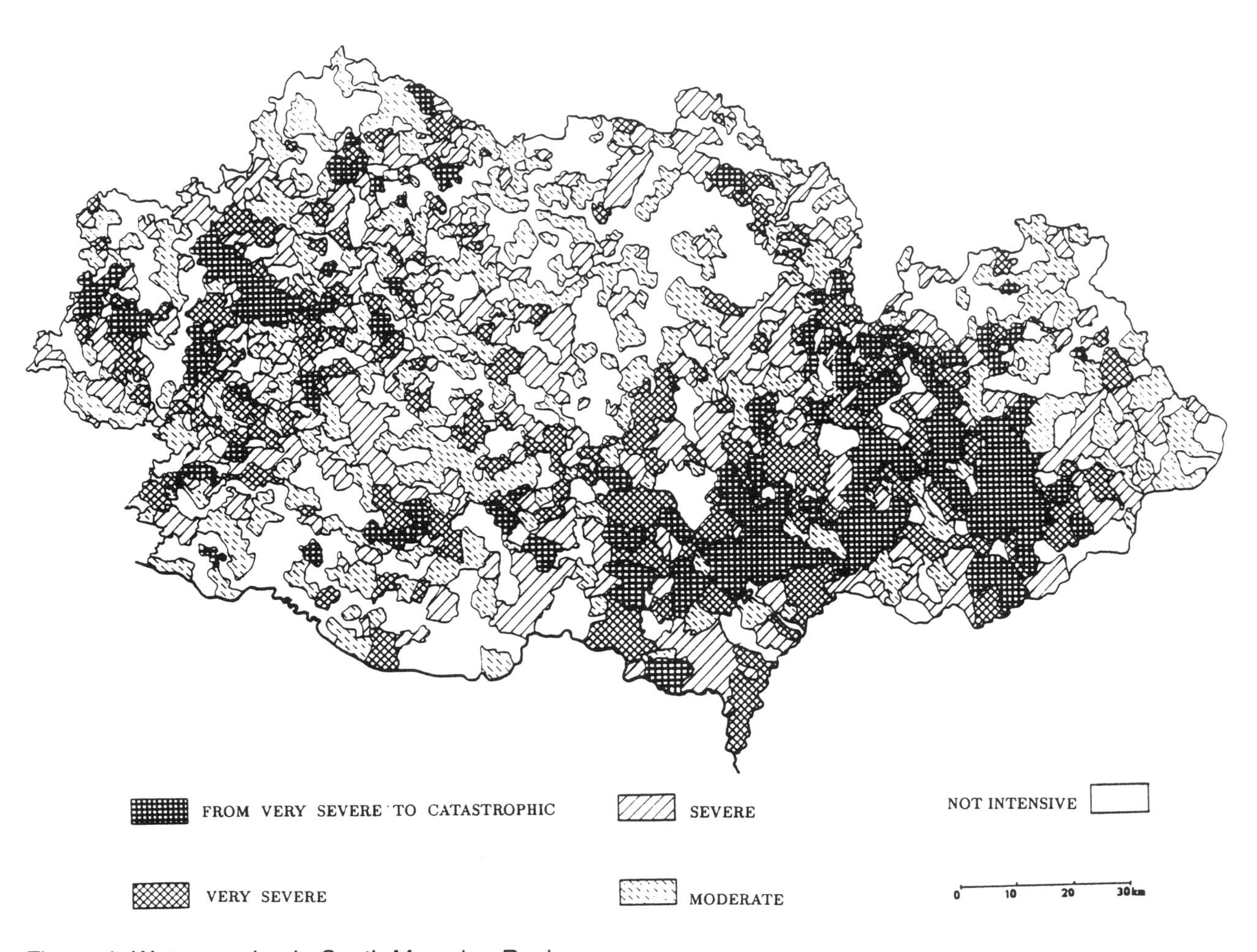

Figure 4. Water erosion in South Moravian Region.

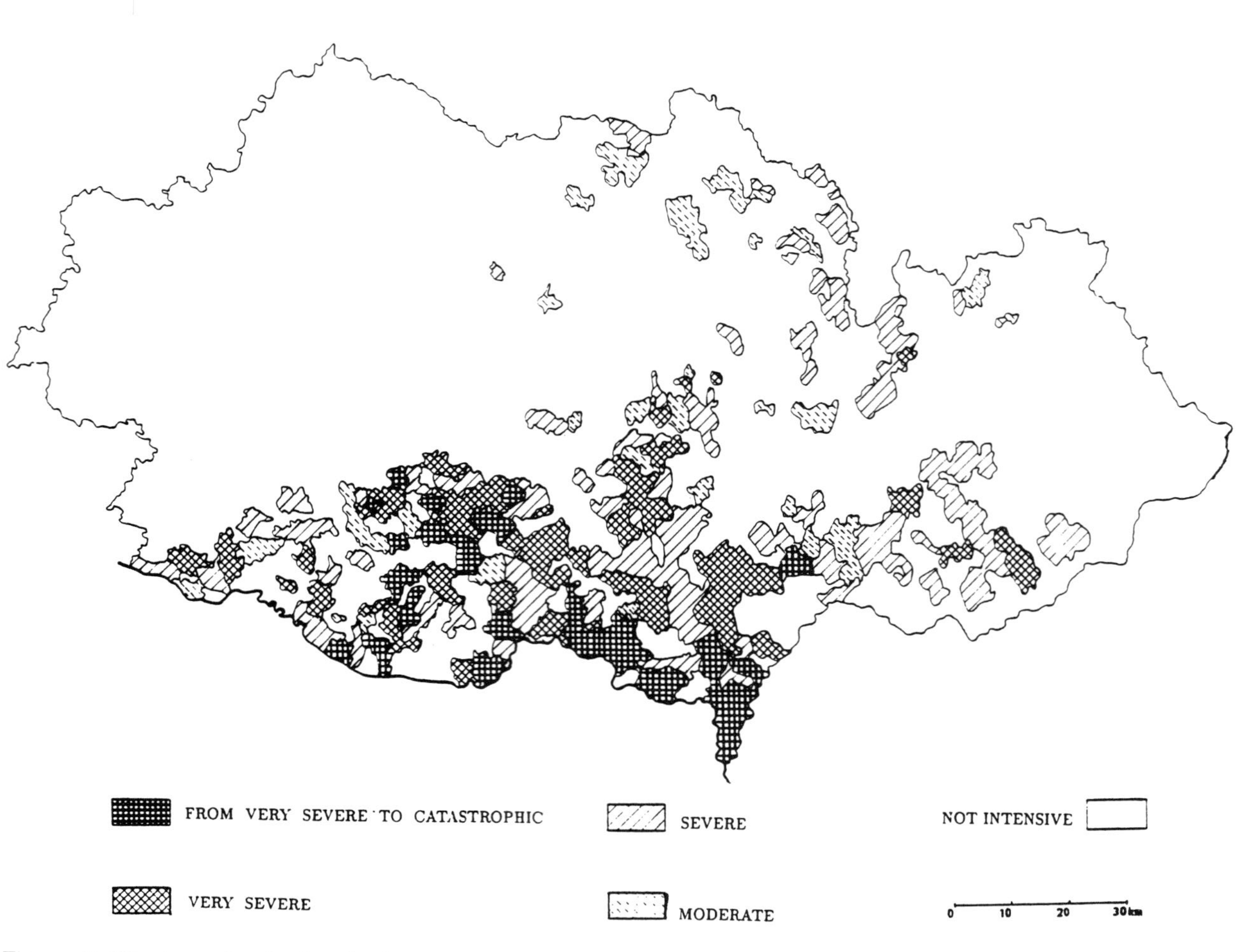

Figure 5. Wind erosion in South Moravian Region.

Research and soil sampling was established to identify relations between the colour of the soil, position on the slope, content of humous, depth of soil and its texture. The aim of the study was to prove the correlation between the shades of soil and the degree of erosion, and to create a method for evaluating this using aerial photography. We took 86 topsoil samples and 65 soil profiles from 20 fields in Litençice hills, Slapanice hills, Zdánice forest, Kyjov hills, and Hustopece hills. Samples were taken from the dry topsoil before the growing season mostly in April and the beginning of May in 1988 and 1989. Typical slope profiles with a mosaic of colour shades were chosen. The final representative sample was taken by sampling inside a circle of diameter of a several metres, always from the places which were relatively light, dark, and intermediate on the slope. The contents of organic matter Cox was analysed, and a correlation matrix of the soil properties was created. We found a correlation between the shade of soil colour and the contents of humous (Cox). Only in 4 cases from 86 analyses (4.65 %) was the Cox content higher in the intermediate zone than in the dark zone. The differences are only in the order of 10-1 % of Cox. The colour of soil is also influenced by other parameters e.g wetness, but they are also a function of the organic matter in the soil and we chose the dry season for sampling. The analyses proved large differences in the contents of Cox along the slope profiles (even 7 times higher). The differences correspond with micro-relief and variation of the distance of a several metres, on less dissected slopes a maximum of tenths of metres was observed. Correlation analyses proved the relation between the shape of the slope and the shade of the soil (60.5, 55.8, 1.1 - hypothesis proved on the probability of 99.9 %, 50 %, rejected, respectively) ; between the depth of the soil and the shade of the soil (75.4 (99.9 %), 23.1 (50 %), 1.5 (rejection)) ; the correlation between the Cox and the shade of the soil (65.5 (99.9 %), 32.1 (50 %), 2.4 (rej.)) ; and between the shade and the texture of the soil (80.0 (99.9 %), 20.0 (50 %),0.0 (rej.))

Field research proved that the soils on slopes are very shallow. From 65 soil profiles 29 cases (44.6 %) we found that the substrata appeared 40 cm below the surface. There is the argument whether this negative fact is created by natural geomorphological and pedogenetic conditions, or by man-made enhanced erosion (comp.10 - 11). After our observations the man- made enhanced erosion is the dominating factor which is responsible for the destruction of the soil profile in the recently ploughed landscape. Besides the high amounts of soil loss described in theses 1 and 2, this fact is supported also by the following results from the samples:

1) The threshold between the top soil and substrata is very sharp in the most shallow soils. The eroded soil is replaced by ploughing of the ground (clays,sands and in better cases loess). Horizon A0 sits directly on the bedrock.
2) On foot-slope accumulation positions, the shade inversion has been observed. In other words the original black soils with high contents of organic matter are being covered by the eroded substrata bedrock from the upper part of the slopes (see Fig. 1).

On fields in dissected hilly lands of the central Moravian Carpathians, the soil losses are being enhanced in erosion positions and the accumulation accelerates on the foothills, in riverbeds and dams. The contents of humous and the soil depth rapidly varies over short distances. The effects of erosion are apparent especially in the upper parts of the slopes and sparse structure of crops (esp. maize or sugarbeet) are visible. The effects of these differences influence the maturing process and the harvest yield. This reality will be difficult to include in the new transformation of fields. Some eroded slopes would be better planted with grass even though they are in the middle of the most fertile traditional agriculture region of Moravia.

REFERENCES

1. Kasprzak, K. (1987) : Contribution to hydrologically important changes in the ploughed soil infiltration, Vodohosp. Ças., 1, 62-80.
2. Cihlar, J. (1987) : A Methodology for Mapping and Monitoring Cropland Soil Erosion. Canadian Journal of Soil Science, 67, 433-444.
3. Wischmeier, W.H. and Smith, D.D. (1965) : Predicting rainfall-erosion losses from cropland east of the Rocky Mountains - guide for selection of practises for soil and water conservation. Agricultural handbook, 282, ARS USDA, Washington, 45 p.
4. Kundrata, M. and Ungerman, J. (1992) : Increase in soil erosion in Czechoslovakia. A case study, Bulletin Association Geography France, Paris, 155-160.
5. Dickinson, W.T. and Rudra, R.P. and Wall, G.J. (1986) : Identification of soil erosion and fluvial sediment problems. Hydrological Processes, Vol. 1, 111-124.
6. Edwards, L.M and Burney, J.R. (1987) : Soil erosion losses under freeze thaw and winter ground cover using a laboratory rainfall simulator. Canadian Agricultural Engineering, 29.2, 109-115.
7. Kirkby, P.C. and Mehuys, G.R. (1987) : Seasonal variation of soil erodibilities in southwestern Quebec. Journal of Soil Water Conservation, 42.3, 211-215.
8. Petrûjová, T. and Dostál, I. (1988) : Regime of sediments on 6 river profiles in Southern Moravia during the period 1985-87, CHMU Brno, unpublished.
9. Walling, D.E. (1988) : Measuring sediment yield from a river basin, in Lal, R. editor, Soil erosion research methods, Soil and Water Conservation Society,Iowa, 39-73.
10. Daniels, R.B. (1987) : Quantifying the effects of past soil erosion on present soil productivty, J. Soil and Water Conservation, 42.3, 183-187.
11. Lipkina, G.S. and Reznikova, D.L. (1987) : The soil formation under the forest grassland in different conditions of relief, Poçvovedenie, 3, 82-93.
Kundrata, M. : Geographical evalution of man-made soil changes, thesis, Institute of Geography Brno, 1990. unpublished.

Farm Land Erosion: In Temperate Plains Environment and Hills
S. Wicherek (Editor)

A specific strategy set up with farmers to succeed erosion control : The experience of a French region "Pays de Caux".

J.F. Ouvry and L. Ligneau

A.R.E.A.S. , 25 rue de Dieppe, 76460 Saint Valery en Caux, France

SUMMARY

In "Pays de Caux", during the 80 's, elected communes representatives and farmers involved themself in an anti-erosion action. A four stages strategy had been defined as following :
- To characterize regional erosion and to define parameters which to act on.
- To investigate cropping systems in other to integrate anti-erosion actions.
- To research and to experiment agronomical solutions and solutions concerning land management.
- To extend and to perpetuate the action.

This strategy involved all partners from regional rural world. The action is basicly set up with farmers and not for them. Its vitality is partly based on a regional association (A.R.E.A.S.) completely in charge of soil erosion and flooding.

RESUME

Dans les années 80, élus des collectivités locales et agriculteurs du Pays de Caux se sont engagés dans la lutte contre l'érosion des terres. Une stratégie en quatre étapes a été définie :
- Connaissance du phénomène, détermination des paramètres sur lesquels agir.
- Connaissance des systèmes de culture pour pouvoir y intégrer la lutte contre l'érosion.
- Recherche et expérimentation de solutions au niveau agronomique et au niveau de l'aménagement du territoire.
- Généraliser et perpétuer l'action.

Cette stratégie implique la participation de tous les acteurs régionaux du monde rural. Le principe de base est d'agir avec les agriculteurs et non pas pour eux. La vitalité de cette opération repose en partie sur l'existence d'une structure régionale (A.R.E.A.S.) entièrement vouée à la lutte contre l'érosion des terres et les inondations.

Every winter, erosion expands in "Pays de Caux". Erosion is typical in that region and it's mainly represented as ephemeral gullies. A specific process is revealed, it can be called concentrated runoff erosion or linear erosion of thalwegs. (Figure 1).

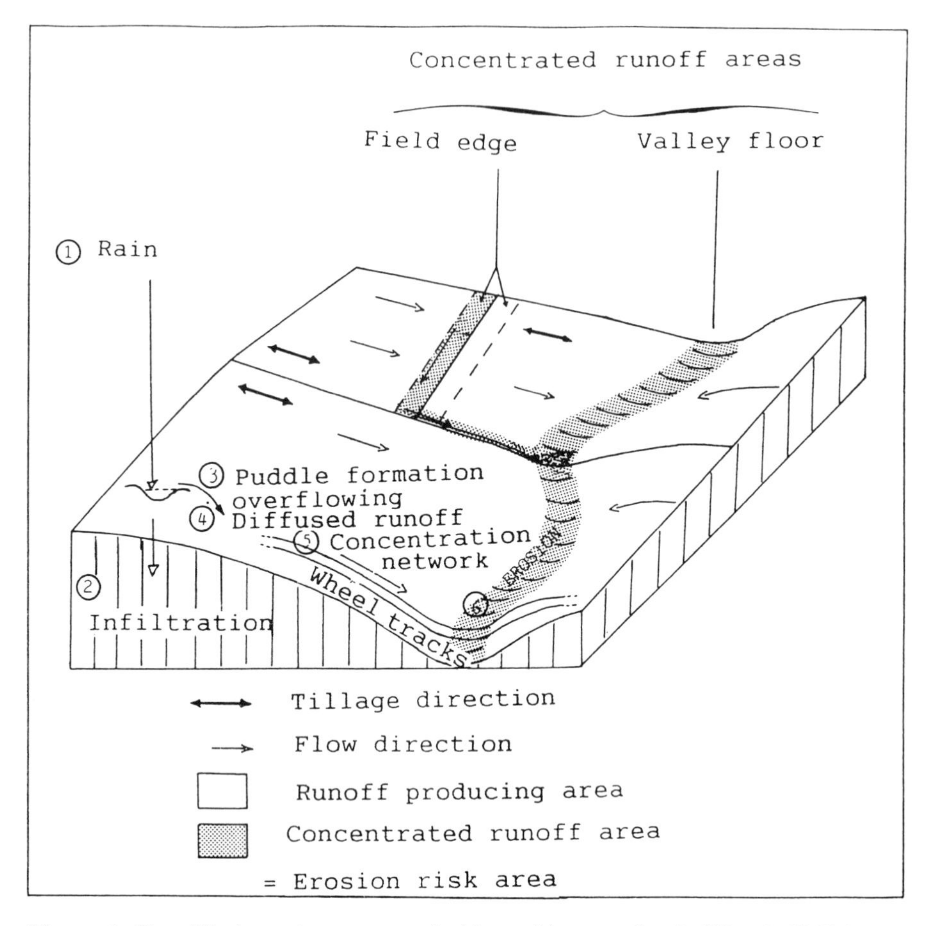

Figure 1. Simplified erosion process in Haute Normandie, Boiffin, J. (INRA).

In "Pays de Caux" this erosion is particularly effective because there is no hydrological network in small valleys.

During stormy summer season, or exceptionally rainy winter, rill-inter rill erosion becomes visible on hillslopes with more than 5% slopes.

This kind of erosion was reported in "Pays de Caux" (1 - 2 - 3 - 4 - 5 - 6 - 7 - 8- 9 - 10 - 11 - 12 - 13 - 14) as well as in the Paris Basin (15 - 16 - 17), the erosive process, the physical and human parameters have been abundantly described, analyzed and quantified.

A STRATEGY TO REDUCE RUNOFF AND EROSION

During the 80's, communes and farmers representatives decided to fight erosion. A strategy at the functionnal watershed scale was developed with the participation of the farmers joining both agriculture development and land management schemes, as it had been suggested by Roose E. (18).

This strategy includes several stages and concerns a lot of partners :

** The first stage consists in the characterization and quantification of runoff and erosion process.*

The aim is to define the physical parameters which are possible to act on now and in future, and to understand the limit of this action. For instant :

. To point out at the watershed scale a disjunction between runoff producing areas and areas affected by erosion (thalweg).

. To define the limits of a possible action to delay crusting (the formation of sealed soil profiles) in winter crops.

. To quantify soil roughness effect on runoff, and according to this criteria, to compare different tillage systems.

. To investigate the positive effect of no breaking up and reinforcement of soil compactness in areas sensitive to erosion (thalwegs). (Photo).

Photo 1.
An example of soil compactness renforcement - an easy anti-erosion action used upstream incatchment.

** The second stage is the investigation of regional cropping systems with farmers.*

Thus, for each systems an assessment, of erosion risk level was employed. Other assessment for farmers needs, farmers sensitivity to erosion problems and for spare time the farmers could spend on the anti-erosion action was done. The technical and economical limits of the farmers action were also defined.

For example, this investigation pointed out that because of the disjunction mentioned before, farmers located upstream in a watershed feel less concerned than farmers located dowstream from which two different ways to motivate them must be defined.

It also have been pointed out that runoff control is not an aim in itself, but it must be completely integrated into agronomical advices at the field scale. The improvement of field productivity and the reduction of runoff must be carried out together. In 1986 it had been the subject of global research on both nitrogen pollution and runoff reduction by cover crops.

An other result of the investigation is that runoff risk in winter increases with the increase in rotation of spring crops. This result is increased because of the expanding of vegetable cultivation.

* *The third stage is carried out at the same time with the second one.* It is a stage of :
. Experimentation of agronomical solutions.
. Sensibilization of farmers and people intervening on a watershed.
. Spreading of concrete and precise advices.

The success of this stage is based on communication, reciprocal trust between farmers and people in charge of soil conservation, and time. As a matter of fact, farmers need time to integrate into their decisions the runoff and erosion control.

It is obvious to recommend easy agronomical solutions since they are accessible for farmers with their own farm equipments. It is solutions such as good machinery adjustment (angle of disks, distance between blades...) reduced-till, good choice of implements, right depth of tillage after potatoes crop...

Thus, farmers understood that runoff and erosion are not fatalities and that they can participate to erosion control. From there, farmers will be able to accept more important constraints such as grassed waterways creation.

In conducting this work, two points were noticed. On one hand the evolution of agriculture is constant. So some solutions we initially put aside are now possible (for example, sowing under a mulch). On the other hand, the new politic of environment can allow innovations such as land retirement.

* *The fourth stage concerns land management.*

The aim is to control unavoidable concentrated runoff in order to avoid disastrous flooding and spectacular gullies.

This land management must be carried out with farmers, land owners, mayors and borough councils, agronomists, ingeneers and competent administrations.

Those actions can be curative or preventive through regrouping of land, for example.

Farmers participation to the organisation of land is concrete. They have responsabilities and decisions to make.

Beside, our intervention is based on sensitization and the technical results are formed to be directly applied by a geometrician who didn't acquired a specific soil conservation training.

* *The last stage is to generalize and perpetuate the action.*

At this stage all partners from rural world are involved in order to work out and applied agronomical and hydraulic solutions. Farmers, communes, department and regional representatives ; specific departmental and regional administrations and councils ; farming advisers, chamber of agriculture, agriculture cooperatives, research centers such as INRA, BRGM, water conservation or environmental organizations (Agence de l'eau, DIREN...) and agriculture colleges are concerned.

Among all those partners, people in charge of erosion control are the necessary foundation for the vitality and success of this action.

In "Pays de Caux", this success is based on a regional association dedicated to soil conservation and flooding control. This association is comprised of agronomists. Communes and region elected representatives and agriculture professionals are administrators. A reason for the good skills establishment is that agronomists are in the business for a long time.

This specialized office makes easier communication between all the partners. It is also a suitable structure to carry out research and to welcome training courses, to work out and prepare concrete solutions for tomorrow.

A strategy to control runoff and soil erosion in a large scale farming area.

CONCLUSION

This action in "Pays de Caux" had been long and difficult to setup in the early 80's.

It had been favourable to proceed from several stages. Investigation in farming system is essential. The creation of a regional association (A.R.E.A.S.) in charge of the action at every stage had been catalytic for all partners and the whole strategy. This permanent structure allows a continual improvement of solutions, according to the evolution of agriculture, and proceeds side by side with the development of agriculture in a way where hydraulic problems are taken into account.

BIBLIOGRAPHIE

1. Auzet V., (1987) : L'érosion des sols par l'eau dans les régions de grande culture: aspects agronomiques. Ministère de l'Environnement et Ministère de l'Agriculture, CEREG, Organisation-Environnement, 53 p. + annexes.
2. Auzet V., (1990) : L'érosion des sols par l'eau dans les régions de grande culture: aspects aménagements. Ministère de l'Environnement et Ministère de l'Agriculture, CEREG, Organisation-Environnement, 39 p. + annexes.
3. Auzet A.V., Boiffin J., Papy F., Ludwing B., Maucorps J., (1992) (in press) : Rill erosion as a function of the caracteristic of cultivated catchment in North of France, Catena vol. 19.
4. Boiffin J., Papy F., Peyre Y., (1986) : Systèmes de production, systèmes de culture et risques d'érosion dans le Pays de Caux. Rapport INA-PG, INRA, 154 p. + annexes.
5. Boiffin J., Papy F., Eimberk J., (1988) : «Influence des systèmes de culture sur les risques d'érosion par ruissellement concentré. I. Analyse des conditions de déclenchement de l'érosion.» Agronomie 8 (8)1, pp. 663-673.
6. Boiffin J., Monnier G., (1991) : Simplification du travail du sol et érosion hydrique. Perspectives Agricoles n°161-162-163.

7. Ludwig B., (1992) (in press) : L'érosion par ruissellement concentré des terres cultivées du Nord du bassin parisien : analyse de la variabilité des symptômes d'érosion à l'échelle du bassin versant élémentaire, thèse.
8. Ouvry J.F., (1982) : Localisation et description des sites d'érosion des sols agricoles du bassin inférieur de l'Yères (Seine maritime). Mémoires DDA, ENSA Rennes. INRA Rennes et Rouen, 72 p. + annexes.
9. Ouvry J.F., (1987) : Bilan des travaux. Campagne 1986-1987. Rapport A.R.E.A.S., 153 p.
10. Ouvry J.F., (1989-1990) : «Effet des techniques culturales sur la susceptibilité des terrains à érosion par ruissellement concentré : expérience du Pays de Caux». Cah. ORSTOM, sér. Pédol vol. XXV (1-2), pp. 157-169.
11. Ouvry J.F., (1992) : L'évolution de la grande culture et l'érosion des terres dans le pays de Caux. Bull. Assoc. Franç., Paris vol. n° 2. pp. 107-113.
12. Papy F., Boiffin J., (1988) : «Influence des systèmes de culture sur les risques d'érosion par ruissellement concentré. II. Evaluation des possibilités de maîtrise du phénomène dan les exploitations agricoles.» Agronomie 8(8)1, pp. 745-756.
13. Papy F., Douyer C., (1991) : Influence des états de surface du territoire agricole sur le déclenchement des inondations catastrophiques. Agronomie 11, pp. 201-215.
14. Papy F., Souchere V., (1992) (in press) : Maîtrise du ruissellement et de l'érosion de talweg : Une démarche d'aménagement de l'espace. Symposium International : In farm lands erosion in temperate plains environnements and hills, 25-29 may 1992, Saint Cloud.
15. Cemagref, (1986-1987) : Les dégâts causés par les pluies intenses dans le bassin du Croult (Val d'Oise). Conseil Général du Val d'Oise, Ministère de l'Environnement, 3 rapports, 236 p.
16. IGN, (1984) : Erosion des terres agricoles. Ligescourt (Somme), Département de télédétection de l'Institut Géographique National.
17. Monnier G., Boiffin J., Papy F., (1986) : Réflexions sur l'érosion hydrique en conditions climatiques et topographiques modérées : cas des systèmes de grande culture de l'Europe de l'Ouest, cahiers ORSTOM, série Pédologie, 12(2), 123-31.
18. Roose E., (1992) (in press) : Is erosion a problem today ? The G. CES : A new strategy to rislve this social problem. In farm lands erosion in temperate plains environnements and hills, 25-29 may 1992, Saint Cloud.
19. Auzet A.V., Boiffin J., Papy F., Maucorps J., Ouvry J.F., (1990) : «An approach to assess erosion forms and erosion risks on agricultural land in Northem Paris Basin (France).» Soil erosion on agricultural lands, Boardman, Foster et Dearing éd., Wiley, London, pp. 383-400.
20. De Ploey J., (1988) : «No tillage, experiments in the Central Belgium loess belt.» Catena, Soil Technology vol. 1, pp. 181-184.

Farm Land Erosion: In Temperate Plains Environment and Hills
S. Wicherek (Editor)

Control of overland runoff and talweg erosion: a land management approach

F. Papy and V. Souchère

INRA, Groupe de Recherche SEFIP, Paris-Grignon, France.

SUMMARY

The silty soils of the Paris Basin loess-belt, dedicated to cash crops, suffer erosion because of runoff concentration in talwegs. Although, in most cases, farmers do not suffer the damage caused by these phenomena, the situation is far more worrying for local communities.

Local communities in danger have built storm basins to protect themselves from disastrous muddy floodings. But this protection below catchments is not really suited to controlling the chronic damage which they are more and more sensitive (cleaning roadways, clearing storm basins, the quality of drinking water, etc). These chronic problems require solutions uphill, in cultivation and spatial planning practices.

However, any proposed solutions have to take the principles of the erosive functioning and the logical reasoning of farmers' actions into consideration. The present economic conditions lead farmers to concentrate on yield, but produce harmful effects. We must therefore try to find technical solutions that can be discussed within a group of farmers to favour as far as possible a community-level management of the agricultural space. The control of runoff and erosion will then be more efficient.

RESUME

Les sols limoneux de la ceinture loessique du Bassin Parisien, voués à la grande culture, sont le siège d'une érosion, due principalement à la concentration du ruissellement dans les talwegs secs de bord de plateau. Si dans la plupart des cas, les agriculteurs ne subissent guère les méfaits de ces phénomènes, la situation est beaucoup plus préoccupante pour les collectivités locales.

Pour se prémunir des inondations boueuses catastrophiques, les collectivités locales les plus menacées se sont équipées de bassins d'orage. Mais cette protection à l'aval est peu adaptée aux dégâts chroniques auxquels elles sont de plus en plus sensibles (nettoyage des chaussées, curage des bassins, qualité de l'eau potable etc). Pour résoudre ces problèmes chroniques, nous devons chercher des solutions, en amont, dans des pratiques de culture et d'aménagement de l'espace.

Aussi, nous pensons que les solutions proposées doivent tenir compte non seulement des principes du fonctionnement érosif étudié, mais aussi des logiques d'action des agriculteurs. Or, les conditions économiques actuelles conduisent les agriculteurs à avoir des logiques individuelles qui visent la production mais induisent des effets néfastes. Aussi, nous devons chercher des solutions techniques susceptibles d'être discutées au sein d'un groupe d'agriculteurs pour favoriser autant que possible une gestion commune de l'espace agricole. La maîtrise du ruissellement et de l'érosion en sera d'autant plus efficace.

1. THE DEMANDS OF LOCAL COMMUNITIES ON AGRICULTURE : NEW MANAGEMENT PRACTICES FOR UPSLOPE AREAS

Specific situations

As related in several papers presented at this Symposium, recent studies have shown the existence of clearly identifiable erosive systems along the edges of loamy plateaus in the cultivated areas of the Paris Basin : talweg erosion due to concentrated runoff (1 - 2). Obviously, other types of erosion may also occur (rill and inter-rill erosion on slopes, sheet erosion) particularly following spring storms. During the winter period however, talweg erosion accounts for 40 to 50 % of the total volume of rills (3). This indicates the extensive damage caused by this particular type of erosion, especially since the sediments detached from the talweg are those transported farthest afield.

Talweg erosion becomes increasingly severe under the following conditions: i) simultaneous occurrence of loamy soils vulnerable to crusting and of cropping systems involving a high proportion of annual crops ; ii) openfield landscape ; iii) valley head located on a plateau edge and overhanging a dry valley floor. This erosional system is relatively better developed than other systems in winter even under low intensity rainfall.

Sediments from the talweg are carried over relatively long distances and deposited both on agricultural land and elsewhere. The water flows on. Where the loamy layer masks a karst relief, a common occurrence in these areas, vertical structures may sometimes connect the surface water with underground waterbodies.

As a result, in situations of talweg erosion more so than for other erosion types, there occurs over long distances from upslope to downslope sites a staggered sequence of processes, i.e. runoff, erosion, transport, sedimentation, overland or deep water movement.

A problem addressed from downslope to upslope areas

The damage to farming is relatively limited, except locally in sites where flow concentration creates deep gullies or in sedimentation areas. But, runoff and erosion do affect local communities.

The severest disasters, those that make the headlines in the press, are the catastrophic floods associated with mudflows (4). Storm basins have been built by the most vulnerable local communities as a preventive measure against such events during the most recent sequence of rainy years which caused natural disasters (5). This solution is costly (about 150 FF/m^2, Ouvry, verbal communication) ; however, it benefits from regional subsidies and has therefore often been given preference over the construction of numerous small storm basins.

Downslope protection is however poorly adapted to recurrent damage to which local communities are increasingly sensitive, such as :

* soil deposits which increase road maintenance costs and the cost of cleaning out storm basins (which involves a yearly 5 to 15 % of the initial building cost, according to Ouvry, verbal communication) ;
* overland runoff water which, once cleared of its solid particle load, carries off dissolved substances and alters water quality.

Severe winter erosion, by creating fast flow routes for water on the soil surface, also appears to contribute to flooding after spring storms by reducing the time of runoff concentration (Meynier quoted in (4)).

Solutions to these recurrent events (elimination of talweg erosion, reduction of flood peaks) must be provided in the upslope zone, as a result of modified cropping practices and farmland planning and management.

The search for compatibilities

The prevailing economic context causes farmers to pursue productive and individual goals. In seeking to be productive, farmers tend to consider erosion damage only insofar as it affects their output: for instance in the absence of specific incentives, it seems difficult to enforce the adoption of practices designed to limit runoff, in the frequent cases where runoff does not trigger erosion on their own land. In being individualistic, these goals result in a land management approach that is limited to the farm territory and does not take into account the continuity of the physical processes involved.

The dual challenge is therefore to develop alternative final objectives for farming and to devise forms of collective land management. For this, the general context will of course have to change. For the time being, however, we shall seek technical solutions which must be compatible with the farmers' rationale and can be submitted to a farmers' group. In this perspective, we seek to :

1. understand the reasons for the farmers' actions and identify what cannot be changed and the possibilities farmers have of introducing anti-erosive measures,
2. propose a range of possible solutions, leaving room for discussion between those who *in fine* will have to implement them.

To collect and structure information for this purpose involves analysing an area of land from two viewpoints : that of the physical units (catchments) and that of the decision units (farms) (6).

2. THE PROCESS OF EROSION : THE CONTINUITY OF PHYSICAL PROCESSES WITHIN A SPATIAL UNIT, THE ELEMENTARY CATCHMENT

In situations of talweg erosion, the functional morphological unit to be considered is not the slope but the elementary catchment, i.e. a first order basin (according to the dry valley network) defined by a sufficiently long valley floor (7). Several functional zones may be identified within this catchment : the talweg and the catchment area consisting of all the land surfaces that will potentially produce runoff. The erosive efficiency of any runoff-prone surface varies however with its location within the catchment area, the severest erosion being generated when the runoff surface lies upslope (Auzet et al. (3) and on-going research). Within the catchment area a distinction should therefore be made between valley sidewalls and valley head, i.e. the area lying upslope from the point where the talweg is clearly formed (Fig. 1).

Let us assume a series of surfaces distributed across the catchment area ; under a given rainfall intensity, these will generate runoff. The runoff water will then travel along a flow network ending in the talweg (Fig. 1). As our objective is to devise protective measures, we shall seek to detect which part of the talweg is vulnerable to incision. A model of erosion functioning is constructed for this purpose; it identifies three structural elements : (A) the runoff contributing areas, (B) the runoff pathways and, (C) the site in the talweg where runoff concentration gives the flow sufficient shear stress to incise into the soil.

TALWEG

VALLEY HEAD

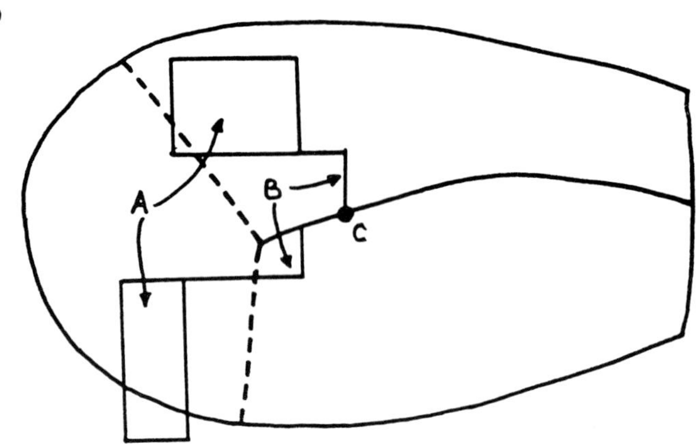

SIDESLOPE

Figure 1. Functional areas in an elementary catchment (valley head : sidewalls ; talweg). The components of the erosion process (A - runoff areas ; B - Water pathway network ; C - Incision point in the talweg.

(A) *The runoff contributing areas.* In the recurrent processes considered here, which are also induced by low intensity rainfall, the overland runoff surfaces are those which have reached the ultimate facies, i.e. the sedimentary crust characterized by very low infiltrability (8, 3). Since our objective is to protect the talweg from erosion, there is no need to construct a model predicting the precise moment when such a state is reached ; this can be done by simply increasing the runoff risk in the model. Empirical knowledge gained from loamy regions with soils susceptible to crusting (clay <14%) enables us to identify the areas that will most probably produce overland flow under low intensity rainfall at various periods of the year (Tab. 1).

(B) *The runoff pathway* depends on the land shape, on the network of roads and tracks, as well as on the field pattern. Within any given plot, runoff flows along the steepest slope and/or along the direction of tillage (Tab. 2) ; it then collects in a headland or in a deadfurrow. Ludwig (in preparation) (9) has proposed a general water flow model for an openfield landscape. In any given catchment, the talweg length to protect increases the further upslope the pathway network carries runoff. General plot orientation in relation to talweg is a determining factor in this respect ; the more perpendicular the plots lie to the talweg, the further runoff will concentrate upslope.

Table 1. Lande-use categories capable of generating runoff at different times of the year under low intensity rainfall (crusting loamy soils).

Land-use category	Early January	Early May
Winter crops (wheat, winter barley, rape)	+	+
Intercrops : stubble or green manure crop (1)	O	(does not apply)
Intercrop with no stubble, uncultivated since harvest	+	(does not apply)
Early spring sowing (spring barley, peas, flax, ...)	(does not apply)	+
Late spring sowing or planting (sugarbeet, potato, maize)	(does not apply)	O

+ = generates runoff   O = no runoff   (shaded) = does not apply

(1) Under given climatic conditions, stubble fields may generate runoff.

Table 2. Relationship between tillage orientation and slope at two sites in the Pays de Caux (Normandy)

Direction of tillage in relation to slope (%)	Blosseville			Fongueusemare		
	slope class			slope class		
	< 2 %	2-5 %	> 5 %	< 2 %	2-5 %	> 5 %
Parallel	32	39	30	47	53	44
Perpendicular	28	27	50	22	19	29
Oblique	40	34	20	31	28	27

(C) *The head of the talweg incision* may be defined theoretically (10 - 11). This will highlight the major determinants in cases where, for a population of given situations, a number of characteristics remain unchanging. In the situations investigated, the following factors may be considered to be determining : talweg topsoil texture (one of the determinants of the critical shear velocity for rill initiation) ; the cross profile of valley bottom (which influences the flow concentration) ; the range of rainfall intensities (which governs flow rates in the talwegs), since the objective sought here is to control recurring damage. The location of the rill headcut is then determined by the area upslope where the "sedimentary crust" stage has been reached, by the local slope gradient at that point along the talweg and, by the compaction degree of the talweg. On the basis of a sample of catchments, we are currently defining the relation between the critical area producing runoff and the critical local slope gradient for loose and compacted talweg floors.

Let us consider runoff-prone areas located in a catchment and the points where runoff water generated by each of these areas discharges into the talweg. Assuming that slope length, gradient and talweg compaction state are known, the point where the rill headcut will occur may then easily be located. Depending on whether this spot is situated in a loose or in a compacted talweg, we advocate creating a compacted strip or a grassed strip from the place of the headcut.

However, changing the distribution of runoff contributing areas on the catchment and redesigning the water flow network so that it discharges more or less upstream along the talweg will modify the length of the talweg segment in need of protection. The available model components can be used to devise an array of solutions which will reduce talweg erosion (on-going research in collaboration with INRA Soil Science Station in Orleans). It should serve as a basis for discussion between the farmers concerned, all of whom have a reason for doing what they do, a fact that must be taken into account.

3. THE RATIONALE OF CROPPING OPERATIONS AND THEIR EFFECTS ON EROSION FUNCTIONING. THE DISCONTINUITY OF SPATIAL STRUCTURES DUE TO THE DECISION UNIT, I.E. THE FARM.

The farm area, managed by a single decision-maker, does not coincide with the catchment area. The catchment area may also consist of discontinuous blocks of fields. Within the farm there are elementary pieces of land managed homogeneously, the arable fields, that form the components of a mosaic which, at a given moment, may be distinguished by the state of their soil surface. On the catchment area, they are characterized by their ability to generate runoff, and in the talweg by their resistance to incision. Another identifiable feature of the landscape is the runoff pathway network which functions are described above.

Let us consider the effect of the farmer's rationale in carrying out cropping operations on the elements of the erosion model. In doing so, we shall distinguish aspects which will be difficult to change from the possible margins of manoeuvre.

(A) Runoff contributing areas

1. The ability of an area of agricultural land to generate runoff depends on the ratio between grasslands and arable land and on the crop rotation within the latter. But the forms of land use are determined by the economic context. For instance, the introduction of milk quotas increased the trend towards abandonment of livestock

production and hence contributed to the decrease in grasslands. Also, the considerable drop in cereal prices favoured the development of pea crops, thereby increasing runoff risks by leaving the topsoil compacted, throughout the winter, from harvest to sowing.

The weight of economic determinants is too compelling to challenge the choice of crops ; we shall consider these as an obligatory factor.

2. The location of land use types within the catchment area influences its susceptibility to erosion. To simplify herd management, grasslands are often sited close to the farmstead. This arrangement is often beneficial when, as is frequently the case, the buildings stand on high ground. For reasons of land suitability, the steep slopes (> 12-15 %) or closed depressions are given over to pastures. In both cases they are located in favourable positions : the former limits runoff and erosion, the latter ensure water decantation, a valuable asset when the depression is connected with groundwater.

Moreover, pastures are sometimes located on valley floors where they help to reduce erosion hazards, sensu stricto.

The farmer's operational logic therefore results in a favourable siting of pastures, which explains why their suppression has such adverse effects.

3. Increasing the size of fields increases the possibility of creating large contiguous runoff surfaces. Moreover when a farmer cultivates fields grouped into larger blocks he will often allocate the same crop to neighbouring parcels for reasons of work organisation ; this is often the case with wheat. The degree of "territorial concentration" of the farm is an indicator of the opportunities the farmer has to do this ; it is linked with settlement type (grouped or dispersed) and with land regrouping. A comparison of field patterns at two sites in the Pays de Caux (Seine maritime, Normandy) which are presented later in detail, highlights this effect. One is a site with dispersed housing (Fongueusemare) and grouped cropfield blocks; in the other, housing is grouped with a dispersed field block pattern (Blosseville). In Fig. 2, the curves of cumulated elementary plot areas (B11 and Fg1) and of field blocks within an individual farm (B12 and Fg 2) show greater concentration at the scattered settlement site than at the other site. However, the decisions of neighbouring farmers may, accidentally in this case, result in larger cropfield blocks ; this is all the more possible as concentration within the farm is already high, as can be seen, from a comparison of curves 2 and 3 in Fig 2.

This change in field pattern is based on a clear objective: to increase work productivity by improving overall organisation. This fact must be taken into account, although there may remain some room for manoeuvre.

4. Crop management techniques will affect the ability of field plots to generate runoff through the effect of cultural operations on soil surface state (fragmentation by machinery, soil surface compaction by wheels) and the timing of operations. Runoff in this case can be limited by making rougher seedbeds, by reducing wheel tracks, by specific tillage operations or by establishing green manure crops during the intercrop period (12). The former measures require specific machinery, the latter call for reorganisation of work planning. Thus when the farmer has to apply fertilizers (in winter before planting the beet crop), he will abstain from tilling the soil to maintain sufficient bearing capacity. Or else, when a work peak occurs in early September (this is frequently the case when farmers have to harvest potatoes and to roll flax simultaneously), the farmer is unable to sow green manure crops early enough to protect the soil against degradation by rainfall.

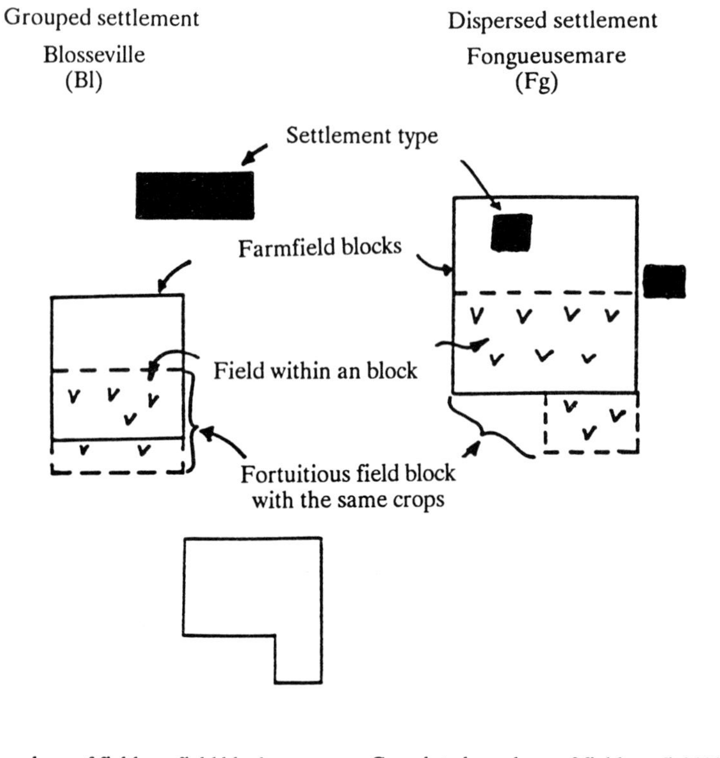

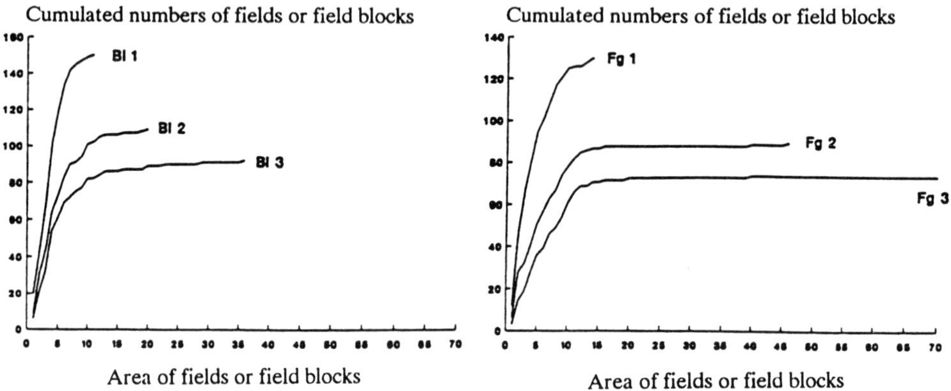

A comparison of the two figures shows that grouping of crops on contiguous parcels is greater at Fongueusemare than at Blosseville, both within the farms (compare curves 1 and 2) and, fortuitiously, on neighbouring farms (compare curves 2 and 3).

Figure 2. Field concentration at the two sites.

Here again the rationale of work planning is not always compatible with the measures advocated ; to find out what room for manoeuvre is available, the links between what these measures require and the way the farm system operates must be investigated (Martin, on-going research).

(B) Water flow network

As shown earlier, the spatial layout of field boundaries has an effect on water movement. The shape of the fields is frequently governed by two major determinants : slope and access. Slopes with a gradient over 10 % will often dictate the direction of tillage operations for reasons of safety, regularity of fertilizer application, crop quality, etc. However, in the regions considered slopes are not steep enough to play this role. This is clearly shown in Tab. 2, for two sites in the Pays de Caux (Normandy). There, the road network is the major factor dictating plot orientation. Access to fields is cheapest when their axis is perpendicular to tracks and roads : orientation of tracks in relation to the talweg is therefore determining for water movement patterns. When track orientation is perpendicular to the talweg, the direction of tillage along the greatest field length channels water downhill along the slope sides. In a region where farm fields are dispersed, tracks must a priori provide access to each group of cropfields ; a well designed land consolidation operation may in this case provide a useful communication network and a general orientation of the fields which will concentrate runoff as far downslope as possible.

(C) Incision

This may be avoided in the talweg through specific management : establishment of compacted waterways and in more severe situations, of grassed waterways (12). The former are less costly and easier to prepare ; the latter require some establishment and maintenance and are only justified in extreme situations.

IV. CONTROL OF RUNOFF AND EROSION WITHIN THE CULTURAL TERRITORY : DIAGNOSTIC ASSESSMENT OF THE MANAGEMENT OF A CONTINUOUS AREA BY A GROUP OF FARMERS

Finding a way to control erosion is rarely possible at the level of a single farm, since the farm boundaries do not generally include the whole catchment area. When, on the other hand, the catchment is the spatial unit for which solutions must be devised this level is rarely adequate, insofar as the allocation of crops to fields and their modes of management are decided and planned on individual farms and often depend on the distance to farmsteads. In addition, the planned measures should take into account neigbourhood effects. The solutions must therefore be devised for a continuous area of land, on which a group of farmers carry out their agricultural activity; this area will often include several contiguous catchments.

This approach led us to consider the landscape features which characterise its erosive functioning and the capacity of the human group cultivating it to control erosion processes. As the soils in the region investigated are homogeneous, and farming systems and crop rotations are similar, the following two spatial characteristics were selected :

1. an indicator of the general morphology of catchments expressing its capacity to concentrate runoff (the valley head area ; the talweg head slope for instance). When this area is small and this slope is steep, the area needed to trigger erosion is reduced. In the reverse case (small talweg head slope and extensive valley head area), the area needed to trigger erosion is greater.
2. an indicator of the concentration of farm blocks (eg, the spatial layout of settlements and the existence or age of land consolidation schemes ; the number of farmers on a given site).

Two sites in the Pays de Caux are compared in Tab. 3 on the basis of these criteria ; they have the same area, the same crop rotation and the same type of soil, and therefore the same general ability to produce runoff. At Blosseville, where slopes are moderate and catchments wide, valley heads cover a large area ; the farms consist of dispersed blocks of fields, since settlement is grouped. At Fongueusemare, slopes are steeper and the valley heads smaller ; settlement is dispersed and each farm consists of a continuous block of land.

Having half the elementary catchments (7 to 14) and gentler slopes, total soil discharge at the Blosseville site is less than at the other site. In addition, talweg incision conditions at Blosseville require larger runoff contributing areas than at Fongueusemare; these conditions are therefore more rarely met and besides, as seen above, concentration of cropfield blocks which could potentially be in the same state is lower. At Fongueusemare, there need only be a few runoff-prone areas to trigger the process : it only needs one or two farmers to group several elementary arable plots. Thus the difference in farm land concentration between the two sites accentuates the effect of morphological differences on the amount of soil discharged.

At Fongueusemare, however, there is a possibility of the farmers controlling erosional processes. Given the small size of the catchments and the size of farm blocks, a farmer will often generate the runoff which erodes his own land lying in the talweg. As a result some farmers apply anti-runoff strategies, in particular by treating the intercrop period (green manure crops). Besides, farm blocks often form a single block, so that farmers have a degree of freedom to reorganize their field pattern (eg, by establishing a field on two catchments, or by orienting it lengthwise along a talweg) or to allocate crops to fields in a more judicious way (eg, avoiding growing the same crop on the same valley head). Collective management of the area may also be possible. In short, at Fongueusemare where erosion risks are high, one may reasonably expect that farmers will reduce runoff and erosion ; cooperation between local farmers and the downslope local communities should concentrate on developing talweg defences (aids for establishing grass strips).

At Blosseville, where erosion risks are low, such steps cannot be expected from the farmers. Consequently runoff will not be reduced, nor will the resulting flood risks for the downslope local communities be lessened. Cooperation between local farmers and local communities should emphasise the adoption of anti-runoff measures and the development of small temporary storm basins.

In both cases, new EC incentives for environmental conservation may be used in a specific way for each situation.

The issues addressed here are not merely concerned with erosion. To change over from one type of farming dictated by a logic of individual production to another, which would include functions of land management and conservation is one of our current major challenges. The above paper seeks to provide an approach for achieving this.

		Blosseville	Fongueuseumare
Common features			
Total area (ha)		490	452
Total spring crops		58	53
of which peas & flax		26	21
Total winter crops		42	47
of which cereals		37	38
Morphological features			
Catchment numbers		7	14
% Valley head areas (TV)		83	69
% Valleys (V)		17	31
Slopes > 5 %	In TV	0	14
	In V	8	44
Field concentration characteristics			
Settlement type		Grouped	Dispersed
Farmers (numbers) on site		30	< 10
Field pattern		Dispersed	Grouped
Average % of farm UAA at site		53	80
Mean size of elementary plots		3.20	4.24
Mean size of cropfield blocks within the farms (for 1992)		4.69	6.07
Mean size of accidental field blocks		5.79	7.55

Table 3. Major characteristics of the two sites in Pays de Caux (Normandy).

ACKNOWLEDGMENTS

The authors thank Anne-Véronique Auzet, Michel Meynier and Jean-François Ouvry for their comments on the text, and Michel Sebillotte for fruitful discussions.

REFERENCES

1. Monnier, G., Boiffin, J., Papy, F. (1986) : réflexions sur l'érosion hydrique en conditions climatiques et topographiques modérées : cas des systèmes de grande culture de l'europe de l'Ouest, Cahiers de l'ORSTOM, série pédologie, 12 (2), 123-131.
2. Auzet, A. V. (1987) : L'érosion des sols par l'eau dans les régions de grande cultrure: aspects agronomiques, Min. Env./Min. Agric., CEREG-UA 95 CNRS.
3. Auzet, A. V., Boiffin, J., Ludwig, B. (1992) : Concentred flow erosion in cultivated catchments : influence of soil surface state. in first international ESSC Congress "Conserving our soil resources" 6-10 april 1992.
4. Papy, F, Govers, C, (1991) : Influence des états de surface du territoire agricole sur le déclenchement des inondations catastrophiques. Agronomie, 11, 201-215.
5. Auzet, A. V. (1990) : L'érosion des sols par l'eau dans les régions de grande culture : aspects aménagements, Min. Env./Min. Agric., CEREG-UA 95 CNRS.
6. Papy, F. (1992) : Effets des structures agraires sur le ruissellement et l'érosion hydrique. Bull. Assoc. Géogr. Franç. (2) 115-125.
7. Auzet, A. V, Boiffin, J., Papy, F., Maucorps, J., Ouvry, J. F. (1990) : Approach to assess erosion form and erosion risks on agricultural land in Northen Paris Bassin (France). In "Erosion on Agricultural Land" (Bordman, Foster and Dearing Eds.), Wiley London, pp. 383-400.
8. Boiffin, J. (1984) : La dégradation structurale des couches superficielles du sol sous l'action des pluies, Thèse, INA P-G.
9. Ludwig, B., Boiffin, J., Masclet, A. (1992) : Origine spatiale des sédiments et contribution relative aux pertes en terre de différentes formes d'érosion au sein d'un bassin versant cultivé. in Farm lands erosion in temperate plains environnements and hills 25-29 may, 1992, Saint-Cloud.
10. Rauws, G., Govers, G. (1988) : Hydraulic and soil mechanical aspects of rill generation on agricultural soils. Journal of soil Science, 39, 111-124.
11. Govers, G. (1991) : The relationship between discharge, velocity and flow area for rills eroding loose, non-layered materials. (doc. reneoté)
12. Ouvry, J. F. (1989/90) : Effets des techniques culturales sur la susceptibilité des terrains à l'érosion par le ruissellement concentré : expérience du Pays de Caux. Cah. ORSTOM, série Pédol., Vol. XXV (1-2), 157-169.
Rico, G. (1990) : Hydrogéologie de la craie : le système aquifère karstique de l'Aubette (Seine - Maritime) : circulations rapides et caractérisation de la turbidité des eaux souterraines. Thèse de l'Université de Rouen.

Farm Land Erosion: In Temperate Plains Environment and Hills
S. Wicherek (Editor)
1993 Elsevier Science Publishers B.V.

Effect of hog manure and fertilizer application on runoff and drainage water quality

A.R. Pesant [a], G. Gangbazo [b], G. Barnett [a], D. Cluis [c], and J.-P. Charuest [a]

[a] Agriculture Canada Research Station, Lennoxville, Québec, J1M 1Z3, Canada

[b] Ministère de l'Environnement du Québec, Québec, G1X 4E4, Canada

[c] INRS-Eau, Université du Québec, Québec, G1X 4N8, Canada

SUMMARY

Annual and seasonal Nitrate losses (NO_3-N) were evaluated from 45 m^2 corn and forage plots which had received the recommended chemical fertilizer rates of 180 and 45 kg N ha^{-1}, respectively, for two consecutive years plus hog manure at twice those rates. Therefore total nitrogen applications were (kg ha^{-1}) : corn 540 and forage 135 except for a check plot receiving only fertilizer. The hog manure was surface applied in three ways : all in the fall, all in the spring or in a split application with half in each season. Corn plots received manure after plowing. Each plot was equipped to collect runoff and drainage water separately throughout the whole year. Annual NO_3-N losses in both runoff and drainage waters were significantly higher for corn with 150 kg ha^{-1} yr^{-1} as compared to only 20 kg ha^{-1} yr^{-1} for the prairie. Due to the residual effect, Nitrate concentration increased significantly between the first and the second year for corn but decreased for forage. Plots receiving chemical fertilizer plus hog manure no matter when the latter was applied had a significant increase on NO_3-N losses as compared to the plots receiving fertilizer only. On forage plots, higher rates of fall-applied hog manure increased NO_3-N losses in runoff and drainage waters.

RESUME

Sur parcelles expérimentales de 45 m^2 nous avons comparé les pertes d'azote total (NT) (azote total Kjeldahl + nitrates) annuelles et saisonnières dues à l'épandage de grandes quantités d'engrais organiques (lisier de porc) et minéraux à celles dues à l'épandage exclusif d'engrais minéraux conformément aux besoins agronomiques de la culture du maïs (180 kg N.ha^{-1}) et d'une prairie (45 kg N.ha^{-1}). Chaque parcelle a été aménagée de manière à recueillir séparément les eaux de ruissellement et de drainage après chaque événement de pluie ou de fonte de neige. Sur une année de végétation, le taux de fertilisation azotée totale apportée par les deux formes d'engrais équivaut à 3 fois les besoins agronomiques de la culture pratiquée soit 540 kg de N.ha^{-1} pour le maïs et 135 kg de N.ha^{-1} pour la prairie. Le lisier de porc a donc été épandu à raison de 360 kg de N.ha^{-1} sur les parcelles de maïs, soit en une seule fois au printemps avant le semis et à l'automne avant le labour ou fractionné également entre l'automne et le printemps, tandis que sur la prairie il a été épandu et laissé à la surface du sol à raison de 90 kg de N.ha^{-1}. Dans l'ensemble, les pertes totales annuelles de NT dans les eaux de ruissellement et de drainage représentent 10 % de la quantité totale d'azote épandue sur les parcelles de maïs et 30 % de la quantité épandue sur les prairies, quel que soit le traitement. Soixante-quinze pour cent de cette perte est sous forme de nitrates. Avec la culture du maïs, les eaux de drainage recueillent 83 % de la perte annuelle en nitrates. Ces pertes sont indépendantes des traitements et varient dans l'ordre suivant : printemps > été > automne > hiver. Ainsi, 45 % des pertes de nitrates ont lieu au printemps en comparaison de 35 % en été, 15 % à l'automne et 5 % en hiver. Par rapport à une fertilisation minérale printanière conforme aux besoins agronomiques, l'épandage du lisier en une seule fois à l'automne augmente significativement la charge totale (NT) de 180 % en hiver alors que le fractionnement de l'épandage augmente cette charge de 190 % au printemps.

1. INTRODUCTION

Within the Quebec farm sector, hog sales totalled $356 million C.D.N. and was the second most important activity after the dairy industry, which had realized total receipts of $1,3 billion C.D.N. (1). In 1979, pork production exceeded consumption by 34 % (2).

While much effort was devoted to increasing the size and improving the economic viability of the hog industry, none was given to the impact on the environment. Hog production generates some 6 million cubic metres of liquid manure or about 24 % of the total manure produced on all Quebec farms (3). Because of limited storage capacity, insufficient available land for application at recommended rates, long winters, and frequent spring and fall rains, large amounts of hog manure are spread in short periods before and after the growing season. Often this is done without taking the nutrient value of the manure, mineral fertilizer applications, and crop need into account, mainly because producers believe that animal wastes have no value (4). It is therefore not surprising that surface water flowing through some agricultural areas is of mediocre quality (5 - 6 - 7 - 8), even though numerous municipal sewage treatment plants have begun operating over the last few years.

Poor liquid manure management, particularly that of the hog industry, was found to be responsible for surface and groundwater contamination in Europe (9 - 10). In Germany and Hungary, Nitrate concentration in groundwater has attained 9.0 and 11.3 mg L^{-1} respectively. In Brittany, potable water exceeds 11.5 mg NO_3-N L^{-1} for 6.6 % of the population (11). It is believed that these high levels are related to the large increase in the size of the hog, cattle, and poultry industries which are not land-based and hence has led to nitrogen availability in excess of crop requirements. It was therefore the objective of this work to evaluate the impact of hog manure and fertilizer applications on groundwater and runoff Nitrate loads.

2. MATERIAL AND METHODS

The experiment was initiated in late 1989 on the experimental farm of the Agriculture Canada, Lennoxville Research Station, 150 km east of Montréal, Québec, Canada (45° 21' N, 71° 51' W) in an area with an average annual precipitation of about 1033 mm. The soil was a Coaticook silty loam (Typic Humaquept) developed on a lacustrine material containing 17 % clay, 80 % silt and 3 % sand, which had 3100 mg total N kg^{-1} (Kjeldahl), 27 mg kg^{-1} Nitrate, and 5.3 %

organic matter. Prior to the experiment, the site had been under timothy clover meadow.

Plots 3 m wide and 15 m long were established on a site with a 6 % slope. They were separated from each other on the surface by a sodded berm, 50 cm wide by 25 cm high along both sides and at the head, and in depth by a black polyethylene plastic sheet installed to 1.2 m on the same 3 sides. Thus plots were isolated from each other and from the surrounding environment thereby preventing runoff and leaching contamination from other sources.

A trough was installed at the foot of each plot to collect runoff and a 10 cm plastic drain at 90 cm depth in the center of each plot to capture leachate. Both systems were connected by pipe to separate barrels which permitted sampling and volume measurement.

The four treatments were as follows : chemical fertilizer only applied in the spring according to crop nitrogen requirement, hog manure applied at twice the crop N requirement in the fall and at the same rate on another plot in the spring, and finally half of the same rate in the fall and the other half in the spring (Tab. 1). The plots to which hog manure was applied also received chemical fertilizer as described for the first plot. These treatments were replicated twice and the 16 plots - 8 in corn and 8 in hay - were arranged in a completely randomized plot design. In addition one corn and one hay plot which had received neither fertilizer nor manure were included to evaluate nutrient uptake.

Table 1
Treatment codes

Crops	Code	Fertilization *			Code
		EP	LP	LA	
Corn	1	0	0	0	000
		1	0	0	100
		1	1	1	111
		1	2	0	120
		1	0	2	102
Hay **	2	0	0	0	000
		1	0	0	100
		1	1	1	111
		1	2	0	120
		1	0	2	102

* EP : Chemical fertilizer applied in spring ; LP : liquid pig manure spring-applied ; LA : liquid pig manure fall-applied. The figures 0 to 2 refer to multiples of the normal recommended rate of application. For example : 2 means twice the recommended rate.

** Hay = Timothy-red and white clovers.

Hog manure and fertilizer (spring only) were applied broadcast on the surface of both corn and hay plots. Corn plots were fall-plowed after manure application and disked before hand seeding at 75 cm inter and 15 cm intra row spacing. The hay was a mixture of timothy, red and white clovers planted after plot establishment before any manure application. The hog manure was obtained from a commercial farm and contained 4.8 % dry matter, 4620 ppm Kjeldahl nitrogen and 24 ppm Nitrates. Fertilizer was applied as ammonium Nitrate, superphosphate, and potassium chloride.

Total runoff and leaching volumes for each plot were measured after each rainfall or snowmelt event and a 125 mL sample to which was added few drops of sulphuric acid, was conserved for analysis. The following analyses were conducted: Nitrate-nitrite on the water (12) ; manure and soil, total Kjeldahl N (13), Nitrate-nitrite (14).

The data was grouped by adding daily volumes and loads according to seasons of 90 days : fall Sept. 15 - Dec. 14, winter Dec. 15 - Mar. 14, spring Mar. 15-June 14, and summer June 15 - Sept. 14. The transformation Log_{10} (x + 1) was used to normalize the data since the variance was proportional to the mean and because of the large variability. Because of the differences in rates of application and management, all treatment effects were evaluated within crops. Inter-seasonal and inter-year variation was measured by difference and subjected to repeated measures analysis of variance (15). Since they were not replicated, the two plots having received no treatment were not included in the statistical analysis. Because of the variability and the fact that there were only two replicates, the data was interpreted at the 10 % level of significance.

3. RESULTS AND DISCUSSION

Weather conditions

During the first year (1989-1990) annual precipitation (as water) totalled 986 mm, of which 814 mm fell as rain and 172 cm as snow. This was normal, since the average totals 1033 $\pm$ 135 mm with 794 $\pm$ 119 as rain and 250 $\pm$ 82 cm as snow. Losses of total precipitation through leaching and runoff totalled 24 % for corn and 21 % for hay. There was somewhat more precipitation the second year (1990-1991) with 1145 mm total and 955 as rain and 190 cm as snow. Of the total precipitation, leaching and runoff losses were 37 % for corn and 33 % for hay.

Annual Nitrate load

Total annual Nitrate load, which includes Nitrate and nitrite losses from leaching and runoff were significantly higher ($P<0.01$) for corn at 151.05 kg NO_3-N ha^{-1} yr^{-1} than for hay at 19.65 (Table 2). Since the total annual fertilization load is a multiple of the crop requirement, which is higher for corn, it therefore follows that the environmental risk is greater than for hay.

Spreading hog manure in the fall in addition to mineral fertilizer in the spring (treatment 102) increased the nitrate-N load from 94.50 for fertilizer-only to160.81 kg NO_3-N ha^{-1} yr^{-1}. Similarly, spreading all the manure in the spring resulted in losses of 175.60 kg N ha^{-1} yr^{-1}. Even when half was applied in the fall and half in the spring nitrate-N losses were 194.88 kg. There was a significant lower amount of Nitrate losses in the fall as compared to spring and split hog manure application with corn. Therefore, it is clear that applying manure at twice the recommended crop N requirement rate, in addition to the regular fertilizer application, results in much higher Nitrate losses.

Table 2
Treatment effects on the annual Nitrate loads in runoff and drainage water during the second year

		Annual load (kg NO_3-N ha^{-1} yr^{-1}) Mean (SD) *
Crop		
	Corn	151.05 (9.81)a
	Hay	19.65 (1.33)b
Treatment		
	Corn : 100	94.50 (12.75)
	Corn : 111	194.88 (25.78)c
	Corn : 120	175.60 (23.25)c
	Corn : 102	160.81 (21.30)d
	Hay : 100	17.24 (2.40)
	Hay : 111	18.36 (2.55)
	Hay : 120	17.11 (2.38)
	Hay : 102	27.31 (3.73)d

*SD : Standard Deviation.
a,b : Means within a column followed by the same letter are not significantly different at the 0.10 level.
c,d : Indicates what the treatment differs significantly from the100 treatment within a crop at the 0.01 and 0.10 levels respectively.

Fall-spreading all the manure on hay plots in addition to the regular fertilizer requirement resulted in a significant ($P<0.10$) increase of annual NO_3-N losses from 17.24 kg ha^{-1} for fertilizer-only to 27.31 kg. The crop effect in the second year is likely related to residual effects from the first year's application as well (Fig.1).

When the first and second year results are compared by means of difference, in effect the residual, the annual load increased in the second year for all corn treatments while they generally decreased for all hay treatments. The decrease for hay is probably explained by a greater sward density in the second year, resulting in better nutrient use. There was no significant difference between all hog manure treatments and the chemical fertilizer (Tab. 3).

Seasonal loads

Loads vary considerably from one season to the next. For example, these decreased from fall to winter for both crops and all treatments (Fig. 2) (Tab. 4) but were much more marked for corn than for hay.

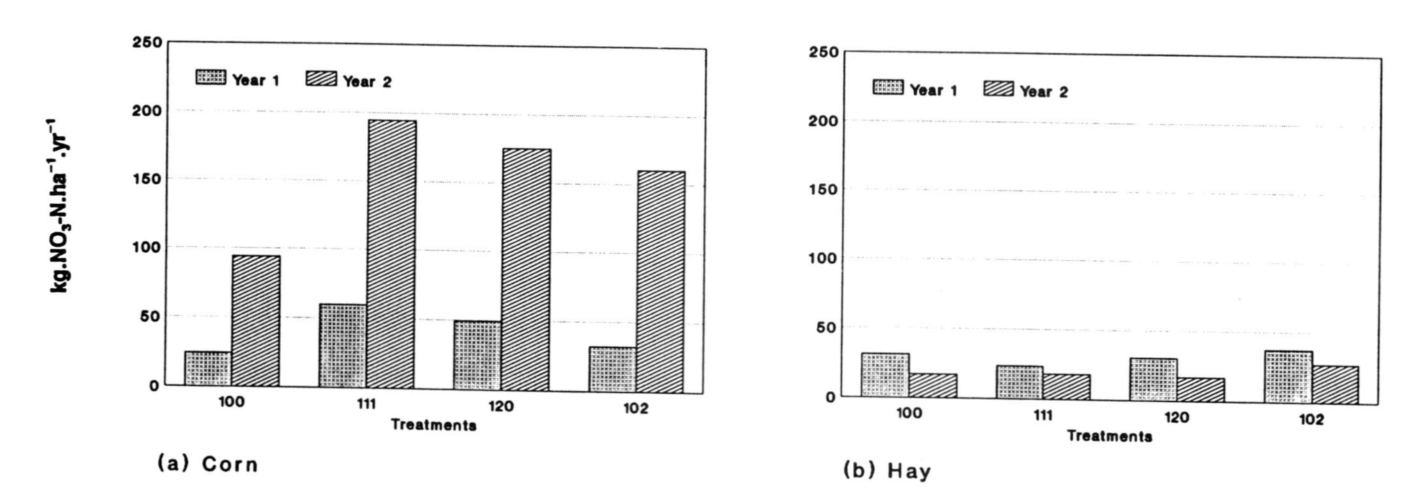

Figure 1. Second year vs first year Nitrate loads in drainage and runoff waters for the two crops.

Table 3
Analysis of variance of annual Nitrate load in runoff and drainage waters : Comparison of the two years

Source	dl	NO_3-N
Mean	1	**
C	1	**
T(C)	6	ns
Error	8	
Contrast		
Corn : 100 vs 111		ns
Corn : 100 vs 120		ns
Corn : 100 vs 102		ns
Hay : 100 vs 111		ns
Hay : 100 vs 120		ns
Hay : 100 vs 102		ns

**, * Significant at 0.01 and 0.10 levels respectively
ns Not significant at 0.10 level

Table 4
Analysis of variance of seasonal Nitrate load in runoff and drainage water during the second year

	Seasonal load (Kg NO_3-N ha^{-1} yr^{-1})		
	Fall vs Winter	Winter vs Spring	Spring vs Summer
Mean	**	**	**
C	**	**	ns
T(C)	ns	ns	**
Corn : 100 vs 111	ns	ns	ns
Corn : 100 vs 120	ns	ns	ns
Corn : 100 vs 102	ns	ns	**
Hay : 100 vs 111	ns	ns	ns
Hay : 100 vs 120	ns	ns	ns
Hay : 100 vs 102	ns	ns	*

**, * Significant at 0.01 and 0.10 levels respectively
ns Not significant at 0.10 level

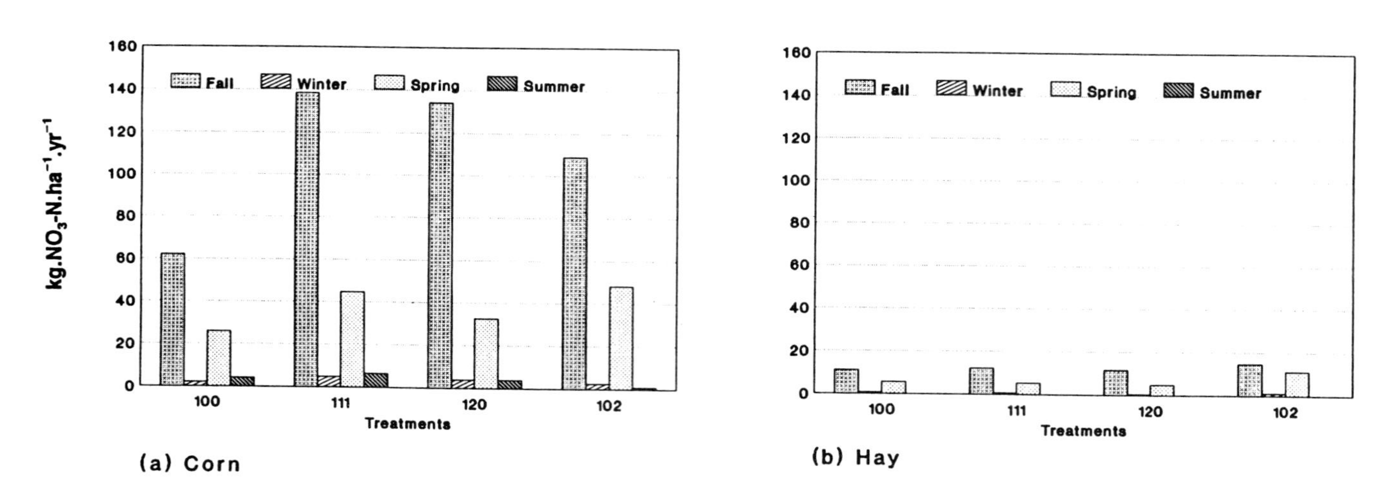

Figure 2. Seasonal Nitrate loads in drainage and runoff waters for the two crops.

Generally, the Nitrate load increased from winter to spring for all treatments on both crops and was again much larger for corn. Nitrate load generally decreases from spring to summer, the greater decrease occurring for the treatment where all manure was fall-applied, no matter the crop. In fact loads dropped from 47.98 to 1.08 kg NO_3-N ha^{-1} for corn and from 11.27 to 0.02 kg NO_3-N ha^{-1} for hay.

Therefore, the Nitrate load was similar for both crops and most treatments. For fall-winter and winter-spring comparisons, seasonal load differences were much greater for a given corn treatment than for hay. By contrast the spring-summer comparison of the difference for the corn receiving all the manure in the fall was much larger than any of the others, no matter the crop.

CONCLUSION

Spreading high rates of hog manure either in the fall or spring in addition to mineral fertilizer for either corn or hay greatly increased Nitrate pollution. On corn, fall-spreading of manure increased the Nitrate load by 70 % over the fertilizer-only treatment, by 86 % when spring-applied, and by 106 % with a split-application of half in each season. The main effects for hay occurred with the fall application which increased the Nitrate load by 58 %. The Nitrate load increased for secondyear corn no matter the treatment, in contrast to hay where loads decreased in the second year.

Seasonal Nitrate loads followed a similar pattern for the two crops. Load differences between fall - winter, which in general decreased, and winter - spring comparisons, which in general increased, were much higher for corn thanfor hay, no matter the treatment. By contrast, for the treatment having received the full manure rate in the fall, the decrease from spring to summer was much greater than for the others. This was true for both crops.

ACKNOWLEDGEMENTS

Financial support for this study was provided by the Ministère de l'Environnement du Québec.

REFERENCES

1. MAPAQ (1989) : L'industrie bio-alimentaire en 1988-Performances économiques et perspectives pour 1989. Ministère de l'Agriculture, des Pêcheries et de l'Alimentation. Direction de la statistique et de la conjoncture, Sainte-Foy, Québec, 87 p.

2. MAPAQ (1981) : Nourrir le Québec-Perspectives de développement du secteur de l'agriculture, des pêches et de l'alimentation pour les années '80. Ministère de l'Agriculture, des Pêcheries et de l'Alimentation. Direction générale de la planification et des études économiques, Sainte-Foy, Québec, 261 p.

3. Theriault, J. (1983) : Inventaire du cheptel Québécois. Dans : Manuel de gestion agricole des fumiers. Ministère de l'Agriculture, des Pêcheries et de l'Alimentation du Québec, Sainte-Foy, Qé, 21-24.

4. Lalonde, Girouard, Letendre et Al. (1984) : Etude de faisabilité d'un programme de transport du lisier dans le bassin hydrographique de la rivière l'Assomption.

Rapport final. Ministère de l'Environnement du Québec, Direction du milieu agricole et du contrôle des pesticides, Sainte-Foy, 108 p + annexes.
5. Simoneau, M. and Grimard, Y. (1989) : Qualité des eaux de la rivière L'Assomption 1976-1987. Ministère de l'environnement du Québec, Direction de la qualité des cours d'eau, Sainte-Foy, Qué, Envirodoq 890318, 234 p + annexes.
6. Primeau, S. and Grimard, Y. (1990a). Rivière Yamaska 1975-1988. Volume 1 : Description du bassin et qualité du milieu aquatique. Ministère de l'environnement du Québec, Direction de la qualité des cours d'eau, Sainte-Foy, Qué, QE--66-1, Envirodon 900060, 136 p + annexes.
7. Primeau, S. and Grimard, Y. (1990b) : Rivière Yamaska 1975-1988. Volume 2 : Résultats complémentaires sur la qualité des eaux. Ministère de l'environnement du Québec, Direction de la qualité des cours d'eau, Sainte-Foy, Qué, QE--66-1, Envirodon 900060, 150 p.
8. Simoneau, M. (1991) : Qualité des eaux du bassin de la rivière Chaudière 1976-1988. Ministère de l'environnement du Québec, Direction de la qualité des cours d'eau, Sainte-Foy, Qué, QE-66-1, Envirodoq 910053, 185 p.
9. W.Q.I. (1987) : Nitrates : A question of time, Water Quality Information (1) : 24-28.
10. Steenvoorden, Jopp H. A. M. (1989). Nitrogen cycling in manure and soils : Crop utilization and environmental losses. In : Dairy manure management proc., Dairy manure management Symp., Syracuse, N.Y., NRAES-31, pp 89-102.
11. André, P., and Dubois de La Sablonière, F. (1983) : Elevage intensif et qualité des eaux souterraines dans un département breton. Techniques Sciences et Méthodes, 5:251-258.
12. MENVIQ (1989) : Eaux-Détermination des nitrates et des nitrites, méthode colorimétrique automatisme avec le sulfate d'hydrazine et le N.E.D. Ministère de l'Environnement du Québec. Menviq 89.07/303, n° 3 1.1
13. Schuman, G. E., Stanley, M.A. and Knudsen, D. (1973) : Automated total nitrogen analysis of soil and plant samples. In : Soil Science Society of America. Proceedings, 37(3):480-481.
14. Mckeague, J. A. (1977) : Manuel des méthodes d'échantillonnage et d'analyse des sols. Soil Research Institute, Agriculture Canada, Ottawa, Ont., 169-171.
15. Rowell, J.G. and Walters, D.E. (1976) : Analysing data with repeated observations on each experimental unit. J. Agric. Sci., Cambridge, 87:423-432.

Farm Land Erosion: In Temperate Plains Environment and Hills
S. Wicherek (Editor)
1993 Elsevier Science Publishers B.V.

Tillage and Crop Residue Management Practices for Soil Erosion Control

J.F. Power, J.E. Gilley, W.W. Wilhelm, L.N. Mielke, and J.W. Doran

Corresponding author : U.S. Department of Agriculture - Agricultural Research Service
University of Nebraska, Lincoln NE 68583, U.S.A.

SUMMARY

Several field experiments have been conducted in Nebraska (400 to 750 mm average annual precipitation) in which we studied the effects of tillage and crop residue management practices on soil properties and the susceptibility of soils to wind and water erosion. Generally we found that by using reduced and no-till systems, we reduced soil disturbance, slowed short-term soil microbiological activity and crop residue decomposition, and increased water storage. These changes subsequently increased crop yields and residue production, reducing soil erosion potential. This increased crop residue production created a somewhat self-perpetuating system to again increase water storage and enhance yields of the next crop. Use of winter cover crops increased crop residue cover, the quantity of organic material returned, and reduced soil erosion potential. Radar backscatter technology is being developed to quantify crop residue cover by remote sensing. Coupled with new generation process-based erosion prediction models that are being developed, this offers possibilities of greatly improving ability to predict runoff and erosion on a watershed basis.

RESUME

Plusieurs expériences de terrain, récemment conduites dans le Nebraska (400 à 750 mm de précipitations annuelles) ont permis d'étudier les effets du labour et des techniques de gestion des résidus de culture sur les propriétés des sols, ainsi que la sensibilité à l'érosion des sols limoneux et des sols morainiques. En règle générale, le retournement par le labour se traduit par une augmentation de l'activité microbienne du sol à court terme et cela a pour effet un accroissement de la décomposition des résidus de culture et une oxydation de la matière organique. Les caractéristiques hydriques du sol sont également modifiées. Par rapport aux sols non cultivés, ceux qui le sont présentent une résistance hydrique moindre et une plus grande sensibilité à l'érosion. L'accroissement des quantités de résidus de surface sur un sol non cultivé, réduit les pertes par évaporation et augmente le rendement dans un environnement peu humide. Ceci engendre une augmentation des résidus de culture pour la saison suivante. On aboutit donc à un système qui s'auto-entretient.

L'utilisation de couvertures végétales d'hiver augmente la quantité de résidus de récolte et la quantité de matière organique réintégrant le sol. Toutefois au printemps, la diminution de la quantité d'eau contenue dans le sol, freine la croissance de la couverture végétale.

INTRODUCTION

Soil erosion by both wind and water is a major problem associated with crop production in Nebraska and surrounding states. With variable topography and a sub-humid to semi-arid continental climate (750 to 400 mm annual precipitation), disturbance of soils by tillage often results in reduction of crop residue cover and subsequent loss of soil by erosion. Not only does soil erosion occur, but productivity potential of these mollisols can also be reduced by loss of soil organic matter through both erosion and increased oxidation (1 - 2). Research has documented that bare tillage practices utilized on most of these soils until recent decades has already resulted in a loss of 50 % or more of the organic C and N originally present in the surface layer of these soils (3).

Early research on soil erosion control in Nebraska pioneered the development of reduced mulch tillage practices (4). An added benefit from reduced tillage systems was improved soil water conservation resulting from the combined effects of weed control, reduced water evaporation, and greater water infiltration (5 - 6). These investigators also showed that with prolonged periods between rewetting (precipitation), total evaporation losses from residue covered soils eventually equalled those from bare soil. Other scientists have verified that these phenomena do indeed occur under field situations (7).

The altered soil water and temperature regimes resulting from different crop residue management practices affect soil biology, and subsequently microbial transformations, and the availability of nitrogen (N) and other nutrients (8). The purpose of this paper is to summarize research in Nebraska that demonstrates to the effects of crop residue management practices on soil property changes and subsequent potentials for soil erosion control, water conservation, and soil productivity.

METHODS

Two field experiments on winter wheat (*Triticum aestivum* L). - fallow systems (400 mm precipitation), established in 1969 at Sidney, Nebraska, were evaluated in regard to effects of no-till, stubble mulch tillage, and conventional bare tillage (plowed) fallow on soil physical, chemical, and biological properties and wheat production. One experiment also contained undisturbed native grass (*Agropyron, Stipa, Bouteloua* sp) plots as controls.These experiments are continuing at this time (9).Soil properties evaluated after varying years of cultivation included soil bulk density, infiltration rates, hydraulic conductivity, air permeability, percent water-filled pore space (WFPS), soil microbial biomass, microbial populations, respiration, total N, N mineralization and immobilization, denitrification, soil nitrate distribution, soil organic C water content, weed growth, wheat above-ground and root biomass, wheat straw, and grain production (10 - 11 - 12 - 13 - 14 - 15 - 16 - 17).

In some years N-isotopes were used as tracers to follow N transformations in more detail. Soils were Alliance (silty, mixed, mesic Aridic Argiustolls) and Duroc (fine-silty, mixed, mesic Pachic Haplustolls) silt loams.

Other field experiments were conducted in eastern Nebraska (700 mm precipitation) using various rotations of corn (*Zea mays* L.), soybean [*Glycine max* L. (Merr.)], and grain sorghum (*Sorghum bicolor* Moench), grown continuously or in different rotations. In some instances, the use of hairy vetch (*Vicia villosa* L.)was evaluated as a winter cover crop (18 - 19 - 20 - 21 - 22). Techniques and measurements were usually similar to those described above for the wheat-fallow experiments at Sidney, although all measurements enumerated were not made for all experiments. Soils were Crete-Butler (fine, montmorillonitic, mesic Pachic Argiustoll-Abruptic Argiaquolls) and Sharpsburg (fine, montmorillonitic, mesic typic Argiaquolls) silty clay loams. Treatments included several tillage practices with variable quantities of crop residues left on the soil surface ranging from little or none (plow, disk, chisel) to those that left all residues on the soil surface (no-till). In one experiment, up to 150 % of the quantity of residue produced by the previous crop (corn or soybean) was applied to the soil surface of no-till plots. Other experiments were conducted to evaluate the effects of various types and quantities of crop residues on hydraulic resistance and runoff (23 - 24), and remote sensing of crop residue cover using radar backscatter coefficients (25). For most experiments, data were analyzed by analysis of variance and least significant differences ($P < 0.05$) or Duncan's Multiple Range Test was used to separate means.

RESULTS

Results from the tillage experiments on winter wheat-fallow rotations at Sidney, Nebraska showed that, compared to plowing (bare tillage), no-till systems resulted in a cooler (3 to 6 °C), more moist soil (Table 1). No-till soils also had greater hydraulic conductivity, water infiltration rate, soil resistance and water-filled pore space. After 19 years total N and organic C was also greater in the upper 75 mm of no-till than plowed soil. Stubble mulching usually resulted in soil properties intermediate between those for plowing and no-tillage.

Table 1.
Effect of fallow tillage practices of Alliance silt loam at Sidney, Nebraska, on several soil properties, May 1980 (12 - 14).

	0-75 mm				75-150 mm			
	NT†	ST†	P†	$LSD_{0.05}$	NT	ST	P	$LSD_{0.05}$
Bulk density, mg m^{-3}	1.29	1.25	1.25	NS	1.30	1.38	1.31	NS
Soil water, V/V	0.28	0.24	0.22	0.05	0.30	0.28	0.27	NS
Water-filled pore, %	54	45	43	5	65	62	56	5
Hydraulic conductivity mm ha^{-1}	32.0	33.0	19.4	NS	21.9	10.1	15.4	10.0
Air permeability, pm^2	2.8	4.1	1.6	NS	--	--	--	--
Water infiltrationn mm ha^{-1}	62	53	51	8	--	--	--	--
Total N, %	.124	.114	.104	.012	.103	.101	.104	NS
Organic C, %	1.08	1.00	0.85	0.11	0.77	0.75	0.83	NS

† NT = No-till ; ST = Subtill ; P = Plow

Table 2.
Tillage effects on microbial populations, microbial biomass, and potentially mineralizable N (8)

Soil variable	Ratio : No-till/plow		
	0-75 mm	75-150 mm	150-300 mm
Total aerobes	1.35	0.66	0.82
Fungi	1.35	0.55	0.69
Aerobic bacteria	1.41	0.68	0.76
NH_4 oxidizers	1.25	0.55	--
NO_2 oxidizers	1.58	0.75	--
Facultative anaerobes	1.31	0.96	0.94
Obligate anaerobes	1.27	1.05	1.01
Denitrifiers	7.31	1.77	--
PMN†	1.37	0.98	0.93
Microbial biomass	1.54	0.98	1.00

† PMN = Potentially mineralizable N.

As a consequence of these tillage-induced changes in soil properties, soil microbiological populations and activity were altered by tillage. In the surface 75 mm of no-till soil, when compared to that for the bare (plowed) soil, populations and biomass of all classes of soil microorganisms were increased (Tab. 2) with greatest increases in denitrifiers. However below 75 mm microbial populations and biomass were usually less in no-till than in plowed soil. Consequently CO_2 and N_2O evolution from the soil surface was two to three times greater for no-till than for plowed soils (data not shown).

These differences in microbial populations and activity reflect the fact that crop residues, the primary source of respirable C for most soil microorganisms, was located at the soil surface for no-till but buried in the plowed soil. Percent WFPS was more often near 60 % in no-till soils, (Tab. 1), the value Linn and Doran (11) concluded was near optimum for aerobic microbial activity (Fig. 1).

As a result of these tillage-induced changes in soil environment and microbial activity, wheat growth and N uptake for no-till equalled or exceeded that for plowed fallow as long as weeds were controlled. Consequently, because of the greater crop residue cover and greater microbial activity (especially fungal activity) in the surface of no-till soil as compared to that of plowed soil, no-till would provide much greater protection from both wind and water erosion as a consequence of greater ground cover and better soil aggregation. However after several years of no-till fallow, control of winter annual grasses (*Bromus tectorum* L.) became more difficult.

In the no-till experiment on the Crete-Butler silty clay loam near Lincoln, both corn and soybean were produced continuously for five years on plots that received 0, 50, 100, and 150 % of the quantity of residues left by the previous crop (averaged 0 to 5.1 Mg ha^{-1} for this range of treatments). Not only did increased crop residue cover reduce soil erodibility, but it also increased available soil water at planting and reduced surface soil temperatures during the growing season.

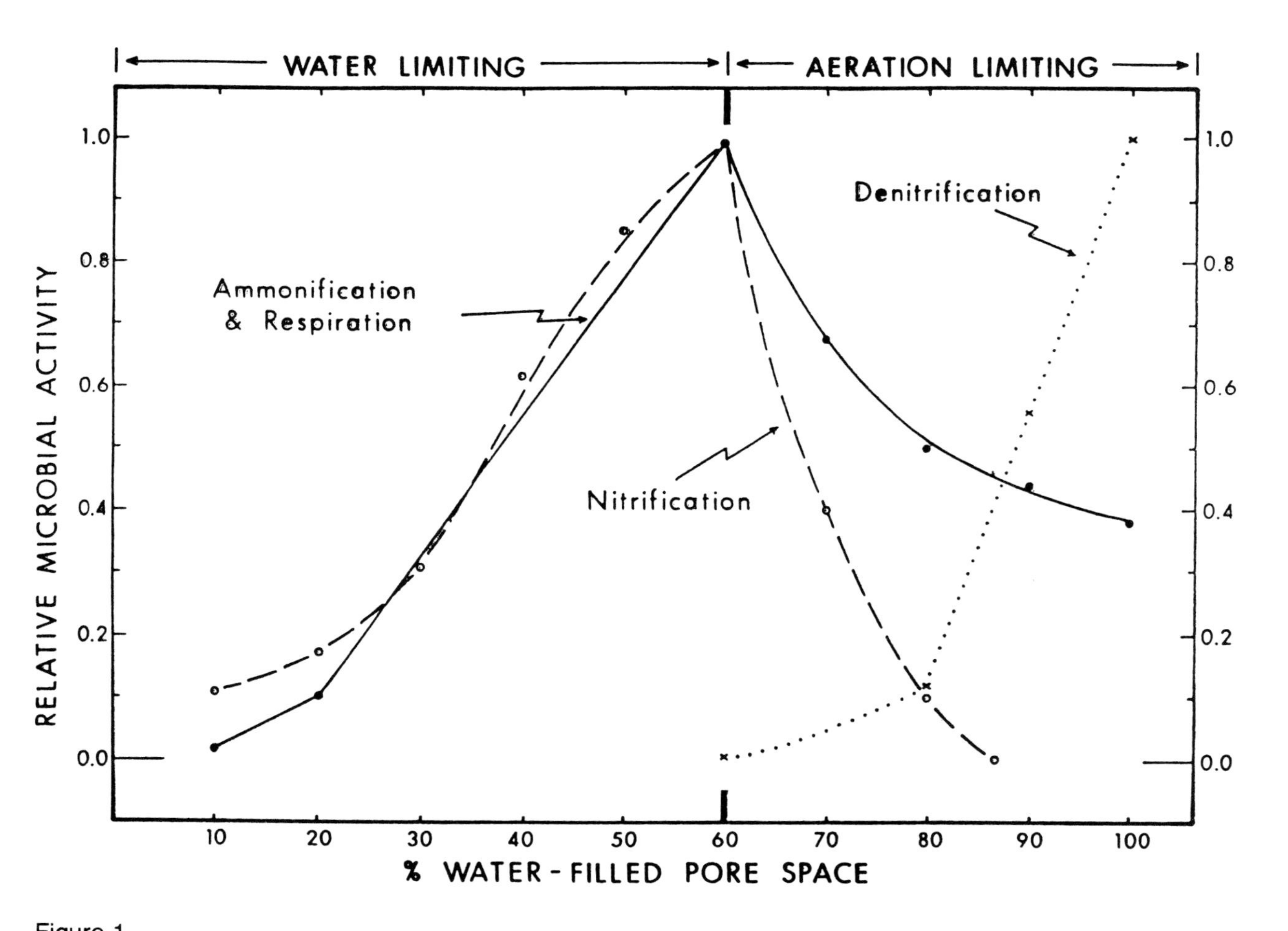

Figure 1.
Effect of percent water-filled pore space in soil microbial activity (11).

This improved environment for microbial and biological activity resulted in increased crop dry matter production with increasing residue rates (Tab. 3).

Table 3.
Regression equation between crop residues returned to the soil surface (X in kg ha^{-1}) with available soil water at planting, grain yields, and stover or straw production (Y) for corn and soybean (15 - 19).

Y	Corn	Soybean
Soil water (mm)	Y = 173 + 5X (r^2 = 0.84)	Y = 175 + 8X (r^2 = 0.71)
Grain (Mg ha^{-1})	Y = 2.91 + 0.12X (r^2 = 0.81)	Y = 1.53 + 0.09X (r^2 = 0.84)
Stover (Straw) (Mg ha^{-1})	Y = 4.34 + 0.29X (r^2 = 0.86)	Y = 2.73 + 0.30X (r^2 = 0.92)

For corn, each Mg ha^{-1} of crop residue on the soil surface resulted in an increase of 5 mm stored soil water, 120 kg ha^{-1} grain, and 290 kg ha^{-1} stover. For soybean, respective values were 8, 90, and 300. Similar effects of crop residues on mineralization and uptake of soil N were observed (data not shown).

Results from this experiment suggest that by returning more crop residues to the soil surface, not only is more grain produced, but also a greater quantity of crop residues is produced for surface soil protection (290 and 300 kg increases per Mg of corn and soybean residues, respectively). Thus, a somewhat self-perpetuating system is created. Although this project was terminated after harvest of the 1983 crop, plots have been maintained but only residues produced by the previous crop were returned (the 100 % treatment). Residual effects of the original treatments on have continued through harvest of the 1991 with crop growth responses of about the same magnitude as that measured during the first 5 years. The increasing residue rate treatment applied the first 5 years increased soil organic C and N contents 10 to 15 % in the upper 75 mm of soil, and these differences are also being maintained through 1991. Consequently these results strongly suggest that by keeping crop residues on the soil surface for soil erosion control, we also benefit in terms of improved water conservation, crop yields, and soil quality.

Several other experiments have been or are being conducted in eastern Nebraska in which various combinations of tillage and cropping systems are being evaluated. Several of these involve studies on the effects of winter cover crops on soil erodibility, water use, and corn yields. Gilley et al. (23) developed equations that related surface cover from crop residues to runoff, sediment concentration, and soil loss, and they also developed equations relating surface cover to residue mass. Using these relationships, Gilley and coworkers (24) evaluated 18 legume and non-legume cover crops planted at different dates for their potential to provide surface cover and reduce soil erosion. Of all legumes evaluated, they found hairy vetch usually grew most rapidly and provided most cover in a given time period. Rye (*Secale cereale* L.) also generally grew rapidly. With July plantings, some warm season species such as cowpea (*Vigna unguiculata* L.) and flat pea (*Lathyrus sylvestris* L.) also performed well.

Collectively these and other experiments emphasize the importance of maintaining crop residues on the soil surface at all times to control soil erosion. Greater crop residue cover reduces raindrop impact and associated soil detachment

(26), and also increases hydraulic resistance by increasing surface roughness. Gilley et al. (27) developed quantitative relationships between surface residue cover and Darcy-Weisbach Roughness Coefficients for a range of Reynolds numbers. He evaluated these relationships for residues from eight different crop species, varying greatly in size and density. Typical examples of the relationships developed are shown in Tab. 4.

Table 4.
Regression Equations for Darcy-Weisbach Roughness Coefficient versus Percent Cover and Reynolds Number for Reynolds Number Less than 20,000 (15 - 19).

Residue type (1)	Regression Coefficients[a] *a* (2)	*b* (3)	*c* (4)	Coefficient of determination, r^2 (5)
Corn	6.30×10^{-2}	1.53	2.34×10^{-1}	0.911
Cotton	8.88×10^{-2}	1.02	7.88×10^{-2}	0.731
Peanut	2.61×10^{-1}	1.56	5.06×10^{-1}	0.924
Pine Needles	8.71×10^{-5}	3.63	6.52×10^{-1}	0.874
Sorghum	5.24	7.96×10^{-1}	4.55×10^{-1}	0.960
Soybeans	9.28×10^{-2}	2.84	1.02	0.919
Sunflower	1.66	8.87×10^{-1}	3.51×10^{-1}	0.916
Wheat	2.98×10^{-4}	3.27	6.28×10^{-1}	0.938
All residue types Combined	1.27×10^{-1}	1.55	3.88×10^{-1}	0.648

[a]Regression coefficients *a*, *b* and *c* are used in the equation:

$$f = a\,(\text{percent cover})^{b}/(\text{Rn})^{c}$$

These algorithms are being integrated into new-generation soil erosion prediction equations such as the Water Erosion Prediction Program (WEPP). Techniques are also being developed to use remote sensing to quantify crop residue cover, using radar backscatter coefficients 25). This technology would greatly improve erosion and runoff predictions on a watershed basis.

4. CONCLUSIONS

Crop residue management is an extremely important practice by which a producer can influence the soil environment, water conservation, nutrient cycling, and ultimately, crop growth and yield. By selecting crop residue management practices that enhance productivity, more crop residues are produced during that growing season, thereby increasing the opportunity for this process to repeat during the next cropping cycle. Thus, a somewhat self-perpetuating cycle is established, thereby greatly advancing the opportunity to control soil erosion and soil degradation.

In general, compared to plowed (bare) soil no-till practices result in a cooler, more humid soil environment. In well-drained soils in sub-humid and semi-arid regions, the soil environment (water, temperature, aeration, and substrate regimes) is usually more favorable for biological activity for more days of the year in no-till than in bare soil. Water and aeration regimes can be defined by calculating percent water-filled pore space. With no-till, values are closer to the optimum value of 60 % more often than in bare soil. Consequently more N is mineralized, more water is available, and there is opportunity for enhanced crop growth. In our studies with corn and soybean, we found that soil water storage increased 5 to 8 mm ha^{-1} per Mg ha^{-1} crop residues on the soil surface, resulting in grain yields increases of 90 to 120 kg ha^{-1}, and stover (straw) production increases of 290 to 300 kg ha^{-1}.

Procedures are being developed to remotely sense percent crop residue cover. Integrating such information over a large area, used with the new generation water erosion prediction models that quantify the hydraulic resistance of surface residues, we can provide greatly improved prediction of runoff and erosion on a watershed basis. With such technology, we can ensure adequate surface cover by use of cover crops or different crop rotations, in addition to use of no-till or reduced tillage practices.

REFERENCES

1. Larson, W.E, Pierce, F.J. and Dowdy, R.H. (1983) : The th reat of soil erosion to long-term crop production. Science, 458-465.
2. Williams, J.R. and Renard, K.G. (1984) : Assessments of soil erosion and crop productivity with process models (EPIC). In : R.F. Follett (ed.) Soil erosion and crop productivity, American Society of Agronomy, Madison, Wisconsin, U.S.A., 67-103.
3. Haas, H.J., Evans, C.E. and Miles, E.F. (1957) : Nitrogen and carbon changes in Great Plains soils as influenced by cropping and soil treatments. U.S. Department of Agriculture, Technical Bulletin 1164, 1-157.
4. Duley, F.L. and Russel, J.C. (1939) : The use of crop residues for soil and moisture conservation. Journal of American Society of Agronomy, 31, 703-709.
5. Black, A.L. and Power, J.F. (1965) : Effect of chemical and mechanical fallow methods on moisture storage, wheat yields, and soil erodibility. Soil Science Society of America Proceedings, 29, 465-468.
6. Bond, J.J. and Willis, W.O. (1969) : Soil water evaporation : Surface residue rate and placement effects. Soil Science Society of America Proceedings, 33, 445-448.
7. Tanaka, D.L. (1985) : Chemical and stubble-mulch fallow influences on seasonal water contents. Soil Science Society of America Journal, 49, 728-733.
8. Power, J.F. and Doran, J.W. (1988) : Role of crop residue management in nitrogen cycling and use. Chap. 6, p. 101-113. In : W. L. Hargrove, B. Ellis, T. Cavalieri, R. C. Johnson, and R. Reginato (eds.) Cropping Strategies for Efficient Use of Water and Nitrogen. Special Publication 51, ASA-CSSA-SSSA, Madison, WI.

9. Fenster, C.R. and Peterson, G.A. (1979) : Effects of no-tillage fallow as compared to conventional tillage in a wheat-fallow system. Nebraska Agricultural Experiment Station, Research Bulletin, 289, 1-28.
10. Doran, J.W. (1980) : Soil microbial and biochemical changes associated with reduced tillage. Soil Science Society of America Journal, 44, 765-771.
11. Linn, D.M. and Doran, J.W. (1984) : Aerobic and anaerobic microbial populations in no-till and plowed soils. Soil Science Society of America Journal, 48, 794-799.
12. Broder, M.W., Doran, J.W., Peterson, G.A. and Fenster, C.R. (1984) : Fallow tillage influence on spring populations of soil nitrifiers, denitrifiers, and available nitrogen. Soil Science Society of America Journal, 48, 1060-1067.
13. Lamb, J.A., Doran, J.W. and Peterson, G.A. (1987) : Nonsymbiotic dinitrogen fixation in no-till and conventional wheat-fallow systems. Soil Science Society of America Journal, 51, 356-361.
14. Mielke, L.N., Doran, J.W. and Richards, K.A. (1986) : Physical environment near the surface of plowed and no-tilled soils. Soil Tillage Research, 7, 355-366.
15. Power, J.F., Wilhelm, W.W. and Doran, J.W. (1986a) : Recovery of fertilizer nitrogen by wheat as affected by fallow method. Soil Science Society of America Journal, 50, 1499-1503.
16. Wilhelm, W.W., Mielke, L.N. and Fenster, C.R. (1982) : Root development of winter wheat as related to tillage in western Nebraska. Agronomy Journal, 74, 85- 88.
17. Wilhelm, W.W., Bouzerzour, H. and Power, J. F. (1989) : Soil disturbance-residue management effect on winter wheat growth and yield. Agronomy Journal, 81, 581-588.
18. Aulakh, M.S., Doran, J.W., Walters, D.T., Mosier, A.R. and Francis, D.D. (1991) : Influence of crop residue type and placement on denitrification and mineralization of N and C. Soil Science Society of America Journal, 55, 1020-1025.
19. Power, J.F., Wilhelm, W.W. and Doran, J.W. (1986b) : Crop residue effects on soil environment and dryland maize and soya bean production. Soil Tillage Research, 8, 101-111.
20. Power, J.F. (1991 : Growth, N accumulation and water use of legume cover crop in a semiarid environment. Soil Science Society of America Journal, 55, 1659-1663.
21. Wilhelm, W.W., Schepers, J.S., Mielke, L.N., Doran, J.W., Ellis, J. R. and Stroup, W.W. (1987) : Dryland maize development and yield resulting from tillage and nitrogen fertilization practices. Soil Tillage Research, 10, 167-179.
22. Zachariassen, J.A. and Power, J.F. (1991) : Growth rate and water use by legumes species at three soil temperatures. Agronomy Journal, 83, 408-413.
23. Gilley, J.E., Finkner, S.C. and Varvel, G.E. (1986) : Runoff and erosion as affected by sorghum and soybean residue. Transactions of the American Society of Agricultural Engineers, 20 (6), 1605-1610.
24. Gilley, J.E., Power, J.F., Reznicek, P.J. and Finkner, S.C. (1989) : Surface cover provided by selected legumes. Applied Engineering in Agriculture, 5 (3), 379-385.
25. Narayanan, R.M., Mielke, L.N. and Dalton, J.P. (1992) : Crop residue cover estimation using radar techniques. Applied Engineering in Agriculture Journal. (In review).
26. Hussain, S.K., Mielke, L.N. and Skopp, J. (1988) : Detachment of soil as affected by fertility management and crop rotations. Soil Science Society of America Journal, 52, 1463-1468.

27. Gilley, J.E., Kottwitz, E.R. and Wieman, G.A. (1991) : Roughness coefficientsfor selected residue materials. Journal of Irrigation and Drainage Engineering, American Society of Civil Engineers, 117 (4), 503-514.

Farm Land Erosion: In Temperate Plains Environment and Hills
S. Wicherek (Editor)
1993 Elsevier Science Publishers B.V.

Influences agro-pastorales sur les caractéristiques physico-chimiques de la rivière l'Assomption (Canada)

C. Pronovost et M. Bouchard

Département de géographie, Université du Québec à Montréal, C.P. 8888, Succ. A, Montréal, Que. H3C 3P8, Canada.

RESUME

Un programme d'échantillonnage hebdomadaire, du printemps à l'automne, de la rivière l'Assomption et ses principaux affluents, dans 14 sites d'amont en aval, a été mené afin d'évaluer l'influence des activités agricoles, urbaines et industrielles sur les teneurs ioniques. L'évolution saisonnière des teneurs était liée à plusieurs facteurs combinés: hausse des températures, lessivage des produits disponibles, dilution par les eaux atmosphériques, activités anthropiques. Si dans la plupart des cas les teneurs en Ca et Mg s'expliquent par la géologie et l'état hydrique du bassin, les eaux résiduaires municipales et industrielles contribuent également à hausser les teneurs en période d'étiage. Il en est de même pour Na, dont la variabilité saisonnière est plus forte dans les milieux plus anthropisés. Le K se caractérise par des pics de courte durée dont plusieurs sont liés à des variables d'activités agricoles (lessivage des fertilisants et des fumiers). Quoique la pollution agricole diffuse maintienne des valeurs élevées en PO_4 dans un sous-bassin où domine l'élevage porcin, les pollutions ponctuelles habituellement révélées en période d'étiage y sont plutôt rares.

SUMMARY

A sampling program on a weekly basis of the l'Assomption river and its main tributaries, in 14 sites from upstream to downstream, was aimed to verify the influence of agriculture, urbanisation and industrialisation on the chemical content of the river. The seasonnal variations in ionic content depended on many combined factors : rise in temperature, flushing of available products, dilution by atmospheric waters, landuse practices. In many cases the content in Ca and Mg can be explained by the geology and moisture status of the watershed, but the influence of municipal and industrial effluents is evident during periods of low flow. The same applies for Na which is more erratic in urban areas. Farming practices produce peaks of K due to flushing of fertilizers and manure. Localised peaks of pollution by PO_4 in the pigsty area rarely occur, but PO_4 contents remain higher here than in the other sites.

1. INTRODUCTION

Les teneurs, exportations et bilans géochimiques des éléments majeurs ont fait l'objet de nombreuses études dans des petits bassins-versants naturels et boisés. D'autres recherches ont mis en évidence des pertes accélérées en cations lorsqu'il y a déforestation (1). Mais peu d'études de ce type ont porté sur des bassins-versants qui ont été passablement perturbés par l'agriculture, l'industrialisation et l'urbanisation. Nous avons voulu vérifier s'il était possible de déceler et d'expliquer l'impact de ces diverses activités dans un bassin-versant varié au point de vue de l'utilisation du sol, le bassin-versant de la rivière l'Assomption. Nous avons considéré le rôle des activités anthropiques associées à l'agriculture et à l'urbanisation, aussi bien que l'influence des sols, des dépôts superficiels et de la roche.

Une étude de Simoneau et Grimard (2), avait pour objectif de décrire l'évolution de la qualité de l'eau de cette rivière et de ses principaux tributaires entre 1976 et 1987, consécutivement au Programme d'assainissement des eaux amorçé en 1979 sous l'égide du ministère de l'Environnement du Québec, qui incluait entre autres des mesures de gestion des fumiers. A partir de1981, de nouvelles politiques ont été mises de l'avant par le Ministère afin de promouvoir l'amélioration ou la construction de structures d'entreposage des fumiers et d'éliminer les déversements de purin en rivière. Quatre sites avaient été échantillonnés sur une période suffisamment longue, une sur la Ouareau, une sur l'Achigan et deux sur l'Assomption. Les auteurs concluaient à une amélioration de la qualité des eaux depuis 1981.

2. SECTEUR ÉTUDIÉ

Le bassin-versant de la rivière l'Assomption a une superficie de 4 220 km^2 et est drainé par la rivière l'Assomption qui se déverse dans le fleuve Saint-Laurent juste en aval de l'île de Montréal (Fig.1). D'amont en aval on retrouve un plateau disséqué en collines rocheuses du Bouclier canadien, ici constitué principalement d'anorthosites et de gneiss, à couverture de till mince et discontinu et à sols sableux et poreux. Presque inhabitée, cette zone est le domaine de la forêt mixte laurentienne, constituée principalement d'épinettes noires, d'érables à sucre, de bouleaux gris. Vient ensuite la plaine des Basses-terres du St-Laurent, développée dans des calcaires et des dolomies et surmontées d'argiles marines. Ici, une partie importante de la forêt a été défrichée pour faire place à l'agriculture qui occupe maintenant 21 % de la superficie totale du bassin de la rivière l'Assomption, alors que la forêt couvre 68 % du territoire, la zone urbaine 2,7 %, le reste étant constitué de lacs. Alors que la rivière l'Assomption est calme dans la plaine et la sillonne en créant de nombreux méandres, l'escarpement qui limite les deux unités physiographiques provoque plusieurs chutes. La rivière l'Assomption a quatre affluents majeurs : les rivières Noire, Ouareau, Saint-Esprit et l'Achigan. Le bassin compte 490 lacs, concentrés surtout sur la plateau.

Le climat d'ensemble est de type continental humide avec étés chauds et hivers frais. Les précipitations sont bien réparties dans l'année et totalisent en moyenne 96 cm/an dont 18 % sous forme neigeuse. Juillet est le mois le plus chaud avec une température moyenne de 20°C alors que janvier est le mois le plus froid avec une moyenne de -11,8°C. Le couvert neigeux dure de novembre-décembre à avril. La crue printannière débute généralement en mars, atteignant des maxima moyens mensuels de près de 70 m^3/s lors de la fonte des neiges, et se termine en mai. Les étiages sont en juillet et août et en janvier et février, la rivière débitant en moyenne un peu plus de 10 m^3/s.

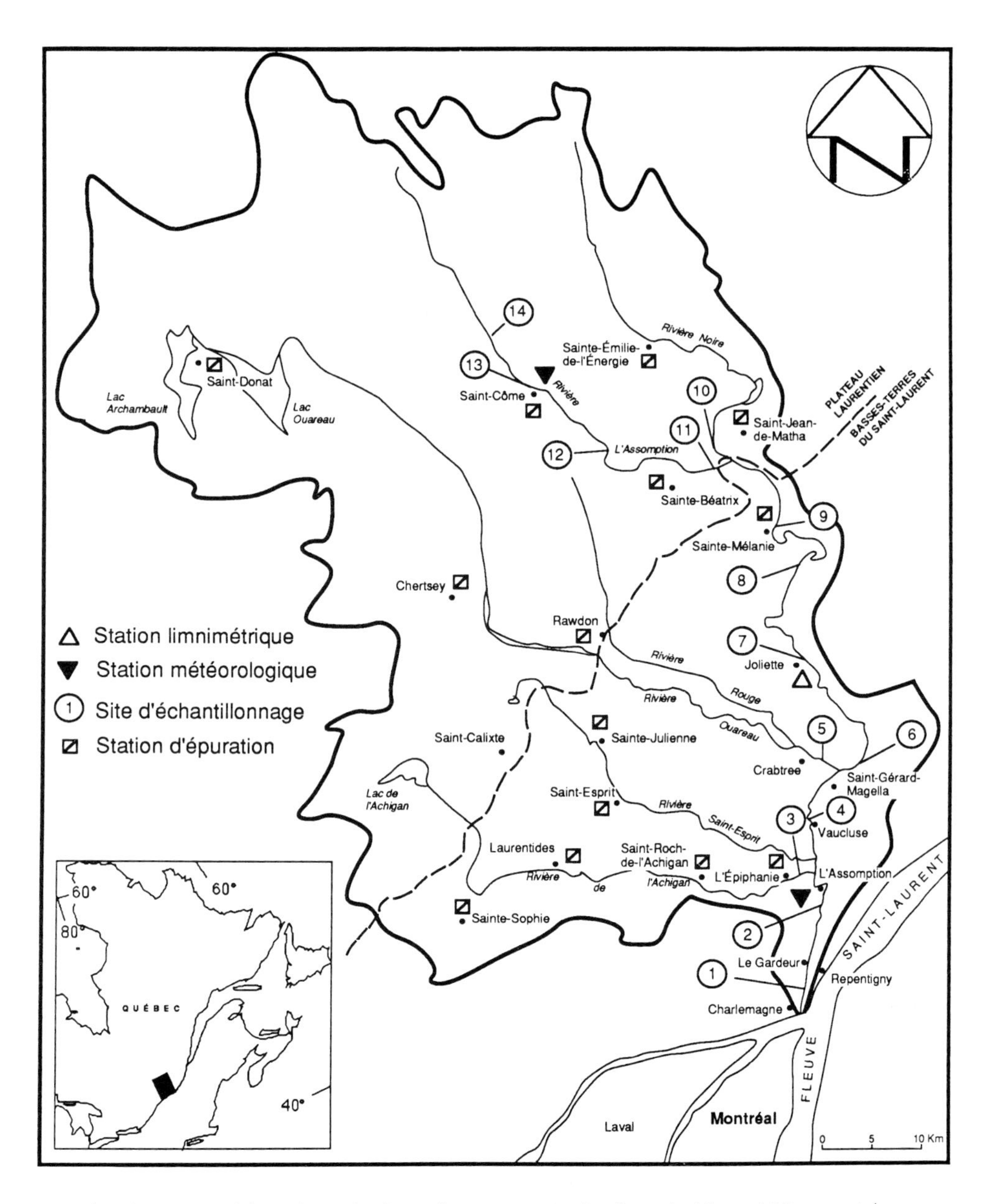

Fig 1 : Localisation du bassin-versant de la rivière l'Assomption.

La partie amont (le plateau dominé par la forêt et les lacs) est surtout consacrée à la villégiature. En été, la population saisonnière vient quadrupler la population permanente du secteur et gonfle à plus de 200 000 habitants. Des problèmes de pollution peuvent résulter des eaux usées brutes des municipalités et des fosses septiques inadéquates des résidences isolées. La population permanente se concentre surtout en aval du bassin, entre Le Gardeur et Joliette ; celle de Joliette était de 133 630 habitants en 1986. En ce qui a trait à l'assainissement urbain, Joliette, la ville la plus importante du bassin, ne possède pas de station d'épuration ; mais 18 des 45 municipalités étaient desservies par de telles stations.

Dans le plateau, les industries, peu nombreuses, concernent surtout l'alimentation et le bois et aucune ne rejète de polluants dans les cours d'eau. Mais à St-Jean-de-Matha, on pratique l'élevage industriel de la volaille et le long de la rivière Jourdain, un affluent de l'Achigan. Selon les calculs de Gangbazo and Buteau (3), la densité animale admissible est dépassée et il y a ici une sur-abondance de fumier. L'activité agricole est importante dans la plaine d'argiles marines Champlain, et elle est dominée par la culture des herbages (et de maïs dans le sous-bassin de l'Achigan); on y retrouve aussi du tabac sur les sols sablonneux. Dans ce secteur, le cheptel est d'environ 100 000 unités animales, majoritairement des porcs, suivis des bovins. L'élevage porcin se pratique de façon intensive le long des rivières l'Achigan et Saint-Esprit donc dans des zones à sols lourds et à faible capacité d'infiltration. Selon Gangbazo and Buteau (3) la production de fumier excède les besoins des cultures dans les bassins de l'Achigan et St-Esprit, mais non dans celui de la Ouareau. Les industries sont majoritairement concentrées dans la portion aval du bassin, surtout à Joliette, soit des industries agro-alimentaires (transformation de produits laitiers), de produits chimiques, métallurgiques et de béton. Une importante usine de papier (papier hygiénique) est localisée à Crabtree, sur la Ouareau. Il y a des abattoirs dans le sous-bassin de l'Achigan.

Il était intéressant de vérifier si ces diverses activités se reflétaient sur la composition chimique de l'exutoire, s'il semblait y avoir un entreposage et un épandage adéquat des fumiers, si les pratiques culturales contrôlaient l'érosion et si les pesticides et engrais étaient utilisés rationnellement, dans le temps et l'espace.

3. MÉTHODES ET ÉCHANTILLONNAGE

Les mesures de débits viennent de la station de jaugeage de Joliette et les hauteurs d'eau précipitées de la station climatique de l'Assomption. Les campagnes d'échantillonnage étaient hebdomadaires et ont débuté lors de la période de dégel de la mi-avril et se sont poursuivies jusqu'à l'automne, soit jusqu'au 1er octobre. Des échantillons d'eau de rivière ont été prélevés sur onze sites d'amont en aval de la rivière l'Assomption, et à l'embouchure de ses trois principaux affluents, les rivières Noire, Ouareau et l'Achigan (Fig.1).

La température et le pH étaient mesurés *in situ* . Les échantillons d'eau de rivière étaient prélevés dans des bouteilles de polyéthylène puis, de retour au laboratoire, filtrés sur filtres Whatman 0,45 microns. Le même jour, les chlorures étaient déterminés par titrage au nitrate mercurique en présence de diphénylcarbazone (4). Ensuite, les échantillons étaient déposés au frais à 5°C jusqu'à analyse (finalisées en moins d'une semaine). Na et K étaient dosés par spectrophotométrie d'absorption atomique avec une flamme air-acétylène, et Ca et Mg avec une flamme N_2O-acétylène. Les anions étaient déterminés selon des méthodes colorimétriques usuelles: les sulfates par la méthode turbidimétrique du sulfate de baryum, modifiée, les nitrates par une méthode modifiée de la réduction du cadmium qui utilise l'acide gentisique, les phosphates réactifs solubles étaient dosés par l'acide ascorbique (4). Seuls les orthophosphates sont réactifs quoique des poly-phosphates et phosphates organiques peuvent être partiellement hydrolysés par les acides réactifs. L'analyse peut donc sous-estimer le phosphate total présent dans l'échantillon. La précision des analyses est estimée ± 0.1ppm pour K^+, Mg^{2+} et NO_3^-, ± 0.5 ppm pour Na^+, ± 1.0 ppm pour SO_4^{2-} et Cl^-.

La pluie était prélevée, après chaque période de pluie, par des techniciens du Ministère de l'Environnement du Québec des stations de l'Assomption et de St-Côme. Les échantillons étaient remisés au frais dans des bouteilles de polyéthylène, jusqu'à analyse, selon les mêmes méthodes que pour l'eau de rivière.

4. RÉSULTATS ET DISCUSSION

4.1. Caractères climatiques de la période d'étude. Chimie de la pluie

Les précipitations d' avril à octobre 1986 se caractérisent par un printemps assez bien arrosé avec maximum de la fin mai au début de juin (Fig.2). Le mois de juillet était le plus sec, alors que août et septembre ont connu de fortes précipitations. Les débits étaient plus forts en avril, à cause de la fonte des neiges. Ils ont par la suite décliné graduellement jusqu'au niveau le plus bas, en juillet (qui est aussi le mois le plus chaud de l'année) et des débits plutôt faibles se sont maintenus jusqu'en septembre. Des orages localisés à la mi-mai et en septembre ont produit des pics de crue assez intenses mais de courte durée. Nous n'avons pas observé un ruissellement de surface de type Horton et estimons que le sous-écoulement apporte une contribution majeure au débit de crue dans ce bassin. Même si on ne peut exclure totalement un écoulement de surface, nos observations suggèrent qu'il ne s'effectue que sur de courtes distances. La microtopographie l'empêche de se produire à une échelle plus grande.

Dans les précipitations, c'est Ca et K qui étaient en moyenne les plus abondants (Tab.1). Les teneurs en Ca, Mg et Na, étaient plus élevées au printemps. K était très variable, étant plus sensible à la contamination par les débris organiques. Les sulfates, très variables, étaient plus abondants à l'Assomption qu'à St-Côme. En moyenne, les pH étaient de 4.6 à l'Assomption et de 4.8 à St-Côme. Les pluies sont donc acides, ce qui s'explique par la dominance des vents du sud-ouest qui amènent des systèmes nuageux qui ont passé au-dessus des Grands-Lacs canadiens et américains et leurs cortèges d'industries sidérurgiques.

4.2. Variations saisonnières et d'amont en aval des teneurs

Dans la rivière l'Assomption et ses principaux affluents, Ca est habituellement le plus abondant des cations, suivi de Na, Mg et K (Fig. 3). Quant aux anions, ce sont les chlorures qui sont les plus abondants en aval de Joliette, suivis des sulfates, des nitrates et des phosphates. Vers l'amont, la proportion des chlorures diminue.

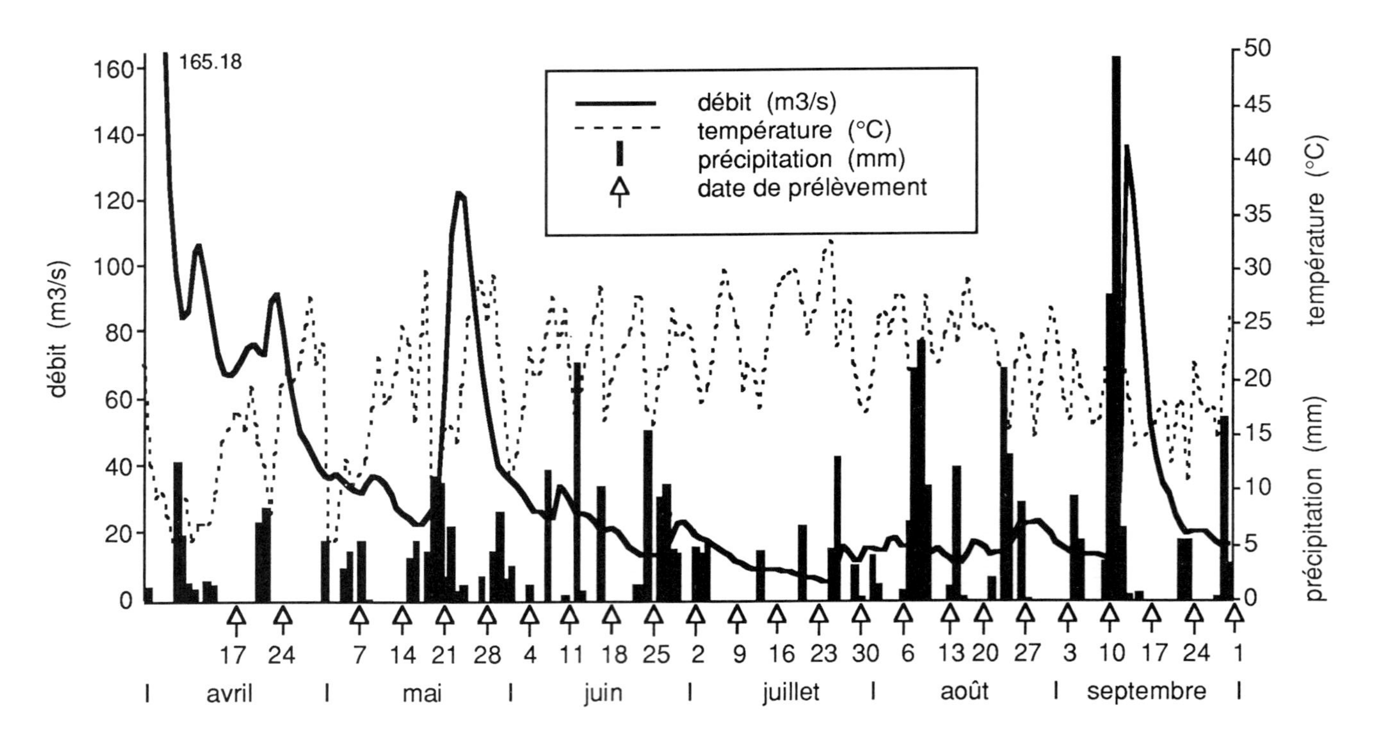

Figure 2. Caractères climatiques de la période d'étude.

	ASSOMPTION						ST-COME					
	pH	SO4	Na	K	Ca	Mg	pH	SO4	Na	K	Ca	Mg
minimum	3.7	0.5	0.03	0.04	0.22	0.00	3.7	0	0.03	0.10	0.20	0.02
maximum	5.7	13.0	2.37	1.42	7.26	0.62	6.8	11.5	1.73	8.22	8.86	0.96
écart-type	0.46	3.72	0.47	0.36	1.70	0.13	0.69	2.83	0.35	1.59	1.65	0.19
variance	0.21	13.8	0.22	0.13	2.88	0.02	0.47	8.04	0.12	2.53	2.74	0.03
moyenne	4.57	4.8	0.32	0.49	1.83	0.14	4.77	2.7	0.34	1.04	1.36	0.17

Tab.1: Composition des précipitations (ppm)

L'examen des fluctuations saisonnières dans chacun des sites, par la suite comparés entre eux, nous permettra d'évaluer l'influence des facteurs naturels et anthropiques sur la chimie de l'eau.

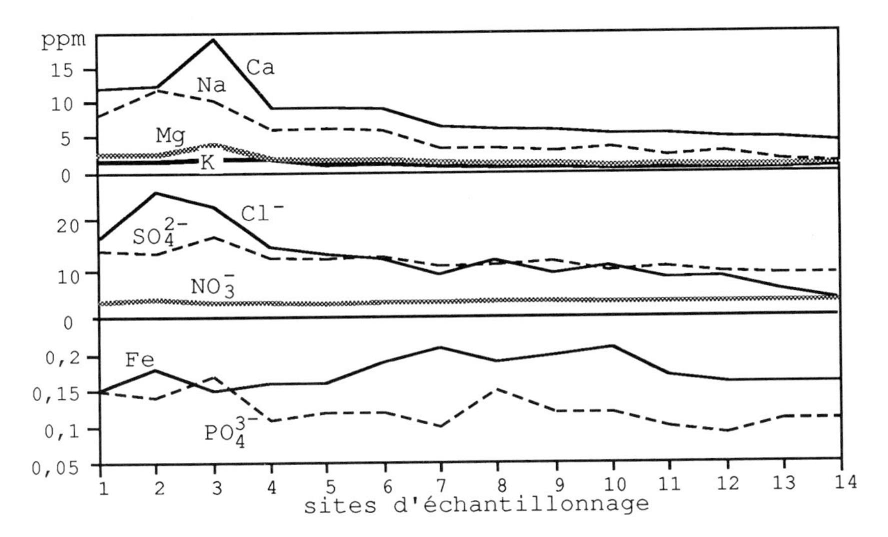

Fig 3 : Moyennes des teneurs ioniques d'amont en aval

On connaît assez bien le modèle général d'acquisition de la composition chimique des eaux dans les bassins-versants naturels. Dans ces bassins, l'écoulement de base, lorsque alimenté par la nappe phréatique, est généralement riche en produits dissous issus de l'altération du substrat, soit en Si. Ca, Mg, Na. Dans ce réservoir où l'eau circule très lentement, les solutions sont mal renouvelées et plus concentrées. En période de crue, les teneurs en ces éléments baissent car le cours d'eau est alimenté aussi par l'écoulement de surface et le sous-écoulement; ces eaux superficielles sont plus diluées, étant donné le temps de résidence plus court, la circulation plus rapide des eaux de sol. Par contre, les constituants qui sont abondants dans les réservoirs de surface (K, NO_3, PO_4) auront des tendances inverses. Il y a même des variations lors d'un événement de crue: les matériaux, provenant de l'altération, qui se sont accumulés avant l'épisode pluvieux seront mobilisés lors de la montée de crue, ce qui amènera une hausse des teneurs. L'état d'humidité du bassin est donc un facteur important. Mais nous verrons que ce modèle peut être perturbé substantiellement et devenir fort complexe dans des bassins-versants plus grands, influencés par l'action anthropique et à utilisation du sol variable.

Dans l'ensemble, les teneurs des cations basiques augmentent d'amont en aval, ce qui dans les bassins naturels et à géologie homogène est attribué à la longueur du parcours de l'écoulement souterrain et donc au temps de résidence plus long. Pour le calcium cette hausse est plus marquée à partir du site 6, ce qui s'explique en partie par la géologie du substratum et en partie par le rejet des eaux usées de Joliette. En effet, en amont du site 7 le sous-bassement est constitué de gneiss et anorthosites, puis, en aval de 7, de dolomies et calcaires. Les plus fortes valeurs de Ca sont dans la rivière l'Achigan, soit le sous-bassin dont la proportion de la superficie drainée dans le calcaire est la plus importante. Mais c'est dans cette rivière aussi que les eaux étaient les plus turbides (érosion et ruissellements importants, typiques des régions à forte vocation agricole). La contribution de cet affluent conserve des valeurs élevées à la rivière l'Assomption, en aval de son embouchure. Un examen des variations temporelles permet de donner des indices sur l'origine géologique probable de Ca; dans ce cas, Ca a généralement une relation inverse avec le débit. Les teneurs étaient faibles lors des forts débits du printemps, alors que les eaux diluées du sous-écoulement contribuaient pour une large part à l'écoulement; elles étaient plus fortes de la mi-juin à la fin juillet alors que la contribution de la nappe phréatique au débit était plus importante (Fig. 4). Cette tendance se retrouve dans tous les sites, quoiqu'elle soit un peu moins nette vers l'amont, où les teneurs étaient plus faibles. Lors d'un événement de crue il y a moins de Ca dans la rivière, étant alimentée aussi par le sous-écoulement, moins riche en Ca que la nappe phréatique (crue du 28 mai). On retrouve aussi cette tendance dans certains sites lors de la crue du 17 sept., mais de façon moins marquée et moins généralisée. Le milieu avait été asséché pendant l'été, et les profils pédologiques ont d'abord dû se recharger avant que l'eau puisse transiter par sous-écoulement (le 17 sept. était la première crue depuis la période plus sèche d'été). Quant aux pics, ils correspondent à un épisode de plus grande sécheresse. Par exemple, l'échantillonnage du 3 septembre, alors qu'il n'avait pas plu au cours des 7 jours précédents. L'origine géologique du Ca dans les cours d'eau est donc grandement probable dans ce bassin, sauf pour un événement isolé : les 16 et 23 juillet, les teneurs dans la rivière Ouareau ont enregistré de très fortes hausses. Comme aucune hausse de cette envergure n'a été notée ailleurs, on pourrait suspecter des déversements des effluents de Crabtree et son usine de papier, un peu en amont.

Les teneurs en magnésium sont nettement plus faibles que celles du calcium mais, elles aussi, augmentent d'amont en aval. L'augmentation est graduelle jusqu'à l'embouchure de la rivière l'Achigan, qui contient plus de Mg. Aussi, le tronçon de la rivière l'Assomption en aval de cette confluence est enrichi en Mg. On retrouve à peu près les mêmes tendances que pour Ca, en ce qui concerne les relations entre les teneurs et les débits, mais les teneurs étant plus faibles, ces tendances sont moins

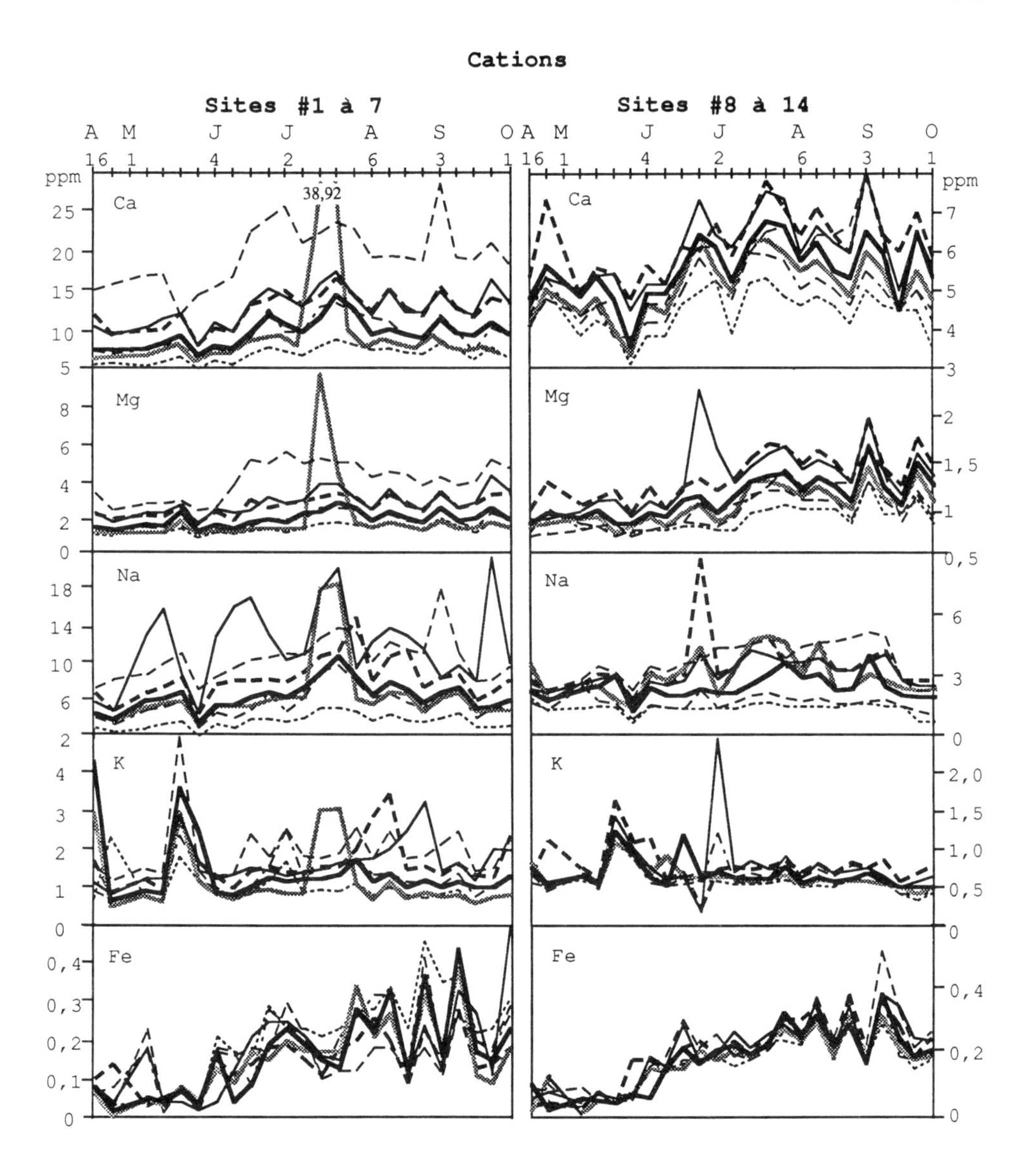

Figure 4. Variations saisonnières des teneurs ioniques d'amont en aval.

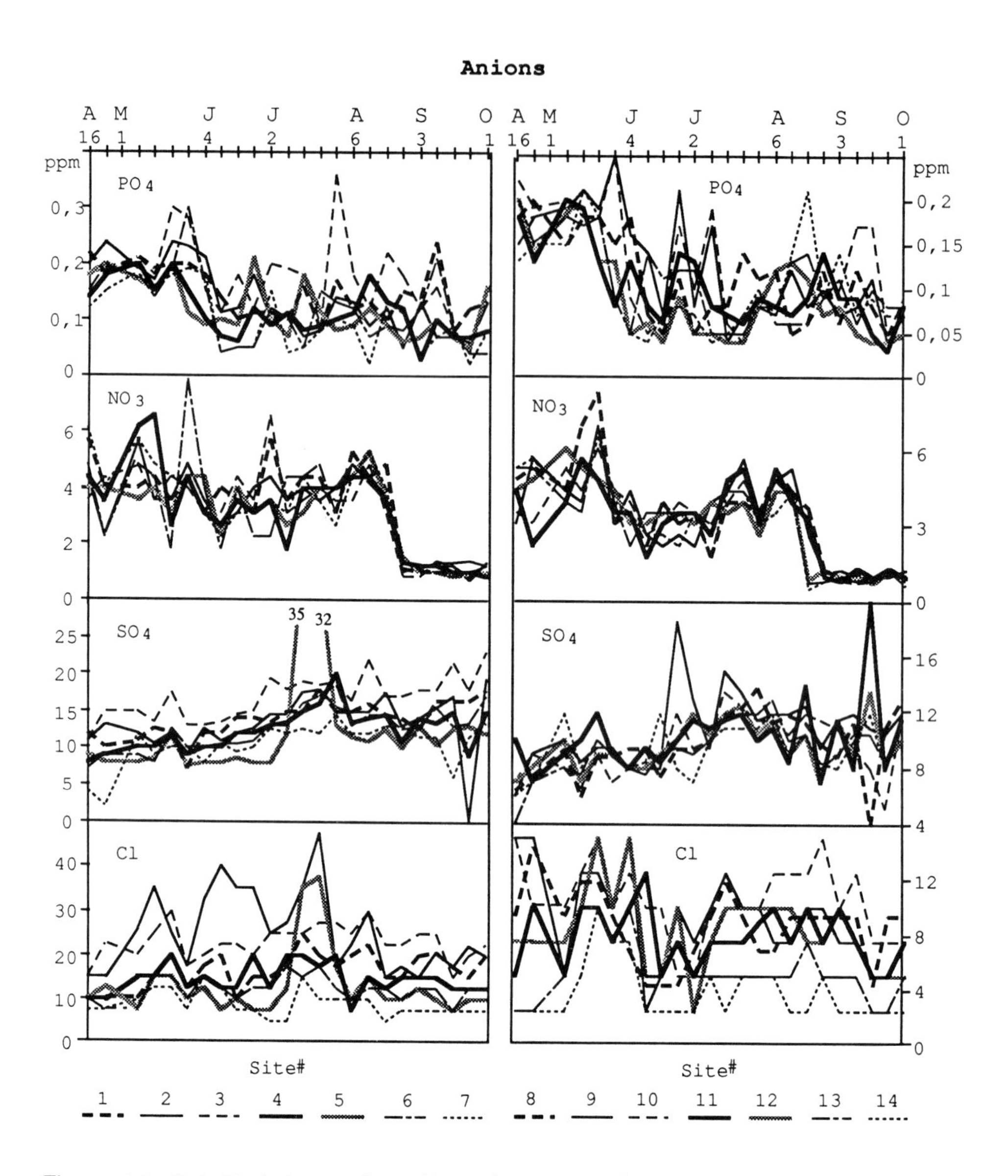

Figure 4 (suite). Variations saisonnières des teneurs ioniques d'amont en aval.

nettes. A prime abord, il semble donc que Mg soit de source géologique. En général, les teneurs sont faibles au printemps, lorsque les débits sont forts, et plus fortes en juillet alors que la contribution de la nappe phréatique au débit était plus importante. Nous n'avons pu expliquer les teneurs élevées au site 9 le 25 juin, ce qu'on pourrait attribuer à une erreur d'analyse. A l'instar du calcium, les teneurs en Mg étaient également très fortes dans la rivière Ouareau les 16 et 23 juillet, influencées par les eaux résiduaires de Crabtree et de son usine de papier. En effet, on est plus susceptible de déceler ces types de contributions lors des périodes de basses eaux.

Le sodium augmente d'amont en aval, graduellement jusqu'en aval de Joliette, puis de façon nettement plus marquée à partir du site 6, avec un palier aux sites 6,5 et 4. Les teneurs deviennent beaucoup plus fortes à compter du site 3 et plus en aval, la plus forte teneur n'étant pas dans la rivière l'Achigan, mais en aval de son embouchure (site 2). La variabilité saisonnière est très faible dans les sites plus "naturels" (13 et 14), mais beaucoup plus forte en aval de Joliette. Dans plusieurs sites, les teneurs sont plus fortes lors des périodes plus sèches de juillet et août, par concentration du Na par évaporation, mais les teneurs ne sont pas directement liées au débit. En effet, même si les crues du 28 mai et du 17 septembre et les pluies de la dernière semaine de juin on dilué les teneurs en Na, on remarque que lors des forts débits d'avril, les teneurs n'étaient pas vraiment plus faibles. Les mécanismes de contrôle des teneurs en Na sont complexes et variés, surtout dans les milieux à forte influence anthropique, ce qui cause la plus grande variabilité saisonnière vers l'aval. On sait que les précipitations sèches et humides peuvent faire entrer beaucoup de Na dans un bassin, surtout dans les régions à proximité de la mer (5) ; mais si Na est d'origine atmosphérique, il ne devrait pas être dilué lors des périodes de débits forts puisque les eaux du sous-écoulement et les eaux de nappe phréatique devraient alors être similaires. Or, ici, on remarque que les teneurs en Na diminuent lorsque les débits augmentent (28 mai), ce qui laisse supposer d'autres sources de Na. Une source géologique est possible, et le Na pourrait être libéré des plagioclases des roches cristallines et des dépôts glaciaires originant du Bouclier canadien. Mais la moins grande variabilité saisonnière des sites amonts laisse supposer un contrôle anthropique vers l'aval. En fait, dans les sites amonts 13 et 14, seul le pic du 7 mai au site 13 semblent suspect. Étant dans le village de St-Côme, il est possible que cette forte teneur passagère soit due à un mauvais fonctionnement temporaire de l'usine de traitement des eaux. Tout comme le Ca et le Mg, les teneurs en Na étaient très fortes dans la rivière Ouareau les 16 et 25 juillet, probablement à cause d'une plus forte influence des effluents municipaux et industriels de Crabtree lors de l'étiage. D'autre part, on sait que le fumier est riche en Na. Pourtant, la rivière l'Achigan est moins riche en Na que le site 2, en aval de son embouchure. L'élevage porcin, pratiqué de façon intensive dans le sous-bassin de l'Achigan, semble, sur la base des teneurs en Na, assez bien contrôlé. Le site 2, immédiatement à l'aval de la ville de l'Assomption, est le plus riche en Na ; cette municipalité ne traitant pas ses eaux usées, nous croyons que les eaux résiduaires municipales contribuent pour une bonne part du Na dans l'eau. C'est aussi dans ce site que les variations sont les plus fortes. La crue de la fin-mai a dilué Na, ainsi que la période plus pluvieuse de la fin juin ; on n'arrive cependant pas à expliquer le pic du 24 septembre, qui ne se retrouve pas dans les autres sites : erreur d'analyse ou déversement plus important d'effluents municipaux ?

Les teneurs en potassium augmentent d'amont en aval, mais très progressivement, jusqu'au site 5. C'est dans la rivière l'Achigan que les teneurs sont les plus fortes, ce que la géologie ne peut expliquer car les calcaires sont sous-jacents à ce sous-bassin et ne renferment donc pas de minéraux pouvant libérer K^+. L'évolution saisonnière des teneurs en K est mal définie, sans tendance précise. K n'est pas plus concentré en période de débit faible et donc la nappe phréatique n'est pas plus concentrée en K que les eaux superficielles : la source en K n'est sans doute pas géologique ici et ses teneurs sont plutôt liées à des variables végétatives et d'utilisation du sol. L'évolution dans le temps des teneurs en K est caractérisée par des pics très forts et de courte durée ; le reste du temps, K varie peu. Il y a eu un pic généralisé à tous les sites lors de la montée de crue du 21 mai, suivie d'une baisse des teneurs la semaine suivante, en période de décrue. Comme le potassium est un constituant important dans le cycle végétal, il peut être emmagasiné dans la végétation, puis lessivé. Ses teneurs augmentent donc lorsque c'est le sous-écoulement et l'écoulement de surface qui alimentent la rivière. Les fertilisants appliqués au printemps peuvent aussi être lessivés vers les cours d'eau lorsque les pluies sont abondantes. Dans une étude antérieure, nous avions souligné que K évoluait par pics dans les profils pédologiques, K étant poussé plus loin avec la progression du front d'humectation (6). K présente parfois, de façon ponctuelle, de très fortes teneurs qu'on ne peut lier à des conditions climatiques particulières. Ainsi, les valeurs 10 fois supérieures à la moyenne au site 9 et 200 fois supérieures à la moyenne au site 8 le 25 juin ne peuvent s'expliquer que par des contributions anthropiques, d'autant plus qu'on ne retrouve pas de tels pics à cette date plus en amont, ni dans les affluents (Ouareau et Achigan). On sait que le fumier contient de fortes quantités de K ainsi que les fertilisants, mais nos analyses ne nous permettent pas d'établir la source exacte de la contamination d'origine anthropique. Ce très fort enrichissement aux sites 9 et 8 s'est propagé vers l'amont et il y a des pics aux sites 7 et 6 le 2 juillet. Cette influence s'estompe en aval de la confluence de la rivière Ouareau, par suite de la dilution de cet affluent moins riche en K à cette date. Les teneurs de la rivière Ouareau étaient plus fortes les 16 et 23 juillet, à l'instar des autres cations à même date, une contribution probable des effluents municipaux et de l'usine de papier de Crabtree. Les hausses des teneurs aux sites 2 et 1 en août sont difficilement explicables sur la base de variables climatiques; comme K est lié à des variables d'utilisation du sol, les milieux les plus fortement perturbés sont les plus difficiles à expliquer, ou du moins si on n'est pas exactement informé des dates d'épandage de fertilisants, de fumier ou de lavage de salles de traite.

Les teneurs en fer varient peu d'amont en aval, mais dans le cas du fer, les moyennes signifient peu de choses car l'évolution dans le temps des teneurs se caractérise par des pics intenses et de courte durée. D'ailleurs les diagrammes des divers sites se ressemblent: les pics et les creux sont aux mêmes dates. Aussi, nous croyons que les teneurs en fer sont liées aux conditions hydriques dans le bassin et non à des sources de contamination diffuses ou ponctuelles. Dans des conditions naturelles, ce sont les milieux de battement de nappe qui fournissent le plus de fer et les horizons B podzoliques. La mise en solution du fer nécessite des conditions humides, mais s'il y a beaucoup d'eau, les teneurs baissent (5). Aussi, les teneurs étaient faibles dans tous les sites au printemps (avril-mai), alors qu'il y avait beaucoup d'eau dans le système. Au cours des mois d'été, les creux correspondent à des prélèvements après une semaine de sécheresse (20 août, 3 septembre). Les pics correspondent à des épisodes plus humides, sans qu'ils ne le soient trop. L'évolution des teneurs lors de la crue du 10 au 17 septembre semble bien indiquer que le fer est libéré très tôt au cours d'un orage (pics du 10 septembre) car il est mobilisé des sols gleyifiés par des processus réducteurs puis est chassé des horizons pédologiques vers le cours d'eau, donc au tout début de l'orage (ou à la montée de crue). Le fer est plus dilué lors de la décrue.

Les nitrates et les phosphates sont souvent liés à des variables d'utilisation du sol.Les nitrates sont peu variables d'amont en aval, passant de moyennes de 3 à 3,8 ppm.Généralement les teneurs sont plus fortes au printemps, une tendance plus évidente vers l'amont. Des études dans des milieux "naturels" ont démontré que les nitrates pouvaient être emmagasinés dans la couverture neigeuse et concentrés par évaporation au cours de l'hiver; ainsi, l'eau de fonte des neiges peut être riche en nitrates (1). Dans plusieurs sites, autant en amont qu'en aval, il y a eu une hausse des teneurs lors de la crue de la fin mai ; comme les nitrates font partie du cycle végétal, ils peuvent être emmagasinés dans la végétation, puis lessivés, ce qui est plus évident dans les sites naturels. Dans les sites à influence anthropique, les engrais répandus au printemps peuvent être lessivés et on en retrouve une partie dans les cours d'eau lorsque le sous-écoulement ou le ruissellement alimentent la rivière.Lors des phases de sécheresse ont ne peut guère mettre en évidence que des pollutions ponctuelles (lavage d'une salle de traite, déversement de purin), ce que n'a pas révélé notre échantillonnage hebdomadaire, mais peut-être que cette pratique est maintenant peu fréquente, mieux gérée ou même absente. On a bien remarqué un pic le 2 juillet exclusif dans la rivière l'Achigan et en aval de son embouchure ; comme cet échantillonnage correspondait aussi à une petite crue, il est également possible qu'il reflète l'impact de la pollution animale par les bêtes au pâturage, ce qui ne peut être révélé que par des prélèvements faits en période humide. Finalement ce genre d'événement est peu fréquent et il semble que l'influence de l'élevage porcin sur les teneurs en nitrates dans la rivière l'Achigan soit moindre qu'il y a dix ans (2). Les teneurs ont chuté à la fin août et en septembre par suite d'une moins grande influence des rejets diffus agricoles et des rejets domestiques des municipalités. On sait que l'influence de ces rejets dans les cours d'eau est plus apparente en période d'étiage. Contrairement à la crue de la fin mai, celle de la mi-septembre n'a pas provoqué une hausse des teneurs, peut-être qu'à ce moment de l'année les sources de nitrates dans les horizons superficiels sont mitigées ou encore que seules les précipitations "utiles" sont susceptibles d'entraîner le lavage des sols. Or ce n'est pas le cas du 27 août au 1er octobre, alors que les profils ont dû se recharger suite à la sécheresse de l'été.

Les teneurs en phosphates de la rivière l'Assomption varient peu d'amont en aval jusqu'à la confluence avec la rivière l'Achigan. Cet affluent, plus chargé en phosphates, maintient des teneurs plus élevées dans le tronçon aval de la rivière l'Assomption. Quoique les teneurs en phosphates dans l'Achigan soient plus fortes que dans les autres sites, elles sont beaucoup plus faibles que ce qui était enregistré avant 1981, donc avant la réglementation gouvernementale concernant la gestion des lisiers de porc. Les tendances saisonnières ne sont pas franches, ce qu'ont également noté Simoneau et Grimard (2). Sauf pour des pics isolés en été et d'autres liés à une crue (fin mai), les teneurs sont plus fortes au printemps (avril-mai) dans tous les sites, et de ce fait, cela pourrait s'expliquer par des causes naturelles. Les pics à l'étiage (25 juin, 9, 16 juillet) dans la partie aval de l'Ouareau et dans deux sites en aval soulignent l'impact des rejets municipaux non traités (St-Liguori et Crabtree) et secondairement, des activités agricoles. Ne se faisant sentir qu'en pérode d'étiage, leur influence est cependant moindre que la pollution agricole diffuse qui maintient des teneurs plus élevées et de façon constante dans la rivière l'Achigan. Dans la rivière l'Achigan, la forte crue de la fin mai a occasionné de plus fortes teneurs en phosphates, ce qui pourrait être lié au lessivage des engrais phosphatés mais aussi à la pollution animale. En plus de l'élevage porcin, on cultive ici le maïs, donc des cultures à grande interligne avec sillons dans le sens de la pente. Il n'est pas étonnant que les eaux de l'Achigan soient si turbides et colorées, surtout en période de crue, reflétant l'érosion, et il aurait été intéressant d'avoir des mesures de phosphore particulaire, sans doute plus abondant que le phosphore dissous lors des événements de crue ici. Le pic du 30 juillet dans la rivière l'Achigan ne correspond pas à un événement climatique particulier; il pourrait s'agir d'une pollution ponctuelle, type de pollution plus évidente en période d'étiage. L'étude de Simoneau et Grimard (2) indiquait une diminution très appréciable des phosphates depuis 1981. En effet, il semble bien que la surveillance exercée par les autorités du ministère a contribué à éliminer les déversements directs dans la rivière l'Achigan, sauf exceptions Il n'en demeure pas moins que les teneurs de l'affluent Achigan sont plus fortes qu'en aval de ce tributaire, et qu'il faudra améliorer encore plus les mesures de bonne gestion agricole.

Les teneurs en sulfates augmentent d'amont en aval et il y a plus de sulfates dans les cours d'eau que dans la pluie, quoique parfois la différence ne soit pas très grande, surtout dans les sites en amont. On sait que les minéraux sulfureux contenus dans les sédiments glaciaires peuvent libérer des sulfates.Cependant, dans cette région grandement influencée par des systèmes nuageux qui ont passé au-dessus des zones industrielles des Grands Lacs canadiens et américains, les précipitations sèches et humides sont chargées en sulfates et peuvent en accumuler sur la végétation. Les sulfates sont ensuite lessivés vers les cours d'eau lors des pluies, ce qui a déjà été démontré par Likens et al. (1) dans le nord-est des Etats-Unis. Mais comme c'est l'Achigan qui en a le plus et comme la Ouareau à son embouchure enregistre de forts pics, des facteurs anthropiques doivent être invoqués. On sait que les eaux usées domestiques et les eaux résiduaires industrielles (pâtes et papiers) fournissent des sulfates. Aussi, c'est en période d'étiage que l'influence des eaux résiduaires est la plus marquée, par exemple les pics des 16 et 23 juillet dans la rivière Ouareau, en aval de Crabtree, une municipalité qui déverse directement ses eaux usées dans la rivière, tout comme l'usine de papier. Quoique la rivière l'Achigan contienne plus de sulfates que tous les autres sites, les variations saisonnières des teneurs sont graduelles, les teneurs étant plus basses d'avril à juin et plus fortes de juillet à septembre. Il n'y a pas de pic marqué ici ; notre échantillonnage n'a donc pas mis en évidence de cas de pollution ponctuelle qui pourrait être une source de sulfates dans ce milieu agricole.

Les teneurs en chlorures augmentent d'amont en aval, avec une hausse très marquée à partir de la rivière l'Achigan. Cependant, ce n'est pas la rivière l'Achigan qui est la plus chargée, mais la rivière l'Assomption au niveau de la municipalité de l'Assomption, sans doute une influence des effluents municipaux. Les tendances saisonnières ne sont pas claires. Dans le site le plus en amont, le seul pic important est associé à la forte crue d'orage de la fin mai. Dans la Ouareau, les teneurs étaient très fortes les 16 et 23 juillet, à l'instar du Na, ce qui est sans doute lié aux effluents de Crabtree et de l'usine de papier; il semble que les chlorures soient ici sous forme de NaCl. Mais pour le reste, le schéma demeure plutôt confus.

5. CONCLUSION

Les résultats précédents suggèrent que les activités anthropiques influencent les teneurs ioniques dans les exutoires. Mais il est parfois difficile de distinguer clairement si tel ion est contribué par des constituants naturels (sols, roches, végétation) ou par des activités anthropiques. Quoiqu'il en soit, l'état hydrique du bassin a une grande importance et explique une grande partie des variations saisonnières des teneurs des cours d'eau. Un travail à l'échelle d'un bassin-versant de plus de 4 000 km^2 fournit cependant des pistes intéressantes sur la contribution des diverses activités agricoles, urbaines et industrielles, ce qu'une étude détaillée des bassins du 1er ordre combinée à un suivi serré des pratiques agricoles ainsi que des rejets d'effluents permettra de mieux préciser.

REFERENCES

1. Likens, G.E., Bormann, F.H., Pierce, R.S., Eaton, J.S. and N.M. Johnson (1977) : Biogeochemistry of a Forested Ecosystem, New-York. Springer-Verlag.
2. Simoneau, M. and Y. Grimard (1989) : Qualité des eaux de la rivière l'Assomption 1976-1987. Direction de la qualité du milieu aquatique, Ministère de l'Environnement du Québec, 234 p.
3. Gangbazo, G. and J. Buteau (1985) : Analyse de la gestion des fumiers dans le bassin versant de la rivière l'Assomption : état de la situation et éléments de solution. Direction de l'assainissement agricole, ministère de l'Environnement du Québec, 83 p.
4. Greenberg, A.E., Connors, J.J. and D. Jenkins (1981) : Standard Methods for he examination of Water and Wastewater. American Public Health Association, Washington, D.C.
5. Bouchard, M. (1983) : Influences stationnelles sur l'altération chimique des sols dérivés de till (Sherbrooke, Que., Canada). Catena, 10/4, 363-382.
6. Bouchard, M. (1984) : Bilan géochimique et origine d'éléments dissous dans un bassin-versant granitique breton (France). 24/4, 363-379.

Farm Land Erosion: In Temperate Plains Environment and Hills
S. Wicherek (Editor)

Erosion hydrique et entraînement mécanique des terres par les outils dans les côteaux du sud-ouest de la France. La nécessité d'établir un bilan avant toute mesure anti-érosive

J.C. Revel [a], M. Guiresse [a], N. Coste [a], J. Cavalie [b] et J.L. Costes [c].

[a] Laboratoire de Physico-Chimie des sols ENSAT 145 Avenue de Muret 31076 Toulouse Cédex, France.

[b] Chambre Départementale de l'Agriculture Allées de Briennes 31000 Toulouse, France.

[c] ITCF Station Inter-instituts 31450 Bazièges, France.

RESUME

Dans les côteaux du Sud-Ouest de la France, les sols ont une répartition très anarchique dans le paysage. Cette mosaïque des sols résulte, en partie, de l'effet cumulatif des pratiques culturales. Un modèle simple est proposé pour tester l'effet de labours réalisés successivement sur une grande échelle de temps. Les résultats auxquels il aboutit se retrouvent bien dans le paysage que l'on observe actuellement.

Une expérience a été réalisée sur le terrain pour quantifier la descente des terres consécutive au labour. Pratiquement, les auteurs utilisent une tranchée de 0,4 x 0,4 x 24 mètres remplie de graviers concassés. Après passage des outils, l'analyse des états de surface montre que la migration des terres provoquée par le labour descendant, est compensée à 60 % environ par le labour remontant. L'analyse pondérale nécessite de définir un flux de terre ou masse de terre traversant une ligne parallèle aux lignes de niveau. L'érosion hydrique a été aussi mesurer sur des rigoles. Mais ce phénomène naturel est très aléatoire car étroitement lié au climat, et aux orages. L'estimation des effets de l'érosion hydrique est comparée à ceux des techniques culturales qui apparaissent prépondérants.

SUMMARY

The patchwork constituted by the random distribution of the soils in the hill landscape of southwestern France actually results partly from the cumulative effect of soil tillage practices. The simple model put forward to test the effect of successive ploughings on a large time scale provided data that accounted for the landscape observed nowadays. A field experiment was carried out to quantify the downward movement of the soil, which was consecutive to ploughing, using a 0.4 x 0.4 x 24 m trench filled with gravel. The analysis of the surface soil after the use of soil tillage implements showed that soil migration caused by down-slope ploughing was compensated for approximately 60 % by up-slope ploughing. Quantitative analysis requires the definition of the soil flux or mass that crosses a line parallel to the contour. Run-off erosion was measured on gullies, but this random natural phenomenon is obviously dependent on the climate, particularly on storms. The effects of run-off were compared to those of tillage, which were stronger.

1. INTRODUCTION

Les côteaux du Sud-Ouest de la France ou Terrefort (Fig. 1) se développent sur une molasse Tertiaire imperméable. Au cours du Quaternaire un réseau hydrographique temporaire très dense a entaillé les formations en une multitude de collines, croupes allongées dont les pentes varient sur les versants de quelques pourcents à 30-35 %. (1 - 2).

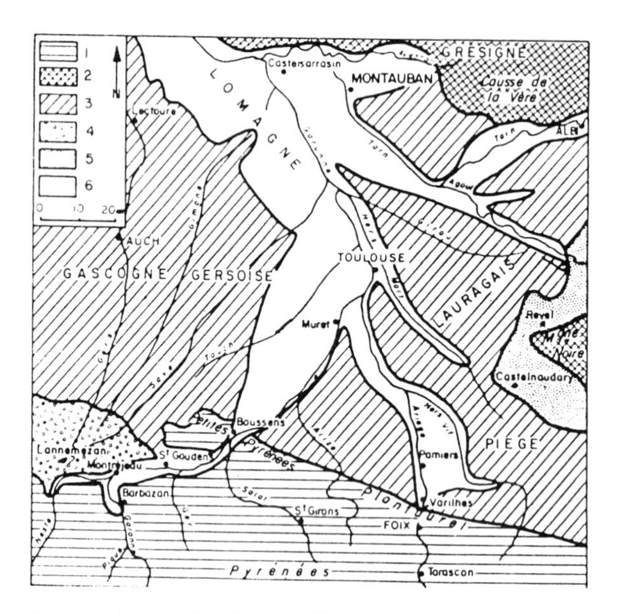

Figure 1. Localisation du Terrefort Toulousain (d'après Hubschman, 1975)
1 : Pyrénées ; 2 : Massif Central ; 3 : Terrefort ; 4 : Plateau de Lannemezan ;
5 : Dépression Montagne Noire ; 6 : Alluvions.

2. LES SOLS : UN INDICE D'EROSION INTENSE

Dans ce paysage de collines, les sols sont d'une extrême variabilité mais correspondent le plus souvent à des troncatures, parfois jusqu'à la roche-mère de deux types de sols seulement. Revel (3) a pu montrer que :

- Les sols développés sont des sols bruns calciques ou des sols bruns lessivés présentant tous deux un horizon d'accumulation du calcaire entre 1,5 et 1,8 m de profondeur. La profondeur actuelle de cet horizon permet de connaître l'épaisseur de la troncature ou du recouvrement.
- Les versants en pente forte (> 12 %) sont décapés et laissent apparaître l'horizon d'accumulation du calcaire sauf au niveau de ravines maintenant comblées.
- Les versants en pente faible (< 12 %) présentent une érosion modérée sauf au niveau de la bordure des ravines ou en limite amont de parcelle.
- Sur tous les versants, il existe des accumulations en aval des parcelles et qui forment des talus même sur des pentes atteignant 35 %.

De cet ensemble il en résulte des sols tronqués : calcaires, bruns calcaires, bruns calciques tronqués, bruns lessivés tronqués et des sols sur colluvions calcaires ou calciques. Cet ensemble résultant de phénomènes d'érosion conduit à la mosaïque des sols du Terrefort.

3. QUELLE EROSION ?

On pourrait penser que l'érosion hydrique sur les versants, en nappe ou en rigoles a permis le décapage des versants avec accumulation au fond des vallons où l'on retrouve 4 à 8 m de colluvions (4). Mais certains faits restent difficiles à expliquer :
- Les accumulations en milieu de pente parfois à 35 % ne devraient pas être présentées dans le cadre d'une érosion hydrique.
- Les ravines de versant, véritables ravins, sont normalement les lieux privilégiés de concentration des eaux, où le flot a son maximum de compétence. Or, ici ces ravines sont un lieu d'accumulation. Barlier (5) y a même rencontré, d'une manière assez générale, un sol en place surmonté par un profil inversé (Fig. 2).

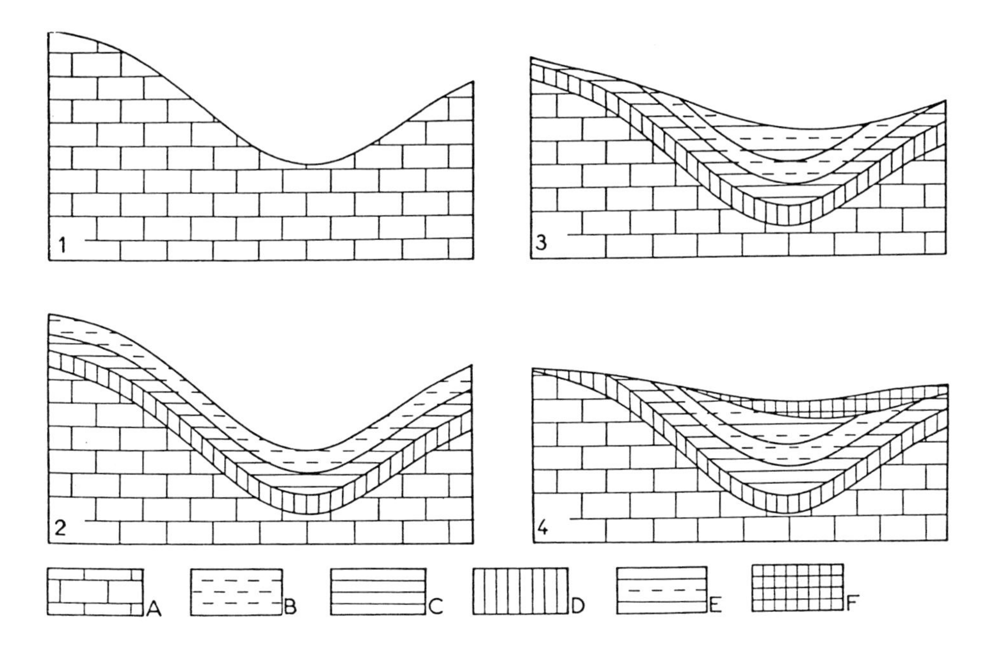

Figure 2. Reconstitution de l'évolution des ravines (Barlier 1977).
1 : creusement de la ravine ; 2 : différenciation d'un sol brun calcique ;
3 : abrasion de la crête et remplissage du fond par le labour ; 4 : état actuel.
A : Substratum molassique ; B : Horizon A du sol brun calcique ;
C : Horizon (B) du sol brun calcique ; D : Horizon Cca du sol brun calcique ;
E : Horizons A et (B) remaniés ; F : Horizons (B) et Cca remaniés.

On ne peut donc pas invoquer l'érosion hydrique pour expliquer tous ces remaniements et transport de matériaux. Par contre, comme le font remarquer Gimpel (6) et Neboit (7), le travail du sol conduit à une modification importante dans le modelé des versants (Fig.3).

4. LE TRAVAIL DU SOL RESPONSABLE DE LA REPARTITION DES SOLS

La mise en culture des sols du Terrefort a débuté au Moyen Age avec la généralisation de la charrue versoir. Des attelages de 4 à 6 boeufs étaient nécessaires pour retourner ces sols riches mais très cohérents. Le labour se faisait dans le sens des lignes de niveau, en versant vers le bas, provoquant un décapage en sommet de parcelle et une accumulation en bas.

Nous avons établi un modèle qui permet de simuler ce type de labour (Fig. 3)

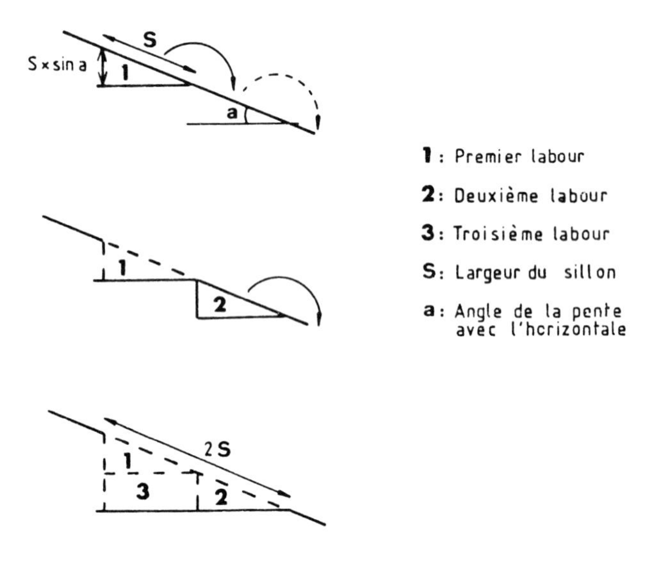

Figure 3. Modèle du décapage d'un sommet de versant par labour.

On considère que la largeur d'un sillon (S) est ramenée à l'horizontale lors du premier labour ; au suivant, par contre, l'horizontalité de cette partie sera rompue. Il faut un troisième labour, pour que la partie correspondant alors à une largeur de deux sillons soit à nouveau ramenée à l'horizontale.

D'une manière plus générale, avant la mise en culture, au temps to, l'épaisseur tronquée est nulle. La première année, lors du premier labour t1, en sommet de parcelle, la largeur (S) d'un sillon est ramenée à l'horizontale et, si a est l'angle de la pente par rapport à l'horizontale, une épaisseur S.sin a est alors enlevée. Pour enlever 2 fois cette épaisseur, ou avoir une largeur horizontale de 2 S, il faut 2 labours supplémentaires. Le nombre de labours t2 est égal à tl + 2, car à chaque labour, on enlève l'épaisseur donnée sur seulement la largeur d'un sillon.

On obtient par conséquent une progression arithmétique de la forme :

$$t_i = 1 + 2 + 3 + ... + i = \frac{i\,(i + 1)}{2}$$

ti : nombre de labours
i : nombre d'épaisseurs S.sin a enlevées.

Pour connaître le nombre de fois où la troncature est ramenée à l'horizontale (i), il suffit alors de résoudre la fonction inverse de la fonction suivante :

$$y = \frac{x\,(x + 1)}{2}$$

avec y : nombre de labours
x : nombre de descentes de terre.

La résolution de cette fonction permet de calculer, en fonction du nombre supposé de labours effectués depuis la mise en culture (à peu près 1 000 ans), et suivant le système de rotation des jachères, la largeur ramenée à l'horizontale et l'épaisseur décapée. Les valeurs obtenues sont réunies dans le Tabl. 1.

La compatibilité de ces valeurs avec ce qui est observé sur le terrain peut valider l'hypothèse formulée en ce qui concerne les épaisseurs tronquées suivant la valeur de la pente et la forme acquise par les versants.

De plus, au niveau des ravines, l'augmentation de la pente du flanc de celles-ci par rapport à la pente générale du versant augmente l'épaisseur tronquée, et corrélativement, le volume de matériaux qui tombe au fond. On arrive donc au comblement des ravines avec troncature plus grande sur les bords. C'est ce que l'on constate sur le terrain. Autrement dit, le travail du sol est bien responsable de la répartition des sols par suite de phénomènes de décapage, transfert et accumulation, c'est à dire d'érosion.

En calculant l'épaisseur tronquée en fonction du nombre de labours pour une pente fixée à 15 %, on constate que la vitesse de troncature n'est pas constante. Elle diminue lentement, sans vraiment tendre vers une valeur limite : il n'y a pas d'asymptote parallèle à l'axe des x. La descente des terrains est donc de plus en plus lente, sans atteindre un équilibre. Il faut souligner que le modèle donne la part la plus importante à l'érosion anthropique et néglige l'érosion naturelle accélérée. Le fait que ce type d'érosion diminue fortement avec le temps paraît rassurant, mais le modèle employé intègre une technique culturale ancienne.

Table 1.
Valeur de la largeur ramenée à l'horizontale et de l'épaisseur décapée en fonction du nombre de labours et de descentes de terre.

Nombre de labours (y)	Nombre de descentes de terre (x)	Largeur ramenée à l'hori-zontale (m)	Epaisseur décapée (m) pour une pente de				
			5 %	10 %	15 %	20 %	25 %
Jachère 1 528 Année/2	32	6,4	0,32	0,64	0,96	1,28	1,60
Jachère 1 703 Année/3	37	7,4	0,37	0,74	1,11	1,48	1,85
Sans jachère 1035	45	9,0	0,45	0,90	1,35	1,80	2,25

Actuellement, le labour effectué suivant la ligne de plus grande pente, modifie les données et a fait l'objet de l'expérimentation suivante.

5. EXPERIMENTATION SUR L'ENTRAINEMENT MECANIQUE DES TERRES PAR LE LABOUR

L'introduction de la mécanisation a permis de travailler le sol dans le sens de la pente. La faible puissance des premiers tracteurs permettait uniquement le labour en descente, ce qui conduisait à un amoncellement à la base de la parcelle. Actuellement, le labour se fait souvent en descendant et en remontant, le transport des terres doit alors être estimé en terme de bilan.

Pour suivre la descente des terres par le labour, il faut pouvoir marquer la terre. Le choix du traceur pose de nombreux problèmes car il doit être incorporé d'une manière homogène sans changer les propriétés des sols. Nous n'avons pas trouvé de traceur malgré de nombreux essais, aussi notre choix s'est fixé sur une tranchée remplie de graviers.

5.1 Conditions expérimentales

Sur un versant en pente régulière de 18 % (10°), une tranchée de 0,4 x 0,4 x 24 m a été creusée suivant une ligne de niveau et la terre portée hors de la parcelle. Elle est remplie de graviers concassés d'un diamètre variant de 4 à 6 mm, bien distincts de tous les éléments contenus dans les sols du Terrefort. Le remplissage s'est effectué avec des bâches pour éviter toute contamination des sols immédiatement voisins et la surface des graviers a été mise en continuité avec celle de la parcelle. Les critiques d'une telle expérience sont nombreuses.

- La tranchée est très étroite par rapport au déplacement probable des particules. Dans cette expérience, nous cherchions à quantifier la totalité de terre déplacée à travers une ligne de niveau. Il nous faudra donc extrapoler les résultats trouvés sur une largeur de 40 cm à une bande dont la largeur sera égale à la distance maximum parcourue par les graviers.
- Les propriétés texturales et mécaniques du sol diffèrent beaucoup de celles des graviers qui ne sont pas mesurables. L'expérience est donc approchée.
- Enfin, la tranchée crée une discontinuité dans la résistance mécanique au passage des outils.

Comme le travail du sol s'effectue suivant la ligne de plus grande pente, aussi bien en remontant qu'en descendant, l'estimation de la descente de terre correspond en fait à la différence : volume de terre descendue - volume de terre remontée. Le labour descendant est fait avec une charrue Huard 3 cors à 5 socs espacés de 0,4 m, tiré par un tracteur Fiat 980 DT de 100 CV à la vitesse de 7 km/h. La profondeur est estimée à 0,27 m. Pour le labour en remontant, seule la vitesse réduite à 6,2 km/h diffère du traitement précédent.

5.2 La dispersion des graviers en surface

Le déplacement des graviers de la tranchée, sous l'effet des outils, a d'abord été observé à la surface du sol. Un comptage visuel et systématique de la surface quadrillée par des carrés de 5 cm de côté a conduit aux résultats suivants :

- La descente et la remontée des graviers sont franches et peuvent atteindre des distances importantes depuis les bords de la tranchée : 2,2 m dans le cas du labour descendant et 1,4 m dans l'autre cas.
- La descente des graviers est plus importante que la remontée, tant par le nombre de graviers transportés, que par l'amplitude de leur transport.

Une étude statistique a permis de déterminer une zone de déplacement moyen sur laquelle sera faite l'étude pondérale.

5.3 La dispersion des graviers à travers la masse du sol

a) Le mode de prélèvement et l'expression des résultats

Pour éviter des différences importantes dans les hauteurs de prélèvement du fait des grosses mottes formées par le labour, il a été procédé à un roulage préalable par le tracteur.

La taille des graviers exclut les échantillons de faible volume. On prélève alors des parallélépipèdes rectangles de 0,1 m de hauteur, 0,1 m de largeur et de 0,4 m de long. La plus grande dimension couvre la distance entre deux outils travaillant, ce qui permet de s'affranchir du positionnement par rapport au passage de l'engin.

Chaque prélèvement est pesé (P_1) puis, après tamisage, le poids P_2 des graviers isolés est déterminé. La densité de la terre et des graviers étant différente à partir du poids des graviers on calcule le poids évuivalent de terre (EqT) par $P = P_2$ (dt/dg) et mesures faites $p = p_2$ (1,63 / 1, 33). Or, les mesures sont pondérales et en tenant compte de la densité des graviers (d_g) et de la terre (d_t), il est possible de calculer l'équivalent en poids de terre (P) des graviers :

$P = P_2 \, (dt/d_g)$. Cette valeur sera notée "EqT".

En faisant la somme de toutes les valeurs P de chaque prélèvement, on obtient le poids de terre qui a été entraîné hors de la tranchée.

Le tamisage pour récupérer et peser les graviers est alors une opération lourde qui, dans un premier temps, exclut toute répétition.

b) Résultats des prélèvements

En examinant les résultats par tranche de sol prélevé 0-10, 10-20, 20-30 cm, Revel et al., (8) constatent une forte dispersion des graviers. Ces particules peuvent migrer horizontalement et/ou verticalement. Ce déplacement dans deux directions peut être à l'origine de la répartition non progressive au fur et à mesure que l'on s'éloigne de la tranchée.

Afin de s'affranchir de la variabilité suivant la verticale, nous avons considéré la somme des graviers trouvés aux trois profondeurs (0-10, 10-20 et 20-30) en fonction de la distance à la tranchée. Ainsi représentés, les résultats montrent une diminution progressive au fur et à mesure que l'on s'éloigne de la tranchée. Les graviers ont migré davantage dans le cas du labour descendant.

c) Interprétation des résultats

La descente des terres par les techniques culturales peut être assimilée à un mouvement de masse sur un versant. S'il est possible de mesurer la masse de terre et la distance qu'elle parcourt sur le versant, la comparaison avec les mesures habituelles exprimées par unité de surface nécessite de connaître la longueur du versant, mais dans un premier temps, on peut considérer la masse de terre (B) qui traverse un segment de 1 m de long et parallèle à une ligne de niveau (LR). Dans cette expérimentation, la ligne de niveau de référence LR est la limite aval de la tranchée, dans le cas du labour descendant, et la limite amont, dans l'autre cas.

Au vue de la figure 4, les particules ont un déplacement variable : certaines migrent loin, d'autres restent plus près de la tranchée. Donc un échantillon prélevé, contient des graviers provenant de la tranchée, et également de la terre provenant de la partie amont ou aval de la tranchée. Autrement dit, en ne tenant compte que des graviers, on commet une forte erreur par défaut. Pour essayer d'approcher la réalité, on a supposé, sur une pente de valeur constante, que la descente des terres suit un régime permanent.

La masse de terre mi qui se déplace sur une distance i est identique en tout point du versant quelque soit sa position. En pratique, mi ne se rapporte pas à un point du versant mais à un rectangle de 1 décimètre de large : ce rectangle correspond à la surface au sol de chaque parallélépipède prélevé.

Au final, le total des particules qui ont traversé LR en effectuant un déplacement de i décimètres est égal à Mi.

Mi = mi1 + mi2 + ... = mii avec i = distance en décimètre.

Comme on a fait l'hypothèse d'un régime permanent, tous les termes de cette addition sont égaux, et on a :

Mi = i x mi

Si on se place à une distance i de la tranchée, les graviers prélevés à cet endroit-là et exprimés en EqT, donne Ni.

Ni = mi provenant du 1er décimètre de la tranchée +
mi + 1 provenant du 2° décimètre de la tranchée +
mi + 2 provenant du 3° décimètre de la tranchée +
mi + 3 provenant du dernier décimètre de la tranchée +

Au total, on a Ni = mi = mi+1 + mi+2 + mi+3.

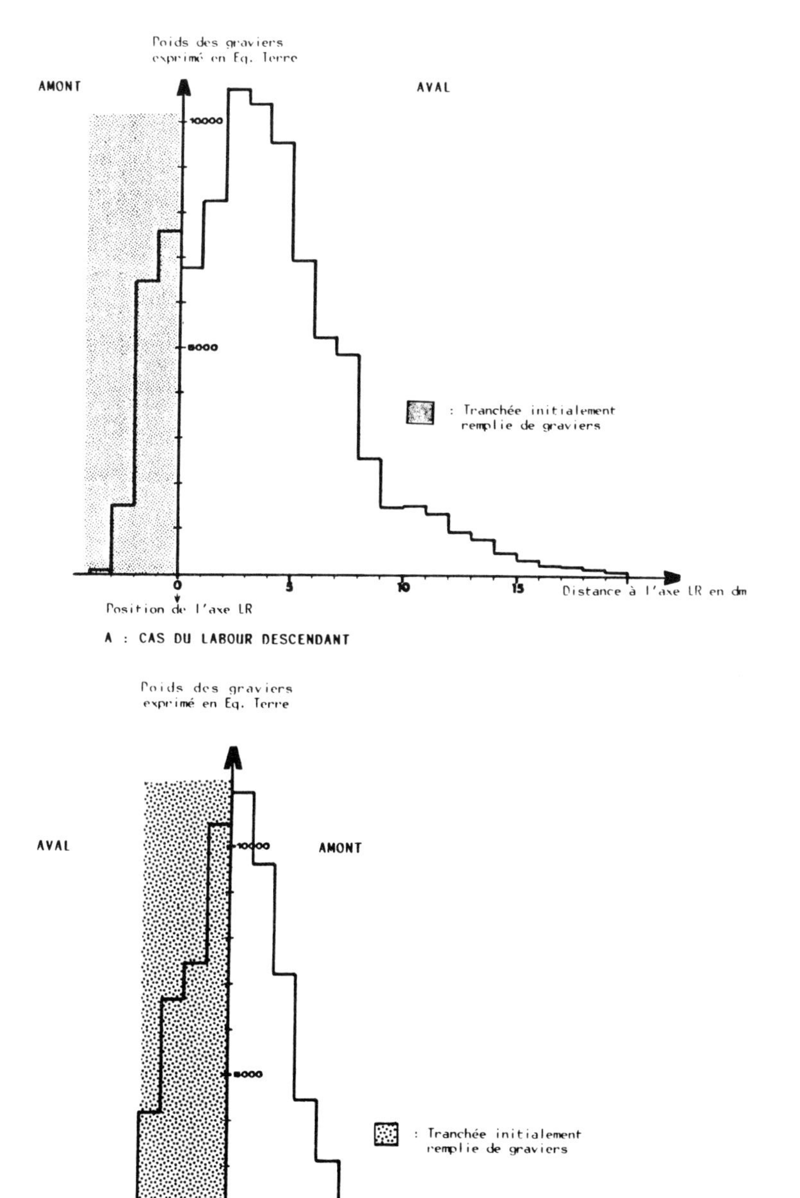

Figure 4. Répartition des graviers de part et d'autre de la tranchée après un labour descendant (A) et un labour (B).

En posant toutes ces équations de i = 1 à i = D, D étant la distance maximale (en dm) parcourue par les graviers, on obtient un système de D équations à D inconnues. La résolution de ce système permet de déterminer tous les termes mi et de calculer la somme S :

$$S = 2{,}5 \sum_{i=1}^{i=D} i \times mi$$

La résolution du système se fait sans problème dans le cas du labour descendant et donne la somme Sd.

Sd = 306,8 kg EqT

Par contre dans le cas de labour montant, la courbe tracée sur la figure 4 A a du être légèrement lissée, de manière à pouvoir résoudre les équations, et déterminer la somme Sm.

Sm = 144,6 kg EqT

Le bilan donne B = Bd - Bm = 162,2 kg EqT

6. COMPARAISON DE LA DESCENTE MECANIQUE DES TERRES ET DE L'EROSION HYDRIQUE

L'hypothèse préalable de régime permanent implique que la terre est déplacée par les outils de la même façon en tout point du versant de pente régulière. Dans la réalité, étant donné la forme convexe concave des versants, la variation de pente conduit à des déplacements différents et même une accumulation en bas. Notre hypothèse s'applique à la partie médiane de pente régulière que l'on a ici extrapolée à toute la parcelle. Par conséquent, la quantité de terre descendue de l'amont vers l'aval reste constante quelque soit la longueur du versant ; mais, ramenée à l'unité de surface, elle dépend de la forme du parcellaire.

Pour ramener les valeurs du déplacement à une expression de l'érosion spécifique (t/ha ou kg/m^2), on doit tenir compte de la forme de la parcelle qui généralement est allongée parallèlement aux lignes de plus grande pente. La quantité de terre déplacée étant mesurée suivant une ligne de niveau, la masse de terre déplacée est proportionnelle à la largeur, inversement proportionnelle à la longueur. Autrement dit cette masse de terre déplacée est d'autant plus faible que le rapport longueur sur largeur de la parcelle est grand. Quelques exemples sont donnés dans le Tab. 2.

L'érosion hydrique a été estimée dans la région par différents procédes : à partir du transport solide des rivières (9) et en tenant compte de l'abaque de Robinson (10) ; par simulation de pluie (11) ; par mesure du volume évidé par les rigoles (4) au cours d'une année où l'érosion a été peu intense.

Toutes ces expérimentations ont donné des valeurs très nettement inférieures à celles résultant de nos mesures (Tab. 2).

Table 2.
Estimation de la descente des terres par le labour et comparaison avec l'érosion hydrique.

D'après ROOSE (1988)	0,25 à 0,001 kg/m^2		
D'après ROUAUD (1987)	de 0,04 à 0,17 kg/m2		
D'après ETCHANCHU (1988)	0,02 kg/m^2		
Bilan de la terre déplacée après un labour descencant et un labour montant sur une pente de 18 % et dans une parcelle de 1 m de large et de longueur : 100 m 250 m 500 m	1,62 kg/m^2 0,65 kg/m^2 0,32 kg/m^2		
	Labour descendant LD	Labour montant LM	Bilan
Flux en kg de terre descendue par le labour sur un mètre linéaire de courbe de niveau	306,8	144,6	162,2

7. CONCLUSION

Dans l'expérience que nous avons réalisée, les outils ont eu un effet érosif très important sur une tranchée remplie de graviers. La structure du sol étant très différente de celle du gravier, il est probable que les valeurs obtenues sont seulement approchées. Mais il faut rappeler que certains agriculteurs ne labourent qu'en descendant. La migration des terres le long du versant peut être alors très importante.

L'érosion mécanique due aux outils à un impact prépondérant dans le modelé du paysage et notamment l'aplanissement des reliefs, la distribution des sols dans le paysage. Les matériaux déplacés restent sur le versant, mais de grandes surfaces peuvent être profondément décapées Revel et al (12) ont montré que, sur le bassin versant du Vermeil (692 ha), 1,08 m de terre ont été décapés sur 58 % de la surface du bassin et 1,47 m de terre ont été accumulés sur 37 % seulement de la surface. Il s'agit d'un remaniement à l'intérieur d'un même bassin versant.

Par contre, l'érosion hydrique a des conséquences irréversibles puisque la plupart des matériaux arrachés au sol quittent la parcelle pour rejoindre les cours d'eaux.

Dans la région que nous avons étudiée, la plus forte valeur de l'érosion hydrique a été mesurée par Roose (11) en simulant une pluie de retour 10 ans (une heure à 40 mm/h). Roose a obtenu 0,25 kg/m^2, ce qui reste inférieur à la quantité de terre érodée par les outils, même dans le cas le plus favorable (0,32 kg/m^2).

Autrement dit, dans le Terrefort, la descente de la terre par les techniques culturales reste largement prépondérante sur l'érosion hydrique. Dans le cadre d'un aménagement visant à protéger les sols, les moyens conservatoires mis en oeuvre devront apporter des solutions pour limiter la descente des terres : travail minimum du sol ou même non travail, vitesse élevée en remontant et réduite en descendant.

De nouvelles expérimentations permettraient de mieux identifier l'impact de cette érosion, notamment en précisant le rôle de paramètres tels que la valeur et la forme de la pente, le type d'outils travaillant et leur vitesse.

REFERENCES

1. Cavaille (1969) : Formations superficielles et sols des côteaux molassiques du Sud-Est de Toulouse. Livret guide excu. A6 Pyr. Cent. et Orient., Roussillon Languedoc, 8° cong. INQUA : 16-32.
2. Hubschman J., (1974) : Morphogénèse et pédogénèse quaternaire dans le piémont des Pyrénées garonnaises et ariégeoises. Thèse Univ. Toulouse Le Mirail, 745 p.
3. Revel J.C., (1982) : Formation des sols sur marnes. Etude d'une chronoséquence et d'une toposéquence complexes dans le Terrefor toulousain. Thèse Sci., INP Toulouse, 249 p.
4. Rouaud M., (1987) : Evaluation de l'érosion quaternaire, des remaniements de versant et de l'érosion en rigole dans le Terrefort toulousain. Thèse 3° cycle, Univ. Toulouse III, 320 p.
5. Barlier J.F., (1977) - Les sols formés sur molasse dans la région toulousaine. Etude des phénomènes de lessivage et de remaniement. Thèse 3° cycle, Univ. Toulouse III, 107 p.
6. Gimpel J., (1975) : La révolution industrielle au Moyen-Age Ed. "points" histoire, 249 p.
7. Neboit R., (1983) : L'homme et l'érosion. Faculté des lettres et Sciences humaines de l'Université de Clermont-Ferrand II Fascicule 17, 183 p.
8. Revel J.C., Coste N., Cavalie J., Costes J.L., (1989-1990) : Premiers résultats expérimentaux sur l'entraînement mécanique des terres par le travail du sol dans le Terrefort toulousain (France). Cah. ORSTOM, sér. Pédologie, Volume XXV, n° 1 et 2, 111-118 p.
9. Etchanchu D., (1988) : Géochimie des eaux du bassin de la Garonne. Transferts de matières dissoutes et particulaires vers l'Océan Atlantique. Thèse Univ. Toulouse III, 178 p.
10. Robinson A.R., (1977) : Relationship between erosion and sediment delivery. Proc. of Paris Symp., AIHS, Public., 122 : 159-167.
11. Roose E., Cavalie J., (1988) : Nouvelle stratégie de gestion conservatoire des eaux et des sols. Comm. Conf. Intern. ISCO 5, Bangkok.
12. Revel J.C., Rouaud M., (1985) : Mécanismes et importance des remaniements dans le Terrefort toulousain (Bassin Aquitain, France). Pédologie, XXXV, 2, 171-189 p.

Farm Land Erosion: In Temperate Plains Environment and Hills
S. Wicherek (Editor)

Soil Degradation in Hungary

P. Szabó

Foundation for Soil Conservation, Zulejka 4, H 1126, Budapest, Hungary.

SUMMARY

Although Hungary is a relatively small country with 9,3 million ha total area, approximately 2/3 of it -i.e. 6,5 million ha - is used by agriculture. In the field of cropland per capita Hungary is among the first European countries.
As far as the natural ressources are concerned Hungary has a medium supply. Besides the soil there are only few resources providing long term quantities.
Soil is our most important natural resource, its value is almost 1/5 of the national wealth and more than double of mineral raw materials. Soil is renewable if efficient conservation is maintained.
Unfortunately several natural and man-made factors have harmful effect on soil productivity :
2,3 million ha of sloping area is endangered by water erosion causing not only soil loss in the mountains and hills but further damages are produced on 200 thousand ha of surrounding walley bottoms and lowlands by sedimentation and flooding.
Some 1 million hectares are endangered by excess water on flat areas due to unfavourable hydrological conditions and relief.
Wind erosion prevails mainly on sands without structure and on peatland ranging totally to 600 thousand ha.
Salt affected soils can be found on approximately 600 thousand ha where the high salt and exchangeable sodium contents are the limiting factors.
Soil acidity restricts soil fertility on 2,3 million ha.

In addition to this because of large scale farming system in the past. 30 years including big fields, overdosage of artifical fertilizers, use of heavy machines, neglecting of manuring and organic farming, further physical, chemical and biological soil degradation can be observed.

Since the mid 80s soil contamination has become one of the most serious problem together with air and water pollution. Drastical reduction of the amount of solid and liquid wastes, their comprehensive reuse or purification belong to the unsolved problems nowadays all over the world. The general handling of wastes - let them flow into rivers, lakes, seas without or with only partly cleaning or land disposal acting as chemical time bombs can cause serious regional or global pollution. The increasing rate of wastes year by year raises the risk.

Joining the European Communities requires from Hungary to face and solve soil conservation problems as soon as possible.
We, Hungarian pedologists and soil conservationists are convinced that these new tasks must be and can be accomplished in the near future.

RESUME

Petit pays (9,3 M d'ha), la Hongrie porte environ 6,5 M d'ha de terres arables ; elle occupe l'une des premières place en Europe pour la surface de terre de culture par habitant. Ressource renouvelable, le sol représente le cinquième de la richesse nationale, cependant divers processus naturels ou anthropiques contribuent à dégrader cette ressource.

Sur les versants 2,3 M d'ha sont la proie de l'érosion hydrique qui entraîne sédimentation et innondation en bas de pentes. 200 000 ha sont ainsi affectés dans les zones basses. Un M d'ha son mal drainés. L'érosion éolienne sévit dans les régions sablonneuses et dans les tourbières et affecte environ 600000.

La présence de sols acides concerne 2,3 M d'ha.

Appliqué depuis 30 ans le système de grandes cultures fondé sur de grandes parcelles, entraîne une dégradation physique, chimique et biologique des sols.

Depuis le milieu des années 80, la pollution des sols est devenue l'un des problèmes les plus grâves au même titre que celle de l'eau et de l'air.

Si la Hongrie devenait membre de la Communauté européenne, cela assurerait par la résolution urgente des problèmes de conservation de sols. Les pédologuqes Hongrois, défenseurs de l'Environnement, considèrent qu'il s'agit là d'une tâche devant aboutir dans un proche avenir.

INTRODUCTION

Although Hungary is a relatively small country with 9.3 million ha total area, approximately 2/3 of it - i.e. 6.5 million ha - is used by agriculture. In the field of cropland per capita Hungary is among the first European countries. In spite of this fact Hungarian soils quality is worsening, their quantity is decreasing. During the last 50 years agricultural area of Hungary was reduced by 1 million ha.

Soil serves as basis and tool for agricultural production and at the same time plays important role in the environmental protection because of its filtering and buffering capacity. As far as the natural resources are concerned Hungary has a medium supply. Besides the soil there are only few resources providing long term quantities. Soil is our most important natural resource, its value is almost 1/5 of the national wealth and more than double of mineral raw materials. Soil is renewable if efficient conservation is maintained.

Unfortunately several natural and man-made factors have harmful effect on soil hindering the achievement of its multi-function tasks (Fig. 1).

2.3 million ha of sloping area is endangered by water erosion causing not only soil loss in the mountains and hills but further damages are produced on 200 thousand ha of surrounding valley bottoms and lowlands by sedimentation and flooding (Tab. 1, Fig. 2).

Water erosion can cause a loss of a 1-3 mm deep soil layer on sloping lands in Hungary as a yearly average in contradiction to the 10-20 mm / 100 years rate of natural soil forming process. Slow and gradual erosion on hilly areas can also be observed in the nature being balanced with other environmental factors. Converting these lands into the agricultural utilization together with intensive cropping system and inappropriate land use raise the erodability risk.

Water erosion producing factors (amount of precipitation, drop size, rain intensity, duration, quantity of thawing snow, thawing time, steepness, length, form and exposure of the slopes), and modification factors (moisture status, moisture regime, structure of the soil and plant coverage) can result in different erosion forms (drop, sheet, rill erosion).

Water erosion impacts can be diminished by comprehensive soil conservation intervention including agronomical and technical erosion control so the soil loss can be decreased below 15 t/ha/year.

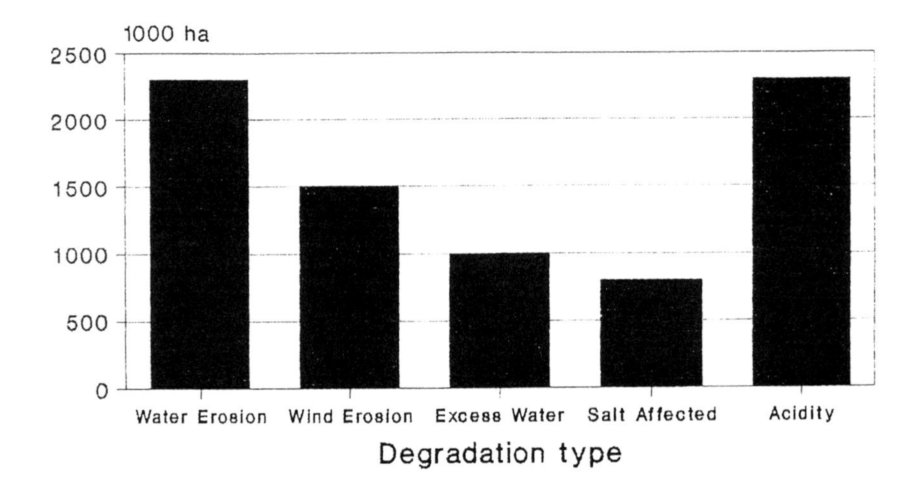

Figure 1. Soil degradation in Hungary.

Table 1.
Degree of Water Erosion in Hungary, (in 1000 ha).

County	Strongly eroded	Moderately eroded	Weakly	Total
Somogy	37	162	121	320
Veszprém	144	52	51	247
Borsod	54	116	54	224
Tolna	40	90	75	205
Fejér	28	46	130	204
Komárom	17	65	100	182
Zala	44	83	47	174
Baranya	24	67	70	161
Nógrád	63	59	25	147
Pest	43	44	52	139
Vas	29	36	45	110
Gyôr-Sopron	12	26	59	97
Heves	19	39	29	87
Total	554	885	858	2297

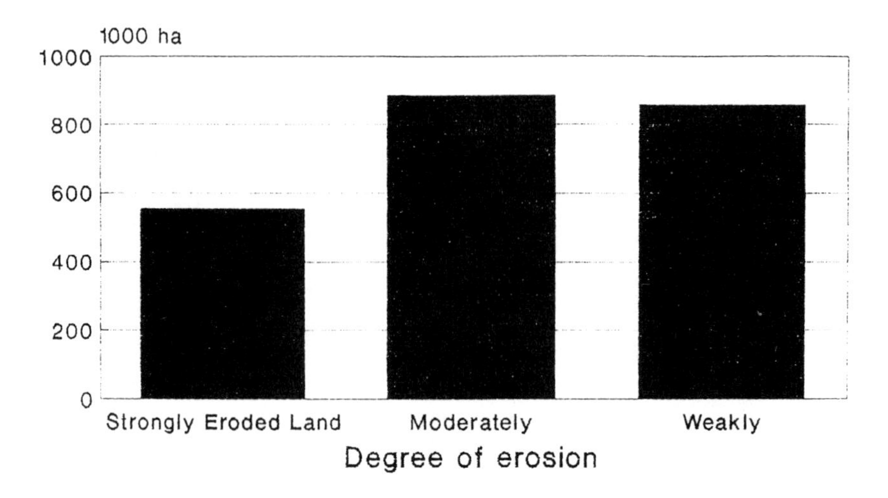

Figure 2. Soil Erosion in Hungary.

Erosion removes the topsoil, rich in nutrients and organic materials and less fertile, bad structured, often very carbonated soil layers can appear near to the surface at the rooting zone resulting in yield decrease and disturbances in water and nutrient uptake. The thinning or completely destroyed humic horizon is not able to perform the filtering and buffering functions.

Sedimentation in the surrounding valley bottoms and lowlands can cause direct destruction of the burried crops or yield decrease by modification of soil properties (anaerob conditions, nitrogen stress etc.).

Water erosion contributes to surface water pollution since soil particles moving down the slope to the rivers and lakes have phosphate content, main reason of eutrophication. Sediments in canals, waterways, reservoires raise the mainatainance costs and decrease their capacity.

Wind erosion prevails mainly on loose sands and on peatland ranging totally to 1,500 thousand ha. Unfortunately recently wind erosion could be observed on originally good structure highly productive chernozem soils as a result of wrong shallow, rotation tillage technology. Too large fields (500-1000 ha), lack of windbreaks and soil coverage also contribute to wind erosion hazard. Wind erosion decrease crop yield directly or indirectly opening infection gates on the plants.

Soil compaction similarly to the erosion belongs to the category of physical soil degradation. Compaction status of the soils are different according to the different soil formation processes and circumstances. Cultivation and tillage can modify the original compaction either positively or negatively. The uniform ploughing depth, shallow tillage, heavy machines application, wrong timing of tillage, improper irrigation result in destruction of soil structure and compaction, erosion and yield loss.

Some 1.0 million hectares are endangered by excess water on flat areas due to unfavourable hydrological conditions and relief.

Salt affected soils can be found on approximately 600 thousand ha where the high salt and exchangeable sodium contents are the limiting factors.

On 200 thousand ha secondary salinization of originally non-saline, non-salty soils took place due to raising salty ground water as a result of human activity (dam and reservoire construction, large scale irrigation etc.)

Soil acidity restricts soil fertility on 2.3 million ha. Soils in Hungary are stronly acid (pH (KCL) < 4.5) on 300 thousand ha, acid (pH KCl = 4.5-5.5) on 800 thousand ha and slightly acid (pH KCL = 5.5-6.5) on 1,200 thousand ha (Fig. 3).

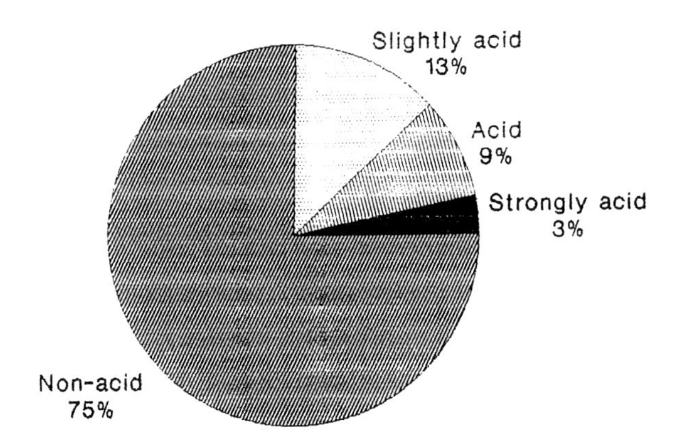

Figure 3. Soil Acidity in Hungary.

Soil acidification proves to be tendency in Hungary similarly to other countries : between 1977 and 1985 area covered by strongly acid and acid soils increased by 80,000 ha while the acreage of slightly acid soils increased by 140,000 ha.

Soil acidity has unfavourable impacts on soil chemical, physical and biological properties resulting in bad structure, compaction, low hydraulic conductivity, degradation of humic substances, nitrogen losses, phosphourus fixation, increased availability of toxic compounds (Fig. 4).

Since the mid 80s soil contamination has become one of the most serious problem together with air and water pollution. Drastical reduction of the amount of solid and liquid wastes, their comprehensive reuse or purification belong to the unsolved problems nowadays all over the world. The general handling of wastes - let them flow into rivers, lakes, seas without or with only partly cleaning or land disposal acting as chemical time bombs can cause serious regional or global pollution. The increasing rate of wastes year by year raises the risk.

One of the prominent problems is the accumulation of toxic materials, especially heavy metals in the soils. High emission rate of heavy metals and other toxic elements is very typical in developing countries because of dirty technologies and lack of effective environmental protection. Pollutants coming from the air and water can accumulate in the soil and instead of using recycling and purification methods, generally speaking, the easiest and cheapest way of getting rid of the solid or liquid pollutants is land disposal. Though buffer capacity of soil can tolerate loading to some extent, uncontrolled disposal finally leads to contaminating of food chain.

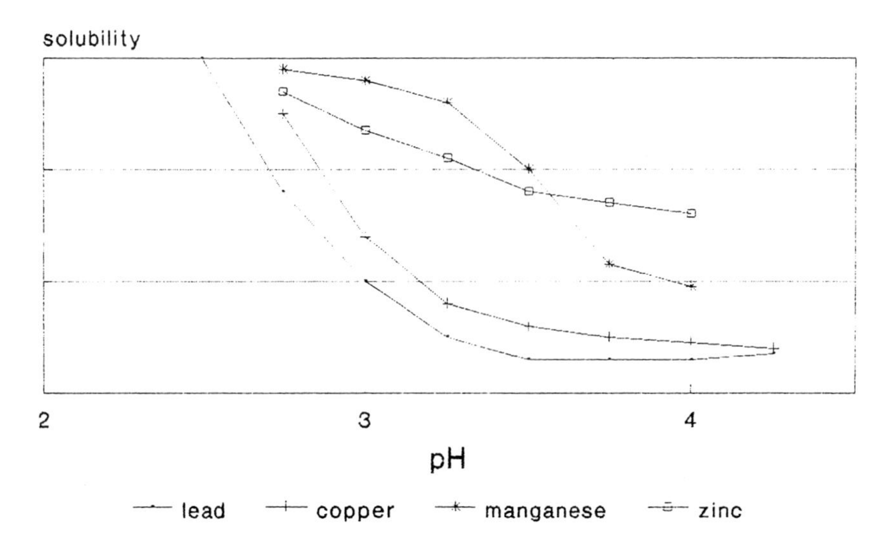

(Swedish Agricultural Ministry 1982)

Figure 4. Solubility of heavy metals in the soil.

In Hungary the first phase of the agricultural intensification took place after the establishment of large scale farming system, at the end of the 60ies, in the beginning of the 70ies. Because of the production-oriented point of view the rate of agrochemicals and the degree of mechanization were significantly insreased and the yield increased too. Since the end of 70ies however harmful impacts of intensive agriculture have been appeared in the environment : soil degradation became more severe and faster (erosion, compaction, acidification, salinization, pollution) nitrate and phosphourus pollution in the surface and subsurface water with agricultural origin, contributing to the eutrophication of our shallow lakes (especially Lake Balaton) and to the decay of groundwater quality making it undrinkable increasing amount of plant protection products, residues, metabolites in the soil, surface and groundwater.

Change of the political system, restriction of the markets and increasing cultivation costs made it clear in the beginning of the 90ies that the Hungarian intensified agriculture must be converted to sustainable agriculture.

Hungarian agriculture has structural changes including ownership, technologies, methods, machines and equipment. At the same time new soil conservation methods and technologies should be developed / adapted and put into the practice in small scale farming system too in order to maintain and increase soil fertility and to avoid, diminish or stop soil degradation and contamination.

The successful accomplishment of this task needs joint activity of governmental and non-governmental agencies in order to be able to bequeath our soil to the following generation.

This was the leading principle of International Peat Society Hungarian National Committee (IPS HNC) in the spring of 1991, when established the Foundation for Soil Conservation (FSC).

The aim of the Foundation : protection and conservation of soil as a renewable natural resource, assistance of practical methods and applied research in the field of soil conservation and soil science, protection of wetland soils, conservation of humus resource of the soil, revealing of soil pollution in order to reclaim the contaminated soils and to avoid further contamination.

For the above mentioned aim the Foundation is : organizing different researches, running independent specialist team, organizing scientific and public relation meetings, making essays and forecasts on pedological problems, publishing different publications, arranging study tours, giving grants and fellowships, collecting and distributing soil information, cooperating with other environmental protection NGOs in solving of soil conservation problems.

Joining the European Communities requires from Hungary to face soil conservation problems and to perform the above mentioned tasks. We, Hungarian pedologists and soil conservationists are convinced that these new tasks must be and can be accomplished in the near future.

Farm Land Erosion: In Temperate Plains Environment and Hills
S. Wicherek (Editor)
1993 Elsevier Science Publishers B.V.

Erosion...a current environmental problem ? The GCES, a new strategy for fighting erosion to resolve this dilemna of a growing society

E. Roose

Director of Pedological Research at ORSTOM, Montpellier headquarters, BP 5045, 34 032 Montpellier Cédex 1, France.

SUMMARY

In Nature, as long as the soil remains covered, geological erosion is no more spectacular than rock weathering (1 meter per 100,000 years). But when population density increases over certain thresholds (40-100-800 inhabitants/km sq.) runoff increases as well and erosion is multiplied by 1000 so that soil is scoured within a few dozen years or is degraded ("worn out") by selective erosion. Accelerated erosion is also the sign of a socio-economic imbalance within the society.

Fighting erosion is more difficult than was formerly believed, because erosion processes are numerous and ecological factors are complicated by socio-economic problems.This can be a linguistic problem as well because various actors may use the same terminology without really understanding the underlying meaning. Erosion processes can vary greatly at different spatial and temporal scales.

During the last centuries, societies have left traces of their attempts at soil and water conservation.

Traditional strategies are linked to socio-economic conditions : slash and burn systems, appropriate where land is abundant and where low-density populations enjoy autonomous subsistence, or Mediterranean bench terracing, where land is scarce and labor is cheap. But numerous traditional systems have been abandonned, higher wages at the factory being more attractive than those paid in the field.

Modern equipment strategies have been developped :

"RTM" Mountainous Terrain Restoration - around 1850 in the Alps and Pyrénée Mountains by civil and forestry engineers to protect roads, dams and valleys through reforestation and torrent correction.

"CES" Water and Soil Conservation around 1930 in the American Great Plains to protect soil productivity and water quality for city-dwellers ; however in 1968 more than 50 % of the researchers in the Soil and Water Conservation Department were doing research on water pollution !

"DRS" Defense and Soil Restoration - was developped in the Mediterranean area between 1940-80 by foresters and engineers in order to restore the foliage cover of the upper valleys and to reduce soil degradation and silting up of reservoirs. Around 1980-87, many researchers have proven the failure of this "rural equipment approach", and have developped a more direct approach, taking the interests of farmers into account. A new booklet, "Land Husbandry"(1) has replaced Bennett's text of 1939 (2), previously the bible of soil conservation manuals, unique in its domaine. In France and its territories, a new approach, GCES (Water and Soil Fertility Management) was developped and experimented on farmers' fields with prime interest accorded to the farmers' problems, such as net income and improving production security, labor valorization, and rural environment management. Soil erosion is no longer the main issue, but rather the consequence of improved water infiltration, soil capacity and increased biomass production - thus better cover - through better water and soil nutrient management.

RESUME

L'érosion est un phénomène de société. Dans la nature, tant que les sols sont couverts, l'érosion est lente (1 mètre en 100 000 ans). Mais, dès que la population dépasse certains seuils de densité, s'agglomère dans des villes ou que les pressions économiques provoquent des changements de systèmes de production, le ruissellement et les divers processus d'érosion s'emballent et les horizons humifères sont décapés en quelques dizaines d'années sur les pentes fortes ou sont vidés de leur substance par érosion sélective sur les pentes faibles (= squelettisation).

La lutte antiérosive est finalement plus complexe qu'on ne croyait car les processus en cause, les facteurs sur lesquels on peut jouer sont très divers et quelques fois contradictoires : par exemple, si on améliore l'infiltration sur les fortes pentes, on réduit les risques de ravinement, mais on augmente les risques de glissement de terrain. De plus, il règne une grande confusion de langage car les différents acteurs utilisent des termes sans en connaître la signification profonde et la philosophie qu'ils cachent.

Depuis des millénaires, les sociétés qui se sont succédées sur les quatre continents ont laissé des traces de leur lutte contre l'érosion. Ces *stratégies traditionnelles* sont intimement liées aux conditions économiques de ces sociétés. Ainsi, les gradins ou terrasses méditérranéennes se retrouvent sur les quatre continents mais toujours là où la terre cultivable manque, en dehors des montagnes, où la population est dense et où le travail est bon marché car on ne va pas investir 500 à 800 hommes x jour pour aménager un hectare si on n'y est pas acculé.

Plus récemment, ont été développées des *stratégies modernes d'équipement rural.* Pour faire face à des problèmes d'intérêt public, l'Etat par l'intermédiaire de ses corps d'ingénieurs (Ponts et Chaussées, Forestiers, Agronomes) a imposé au monde rural des équipements dont l'objectif est la restauration des sols certes, mais surtout la protection des routes, des ouvrages d'art, des barrages et de la qualité des eaux pour les citadins. Il s'agit de la RTM, Restauration des Terrains en Montagne (Alpes et Pyrénées, 1850-1900), de la CES, Conservation de l'Eau et des Sols des plaines américaines (Bennet, 1930), de la DRS, Défense et Restauration des Sols autour du Bassin méditerranéen (1940-1980).

Ces stratégies ont été rejetées par les paysans qui n'y ont pas trouvé leur compte (perte de terrain sans augmentation des rendements) et ont abouti dans plus de 80 % des cas à des échecs, tant aux USA qu'en Afrique et en Asie.

Devant cet échec, les chercheurs ont tenté depuis les années 1980-1985 de mettre au point une nouvelle approche tenant mieux compte des intérêts des gestionnaires des terres (les paysans) et les associant à toutes les phases des nouveaux projets concernant la lutte antiérosive. Une nouvelle bible (Land Husbandry, 1988) remplace le gros manuel de Bennet (2) qui a servi de théorie et de modèle à tous les manuels de conservation des sols. En France, *la Gestion Conservatoire de l'Eau et de la fertilité des Sols (G.CES)* tente de formaliser cette nouvelle approche *de développement rural* où la lutte antiérosive n'est plus une fin en soi, mais un préalable pour développer la production de biomasse, c'est à dire à la fois intensifier la production, améliorer l'infiltration et réduire les pertes en terre et nutriments par les voies biologiques.

INTRODUCTION

Is Erosion still a research problem ?

Erosion is an age-old problem. Today, however, its importance increases along with demographic and socio-economic pressures. As the population increases as do man's needs, fragile land is being cleared for agriculture, excessive grazing, and intensive exploitation of soil without bringing the necessary nutrients and organic matter to compensate this overuse. The result is first of all the accelerated degradation of vegetal cover, then of the soil, then of the hydrological network followed by degradation of the microclimate. Erosion is therefore the sign of an imbalance between the environment's potential resources and the way in which modern society manages these resources. If the way soil is exploited is the reason behind erosion, we should hope to develop better adapted production systems, and thus reduce losses due to erosion.

Seven thousand years ago man showed his capacity to combat erosion : we can thus analyze past history and ask why 80 % of the currect LAE projects end in failure ? First we must stress the diversity of erosion processes, linguistic misunderstanding, temporal and spatial discontinuity and two different approaches to fighting erosion : on one hand, protection or even better, improving soil production potential; on the other hand, water quality conservation.
Next, possible solutions :
- a few research results,
- a new LAE strategy, the GCES, whose average objective is to resolve the problems of both farmers and their environment.

Finally, we will draw a few conclusions and test their validity in function of regional situations :
- in countries with intensive surplus agriculture, the fundamental problem is maintaining water quality,
- in developping countries, not only must soil be preserved, but its fertility needs to be restored and another challenge met: production should be doubled every 20 years !

1. EROSION OR EROSIONS... A LINGUISTIC PROBLEM

1.1. Diversity of erosion processes depending on temporal scales

Normal geological erosion in the natural milieu is very slow (about 0,1 t/ha/year, or 1 meter of eroded soil per 100,000 years) and corresponds to rock weathering : this is why the soil formation is deeper where the climate is hot and humid.

But once men and animals begin to multiply, vegetal cover begins to degrade, runoff develops and erosion is multiplied by 1000 (E=10 to 700 t/ha/year; runoff = 20-70 %) Just a few decades are enough to scour off 1 meter of soil !

Catastrophic erosion occurs when geological causes are exacerbated by careless human management.
Erosion worsens in a most dramatic way, for ex. :
- gully erosion in the Mediterranean reaching *100-300 t/ha in one day.*
- Land slides moving *millions of cubic meters within one hour.*

The example of the Nimes storm on Oct. 3 1988 is spectacular. Within six hours' time, 420 mm of rain fell : torrents swept away the old village areas. Results : 11 deaths and 4 billion francs in damages! Worse still, in Columbia when the Nevado Ruiz volcano spitting vapor brought on an enormous torrent of lava that buried a city of 25,000 inhabitants in one single night !

In conclusion :

It is difficult to measure processes that are so temporally and spatially sporadic. Research is still not steady on its feet.
- The Press and the State only deal with catastrophes that elicit strong reaction from the public.
- As for us, we are particularly interested in the start of erosion processes - sheet and gully erosion - erosion accelerated by Man, because this is the stage where something can be done to prevent catastrophes.

1.2. Diverse logics: different spatial interests

Erosion is the result of three processes : washing-out, transportation and sedimentation. These three processes are ever-present , but their importance varies spatially.
In mountain environments : washing-out and therefore RTM is developed by foresters.
In Piedmont environments : both erosion and transportation are problems: the DRS tries to revegetate the upper valleys and to master torrents.
In hill and plain environments : Washing-out is less a problem than is sedimentation and pollution : the CES and control of sedimentation and pollution are strong imperatives.

In function of the space concerned, the strategies vary, and the actors involved as well :
- *On slopes*, erosion damages soil. The *"uphill logic" or farmer's strategy attempts to increase land productivity,* and seeks lasting results and insured productivity. The best methods seek to modificy poorly balanced production systems. The principal actors involved are the farmers themselves, the only ones capable of managing the rural environment and land productivity.
- *In rivers*, solid transportation damages *water quality,* and threatens downhill plains. The *"downhill logic"* seeks to equip the valley in order to protect the water, reduce silting up of reservoirs, and correcting torrents and gullies. The main parties involved are city-dwellers, engineers and local officials : they usually look to heavy equipment and physical methods to reduce solid transportation.

It is essential that we differentiate these two logics whose objectives are quite different. Only the farmers can manage and improve their lands, but the State should take responsability for major projects such as reforesting summits, correcting gullies and torrents, fixing river banks, desilting, as well as the cost of instruction and research.

1.3. Diverse processes, causes and factors (Tab. 1)

Land degradation can have numerous causes : Salinization of arid zones, compaction, during motorization, acidification by mineral fertilizers, mineralization of organic matter. It develops once land is cleared, before the preliminary symptoms of erosion(but not transportation), but worsens due to erosion.
Sheet erosion is dangerous because it often goes unnoticed (1mm=15 t/ha !) and is selective vis-à-vis organic and mineral colloids as well as absorbed nutrients. It wears out the land, i.e. it prevents the soil from stocking water and nutrients. This process is not clearly understood by peasants : to my knowledge, no African dialect includes a specific word for this pernicious process.
Linear erosion is better understood : even before Bennet, runoff speed was being slowed down by thresholds, waterfalls absorbed energy, damage was reduced , but infiltration and runoff volume related problems were still not under control.
Mass wasting is still not clearly understood (except by specialists) and is difficult to control : the presence of trees seems to reduce its frequency. However, cultivating on slopes causes progressive land sliding.
Dry mechanical erosion is often mistaken for sheet erosion. We don't know much about the intensification or reduction factors for this sort of dry creeping which is very active on overpopulated mountains. Research is starting to attack this problem!
Wind erosion only occurs when wind exceed certain speed thresholds (25 km/hour). Balancing factors are close to those for sheet erosion.

1.4. The underlying nature of the problem : the imbalance within the cultivated "converted" milieu. (Fig 1)

The dense humid forest has a biomass of 850 t/ha and produces 8-15 t/ha/year of litter perfectly protecting the soil from sun and rain. Roots are deep under the soil, but most roots dig into the first 25 centimeters, those richest in organic matter and nutrients. Under foliage, profiles develop over numerous meters, even tens of meters. Erosion is slow.

In the swamp environment, there is less protection : the canopy (50-150 t/ha) and especially the burnt or pastured litter, lets heat and rain penetrate, which crusts the soil surface. Runoff can reach 40-70 %, especially if fires come late. Water penetrates more superficially and soil is leached, and more clayey between 50-100 cm.

But under cultivation, the atmosphere is even dryer : the canopy is less dense (2 to 20 tons depending on agriculture) and litter is often inexistent, burnt, used elsewhere, valorized by livestock or artisans.
From a vegetal point of view, we can observe :
- simplification of the agrosystem,
- reduction of the biomass.

The pedo-climate is hotter, more arid than the forest, especially since rooting is limited to 25-50 cm in depth.
The soil is characterized by :
- reduced biological lifting (feeble rooting)
- decrease in litter supply, and therefore decrease in organic matter in the soil and biological activity.
- structural degradation: crusts, damage to microporosity and decrease in infiltration capacity.

TABLE 1 - Diversity of Erosion Processes, Causes, Factors and Consequences.

Process	Causes	Factors	Consequences
Soil degradation	- Mineralization of organic matter - Salinization, motorization, etc...	- Temperature - Humidity - Litter turn over	**decrease** : ↓ OM Content decrease ↓ Water + nutrient storage capacity ↓ Porosity, infiltration **increase :** ↑ of runoff and Erosion risk
Sheet Erosion	Splash = - setting - shearing - projection	- Vegetal cover 1000 - Slope 200 - Doil 30 - Structure A.E 10	Sealing crust + Setting Runoff - Selective Erosion - Scouring
Dry mechanical	Tillage practices	- Frequences - Intensity - Slope steepness - Soil fiability	- Scouring - Humiferous - Horizon
Gully Erosion	Runoff energy $E = \frac{M.V2}{2}$	- Runoff - volume = f {surface, rain, intensity - Speed = f (slope, roughness) - Resistance of the soil × vegetation - A.E structures : weir, etc.	- Deep gullies - Imbattance of slopes - Alluvial fans
Mass mouvements (Sliding) on hillslopes	Gravity > Cohesion of the soil	- Cover weight {soil+water+vegetation - Humectation of sliding plane - Slope + drainage	- Hillslope scouring - Mud slides.

CONCLUSIONS :

1 - Diversity of forms, causes, factors and means for fighting erosion.
2 - Temporal and spatial variability of erosion intensity.
3 - Great importance of the soil surface state.

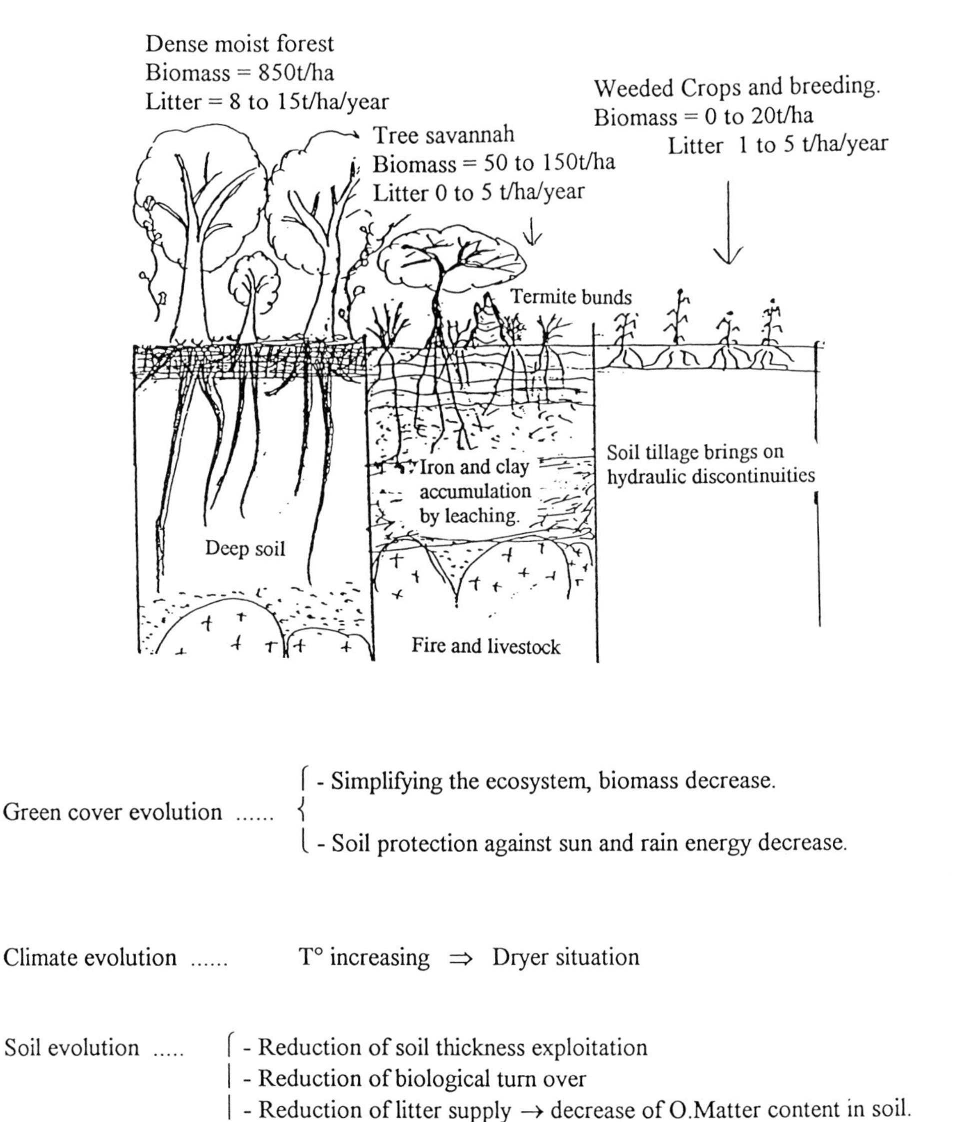

Green cover evolution ...... { - Simplifying the ecosystem, biomass decrease.
- Soil protection against sun and rain energy decrease.

Climate evolution ...... T° increasing ⇒ Dryer situation

Soil evolution ..... { - Reduction of soil thickness exploitation
- Reduction of biological turn over
- Reduction of litter supply → decrease of O.Matter content in soil.
→ decrease of biological activity.
- Soil structure degradation : - sealing crust, compaction + clay pan,
- decrease of macroporosity,
- reduction of infiltration capacity.
- Runoff + Erosion + Leaching risks increase.
- Nutrient losses increase ↑
- Decrease water + Nutrient storage capacity → "TIRED SOIL"
- Decrease water disposal for biomass production.

Figure 1. The soil degradation causes : The imbalance of cropping systems.

Risk of runoff increases, as erosion and leaching of nutrients by drainage and runoff water increase. Nutrient and organic matter losses are accelerated, especially in forest soils rich in surface litter : water and fertilizer storage capacity is reduced. The soil becomes "worn out" and is incapable of valorizing the water and nutrients it receives : productivity decreases to a stable limit, taking into account the production system of biological lift and supply hindered by rain and dust. Traditionally, without fertilizer, 4 to 8 quintals of cereals is produced! Just enough to maintain a state of relative survival !

1.5. Economic consequences of erosion

Erosion brings on soil production losses and downhill damages. This economic aspect is currently the object of extensive research.

a) On-site loss due to erosion.

Water loss :

- runoff brings on decrease in drainage in humid zones and increase in river debit, but little change in production.
- however, in semi-arid zones (if only during dry season) runoff brings on a decrease of "ETP" and of the biomass.

The anti-erosion combat (LAE) should be even more spectacular concerning productivity in semi-arid areas.

Loss of fertilizer : is calculated in terms of tens of kilos of "NPK", by hundreds of tons of erosion. It represents millions in payment to compensate fertilizer loss for each country. (3 - 4).

Loss of immediate production is easily compensated by bringing in fertilizer on the regional level (2 to 10 %) but this loss can be catastrophic for certain farmers who live in high-risk zones as the annual losses represent large portion of their net income.

Loss of potential surface reaches 7 to 10 million hectares per year. On the world-wide scale, in three centuries' time, all arable land could be destroyed. On the regional scale, from 2 to 5 % of land is affected each year; however, certain fragile plots lose up to 20 or 100 % of their arable land due to certain serious events.

Loss in long-term productivity is highly variable in function of the type and depth of the soil *(see diagram).* Soil memory is inscribed in the loss of thickness of the humus horizon, loss of infiltration and water/nutrient stocking capacity, loss of water's efficiency and "intrants" (= wearing out of the soil) and finally, the plunge in productivity that results. Stocking and Biot modeled the duration of soils in function of erosion rhythm and minimum soil thickness needed to maintain production system profitability.

Note that certain deep, rich, even soils never see their productivity decrease. This is the case for brown deep loess soils.

However, the majority of African soil has fertility concentrated on the surface. The slightest soil loss, by stripping, but even moreso by selective erosion, brings about a drastic drop in productivity, up to stability thresholds of very low levels (4-8 Q : ha/year for cereals).

b) Off-site problems

Water from runoff mixes with discharge from the base of ground waters, pollutes the water sources and accelerates their eutrophization. This pollution jacks up the price of drinking water and electricity, reduces the growth of fish and brings on silting of reservoirs, canals and ports. Mudslides encumber ditches and poorly placed village roads. The cost of damage from erosion during storms of rare frequency (1/1 to 1/10) is increased, and attracts the media, provoking reactions from the government more often than does soil degradation. This explains the traditional approach consisting of stopping first the erosion from the badlands, which are worn-out lands.

c) Major consequences of economic analysis of the LAE (Fig 2)

- If investment is made in the fight against erosion on degraded soils, whether or not the soil is deep, productivity of the land is not altered (thus this does not interest the farmers) but transportation of solid matter is reduced. This is the case for RTM and DRS.
- If investment is made on good land being damaged :
- if the soil is deep, no improvement in profitability can be reached,
- but if the soil is superficial, productivity can be restored (in the interest of the farmer).

This is revolutionary in relation to the classic LAE which deals with damaged land, while farmers prefer to invest in better land...except if they can recuperate acidentally damaged land. This was experimented in very different ecological conditions : the Nord-Pas de Calais region (5) Mali (6) and Yatenga (7) Rwanda (8) and especially Algeria (9).

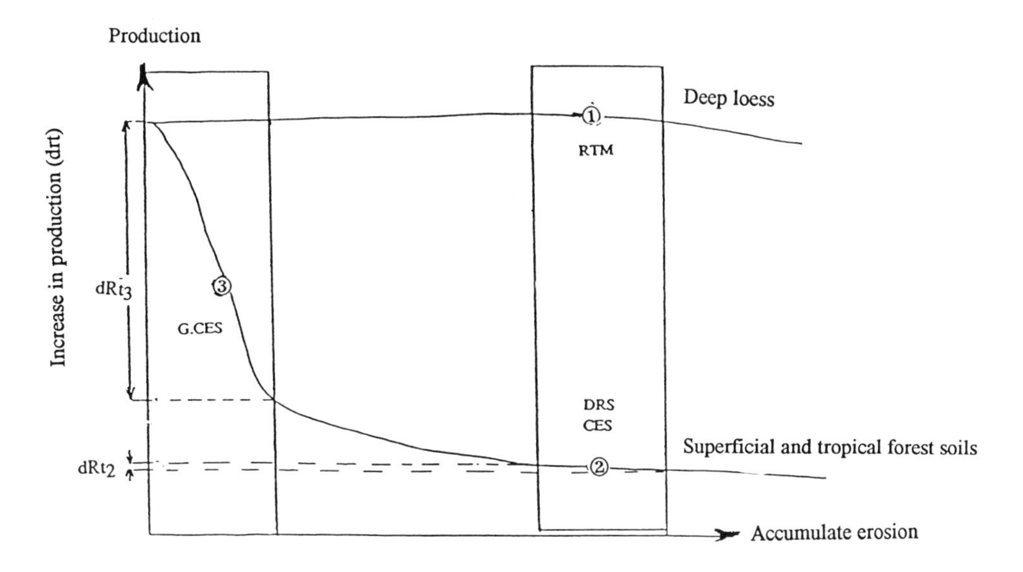

Figure 2. Effect of cummulated erosion or soil scouring on their productivity. Interest of rural equipment strategies (RTM, DRS, CES) and rural development strategies (G.CES) in relation with two different soil situations.

1.6. Conclusions

a) Inconsiderate use of the same terms by different disciplines in speaking of erosion while there are at least five different processes at work, with varying gravity in function of different factors and thresholds according to temporal and spatial scales, has brought about quite a bit of confusion, both on the research level (where links between lab tests, plots, fields, on slopes, on catchment areas are desperately sought) and between those involved in fighting erosion.

In spite of thousands of years of empirical practice fighting erosion, research is still not progressing in certain domaines such as that of dry mechanical erosion, fighting crusting, simultaneous management of water and nutrients, socio-economic aspects of the LAE, acceptation by farmers and cost of LAE techniques.

b) Analysis of economic policy of anti-erosion combat clearly distinguishes two objectives and two different actors involved :

- if the problem in conservation - or better still, restoration of soil fertility - means of modifying production systems should be proposed to farmers in order to resolve immediate problems of valorizing work, improvement and security of production of the best lands (this is the GCES) : only the farmers can manage the rural environment ;
- if it is a question of conserving water and reducing solid transportation from rivers, only the State can take charge of major mechanical and biological projects for stabilizing gullies, torrents, land slides and rivers, reforestation of upper valleys, protecting banks, routes and other works.

2. ELEMENTS FOR A SOLUTION

2.1. Traditional or modern strategies

Man has not given up on handling erosive phenomena which hinder his projects and development: we have found traces of his ingenuous attempts in this domaine over thousands of years (ex. stepped terraces in 1800 around the Mediterranean basin - in 1200 in Crete - in 1400, by the Incas in Peru and even earlier in China, Bali and Latin America).

These traditional methods are often described, but we don't understand their functioning very well or their limitations. Engineering studies have considered them inefficient and not worthy of modern times. However, many of these methods, even if they have been abandoned for economic reasons (the factory pays more) are quite efficient, and inform us about the socio-economic conditions of a certain age in history, indicate extreme ecological conditions in specific regions and can be improved and serve as a starting point for farmers in taking charge of their environmental problems. (ex. the Zaï and stone bas in Yatenga, Burkina : 7).

Since 1850, "modern methods for rural equipment" have been developed :

- the RTM (1850) "Restoring Mountain Land" in the Alps and Pyrénées by foresters who restore vegetation in degraded mountain slopes and remedy torrents ;
- the CES (born in 1930 in the USA) "Conservation of Water and Soil" by agronomists charged with providing both financial and technical support. for farmers wishing to protect their lands' productivity - and the quality of water indispensable to city-dwellers (thus explaining the State support for the project) ;

the DRS (developed in the Mediterranean area between 1940-1980) "Defense and Restoration of the Soil" (damaged by erosion) proposed by State forestry engineers to protect land, restore vegetation on damaged land and more importantly, to reduce the silting up of dams, water being a precious asset in arid areas.

All these approaches, carried out by the centralized power of different States, have often succeeded in realizing admirable projects, but most often have met with failure at long-term, because farmers do not feel concerned. This equipment does not improve productivity, on the contrary, it reduces exploitable surface without attacking the heart of the problem: sheet erosion and degradation of organic matter in the soil.

Since 1987 and the Puerto Rico (WA.SWC) (10) , Niamey (ICRISAT) Medea (INRF) Butare (ISAR) and Bujumbura (ISABU) seminars, a new strategy is developing which attempts first of all to resolve farmers' problems, double profitability over 20 years, improve stability of production systems, and intensify agriculture without destroying the environment. These methods have been described in a pamphlet entitled "Land Husbandry" or the art of caring for land, and a series of articles and conferences on the GCES (11). The FAO should be putting out a volume on the application of this strategy tied to lasting development of agriculture in different countries of Europe and Africa.

In short, the process is the following :

a) *Diagnostic :* Where does run off and erosion start ?

- Is it due to degradation of the soil? When ? How ?
  If so, the production system should be improved with the help of farmers involved.
- Is it a case of untimely, excessive water input from uphill ?
  If so, then the drainage and water management system should be improved with the help of engineers.

b) *Water management function of the region hydric balance*

There are four means of managing water corresponding to anti-erosive structures and adapted agricultural techniques.

c) *Managing the biomass and nutrients*

Conservation of the soil often discourages farmers for it demands great effort, investment in labor and "intrants" without providing immediate improvement in profitability since soil is often already damaged.

Restoration of physical and chemical qualities must therefore be envisaged.

A means of exploitation of the available biomass must be chosen, through breeding and manure production, (foresee and coefficient of profitability of 30 % and heavy azote waste), through composting (much labor and same profitability as manure, but without animal production), direct removal of agricultural residue (much work and slower degradation of the soil) or straw protection of the surface. Production systems increasing the available biomass are imperative (agroforestry, cultivated fallowland, concealed green fertilizer etc). In any case, airation of the soil has to be provided, as well as correction of the pH factor (aluminic toxicity if pH is over 4.8) and soil deficiency (mineral supplements directly brought in to valorize labour and available water).

Finally, infrastructures must be developed (routes, markets, schools, dispensaries) in order to valorize the surplus in rural production by using it to feed cities.

We often use the individual case as a starting point for developing an exploited slope by a small rural community before carrying the projects to the level of a field, and then to that of a catchment area. This involves more than does a simple equipment project because it also a question of changing mentalities, of liberating the creative spirit in peasant farmers, but this is the only way to avoid imminent disaster: farmers must take their environment into their own hands.

Table 2.
Antierosive structures and cultural practices in relation with running water management.

Management type	Antierosive structures	Cultural practices
Runoff farming in arid and semi-arid areas	- Impluvium, sistern, - Soil dykes on wadies - Discontinued terraces	Deep plowing,pitting Microcatchment
Total infiltration in semi-arid (less 400 m) or sub tropical areas on highly permeable soils	- Pitting - Beuch terracing	- Rough plowing - Tied-ridging - Mulching
Water diversion Humid areas on slow permeable soils	- Channel terraces with lateral drainage	- Oblique ridging - Ridging paralelly to the slope
Runoff energy dissipation in all climat, permeable soils on slopes < 60%	- Stone bunds or walls - Grass lines, grassed embrankment - Hedges	- Agroforestry - Rough tillage - Crops alternating with meadows - Mulching

2.2. A few solutions have emerged through research

a) The major importance of biological methods, vegetal cover and in particular, at the soil surface, litters, self-propagating vegetation, even cover plants introduced under agriculture : canopies are less efficient than litter.

b) The ambivalent role of working the soil which, on one hand, temporarily improves infiltration, rooting and profitability, but which, on the other hand, speeds up mineralization, imbalances and degradation of soil. To be used with a certain carefulness: minimal labor along with mulch permits a serious reduction in degradation risks.

c) The role of inclination of the slope seems more important than its length except when linear erosion is developing.
The topographical position is especially important: the bottom of the slope is often more quickly saturated and eroded than the upper segments. Complete revision of classic plans for installations must be done and use of mathematic formulae defining - empirically - the space between terrasses, should be left aside. Only observation of the land on early rills and gullies and discussion of economics with farmers will lead to definition of the spacing between anti-erosive structures.
d) major effect of soil surface state upon runoff - particularly litter biological activities and crustings : must mode us revised the notions of deep/superficial work, heavy clods or powdery. Function of runoff generation we must try to improve infiltration by a production system or to drain water excess with out increasing solid transportation. Cultural profile often makes clear the plant development but rarely elucidates infiltration ! Soil surface state is much more significant.
e) Treating gullies is costly, but is not complicated if one respects a dozen or so elementary rules. Costs can be reduced and water and lost sediments valorized by transforming gullies into a kind of " linear oasis" so that each ecological nitch is used by well adapted productive species.
f) Modelization of runoff has made great progress using surface states, evolution of infiltration capacity and water stockage in diverse horizons. We still have to try to better understand the redistribution of water at the catchment area level during different seasons.
g) Modelization of erosion is much more complicated. There is no universal model. The USLE model is an empirical approach adapted to field engineers who need markers to choose different modes of LAE. It can be useful if rainfall energy and danger of sheet erosion are greater that runoff and gully or mass erosion. Studying the functionning of slopes is a prerequisite to using the model.
h) Sociological aspects of erosion (property problems, typology of exploitations, motivation and available resources for undertaking property investments) has hardly been approached. Economic problems are handled first : economic conditions for erosion, influence of the regional and world markets, cost of erosion, loss in short and long-term productivity of eroded land, cost of fighting erosion. We have trouble classifying different soils according to their erodability for each type of soil is a dynamic entity, evolving over time in function of the production system applied and the type of erosion.

We are advancing in the domaine of theoretical and practical knowledge... but the next generation will still have a lot left to deal with.

3. GENERAL CONCLUSIONS

a) *Erosion is a societal problem, in constant growth.*
- In developping countries, water has to be managed along with soil fertility in order to meet the challenge of doubling productivity over 20 years' time, along with the increase in population.
- In industrialized countries, quality of water, a natural resource as precious as soil, must be conserved, or restored, - water is a resource whose renewal, and restoration are often possible, but costly.

b) *Today this complex problem interests many people*
- Researchers from diverse disciplines who seek to understand and modelize the process.
- Developpers, because mastering erosion is one of the keys to lasting agricultural development.
- Politicians... because environmental problems are a contemporary issue: those who pollute should pay, which is not the case today.

c) Today, possibilities for action are opening up
- The problem is evident, urgent : we're backed into a corner, there is no way out. Everyone is concerned.
- Demographic pressure is enormous.. but production is stagnating.
- We're starting to find a way out: there are already a few results showing that it is possible to intensify agricultural production without degrading the environment.

This is a fascinating area of study because it integrates physical and human environments.

However, French research institutes do not have any major research program on fighting erosion. Technical higher education on erosion processes is not plentiful and is more or less nonexistant concerning ways of fighting erosion. It is difficult to do serious doctoral research on quantified erosion in the field and especially so concerning development of anti-erosion strategies, because erosion varies so greatly according to space and time.

Evidently, the problems are many and frequent... but devoting a whole career to this area is difficult.

I would like to conclude on an optimistic note - once the problems of world security have been resolved, we can hope that Man will attack the real problems of the Earth - and that, by hoping to rule the moon, he will learn to master Mother Earth, for a long time to come.

REFERENCES

1. Shaxson, I.F., Hudson, D.W., Sanders, D.W., Roose, E.J., Moldenhauer, W.C., (1989) : Land Husbrandry : a framework for soil and water conservation, S.W.C. Soc., WASWC, Andeny, Iowa : 64 p.
2. Bennet, H.H., (1939) : "Elements of Soil Conservation", 2d Edit. New-York, Mac Graw-Hill : 534 p.
3. Roose, E., (1973) : Dix-sept années de mesures expérimentales de l'érosion et du ruissellement sur un sol ferrallitique sableaux de basse Côte d'Ivoire. ORSTOM, Abidjan : 125 p. Thèse Doct. Ing. Fac. Sci. Abidjan n° 20.
4. Stocking, M., (1986) : The cost of soil erosion in Zimbabwe, in terms of the loss of three major nutrients. FAO. AGLS. Rome : 164 p.
5. Roose, E., Masson, F.X., (1983) : Conséquences of heavy mechanization and new rotations on runoff and on loessial soil degradation in the North of France. Communication ISCO 3, "Preserve the land", Honolulu, Edit. S.C. Soc. America, Andeny, USA : 24-33.
6. Roose, E., (1985) : Rapport de mission auprès de la DRSPR dans la région Sud du Mali (3-17 décembre 1984). IER, Bamako, KIT Amsterdam : 42 p.
7. Roose, E., Rodriguez, L. (1990) : Aménagement de terroirs au Yatenga (N.O. Burkina Faso). Quatre années de GCES : bilan et perspective. CRPA Ouahigouya, ORSTOM Montpellier : 40 p.

8. Roose, E., Ndayizigiye, F., Nyamulinda, V., Byiringiro, E., (1988) : La GCES, une nouvelle stratégie de lutte antiérosive pour le Rwanda. Bull. Agricole Rwanda, n° 21 (4) : 264-277.
9. Arabi, M., Roose, E., (1989) : Influence de quatre systèmes de production et le ruissellement en nappe en moyenne montagne algérienne. Bull. Réseau Erosion n° 9 : 30-51.
10. Roose, E., (1987) : Water efficiency and soil fertility conservation on steep slopes of some tropical countries. Communication Workshop WASWC, Puerto ¤ico, 22-27 March 1987. Edit. Moildenhauer (W.C.) and Hudson (N.W.), S.W.C. Soc. Ankeny, USA : 296 p.
11. Roose, E., (1991) : Introduction à la Gestion Conservatoire de l'eau et de la fertilité des sols (GCES) : bilan et perspective. CRPA Ouahigouya, ORSTOM Montpellier : 188 p.

AUTHOR INDEX